Synergetics of Measurement, Prediction and Control

Springer
Berlin
Heidelberg
New York
Barcelona
Budapest
Hong Kong
London
Milan
Paris
Santa Clara
Singapore
Tokyo

Springer Series in Synergetics

Editor: Hermann Haken

An ever increasing number of scientific disciplines deal with complex systems. These are systems that are composed of many parts which interact with one another in a more or less complicated manner. One of the most striking features of many such systems is their ability to spontaneously form spatial or temporal structures. A great variety of these structures are found, in both the inanimate and the living world. In the inanimate world of physics and chemistry, examples include the growth of crystals, coherent oscillations of laser light, and the spiral structures formed in fluids and chemical reactions. In biology we encounter the growth of plants and animals (morphogenesis) and the evolution of species. In medicine we observe, for instance, the electromagnetic activity of the brain with its pronounced spatio-temporal structures. Psychology deals with characteristic features of human behavior ranging from simple pattern recognition tasks to complex patterns of social behavior. Examples from sociology include the formation of public opinion and cooperation or competition between social groups.

In recent decades, it has become increasingly evident that all these seemingly quite different kinds of structure formation have a number of important features in common. The task of studying analogies as well as differences between structure formation in these different fields has proved to be an ambitious but highly rewarding endeavor. The Springer Series in Synergetics provides a forum for interdisciplinary research and discussions on this fascinating new scientific challenge. It deals with both experimental and theoretical aspects. The scientific community and the interested layman are becoming ever more conscious of concepts such as self-organization, instabilities, deterministic chaos, nonlinearity, dynamical systems, stochastic processes, and complexity. All of these concepts are facets of a field that tackles complex systems, namely synergetics. Students, research workers, university teachers, and interested laymen can find the details and latest developments in the Springer Series in Synergetics, which publishes textbooks, monographs and, occasionally, proceedings. As witnessed by the previously published volumes, this series has always been at the forefront of modern research in the above mentioned fields. It includes textbooks on all aspects of this rapidly growing field, books which provide a sound basis for the study of complex systems.

A selection of volumes in the Springer Series in Synergetics:

Synergetics An Introduction
3rd Edition By H. Haken

Handbook of Stochastic Methods
for Physics, Chemistry,
and the Natural Sciences
2nd Edition By C. W. Gardiner

Noise-Induced Transitions
Theory and Applications in Physics,
Chemistry, and Biology
By W. Horsthemke, R. Lefever

The Fokker-Planck Equation
2nd Edition By H. Risken

Chemical Oscillations, Waves, and Turbulence By Y. Kuramoto

Advanced Synergetics 2nd Edition
By H. Haken

Dynamics of Hierarchical Systems
An Evolutionary Approach
By J. S. Nicolis

Complex Systems – Operational Approaches in Neurobiology, Physics,
and Computers Editor: H. Haken

Information and Self-Organization
A Macroscopic Approach to Complex
Systems By H. Haken

Synergetics of Cognition
Editors: H. Haken, M. Stadler

Relative Information Theories
and Applications By G. Jumarie

Neuronal Cooperativity
Editor: J. Krüger

Synergetic Computers and Cognition
A Top-Down Approach to Neural Nets
By H. Haken

Foundations of Synergetics I
Distributed Active Systems
By A. S. Mikhailov

Foundations of Synergetics II
Complex Patterns 2nd Edition
By A. S. Mikhailov, A.Yu. Loskutov

Synergetic Economics By W.-B. Zhang

Ambiguity in Mind and Nature
Multistable Cognitive Phenomena
Editors: P. Kruse, M. Stadler

Modelling the Dynamics of Biological Systems
Editors: E. Mosekilde, O. G. Mouritsen

Igor Grabec Wolfgang Sachse

Synergetics of Measurement, Prediction and Control

With 153 Figures

Springer

Professor Dr. Igor Grabec
Faculty of Mechanical Engineering
University of Ljubljana, pob. 394
1000 Ljubljana, Slovenia

Professor Dr. Wolfgang Sachse
Theoretical and Applied Mechanics
Cornell University
Ithaca, NY 14853-1503, USA

Series Editor:

Professor Dr. Dr. h.c.mult. Hermann Haken
Institut für Theoretische Physik und Synergetik der Universität Stuttgart
D-70550 Stuttgart, Germany
and
Center for Complex Systems, Florida Atlantic University
Boca Raton, FL 33431, USA

Library of Congress Cataloging-in-Publication Data

Grabec, Igor, 1939–
Synergetics of measurement, prediction and control / Igor Grabec, Wolfgang Sachse.
p. cm.
Includes bibliographical references and index.
ISBN 3-540-57048-9 (hardcover : alk. paper)
1. Self-organizing systems. 2. System theory. 3. Information storage and retrieval systems. 4. Mensuration. I. Sachse, Wolfgang, 1942– . II. Title.
Q325.G73 1997
003'.7–DC20 96-36381

ISSN 0172-7389
ISBN 3-540-57048-9 Springer-Verlag Berlin Heidelberg New York

Typesetting: Camera-ready copy from the authors using a Springer TEX macro package
Cover design: *design & production* GmbH, Heidelberg
SPIN 10080785 55/3144 - 5 4 3 2 1 0 - Printed on acid-free paper

"When you can measure what you are speaking about, and express it in numbers, you know something about it; but when you cannot measure it, when you cannot express it in numbers, your knowledge is of meager and unsatisfactory kind: it may be a beginning of knowledge, but you have scarcely, in your thoughts, advanced to the stage of science."

William Thomson – Lord Kelvin
Popular Lectures and Addresses [1891-1894]

"Measurement demands some one-to-one relations between the numbers and magnitudes in question – a relation which may be direct or indirect, important or trivial, according to circumstances."

Bertrand Russell
The Principles of Mathematics [1937]

Preface

In the last century the description of natural phenomena in terms of physical laws became well established in the natural sciences. The principal advantage of such a description is that it permits an exact prediction of some quantity, provided that information about other quantities related to the phenomenon is provided. However, there are many complex natural phenomena whose exact description in terms of natural laws is difficult, if not impossible. In contrast, intelligent beings are often able to predict an unknown quantity or property based solely on learning from previous occurrences of a phenomenon. In order to determine whether such an approach can be implemented in electronic information processing systems, it is appropriate to explore the physical aspects of the ability to learn from past experiences. In striving towards this goal, we consider a system comprised of sensors, actuators and memory cells and investigate the synergy of these elements in the system so that it can operate as a modeler of natural laws. The ultimate goal of our investigation is the establishment of a link between a quantitative physical description of natural phenomena based on analytically expressed physical laws and one that is a less strict empirical description which is based on learning from examples.

The past decade has seen an explosive growth in research into artificial neural networks and similar information processing systems. One impetus was the discovery that multilayer perceptrons can operate as adaptive modelers of relationships between empirical data. This is a basic characteristic of an intelligent device and one might therefore expect that our goal could be most easily realized by applying such perceptrons. But a multilayer perceptron is a complex system that possesses intricate behavior that is a consequence of the nonlinear interactions between neurons. Most of the literature dealing with the dynamics of the perceptrons begins with a description of biological neurons and their associated networks. In such a connectionistic approach, however, the modeling capability and related primitive intelligence of artificial neural networks is thus often represented as being an almost incomprehensible, a nearly magical property. Such a treatment is often difficult to accept for scientists and engineers who are accustomed to working with quantitative data that has been obtained from measurements and is then used to describe the properties of natural phenomena in terms of exact relationships between these data.

In this book the physical description of natural phenomena is based on those methods of empirical modeling of physical laws that can be described using the methods of probability and statistics. The fundamentals of these two disciplines are well understood by many of today's engineering graduates who have learned them in courses on applied mathematics, probability and statistics, stochastic processes, measurement science, system dynamics and control. These fundamentals serve as the starting point for our description of the tasks needed to obtain an empirical modeling of natural phenomena. In this context, a sensory-neural network is viewed only as a special kind of information processing system which is capable of performing the specific tasks needed to carry out adaptive empirical modeling and its application. The performance of the sensory-neural network is then explicable as the synergy of measurement, modeling, prediction and control.

This monograph is written mainly for experimentalists and, in particular, students who are interested in experimental work related to the adaptive modeling of natural laws, informatics, sensory-neural networks, intelligent control and synergetics. No knowledge of advanced topics of mathematics is presumed. On the other hand, where it is advantageous, unavoidable or required, we make use of the language of mathematics to provide a generalized explanation of particular topics. Our main thrust is an emphasis on the relationship between rigorous quantitative modeling of natural phenomena that is based on physical laws and empirical modeling that is statistically based. Our focus is on the development of a general information processing system capable of automatically modeling the relationships between quantitative sensory data that can subsequently be used in applications of the model to solve real measurement problems. We hope, therefore, that this book will be of general use to experimentalists, applied scientists and control or process engineers who are interested in the modeling, prediction and control of natural phenomena and technical processes based on the information in sensory signals.

The examples presented in this monograph have been drawn, in large part, from our experimental work and from the diligence of our present and former students and research associates. Many examples are related to the empirical modeling of acoustic emission and ultrasonic phenomena which have been investigated over the past decade in our laboratories during our development of adaptive non-destructive testing techniques and procedures for monitoring, characterizing and predicting chaotic manufacturing processes. But clearly, the ideas we describe are also directly applicable to measurement systems relying on other kinds of sensory signals, be they electromagnetic, optical, radiative or any other.

We acknowledge the cooperation of our respective research groups, the Laboratory for Technical Physics in the Faculty of Mechanical Engineering, University of Ljubljana and the Ultrasonics Group in the Department of Theoretical and Applied Mechanics at Cornell University. The students and collaborators whose work we have cited include Edvard Govekar, Dušan Grošelj, Peter Mužič, Iztok Peruš and Egon Susič. We also appreciate many useful comments and suggestions received from colleagues and students. Of special note are the contributions of Boris Antolovič, Richard Baker, Miran Kokol, Nelson Hsu, Jennifer Michaels, Primož Potočnik, Rom Janez, and Aleš Rukavec in preparing the text or their critical reading and review of the manuscript or portions thereof.

The preparation of this book was facilitated by a number of our exchange visits between Cornell University and the University of Ljubljana. Our research programs have been an essential incubator for many of the ideas described in the text. The research work at the University of Ljubljana has been supported by The Ministry for Science and Technology of Slovenia. And the work at Cornell University has been supported by the Office of Naval Research and the Materials Science Center which is funded by the US National Science Foundation. Recent support has also been obtained from the US Air Force Office of Scientific Research. The authors are deeply grateful for this support of their work.

Ljubljana and Ithaca
April 1996

Igor Grabec, Wolfgang Sachse

Table of Contents

List of Symbols

Symbol	Meaning
	General notations
dx	differential of x
∂	partial derivative
∇_q	gradient in parameter space
$L(\partial_{\boldsymbol{r}}, \partial_t)$	linear differential operator
δx	variation of x
$\eta(s, s')$	arbitrary variation
$\triangle x$	spacing of sample points
	Fundamental variables
L	length
[m]	unit of length
V	volume
m	mass
ρ	density
t	time
ω	angular frequency
Hz	frequency unit
	Functions and transformations
$\sup(x)$	supremum of x
$\min(x)$	minimum of x
$\exp(x)$	exponential function of x
$\log(x)$	natural logarithm of x
$\tanh(x)$	hyperbolical tangent function of x
$\mathrm{sinc}(x)$	sin(x)/x
$\mathrm{U}(x)$	unit step function
$\delta(x)$	Dirac delta function
$I_A(X)$	set–indicator function

$I_k(\boldsymbol{x})$	cell indicator function
$g(x)$	Gaussian function
$w(x)$	triangular basis function, window filter function
$C_i(x)$	normalized basis function, similarity measure between x and q_i
$\Psi(x)$	sigmoidal basis function, neuron response function
$G(\boldsymbol{r},\, \boldsymbol{r}';\, t,\, t')$	Green's function
G^{-1}	inverse Green's function
$G(x_1, \ldots, x_N, h_i)$	generating function
$W(\omega)$	Fourier transform of function $w(t)$
$S(\omega)$	spectral density
$x * y$	convolution of $x(t)$ and $y(t)$
$\mathcal{F}$	functional
λ	Lagrange multiplier

Vectors and matrices

$\boldsymbol{a}$	vector
$\boldsymbol{a} \cdot \boldsymbol{b}$	inner or scalar product
$\Vert\boldsymbol{a}\Vert = \vert\boldsymbol{a}\vert$	norm of a vector
$\boldsymbol{a}\boldsymbol{b}^T$	outer or matrix product
$\boldsymbol{r}$	space vector
$s = (\boldsymbol{r},\, t)$	four-dimensional space-time vector
$w = (\boldsymbol{k}, \omega)$	four-dimensional wave number
$\boldsymbol{R}^n$	Euclidean space
$D(f, f')$	Euclidean distance between f and f'
$\vert B\vert$	determinant
$\boldsymbol{T}$	rotation matrix
$\boldsymbol{I}$	identity matrix
$\boldsymbol{e}$	eigenvector
$\boldsymbol{W} = \boldsymbol{e}\,\boldsymbol{e}^T$	outer product of eigenvectors
λ_k	eigenvalues

Signals and dynamics

$\boldsymbol{X} = (x_1, x_2, \ldots x_n)$	input signal
$\boldsymbol{Y} = (y_1, y_2, \ldots y_n)$	output signal
$\boldsymbol{Y} = \boldsymbol{F}(\boldsymbol{X})$	input-output relation
$\boldsymbol{p}$	system parameters vector
$\boldsymbol{F}(\boldsymbol{X};\, \boldsymbol{p})$	system dynamics generator
$\boldsymbol{\Phi}(\boldsymbol{X})$	continuous chaos generator
$\boldsymbol{\Psi}(\boldsymbol{X})$	discrete chaos generator
$\boldsymbol{F}(\boldsymbol{s}(t),\, \boldsymbol{v}(t))$	driven chaos generator

$G_i(x(t), x\ (t-1);\ \boldsymbol{v})$	cascade expansion term of driven chaos generator
$x_k(t+1)$	cascade expansion term of chaotic state vector
$\triangle V'(t)$	volume changes
Λ	average rate of change of a volume element
$d_{12}(t_i) = \lvert \boldsymbol{X}_1(t_i) - \boldsymbol{X}_2(t_i) \rvert$	distance
M	space dimension
$\{D_n\}$	set of dimensions
D_o	Hausdorff dimension
D_1	information dimension
D_2	correlation dimension
$R_{er}(s)$	relative error
L_{ex}	characteristic exponent

Probability and statistics

s	sample point
S	sample space
$A = \{s_1, s_2, \ldots\}$	elementary event
$\emptyset = \{\}$	non-realizable event
$A \subset B$	implication or inclusion
$C = A \cup B$	union of events
$C = A \cap B$	intersection of events
A^c	complement of A
Φ	relative frequency
$P[A]$	probability of event A
$P[A \vert B]$	conditional probability of event A at condition B
p_i	probability of x_i
$F_X(x) \equiv P[X(s) \le x]$	cumulative probability distribution function of variable X
$f_X(x)$	probability density function; PDF
$f_{X,Y}(x, y)$	joint probability density function of X and Y
$f(x \vert y)$	conditional probability density function
$\Delta f(\boldsymbol{X})$	perturbation of probability density function
$\psi(x; \Delta x) = \frac{\Phi_{\Delta x}}{\Delta x}$	estimator of PDF
η	bias
$\{X(n, s);\ n = 1, 2, \ldots\}$	discrete parameter random process
$\{X(t, s);\ t \in (0, T)\}$	continuous parameter random process
$x(t, s)$	a sample record of random process
$\boldsymbol{X}(t)$	representative vector of random process
$\mathrm{E}[X]$	expected or mean value of random variable X
$\mathrm{E}[X \vert y]$	conditional average of X at given condition $Y = y$

m	mean value
$\langle x \rangle$	sample mean
$m_X(t)$	mean value of random process at t
$\mathrm{var}(X)$	variance of X
R_{XY}	cross-correlation of random variables X and Y
$R_{XX}(t_1, t_2)$	auto-correlation function
$R_{XY}(t_1, t_2)$	cross-correlation function
$\mathrm{cov}(XY)$	covariance of x and Y
σ_{ij}	covariance of components X_i and X_j
$\boldsymbol{\Sigma}$	covariance matrix
$\boldsymbol{B} = \boldsymbol{\Sigma}^{-1}$	inverse covariance matrix
A	correlation coefficient
τ	correlation width
$\Sigma(t)$	covariance function
$C_w(h_i)$	correlation criterion
$C(s)$	correlation integral
N	number of samples
K	number of prototypes
$k(N)$	number of neighbors
q_i	prototype
σ	window or transition interval width
$h(x)$	kernel width

Information and entropy

k	unit of information
H	entropy of information
K_n	entropy of joint information
K	Kolmogorov entropy
$\mathcal{K}$	complexity measure
ΔI	information change, Kullback information
Ψ	information functional
G_k	constraint
$g_k(x)$	function defining constraint
$D(f, \phi)$	mean square difference between f and ϕ
$D_i(f_1, f_2)$	information divergence between f_1 and f_2

Neural networks, modeling and control

a	activation
U	electrical membrane potential
θ	threshold value
m_{ij}	weight, memory element
$\boldsymbol{M} = [m_{ij}]$	weight, memory matrix
$\boldsymbol{N}$	interconnection matrix of neurons
c_i	weight memory element

τ	adaptation time
I, L	in-flow and loss term
β	forgetting constant
ϕ	forgetting rate
η_i	adaptation rate
$\boldsymbol{\Phi}$, $\boldsymbol{\Omega}$, and $\boldsymbol{\Theta}$	generators of NN dynamics
$\mathcal{E}$	error measure of NN response, reconstruction error
y_i^l	response of i-th neuron on l-th layer
a_i^l	activation of i-th neuron on l-th layer
m_{ij}^l	synaptic connections of i-th neuron
δ_i^l	sensitivity terms in back-propagation rule
$\boldsymbol{g}, \boldsymbol{h}$	truncated input and output vectors
$\widehat{\boldsymbol{h}}$	estimator of truncated output vector
$\boldsymbol{\Sigma} = \triangle \boldsymbol{h}_p \triangle \boldsymbol{g}_p^{\mathrm{T}}$	covariance matrix of a linear regression
$\overline{\boldsymbol{g}}, \overline{\boldsymbol{h}}$	central values of linear regression
$z_k(\boldsymbol{x})$	reference function
$\boldsymbol{q}_k$	centroid
$\boldsymbol{Q} = (\boldsymbol{q}_1, \ldots, \boldsymbol{q}_i, \ldots, \boldsymbol{q}_K)$	parameter vector
$\delta \boldsymbol{Q}$	change of parameter vector
$\boldsymbol{Y} = \boldsymbol{H}(\boldsymbol{X}; \boldsymbol{Q})$	system response
$\mathcal{D}$	discrepancy, performance measure
α_i	adaptation constant
$\boldsymbol{G} = \nabla_q \otimes \nabla_q \mathcal{D}(\boldsymbol{Q}; f)$	second term in the series expansion of D
γ	random fluctuation term
$\{C_k;\ 1 \leq k \leq K\}$	partition to cells
$\boldsymbol{B}$	driving term of perturbation
$\boldsymbol{C}$	adaptation matrix
$\boldsymbol{c}_i$	evolution coefficients
η	adaptability of neurons
$\alpha(t)$	adaptation constant
$\langle g \rangle_e$	empirical average
$\langle g \rangle_r$	representative average
$\epsilon = \langle g \rangle_r - \langle g \rangle_e$	difference of averages
$\overline{\epsilon^2}$	mean square difference of averages
C_{lmki}	adaptation matrix elements
B_{lm}	driving term elements
$\Delta \boldsymbol{q}_l$	change of prototype vector
$\boldsymbol{V}$	excitation state vector
$\widehat{Y}$	predictor of random variable Y
$\widehat{\boldsymbol{Y}} = E[\boldsymbol{Y} \mid \boldsymbol{X}]$	conditional average predictor
$C_n(\boldsymbol{x}), C_n(j,k)$	similarity measure coefficients in conditional average
$\boldsymbol{Z}_c = \boldsymbol{X} \bigoplus \widehat{\boldsymbol{Y}}(\boldsymbol{X})$	concatenated vector

$z^{(i)}$	iterated variable
ζ	limit point of iteration
$\boldsymbol{C}(\boldsymbol{z})$	vector of relative excitations
E_1, E_2	error measures
A_i	iteration correction constant
$\boldsymbol{v}$	correction vector
$\boldsymbol{u}$	projection to regression line
$u_i(\boldsymbol{r},\, t)$	displacement field vector
$\gamma(\boldsymbol{r},\, t)$	source force density
$\boldsymbol{Z}(\boldsymbol{r}, t, i)$	pattern field vector
$\boldsymbol{z}(t)$	vector in embedding state space
$\boldsymbol{z}(t) = [\boldsymbol{x}(t), \boldsymbol{y}(t)]$	state vector concatenated of control and response variables
$\boldsymbol{v}(t) = [\boldsymbol{r}(t), \boldsymbol{s}(t)]$	environment state vector concatenated of reference and disturbance variables
$u(t)$	instantaneous utility
J	performance measure
$\boldsymbol{Z}_t$	joint state vector
$\boldsymbol{G}(t)$	joint condition vector
$\overline{J(t_i, t_f)}$	statistical expected value of performance measure
$\mathcal{Q}$	set of representative prototypes
$\widehat{u}(t)$	predicted utility
$\widehat{J}(t_i, t_f)$	predicted performance measure
$\overline{J^*(t_r, t_f\,;\, \boldsymbol{y}(t_r))}$	optimal performance measure
$\mathcal{X}_{ad}$	set of admissible controls
$\mathcal{Q}^*$	set of optimal representative prototypes
$\boldsymbol{S}(t) = [\boldsymbol{z}(t), \boldsymbol{v}(t)]$	joint organism-environment state vector

1. Introduction

1.1 Goal

The progress of natural sciences and technology is closely related to quantitative research work. Its methodology is well established and can be described as follows: An object or phenomenon of interest is first quantitatively explored by various measurements and then the experimental results are represented in terms of empirical relations which may subsequently be analytically described in terms of physical laws. The research work in various fields of natural sciences and technology is performed by scientists, utilizing similar experimental techniques and information processing methods. Therefore, a question arises whether it might be possible to develop a general type of machine that could perform such work autonomously, analogous to robots that do mechanical work on industrial production lines. In order to be able to design such a machine, the general procedures of quantitative research work must first be formalized so that such work can be described as a process similar to that which might be used as an industrial test procedure. And then, this process must be implemented using an appropriate technique.

Our goal in this monograph is to provide a description of the general properties of quantitative research work that is based on measurements and the modeling of physical laws. We further wish to establish the foundations needed for their implementation and automatic execution via electronic devices. Our principal idea is that an empirical physical law is nothing more than an encoded presentation of compressed quantitative information about some general property of nature that can be acquired, organized, stored and presented to a user using an information processing device. [5] A system capable of synergetically performing these tasks will include elements called sensors, processors and actuators and these can be realized either electronically or in some other way as is the case, for example, in biological systems. [2, 3, 7, 8, 10, 11] In fact, the biological world offers a broad diversity of systems that have been developed by natural selection during evolution for the purpose of sensing, information processing, and control and these systems can serve as examples for the development of artificial systems that are designed for similar tasks. While our interest is towards a quantitative physical treatment of natural phenomena, it should be self-evident that our approach is quite naturally related to research of neural networks. Our approach, addressing only empirical information about a natural phenomenon, really represents just the first step towards an automation of scientific work. But in science, analytical work is equally important, as it provides a basis for understanding and methods for formulating natural laws. The related problems are presently studied in the field of artificial intelligence and will not be considered here. One might ask what could we achieve in a scientific field if we exclude the analytical descriptions of natural phenomena from consideration? That is, would it be at all possible to completely avoid analytical work in developing a scientific description of natural phenomena?

It is well known, that the ultimate goals of fundamental and applied quantitative sciences are somewhat different. In the fundamental sciences, one seeks to describe natural phenomena in terms of analytical physical laws which are not contradictory to empirical observations. In this way, one obtains a firm basis for understanding the phenomena that provides directions for further analytical research. In contrast, the ultimate goal of the applied sciences is to provide a description of natural phenomena that will permit a quantitative prediction of some of the characteristics of a natural phenomenon provided that some complementary information is given *a priori*. In this application, empirical relations are equally useful as are analytical physical laws. [5] Even more, one does not need a mathematical relation but merely a model in order to make a quantitative prediction. In this regard, a mathematical relation is equivalent to a mechanical, biological, electronic or any other model that permits a transformation of information. Our goal is to formulate a general basis that will permit a quantitative empirical modeling of natural phenomena and thus also provide a common basis for its implementation in various techniques and common applications in the applied sciences. While an analytical treatment is utilized to obtain a description of the modeling process, the model itself can be implemented using any technique based on electronic, biological or numerical means.

The benefits which result from a quantitative description of natural phenomena in terms of physical laws are most significant in those engineering applications in which such a description leads to a reliable solution of problems related to the prediction of the characteristics of manufactured products, the description, forecasting and control of technical processes and to the planning of production, etc. There are usually two components in the specification of an applied problem. First, the natural laws must be specified that govern the properties of the phenomenon under consideration and second, there must be given some specific parameters and initial or boundary conditions that specify in which particular environment the phenomenon occurs. As a typical example, we consider the prediction of the temperature in the interior of a specimen that is heated in an oven. In this case, we apply the heat diffusion equation and specify the diffusion coefficient of the material as well as the initial and boundary conditions. The problem can be further solved using the principles of mathematics. Although this route appears advantageous and straightforward, unfortunately, it often cannot be followed. First of all, there are many natural phenomena which, because of their complexity cannot be simply quantitatively described by physical laws. Typical examples are processes taking place in biological systems. Second, there are many phenomena where the natural laws are known but the boundary or initial conditions are either not known or cannot be specified because of their complexity. As a typical example, we point to the problem of describing the fluid flow in an irregular river bed. Beside this, the execution of mathematical procedures involved in a quantitative solution of a well-specified problem may not be economic if they require a cumbersome analytical treatment, tedious programming, or time-consuming numerical calculations. As an example, we can mention the description of the deformation of a piece of an inhomogeneous material of geometrically complicated form under dynamical loading. The question thus arises how to best describe such phenomena. That is, is there a procedure by which the aforementioned problems could be efficiently solved without losing most of the benefits obtained by using a quantitative physical and mathematical description of natural phenomena?

In order to obtain an answer to the posed question let us consider an example from the field of non-destructive materials testing. An inspector who is to ascertain the quality of a certain manufactured product does not calculate the distribution of internal stresses resulting from a particular loading and thermal history. Nor can an inspector load a specimen to failure in order to estimate if an existing defect in the specimen will result in failure of the material under a prescribed loading. For this purpose, it is often sufficient that the inspector learns to estimate the severity of a defect from measurements of the mechanical strength of similar samples. Such an estimation is based on a recognition of defects and an associative reasoning by the inspector. This approach is far less exact than a rigorous, mathematically based treatment, but it is less demanding and it is applicable in a wide variety of situations. The recognition relies on sensing which is used instead of quantitative measurements so that the main difference is, in fact, in the processing of the information which differs significantly in a mathematical treatment from that in the associative reasoning. Consequently, the question arises as to how one can properly relax the need of a mathematically strict, quantitative description of natural phenomena to a process that relies on recognition and association-based processing that includes sensing from given examples. Learning from examples is a typical attribute of natural intelligence and it is our goal to demonstrate how one can quite formally proceed to obtain a description from the fundamentals of quantitative sciences. Such a demonstration is done intentionally because the methods of quantitative science are widely accepted in engineering and we expect that the transition from a rigid, physically-based description of technical problems to a soft or fuzzy one that is based on learning from examples, can best be introduced by treating both descriptions in common. [11]

1.2 Relation to Other Scientific Fields

Artificial learning systems are the object of intense investigation in the fields of neural networks (NN) and artificial intelligence (AI). [1, 6, 8] In the former, one usually begins with a specification of simple adaptive elements called formal neurons. The main task is then to find such an inter-connection of neurons in a network that will transform the input into output signals in accordance with a set of presented examples. In contrast, in the field of artificial intelligence one begins with a set of logically well-defined rules and attempts to construct, based on given examples, a logical system that is capable of finding a solution of the posed problem from available information. In fact, both approaches deal with information processing that is needed for the modeling of natural phenomena, although this is usually not explicitly pointed out. The processing in both NN and AI approaches is based on specific methods that are not yet widely adopted in most technical disciplines. Therefore we try to proceed to a description of learning systems by following the general route of quantitative scientific exploration.

We begin with a statistical treatment of measurements and describe the fundamental steps of empirical modeling of natural laws in such a way that they could be performed automatically on an electronic information processing system. We call such a system an *empirical modeler of natural laws.* [5] Our fundamental problem is then to specify the structure of a modeler that optimally stores and recalls the empirical information provided by measurements. An analytical treatment of this optimization problem indicates that the structure of the corresponding optimal modeler resembles a self-organizing neural network. [4] For this reason, we need not begin with a description of an artificial NN. The optimal storing of empirical information is achieved by a self-organized synergetic operation of modeler memory elements while an optimal utilization of stored information is provided by non-parametric regression. The complete procedure of empirical model formation and its application corresponds to a formalization of the fundamental tasks of quantitative applied sciences that are included in sensing, storing and recalling of information. Our method is complementary to the connectionistic one utilized in the field of neural networks as well as to the logical approach that forms the basis of artificial intelligence, although there are many common properties of all three fields. In fact, the method we shall describe in this monograph corresponds to a top-down approach from quantitative science to neural networks.

In current measurement systems, the acquisition and processing of data, is performed by standard sensors and instrumentation controlled by sequential digital computers. But the algorithms stemming from our approach suggest implications for automatic modelers based on analog, parallel processing techniques. What is the basis for this suggestion? In an analytical approach for solving problems based on the application of natural laws, we usually employ precisely specified arithmetical operations performed on accurately specified quantitative data. For this purpose, a digital computer is well-adapted. Its sequential operation is often appropriate for a complex programming of algorithms relying on step-by-step operations. In contrast, a solution of similar problems based on learning from examples relies on a comparison between some given, most often multicomponent data with a vast amount of memorized data. Such an operation generally does not require great precision but rather, speed. A sequential digital processing of the data is, in this case, not as advantageous as an analog and parallel processing. At the same time, the utilization of stored information that is related to the prediction of data based on non-parametric regression does not require complex or very accurate operations. It also appears that the self-organization which is needed in an optimal storage of empirical information can be simply implemented using analog and parallel techniques and a proper connection between memory elements without any complicated programming.

There is yet another reason in favor of analog and parallel processing which is not immediately obvious. When dealing with complex natural phenomena where many different parts synergetically contribute to the observable properties, one needs to use many sensors of different types to better specify a particular phenomenon. In this case, one needs to deal at the outset with data comprised of huge numbers of components so that sequential and digital performance of even rather simple operations can essentially delay the determination of the final result. A typical example of this kind is the control of a biological system using visual information of a dynamically changing scene. In this case, an eye provides the signals of millions of sensors operating in parallel. As a consequence, the advantage of systems resembling biological sensory-neural networks that utilize analog and parallel processing of information becomes outstanding for the empirical modeling of natural phenomena and its application dictates development of corresponding electronic devices like networks of sensors and analog processors connected in parallel. [9] The trends of development in the advanced electronics industry indicate that this will soon be the case and we need to prepare fundamentals for their efficient application. However, there are many other characteristic properties of biological information processing systems that seem to be related to analog and parallel operation but are at present still not completely explained and understood. As a typical example we mention the growth or decay of a sensory-neural network without reprogramming its operation.

With all above mentioned properties in mind, one might ask: "Why is this research at all needed and why are we not satisfied with so well-developed quantitative science? Or, wouldn't it be sufficient to develop more powerful processors, that would eventually operate in parallel and in an analog mode and then apply existing methods of quantitative and analytical science?" The answer is, "No!" The argument is very simple, although it appears philosophically motivated. The most sophisticated systems in nature that are also capable of coping with very complex phenomena are sensory-neural networks of living beings. Nature has developed them mainly in relation to the optimal control of biological organisms and has not incorporated into their operation the execution of arithmetic operations in such a manner as we perform them in computations. A bird does not solve the equations of aerodynamics in order to fly. Mathematics and a quantitative physical description of natural phenomena have emerged as a result of the evolution of human civilization and the development of related instruments and measurement techniques only in the period of last few millennia which is negligibly short in comparison with the time of existence of living beings on earth. Prior to this, the origin of all knowledge and its utilization was solely based on direct observation and it was sufficient to support life and related control of biological organisms in various environments. This indicates that there must exist an entirely empirical and operational means that permits an efficient modeling and prediction of natural phenomena. Unfortunately, we do not yet know how to develop a processing system similar to that of a brain. [8, 10] Our hope is that the success of quantitative sciences when analyzed and followed in detail, will permit a general description of the process of exploration and thus open the door to an operational description of nature which will lead to an understanding of the fundamental properties of natural intelligence. We also expect that this will suggest directions leading to the development of systems exhibiting similar properties as biological sensory-neural networks. Without doubt, we are only at the beginning of this journey and we do not want to start with highly developed systems that are operating on the basis of sophisticated logical rules. We anticipate that such rules will quite naturally emerge from the description of the properties of optimal modelers. Further, we also do not wish to start with a description of the evolution of biological systems on a molecular scale which would be enormously complex. [2, 7] Our intention is rather, to begin with the simplest systems including those which are capable of sensing, storing and a simple processing of information.

We suppose that such systems have developed in the biological world because they result in an optimal behavior of living beings in a given environment. [7, 11] An optimal behavior corresponds to an optimal self-control, for which a modeling of natural phenomena is needed. In the formulation of optimal modeling, we do not follow the evolution of living beings which we do not know in detail, but rather, we try to obtain suggestions from the analysis of the methods of quantitative sciences and we turn first to sensing that is according to our opinion the basis of the quantitative exploration of natural phenomena. Following this, is the processing of information, which we can explore step by step by utilizing analytical methods as well. And finally, there is an efficient utilization of information which can be analyzed based on the methods of quantitative system science and control. By proceeding along this route, we are, in fact, presuming that one can obtain an explanation of natural intelligence based on the knowledge of basic properties of natural phenomena that are describable in terms of quantitative physical laws and thus to fill the gap that exists between quantitative natural sciences and humanistic ones. At present we still do not know how and where along this journey will emerge the typical characteristics of intelligence that can be described as thinking, concept formation, abstraction, consciousness, etc. We also do not know how to describe the characteristics of behavior of intelligent beings that can be related to emotions. But we suspect that the answers to these and related questions will be found near the end of this journey.

1.3 Plan of the Monograph

It is our plan in this monograph to follow and to analyze the fundamental steps of quantitative empirical description of natural phenomena and to show how they can be automatically performed by existing information processing systems. In Chap. 2 we first propose an explanation of the development of structures and the evolution of biological organisms on the basis of synergetics. Although we are not yet able to exactly model how life can emerge from the fundamental interactions between atoms and molecules, the concepts of synergetics do provide a basis for our rudimentary understanding of many of its characteristic features. The synergetic view also helps us to understand why it is advantageous that the controlling organism of a living being operates as an empirical modeler of natural phenomena. Chap. 2 also includes some fundamentals of quantitative science and a description of the basic components needed for making systems capable of automatic measuring and modeling of natural phenomena. The input unit of such a system is an array of sensors. Therefore, Chap. 3 is devoted to the characteristics of transducers and basic properties of the biological sensors and neurons. By interfacing between a physical phenomenon and a measurement system, the transducers provide the basic information about a natural phenomena in terms of signals which are treated as representatives of various physical variables. The total, combined output from an array of sensors is treated as a multi-component vector variable which represents the observed state of nature. The signals excited in sensors by various states of the real world are generally of random character and therefore we utilize a statistical treatment for the description of physical variables. As a basic tool for the quantitative description of natural phenomena, we introduce the probability distribution of a vector random variable. Its characterization, in terms of empirical samples, represents the basic step of modeling empirical natural laws and it is elaborated in detail in Chap. 4. This explanation is supported by a description of some fundamental concepts of probability and statistics which is given in Appendix A.

The acquisition, modeling and application of natural laws is intuitively understood as a processing of information. Therefore we include into the description also the mathematically formulated concept of information which is presented in Chap. 5. This leads us in Chap. 6 to a formulation of the maximum-entropy principle which is applicable for the optimal estimation and storage of the density of probability distributions of sensed variables. In real world problems, the estimation of the probability density and the corresponding modeling of natural laws proceeds adaptively based on the acquisition of empirical data by the measurement system. The basic problems related to adaptive modeling are described in Chap. 7.

A statistical treatment of variables and the presentation of relations between measured quantities by models leads quite naturally to the problem of maximal preservation of empirical information and to the formulation of an optimal modeler. Solution of the corresponding optimal problem includes two characteristic phases which can generally be described as the self-organized storage of data and the optimal recall of stored information. [5, 4] Both phases represent a process of learning and estimation from examples and they are treated separately in Chaps. 8 and 9. In Chap. 8 the maximum-entropy principle is first utilized for the derivation of an algorithm that describes an optimal storage of empirical information. It corresponds to a general modeling of natural laws by a self-organizing memory. In the following chapter, the problem of the optimal utilization of information that is stored in the model for the purpose of prediction is solved by deriving a general non-parametric regression.

The complete problem of learning from examples is currently under intense investigation [8, 10, 11] and we do not expect that our approach based on a self-organized neural network and non-parametric regression will immediately provide a full explanation of natural intelligence but rather only some of the basic properties of learning systems. Our approach is supported by results of experiments in a number of diverse fields ranging from an analysis of acoustic emission and ultrasonic phenomena in non-destructive testing to the prediction of seismic capacity of structural elements in civil engineering and to the modeling of healing processes in medicine. Together, they demonstrate the applicability of the information processing systems constructed in accordance with some general properties of modelers of natural laws.

In some measurement situations, one observes that the measured data are invariant with respect to certain transformations. Typical examples include changes in signal amplitude, shifts in time, and translations in space. Hence, any modeling of natural phenomena needs to consider the problem of how one includes such invariances into its execution. This problem is considered in Chap. 10 which also presents some examples of the application of linear models.

Principles identical to those used for the automatic presentation of empirical laws are also applicable for the modeling and prediction of chaotic processes. The corresponding derivations are presented in Chap. 11, while the fundamentals of deterministic chaos description are reviewed in Appendix B. The underlying mathematics of the applications appearing in Chaps. 8–11 demonstrate that for an optimal application, one needs an information processing system comprised of synergetically operating and self-organized units whose operational characteristics mimic the functional properties of biological neurons. For this reason, we review some of the characteristic properties of artificial neural networks in Chap. 12. There we also describe the connection between a conditional average estimator and the three-layer perceptron model.

By modeling chaotic dynamical phenomena, one can predict their occurrence. Further, by modeling the operation and performance of driven dynamical systems one can form the basis of optimal control procedures. The route is thus open for the development of intelligent controllers that will be able to learn from examples presented to it, how to control a system in a given environment which itself might even also be chaotic. In Chap. 13 we present some of the fundamentals that are needed for this purpose and we demonstrate the applicability of empirical modeling to tracking and cloning, and we describe an empirical approach to optimal control that could be further generalized to obtain an intelligent controller. Research of the latter control appears very promising for applications in a number of fields. During biological evolution, natural intelligence developed mainly in relation to the control of biological systems in various environments. Therefore we expect that the corresponding research of artificial self-controlled systems could lead to an explanation of the fundamental properties of natural intelligence as well. [11] Some basic problems related to the development of self-controlling systems are summarized in Chap. 14.

What is the long-term goal of this approach? It is nothing less than finding answers to questions related to the building of information processing systems which are capable of self-organized extraction and presentation of general properties of natural phenomena, their representation in terms of abstract objects and the creation of new concepts for optimal control and eventually also for autonomous analytical scientific work. Although this goal seems like science fiction, we expect that this research will finally lead also to a quantitative or physically-based explanation of consciousness. An idea for an operational description of consciousness is also presented in Chap. 14.

2. A Quantitative Description of Nature

2.1 Synergetics of Natural Phenomena

One of the most important characteristics of Nature is its non-uniformity. If it were homogeneous and stationary there would be no distinct objects or events and no life. The inhomogenity of Nature extends from the subatomic to the cosmic scales and the nonstationarity includes events whose characteristic times range over all time scales. The non-uniformity of Nature is observable in the variations of different properties of Nature in space and time. The most widely used variables for their physical description are the mass density, the chemical composition of substances, pressure, temperature, and mass flow distribution, among others. Because of the spatial and temporal fluctuations of these variables, various dynamic phenomena occur in Nature. Weather phenomena and the growth of living systems are just two examples. Any dynamic phenomenon in Nature can be related to a flow or exchange of energy between parts of a system. For some natural phenomena, this exchange of energy occurs smoothly and quasi-statically, but in many cases, when the exchange of energy is increasing, a sudden transition in the behavior and structure of the system is observed. [5, 11, 17, 23] In this new state, the dynamic behavior dominates the structure of the system. In those cases in which the exchange of energy is high, the exchange may occur with strong fluctuations.

In the past two decades a new interdisciplinary research field called *synergetics* has emerged. [11, 12] It deals with the exploration of natural phenomena for which a synergy or cooperation of many coexisting units is characteristic. These units may be atoms, molecules, cells, electronic elements, organisms, plants, animals, computers, societies, etc, and comprise a set that can be generally called a population. The main goal of synergetics is to find and explore those properties of underlying phenomena which are consequences of cooperation between members of the observed population and could therefore be treated as generic and common for many different populations with similar properties of interaction. Most common and characteristic for synergetic systems is that the cooperation between the constituents may lead to the formation of expressive structures in space and time. The state of a synergetic system is usually characterized by densities and interactions of constituents in space. Its evolution in time can be quite generically described by a nonlinear equation of generalized reaction-diffusion type. It is helpful in a quantitative exploration and explanation of the properties of synergetic systems, especially in the study of their self-organization and emerging structures. Here we shall not review in depth this research but rather we shall only mention some characteristic, qualitatively expressible properties of synergetic systems that could lead to a physically based explanation of natural evolution. [17]

A simple, well-known example is the streaming of water in a pot in contact with a hot plate. If the temperature of the heater is just above the temperature of the surroundings, then the layer of water conducts heat from the plate to the upper surface of the water and to the environment with little observable macroscopic fluid flow. But as the temperature of the heater is increased, this steady state becomes unstable and an observable fluid flow called Rayleigh-Bénard convection abruptly develops in the pot. [10, 11, 12] If the height of the water is comparable to the radius of the pot and the heater temperature is approximately constant, then water streams up along the central axis of the pot and flows down at the outside walls. This anuloidal flow resembles a rotating motor driven by the temperature gradient. If the fluid level is decreased, then the single roll disintegrates into several coexisting smaller rolls or cells. The spatial ordering of the cells in the developed state need not be regular or periodic but also quite irregular which indicates the presence of chaos. In spite of the chaotic interaction between the various cells of some population the structure of individual cells is rather stable and emerges as a consequence of *self-organization* in the system. It is interesting that this phenomenon resembles some of the characteristic properties of living organisms. First of all, it is a consequence of a dynamic phenomenon related to the transfer of energy from one part of the environment to another. If the properties of the environment are slowly changing, then changes in the structure occur which correspond to the evolution of the system. If proper conditions for the existence of rolls develop in the surroundings of the structure, then the existing structure causes a synchronized development of new cells, analogous to the replication or growth of an organism. And finally, when the source of the energy disappears, the structure dies.

Qualitatively similar natural motors can also be observed in quite different environments, as for example, in chemical reactors, lasers and discharge tubes, solid state devices, spin glasses, etc. [3, 4, 11, 17, 23] They can also appear at certain intervals during transient phenomena, as for example, during an explosion or the failure of a material, during the magnetization of a ferromagnetic material or the polarization of a dielectric, etc. In all of these phenomena, it is common that the characteristic physical variable develops as a growing fluctuation that is a consequence of cooperation of many constituent parts of the system. This instability is driven by an inhomogeneous distribution of energy in the system and it generally results in the development of a structure, which may be either ordered (i. e. periodic) or disordered (i. e. chaotic). [10] It is characteristic for unstable dynamical phenomena that they usually sensitively depend on the initial conditions or disturbances from environments. This further means that a natural motor with a proper construction can drive an unstable phenomenon in such a direction that the motor obtains more energy than it needs for its motion. If the energy source of a natural motor is suddenly disrupted, then some of the characteristics of the dynamical structure may become frozen and preserved in the environment. Many objects found in Nature exhibit tell-tale signs reminiscent of previously existing natural motors. Fossils are a good example. At the same time, other objects in Nature, such, as for example, riverbeds, mountains, volcanoes, and even continental plates are the result of still operating natural motors. Specific forms which are comprised of ordered sub-units or cells are most characteristic for the living world or animals and plants which can be treated as chemical natural motors that operate because of the inhomogeneous distribution of the free energy in the environment. [3, 17]

One of the characteristic properties of Nature is that the stationary states of its constituents are often only metastable. [5, 11, 23] Under the action of a particular disturbance, which we term a *proper* disturbance, a metastable state can be destroyed and changed into a new, more stable state. In this process, the energy difference between the two states is released and transferred to the source of the disturbance. This property is essential for the existence of perpetual dynamical phenomena which form the basis of a number of oscillators as, for example, a clock. In a clock, a weak disturbance generated by the pendulum results in a transfer of energy from the weight, spring or a battery, back to the pendulum where it is spent to compensate for the energy losses caused by friction. In such a system, the energy flow is triggered by that portion of the system which obtains the energy. Therefore, the properties of the system must be adapted to the metastable energy source. The bound energy of various metastable states thus represents a potential possibility for the development of natural motors. This occurs only if the construction of the corresponding motors is properly adapted to the source.

Biological organisms are typical examples of natural motors that operate because of the energy stored in metastable chemical compounds within them. Their structures have been adapted to various possible sources of food or energy in the environment by a process of natural selection during the evolutionary development of the organisms. The survival of a living being depends on its ability to find food. During evolution the abilities of movement and recognition have evolved in animals. For the survival and development of a species, the level of development of these abilities has presumably played a decisive role in the process of selection.

How biological evolution has proceeded from simple organic compounds to highly developed organisms is not entirely clear. [17] But from studies of the properties of available examples, it has been found that the most common feature of all living beings is their *expressive structure*. By this we mean those identifiable features which differentiate one species from another. An animal's expressive structure is replicated during reproduction. The similarity of species in successive generations assures that they react similarly to similar situations. The result is an optimally adapted pattern of behavior of a species to a given environment which further assures the continuation of the species. The pattern of behavior can be represented by means of various prototypical coordinated actions. In order to preserve an optimal behavior of the species in an environment, the structure of individual organisms must be transmitted from parents to their offspring. This can be described as a transmission of information. In actuality, this occurs by transmitting the fundamental building blocks, represented by DNA and RNA molecules, or genetic code from generation to generation. The survival and reproduction of optimally adapted beings in the process of selection has given *significance* to the information represented by the structure of DNA and RNA molecules and the corresponding genetic code. [17] The complete process of selection and preservation of information about an optimal structure has proceeded from simple organic compounds to the organisms living today by continuation of species as an entity. In this aspect, such transfer of information is crucial for the existence of life.

But in addition, the concept of information is needed to describe how a living system interacts with its environment because the survival of an animal depends on information from this environment. The source of food, or energy available to an animal, is obtained from species found in its surroundings. Because of the well-expressed form or pattern of behavior of a species, a predator animal obtains indications of encoded information about the free energy available in the prey or plant living in its surroundings. Therefore, there exists the possibility for an animal to obtain food if the properties of the objects in its surroundings are somehow transformed into actions of the animal. For this purpose, sensing and recognition of patterns in the surroundings as well as the self-control of organisms based on the processing of information have developed. It is clear that the principal purpose of an animal's sensory–nervous system is to find a proper, eventually an optimal behavior of the animal in its environment. This behavior includes the tasks of feeding, reproduction and survival.

It would be an extraordinary achievement in the natural sciences if an analytical, physically based quantitative description of biological evolution could be formulated. For this purpose the theory of adaptive evolution of Darwin alone is not sufficient because it does not explain what kinds of biological systems can evolve in Nature but there is needed a physically and chemically based approach to the research of this problem. [3, 17] Unfortunately, the diversity of substances and their complexity make it impossible to complete this task today even with all the tools of modern science. However, simple numerical simulations of the behavior of cellular automata explicate the evolution of primitive organisms in terms of elementary processes which occur between ever more complex constituents of an artificial environment. [17, 20, 29] According to such models, a transition from an inanimate to an animate state in Nature is similar to the development of complex natural motors. With increasing changes of an unstable environment, conditions may arise in which simple motors first evolve from primary fluctuations similar to the phase transitions that occur as a material is heated. From the interaction and self-organized cooperation of simple motors, a cascade of increasingly complex dynamical structures, resembling primitive organisms, then evolve. The evolution of living organisms is thus explained as a fundamental dynamical property of an inhomogeneous and unstable environment. According to this view, biological evolution in Nature consists of a sequence of collective or self-organized transitions occurring at various levels of complexity of synergetic dynamical chemical systems. [3, 17] Because of the finite free energy that is available in an environment, a competitive selection favors a particular combination of constituents such that the surviving species possess distinct properties. By this process, instinctive as well as intelligent behavior of a species evolves.

For the evolution of the living world, one individual is of little importance. Rather, selection is carried out by variations of individuals occurring with the interchange of genetic information at replication during reproduction. A species of more or less interacting individuals is thus the fundamental unit by which evolution proceeds. However, various species are normally interacting. In fact, the entire biosphere is involved in the process of coevolution. For the survival of an individual, not only its behavior, but also the properties of the other individuals in its and other species present in the environment are important. As a consequence of this interaction a synergetic behavior has emerged in species which determines the characteristic patterns of behavior or relations between individuals and their environment. Synergetic behavior is mainly the consequence of the communication between individuals of species. In this way good or bad experiences can be transmitted which again can increase the possibility of survival.

One of the most characteristic properties of highly developed societies, as, for example, families of bees or ants, flocks of birds, herds of animals or even human society, is a well-developed system of communication between individuals. [3] For this purpose, various communication modes have been developed which utilize the senses and activities possessed by the individuals in a family. Among these, the chemical transmitters of smell and the mechanical transmitters based on gestures and voices are the most developed. For an effective reception and utilization of information, an individual must possess a proper sensing system, a decoder of the received messages and an activator which transforms a message into a proper action. The ability of information utilization can either be inherited or learned. The first is instinctive while the second represents intelligent communication.

The development of the capability to communicate based on learning has enabled an efficient transmission of experiences between generations of species. In human society, this has led to the development of language, abstract reasoning and the development of science. In the context of a synergetic explanation of evolution of intelligent beings, the emergence of science can be treated as the latest level of the self-organization process in human society. Scientific activity based on the systematic search, encoding, storage and utilization of information increases the possibility of survival of a society.

It is interesting to note that not only human society but also science has undergone a marked evolution. By following this evolution, at least in its most critical steps, one may even hope to obtain ideas for forecasting future trends in science. Among the notable breakthroughs in the evolution of science, we may include: The application and development of tools and weapons, the discovery of fire, the development of agriculture and food preparation, application of signs and the development of writing for the transmission of information, the invention of numbers and their use in describing natural properties from measurements, the systematic collection and storage of knowledge and discoveries from Aristotle to today's data bases. Also included should be the discovery of natural laws, such as the theorem of Pythagoras, the formulation of logical structures such as the axioms of Euclid, the differential calculus of Leibnitz, the emergence of the theoretical sciences and the mapping of observations to models, and their utilization for the prediction of unknown properties of Nature such as electromagnetic waves by Maxwell, gravitational singularities or black holes by Einstein and anti-particles, like positrons, by Dirac. Complementary to this is the discovery of the elements and the structure of chemical compounds, the invention of mechanical, steam and electrical machines, the discovery of electricity and electronic devices including the invention of electronic information processing systems, and the discovery of natural evolution, genetic code and the possibility of induced mutations.

It is possible to represent the evolution of science in the form of a tree. One might then ask what are the common concepts in the tree of scientific evolution that might be thought of as being the principal parts? The roots may be used to symbolize the evolution of living beings, involving: 'self-organization → adaptation → sensing → control → movement → optimal self-control'. In this tree, the trunk would symbolize the evolution of intelligent beings involving: 'recognition → thinking → communication → work → discoveries', while the principal branches which symbolize mechanistic activity involve: 'tools → machines → robots → intelligent machines'. Other technological, material-related activities are: 'fire → food processing → chemical analysis and synthesis → complex chemical and biological processes'. And yet another branch symbolizes scientific exploration involving: 'measurements → numbers → experimental sciences → models → natural laws → theoretical sciences'. It is characteristic of the evolution of science that bifurcations of a trunk into its branches were already achieved by the ancient societies and that many of the achievements of modern science are often only further bifurcations of the branches which result as a consequence of the systematic repetition of well-established procedures of scientific exploration.

What is the fruit of this tree? The most characteristic property of the fruit is that it carries encoded information according to which a new tree can develop. What constitutes the fruit of scientific evolution is still a matter of conjecture. We recognize, however, that the development of the sensory nervous system has played a decisive role in the evolution of life on earth. It has enabled humans to work and to build machines that perform mechanical work and to develop science. This altogether increases the chance of human species to survive. We can imagine the fruit of the scientific tree as a replication of the development of intelligent systems resembling the capabilities of the humans. Is it then naive to expect that a similar transition may also occur in the field of scientific work? Related to this, one might predict that the development of information processing systems capable of autonomous scientific exploration could play a similar role in the future evolution of science as machines previously did in technology. That is, one might expect that the procedures of scientific inquiry could become automated and be performed autonomously by some kind of robot endowed with a capable processor. Many inventions such as electronic sensors and information processing systems that are already available today, indicate that technical evolution proceeds in this direction.

2.2 A Description of Nature

Our goal of developing a system capable of performing measurements that can be used to model natural phenomena or even to forecast their evolution requires that the fundamental characteristics of scientific inquiry in the quantitative sciences be understood. To do this requires a description of Nature. It is commonly accepted that the quantitative sciences rely on a physical description of natural phenomena. Various introductory physics texts state that measurements and experiments form the basis of all physical descriptions of natural phenomena. However, a detailed discussion about the common properties of the various measurements is usually avoided in these texts in favor of an enumeration of specific characteristics of particular measuring procedures or devices. Our goal here is to focus on the common properties of measurements because these suggest the requirements which need to be met if one wishes to develop systems capable of imitating scientific work on information processing devices. Therefore, we first point out those properties of the physical description of natural phenomena by which the description differs from others, as for example, literary or artistic. With this goal in mind, we first analyze the human activity which leads to a physical description of Nature.

Humans as other living beings perceive their environment through their organs of sense called *sensors*. On the basis of these, our perceptions or notions about Nature are established. We are able to concentrate our observation and focus our attention on particular objects in space and events in time and to specify or to recognize those objects and possibly also their composition that make up our environment. Similarly, we can identify or recognize sequences of states as events. But in addition, we are able to influence Nature by effecting changes in it.

Among the various actions that humans can perform, let us focus here on those by which we create new objects that excite in us perceptions similar as those excited by some other object. As a typical example of such an activity we mention a sculptor who creates a statue of a person or an artist who draws a picture. In this case, we say that the created object is an imitation or a copy of the original one. We can look upon copying as being a direct, although usually not the simplest description of an object. In addition to objects, we can recreate events in our surroundings. As a characteristic example, we mention the performance of an actor in a theater who imitates movements and repeats words in order to recreate a particular event. Copying can also be performed by other living beings, or even by special instruments designed for this purpose. Examples include a photographic camera, a copier, or an audio recorder. Common to these examples is the existence of a certain relation between the observed object and its copy. [25] Such a relation is generally understood to correspond to a transfer of information from the source to the storage medium. Any system for which the transmission of information is essential, can, in principle, be considered for providing a description of natural phenomenon. Consider, for example, the *flip-flop* circuit. In its operation, an input impulse results in a transition of the element in the circuit to a corresponding output state which, in effect, corresponds to a memorization of the input signal. There are many other so-called self-organizing dynamical systems, including biological ones, for which reproduction or duplication is an essential characteristic. [3, 4, 23] Such systems could be utilized in the description of natural phenomena.

Although the duplication of objects is essential for a number of technological applications, it is not the most appropriate for scientific work. For this, we must turn to a symbolic description. Let us consider the case of modifying a copy. For example, it may be reduced or enlarged in size, or it may be simplified such that only some of its features are reminiscent of the original. As an example, we can imagine that a circle, two points, and a line can be used to represent a person's head. In this case we recognize that the created object only symbolically describes the original one. If the modification is carried out to such an extent that the modified object or symbol no longer resembles the original one, although we might still wish to relate them, then we term this a *symbolic description*. There is a characteristic relationship between an observed object and its representative symbol in a symbolic description, but it need not be based on the similarity of sensory excitation by the original and its descriptive symbol. Therefore we can correctly interpret a symbolic

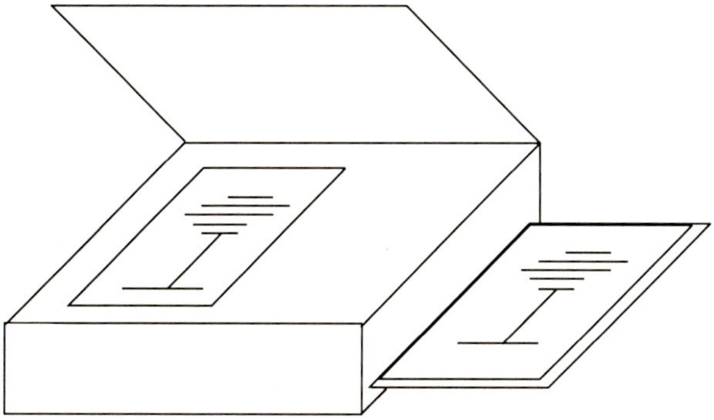

Fig. 2.1. A scheme of reproduction

description of Nature only if we have available all the proper relations between the original objects and their corresponding descriptive symbols. Such a transition from objects to corresponding symbols is generally called *encoding*. It corresponds in human activity to the mental ability to create a set of mental associations. We can thus conclude that for any description of Nature, sensing is needed which is then associated by a specific action to another sensing. Symbolic descriptions are generally a characteristic of communication between living beings and can include various senses. But they can also be performed by various devices, as, for example, an automatic weighing device or a scale which associates with an item, the corresponding display of a register representing the price of the item. A symbolic description can also be applied to the evolution of states in Nature or events. A well-known example is the description of speech by letters or music by notes.

Fig. 2.2. A transition to symbolic description

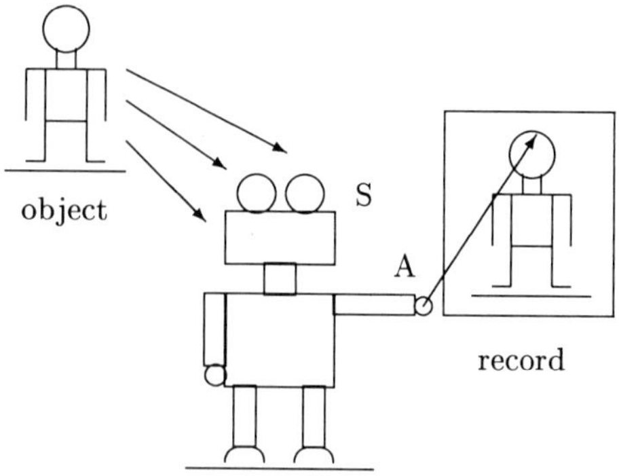

Fig. 2.3. A scheme of a system capable of describing natural phenomena in which 'S' denotes the sensors and 'A' the actuators

Science is based on observing and describing natural phenomena. If one wishes to develop a system capable of automatic scientific work, one must include in it elements called *sensors* which are sensitive to particular influences from the environment. Also needed are elements, called *actuators*, which are capable of generating responses in their environment in order to generate a description of natural phenomena. The operational characteristics of sensors and actuators, which together, are called *transducers*, will be discussed in the following chapter. But in addition to the transducers, there is needed an element by which the sensed influences are transformed into input signals to the actuators. One generally represents such a system by a "black box" with sensors at its input and actuators at its output. A description of natural phenomena can generally be treated as a transformation of the influences from the surroundings into an output record by actuators in a recording medium. The *influences* transmitted from an environment through the sensors into the black box and then through the actuators into an environment are generally called signals. We symbolically denote signals by a set of letters

$$\boldsymbol{X} = (x_1, x_2, \ldots x_n) , \tag{2.1}$$

where $\boldsymbol{X}$ denotes the complete set of signals while x_n denotes one particular signal from this set. The transformation of input signals $\boldsymbol{X}$ into output signals $\boldsymbol{Y}$ is also symbolically represented by some function $\boldsymbol{F}$

$$\boldsymbol{Y} = \boldsymbol{F}(\boldsymbol{X}) . \tag{2.2}$$

Here $\boldsymbol{F}$ describes the response of the complete system while the black box can be treated as a transformer whose operation is characterized by this response. The operation of most scientific instruments can be represented by such a scheme. For example, in a seismograph [9, 8, 26], the detected motions of the earth are transformed into traces on a recording chart. Similarly, the observations, thinking and actions of a scientist can also be represented by such a symbolic description.

One of the greatest advantages of a symbolic description of natural phenomena is that it also can be used to describe human mental activity. In this case, the natural objects or events are first associated with thoughts or abstract objects and these are further represented by symbols. Literary or artistic work is often related to such a description, while the most characteristic example is the description provided by mathematics. Thoughts or abstract objects can be related to internal states of the black box to autonomously create output signals. However, this output cannot be treated as a direct response to an input signal. In that case, the black box is not a transformer but rather a generator. From a scientific point of view, a transformation of signals appears to be of primary importance in empirically-based sciences because it is related to the direct description of natural phenomena which corresponds to the acquisition of knowledge. In contrast, an autonomous generation of signals can form the basis of an analytical or a theoretical procedure which is the successor to the empirical approach for providing an understanding of natural phenomena. In this monograph we are more interested in the direct description of natural phenomena and therefore our attention is focused principally on the transformation and storage of input signals.

2.3 Fundamentals of Quantitative Description

In the technical world we often encounter the task of reproducing as accurately as possible an object or a process, or a particular state of an environment. To do this, we generally need an appropriate method for comparing the original and the copy because our senses are not always a reliable indicator of the properties of Nature. For this, we select a set of objects or events which are called *measures* and form from these a basis for comparison. We further select a set of processes, called measurements, by which the measures of the objects or events under observation are related according to well-specified procedures with some abstract object. This abstract object is usually denoted by a symbol which is called the result of a measurement. Most often it is expressed as a number, although this is not always suitable.

As an example of such a measurement, we consider the weighing of an object. First we select a set of weights, then we select a balance beam pivoted about a particular point along its length. We place the sample under observation at one end of the beam and the weight on the other end. If the beam is balanced, then the weight of the sample corresponds to the weight of the balancing mass. The result of this measurement can then be symbolically described by the index which is associated with the measure. The symbolic description of the results of measurements is the general basis of the physical description of the properties of Nature.

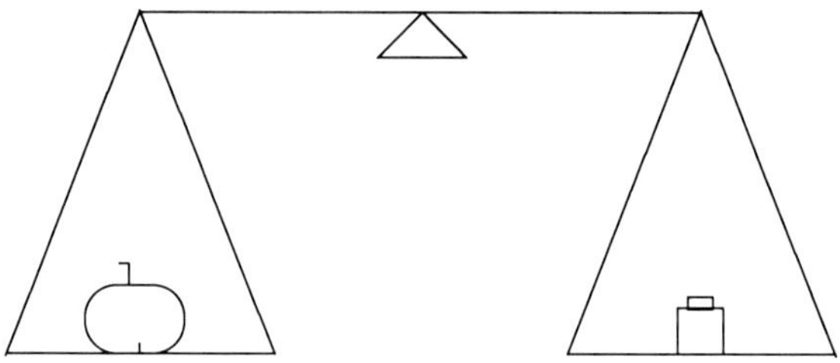

Fig. 2.4. Representation of a measurement

The aforementioned example of weighing shows that a specification of a measuring procedure includes comparison or specification of a relation of equivalence. Two samples being compared are equivalent with respect to the selected set of measurements if the results of measurements on both samples are equal. By establishing the equivalence of two samples, one can decide if a sample under observation represents a reproduction of another one. This is of practical importance to applications in many fields of human activity, especially in manufacturing and science. The possibility of determining the equivalence of objects or events leads to the most important step in a quantitative description of Nature, that is, to the formation of measurement scales. For this purpose, an object or an event, called a *unit*, is arbitrarily selected as a representative of a specific property of Nature under consideration. The comparison procedure permits the selection or formation of a set of equal units. Using the elements of this set, we can then compose new elements by using a proper procedure of scale formation which is usually called *addition*. A repetition of this process leads to a broader set of generated elements which is called a scale. Each element of the scale is related to a *symbol* or an abstract element representing the number of repeated additions of equal units. This symbol represents the operation needed to create an element of a scale from a unit, which at the same time describes the relationship between units and scale elements or the relationship between their properties. The formation of the scale is thus operationally based.

A specimen property which is not an element of a scale, can still be described symbolically, provided that the specimen is compared with the elements of the scale. The comparison then yields the element of the scale which is most similar or equivalent to the observed specimen. Its associated characterizing symbol is further used to represent the result of a measurement or the observed quantity. This means that the description of an observed property in terms of properties of the elements of the scale is also of operational character. We cite as an example, the quantitative description of the length of an object. For this purpose, we first define a measurement procedure for

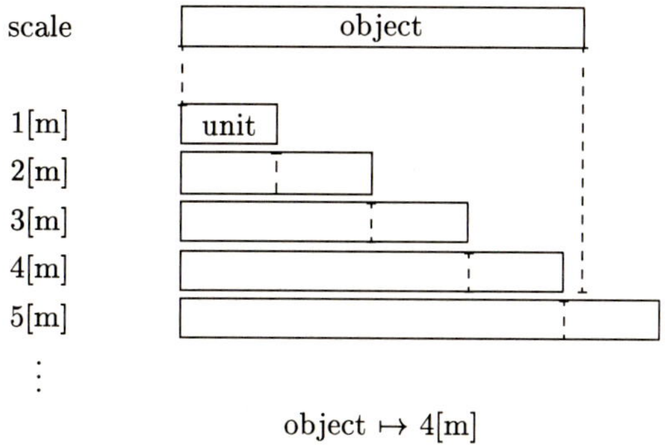

Fig. 2.5. A scheme for obtaining a quantitative description of an object

comparing the lengths of two arbitrary objects. We then select an unit, called a *meter* and define an operation by which we can compose such equal meters into elements of the scale, denoted by 1 [m], 2 [m], 3 [m], etc. In order to measure the length of an arbitrary sample, we must compare it with the elements of the scale and find that one which is equivalent to it. Suppose that in one example it is 3 [m]. The result of the measurement of the specimen length is then expressed symbolically as $L = 3[\mathrm{m}]$. Here the first element denotes the observed property, the second describes the comparison or the operation of the equivalence estimation, the third is the number denoting the repetition of composition or addition of units needed to form the corresponding element of the scale, and the last element in brackets denotes the measurement unit. This example illustrates that a quantitative description of a particular property is nothing other than a symbolic description of a series of operations included in the complex process of a measurement. In order to define a physical variable, one must specify an operation of comparison or an estimation of similarity, to select a unit and to specify how a scale can be created from it. Because of the operational character of measurements, they can be autonomously carried out by measurement devices.

However, all the physical variables need not be determined by specific measurements. For example, let us consider how we might describe a volume of a rectangular object such as a cube. If we define the unit of a volume by a cube with the side equal to the unit of the length, then we can establish on the basis of geometrical observations that the volume of a rectangular object can be expressed in terms of the lengths of the sides a, b and c by using a mathematical product: $V = a \cdot b \cdot c$. In this case, the value of the variable V is defined as the product of the values of the sides and the unit is described as a cube whose edges are of unit length. A measurement of a volume can thus be composed of three separate equivalent measurements of a single variable. A variable that can be specified by means of measurements of more basic variables which may be modified by mathematical operations or derivations is therefore called a *derived variable*. The corresponding mathematical expression can thus be interpreted as a symbolic description of the operations included into the specification of a derived unit.

A quantitative or operational description of the properties of objects or events in Nature is especially appropriate for industrial applications because it permits a symbolic description of the properties of the operations or processes included in manufacturing. [8] But an essential characteristic of any quantitative description is its reproducibility. By a proper selection of operations included in a specification of a measurement, one can exclude to a certain extent the unreliability of human perception. Furthermore, the operations can be performed by properly constructed devices or instruments. If a certain property of an object is characterized by similar measurements, then it should be possible to obtain equivalent results from a measurement, independent of the operator who is performing the measurement or the measurement device. This is the basis of objective experimental work and it leads to the creation of science which is, in turn, objective. [26]

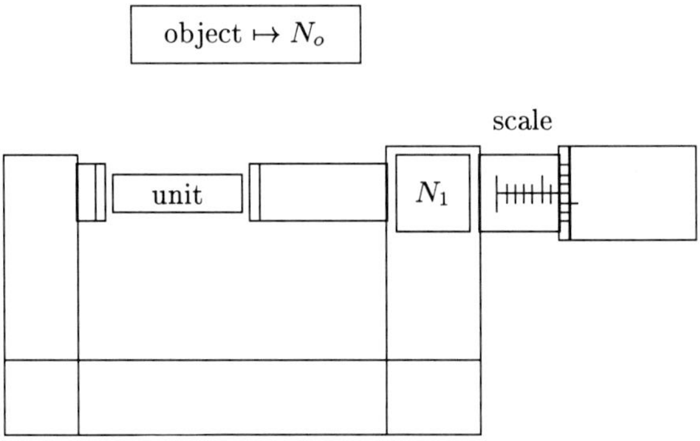

Fig. 2.6. A micrometer as an instrument designed for the quantitative description of length and the process of calibration

In contrast to human observations, measurements performed by measurement devices because of their quantification and often higher precision, usually show a greater reliability. This is a consequence of the inclusion of devices into the process of comparison and the operational definition of units, which, in turn, can also be performed by devices. In order to elaborate this, let us again consider the task of a quantitative length measurement. To do this, we construct a *micrometer* for comparing the dimensions of objects. The frame of the micrometer has two points or small plattens, one of which is fixed and the other movable by a micrometer screw. In order to compare the length of two objects, we must insert each of them separately between the points and then turn the screw until the points just come into contact with the selected points on the surface of the specimen being measured. To facilitate the process of comparison, we utilize the number of rotations of the screw N. If the number of rotations corresponding to two lengths being measured are the same, that is, if: $N_1 = N_2$, then these lengths are the same. The use of the micrometer screw results in a more reproducible result of the length comparison than might otherwise be obtained by a visual comparison of the two lengths or even a visual estimation of final position of the movable measuring point. Furthermore, the operational comparison makes it possible to include mathematical operations into the measurement. Let us assume that the pitch of the micrometer screw is uniform along its length and we begin by counting rotations when the measuring points of the micrometer come just in contact. Let the standard unit of length correspond to N_1 [turns/length]. If the dimension of an object is found to be equal to N_o rotations of the screw, then the length of the object can be expressed as: $L = N_o/N_1$ units. In this example, the screw plays the role of the scale which is calibrated by determining the number N_1 of screw rotations per unit length. This approach permits the sub-division of units into smaller divisions as well as their multiplication into larger elements on a scale, which results in a direct advantage when a measuring instrument is used in a measurement process. A further advantage is realized when a device called a counter is constructed which can also record or store the numbers N_1 and N_o. In this approach, a human operator can be excluded from the measurement process. And in this case, the complex measuring device or instrument begins to correspond to the previously mentioned scheme of a system that can be used to obtain a description of Nature. However, the observed property is now expressed symbolically by the state of the counter as a physical quantity.

An essential property of a quantitative description of natural phenomena should be mentioned here. A quantitative description of a variable is a symbolic presentation of a series of operations included in the process of the measurement. When writing $L = 2\,[\mathrm{m}]$, we denote a comparison of the property of a sample with the same property of the elements of the length scale. The comparison can often be performed by a sensor, as, for example, a point and a counter on a micrometer screw, which responds to an object presented to it by a change of its state. The quantitative specification of a variable thus generally includes two steps: The first corresponds to a specification of the response of the sensor to the sample under examination and the second corresponds to a specification of the element of the scale which results in a similar response. The latter operation is, in fact, an inverse procedure of the former. A quantitative measurement therefore cannot be represented only by a single response. This is an important feature that must be noted when considering the development of an automatic quantitative description of natural phenomena.

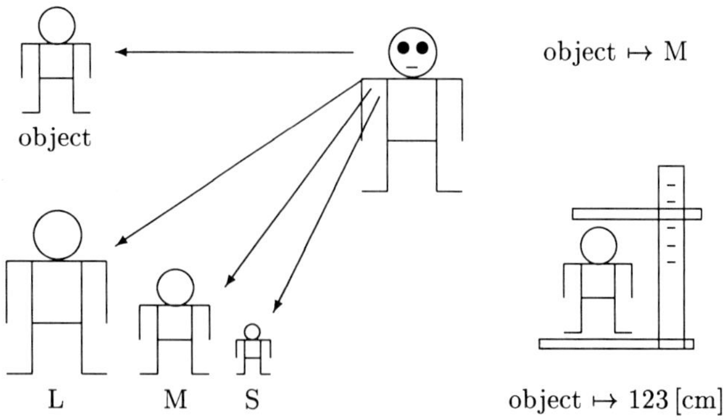

Fig. 2.7. Analogy between a quantitative and qualitative description of size

Although a measurement procedure appears as a complex mechanical operation, it resembles the process of human recognition. Recognition is based on an association of the observed specimen with a set of notions which are prototype elements in the memory. The prototypes generally represent particular properties, as do the elements of a scale. Recognition may correspond to the process of comparing a specimen with the elements of a scale to obtain a quantitative description. Because the process of comparison is usually less strictly determined, recognition can generally be considered as a basis of a qualitative description of Nature. However, there is a profound difference between the properties of the set of prototypes we are using in recognition and the properties of the elements of a scale. Each element of a scale is specified by a reproducible and repeatable composition of a unit. This is generally not the case with abstract prototypes. In general, only a small number of prototypes are required to obtain a description of the variation of a certain property but many prototypes are needed to span the broad range of properties of Nature. In contrast, a scale is a large set of elements all describing variations of a common property or a physical quantity, but the number of all fundamental physical quantities is relatively small. It is well known from elementary physics courses that altogether, only four fundamental quantities are needed to form the basis for a physical description of Nature. These are: length, time, mass, and electrical charge. Because of this difference, a qualitative description that is based on a person's notions, such as a literary one, is in a sense, soft, i.e. "fuzzy". Such a description does not permit great reproducibility as would be possible from a more rigid physical description that is based on more precisely specified elements of scales. In contrast, a qualitative description of natural phenomena in terms of prototypes is much more adaptable to complex situations as those which may be encountered in everyday life. A question therefore arises how one could put the physical as well as the qualitative descriptions on the same basis in order to realize the advantages of each and so to permit a unique treatment of physical phenomena. A qualitative description is often interpreted as the foundation of a soft science in contrast to the quantitative description that is a basis of the exact sciences, such as physics, chemistry, etc.

The properties of a composition of units into elements of a scale are usually describable by a mathematical structure. The quantitative description of natural phenomena, and with it the associated physics, is thus from the very outset, intimately related to mathematics. The most important advantage of the application of mathematics and physics to the description of Nature stems not only from the reproducibility and accuracy of the quantitative treatment but also, and most importantly, from the formulation of natural laws which lead to an abstract analytical treatment of natural phenomena and the prediction of indirectly observable properties of these phenomena.

2.4 Fundamentals of Physical Laws

After carefully observing a natural phenomenon, we often discover that several properties are mutually related. A quantitative description makes it possible for us to symbolically express the properties of natural phenomena in terms of mathematical relations between the results of various measurements. Let us for example imagine how we might quantitatively express the well-known property that the weight or the mass of a cube of a homogeneous material depends on the length of one of its sides or its volume. For this purpose, we first perform a set of simultaneous measurements to determine the mass and the size of a number of specimens. Let us assume that the experiments have been carried out on solids in the form of a cube with sides whose dimensions are denoted by a and expressed in units of [cm]. The results of four measurements are tabulated in the first row of Table 2.1 and they represent a set of sample values. The measurements of the mass m of each specimen is listed in the second row of the table. It is evident that the quantity of the mass corresponds to the value of three times a^3. This observation can be mathematically expressed by the relation

$$m = \varrho a^3 , \tag{2.3}$$

in which ϱ denotes the factor of proportionality between the values of mass m and the length a of the edges of the cube. Although the measured values

Table 2.1. Results of a simple experiment

a [cm]	1	2	3	10
m [g]	3	24	81	3000

will differ when we use different units, the equations relating the mass and the third power of the length of a side will be valid, provided that we include also a symbolic multiplication and division of units. In this case, we must assign to the factor of proportionality ϱ, a derived unit $[\varrho] = [\mathrm{g \cdot cm^{-3}}]$. Now we have defined a new physical variable called the *density* of the material which, in this example, is equal to $\varrho = 3\ [\mathrm{g{\cdot}cm^{-3}}]$. This variable represents a property of a material. Its definition is indirect or derived because one needs to measure two physical variables of a specimen – its mass and the length of a cube edge – in order to determine the density using the equation

$$\varrho = m/a^3 . \tag{2.4}$$

The units of this quantity are not arbitrarily selectable but rather, they are derived from the units of mass and length. The definition of density and the relationship between mass and the length of the cube edge become still more transparent if we introduce the volume of the cube $V = a^3$. Then,

$$m = \varrho V . \tag{2.5}$$

The relationship between these three variables quantitatively expresses the well-known property of objects made from a homogeneous material, that is, that the mass is proportional to the volume of an object with the constant of proportionality given by the material's density.

It is characteristic that the relationship between the mass of an object and the volume it occupies does not depend on a particular sample under observation and it is preserved when evaluating a large class of similar samples. It is the form of the equation which is representing a common property of the class of samples under observation. Such a relation is more general than a particular measurement and it can therefore be interpreted as a natural or physical law. As such, it can also be used as the basis for a description of the relationship between the mass and volume of other, similar classes of objects made, for example, from other materials. The property of an examined particular class is then characterized by only one quantity of a common specific variable or parameter called the density. Similar situations are often encountered when observing the relations between other physical variables. It is for this reason that physical laws are particularly useful in providing a general, quantitative description of the properties of Nature.

There is a significant difference between a presentation of the observed relationship between physical quantities comprising a table of numbers and its expression via an equation such as Eq. (2.3) or (2.5). A table is compilation of discrete data values that are the result of measurements made on samples of particular sizes. In contrast, the equation relating mass and volume that is expressed by Eq. (2.5) holds for any value of volume and in every instance, the mass is related to the volume in exactly the same manner as in the specific cases that had been used to formulate the law. Hence, the formulation of a physical law results in a generalization of a finite, discrete set of measured data into a continuous set.

This immediately raises the question as to how one knows that for a cube of a certain size, it has a certain mass. For example, if the cube has volume 2 [cm^3], how can we know that its mass will be 6 [g] when we have not measured it? This question is indeed so profound that children usually ask it, if a transition from measurements to physical laws is properly explained to them. In fact, the correct answer is that we do not know for certain, although we can prove it, if the result obtained from measurements made on a particular cube confirm the value predicted from the expression. And yet, even when the data indicate that the mass–volume relationship holds, then we may still not know whether the expression will be valid for additional data values that are not yet members of the set of sample values.

Quite different is the situation if another experiment yields a value of mass that was not predicted by the expression. We must then conclude that the relation is, in fact, not correct or that the hypothesis on which it is based is false. The problem is the following: In making the transition from a table representing the results of discrete measurements or empirical data to formulate a natural law, we include our knowledge of similar relations in mathematics and we assume that some of them are applicable for the expression of the relations between measured variables. A law such as given by Eq. (2.3) or (2.5) is a hypothetical generalization or even an assumption. The hypothesis that the generalization is applicable to all data can be made false if the prediction of the generalization is not consistent with the measurements. But the validity of the law for all possible examples cannot be confirmed by a finite set of measurements. However, many examples have shown that by using such hypothetical generalizations or assumptions, we can correctly guess or predict the values of physical variables for a variety of properties of Nature. It is for this reason that the relationships between physical variables are usually expressed in terms of mathematical relations. It is the set of such hypothetical relations which forms the basis of the theoretical physical description of Nature. The benefit is evident. Instead of presenting tables with particular values, we are using general symbolic expressions that are applicable to a broad range of examples. We are thus equipped with a powerful theoretical tool which provides a general, symbolic description of the properties of Nature although we can never be completely certain that it will always be correct or applicable. The main task of analytical science is thus reduced to searching for appropriate mathematical expressions which correctly represent the relations between physical variables that are used to characterize natural phenomena. It is one of the goals of this book to determine to what extent the presence of a scientist can be excluded from this effort so that the natural laws can be modeled autonomously by devices.

The application of mathematical relations for the description of natural laws leads to an interesting possibility. Let us consider again the example of the relationship between the volume and mass of an object which has been estimated from a finite number of observations. From a mathematical point of view, there is no difficulty in considering negative values of the variable V or magnitudes of it which, say, are only a few nuclear dimensions or which exceed the size of the galaxy. But we cannot find a corresponding sample in the real world. We are thus faced with the problem of the applicability of the natural law to particular situations, yet the law does enable us to predict the corresponding property of an imagined object. This capability is of great advantage for the prediction of indirectly observable or hidden properties of Nature. The history of science has recorded several examples of predictions of natural phenomena that were subsequently confirmed. A notable example is the prediction of the existence of electromagnetic waves by Maxwell.

Related to the application of mathematical relations to express natural laws, one might ask why the properties of Nature should at all be expressible in terms of natural laws. Mathematical relations are constituents of mathematical structures which have been formulated in connection with an analysis of regularities or observable relations between various properties of Nature. The problem thus becomes a question of why Nature at all exhibits a structure which is expressible in terms of mathematical relations. This is the philosophical question of the origin of Nature which cannot be answered based on an operational description of natural phenomenon. In such a description, one accepts the properties of Nature as given facts and only tries to express one property in terms of one or more other properties.

2.5 The Random Character of Physical Variables

It is well-known from experience, that a repetition of experiments carried out under equivalent testing conditions, does not always yield exactly the same results. This variability is usually described in terms of the random nature of physical variables. Because of this, any proper presentation of experimental data should include the empirical distribution functions by which the relative number of experiments yielding equivalent results is described. The relevant procedures are detailed in Chap. 4. The distribution function depends on the value of the measured variable. When we wish to develop an empirical basis for discovering a natural law, then we must observe a number of variables simultaneously and describe the measured data in terms of a multivariate distribution function. The fundamental problem is then to make a transition from a distribution function to an analytical expression of the relation among the variables imbedded in a natural law. With few exceptions, the formulation of a natural law usually does not include a description of the variability of the results that is obtained in repeat experiments. Thus, a natural law, that is expressed in terms of relations between physical variables only, cannot completely describe situations which are inherent in experiments. This deficiency is most frequently justified by an argument that the variability of the results obtained in successive experiments is a consequence of imperfect measurements or imperfectly prepared experimental conditions, which altogether can lead to experimental errors. Based on this argument, it is further hypothesized that there must exist a perfect relation between those variables which corresponds to the property of Nature that is under study. This point of view stems from the beginnings of quantitative science in ancient times and it continues to be widely accepted in modern scientific society, principally because of the many simple and successful applications of natural laws. However, when the variability of results in repeat experiments is real then one must include it into a proper description of the phenomenon and to apply the probability distribution as a basis of a more general description of experimental results. Such a generalization must also allow the application of natural laws and we must find a proper procedure by which the information hidden in the corresponding distribution function can be reduced to a less general form that can be represented by a natural law. For this purpose the methods of statistical estimation are applicable. These will be described in Chaps. 8 and 9. We shall see that these methods form the basis for the solution of various problems based on learning from examples.

2.6 Expression of Natural Laws by Differential Equations

As stated earlier in this chapter, one of the principal purposes for introducing a natural law is to reduce the number of experimental data that is needed to describe a phenomenon and at the same time, to permit a generalization of the description of it. However the actual extent of the reduction is left to the observer. Let us recall the example of the relation between the volume and the mass of an object. The corresponding set of data that were obtained from experiments on a set of cube-shaped specimens made of a homogeneous material could be reduced to the expression $m = \varrho V$. The value of density was introduced to describe the common property of the mass per unit volume that is one characteristic of an object made of one material. In a similar fashion, one can try to express the common property of all objects made from a homogeneous material. One relates homogeneity with a translational invariance of material properties. That is, a measured property is the same regardless of material volume under observation. One might therefore expect that the common property of all homogeneous materials could be described by including the changes of the volume into the expression of the natural law. If V_1 and V_2 denote the volumes of two observations, then the change of the observed volume when we transfer our observation from the first volume to the second one is $dV = V_2 - V_1$. The corresponding change of mass is $dm = m_2 - m_1 = \varrho(V_2 - V_1)$ which is ϱdV. Thus, from the fundamental law, it follows that $dm/dV = \varrho$. For all homogeneous materials, the density

ϱ, and hence also this differential quotient, is independent of the volume of material under observation. In accordance with this example, we expect that any invariance of a natural law can be expressed by some constant parameter occurring in the corresponding differential form of the natural law.

By differentiating the relation $m = \varrho V$ with respect to V, we obtain the new equation $dm/dV = \varrho$ in which the material parameter appears. We show how we can exclude it from further treatment. By using the first equation, we express the density $\varrho = m/V$ and insert it into the second expression to obtain the following differential equation

$$\frac{dm}{dV} = \frac{m}{V} . \tag{2.6}$$

The parameter ϱ describes the property of a particular material sample. By excluding this parameter from the description of the natural law, we have obtained a differential equation which is independent of the particular sample under observation. Therefore, this differential equation describes a common property of all examples dealing with a family of homogeneous materials. This means that the equation is invariant with respect to the choice of material.

A similar situation is encountered when one attempts to express experimental results that have been observed on a family of similar examples that differ in one or more parameters. An example of a two-parameter family arises in the description of a simple oscillator that is oscillating at frequency ω. The motion can be expressed in terms of a family of time-dependent signals of the form $x(t) = A\sin(\omega t + \phi)$. Here A and ϕ play the role of parameters by which particular examples of the phenomenon can be characterized. In order to exclude these two parameters from the description, we perform differentiation of the signals twice with respect to time t. By excluding both parameters from the expressions, we obtain the differential equation

$$\frac{d^2x}{dt^2} + \omega^2 x = 0 . \tag{2.7}$$

A generalization of the description that is obtained by excluding parameters through differentiation is quite generally applicable and it indicates which derived variables are convenient for obtaining a general description of the phenomena considered. Therefore, the description of natural phenomena in terms of differential equations is commonly accepted as a basic scientific tool in practice. Consequently, a question that is of fundamental importance arises: How can one build an information processing system that is capable of expressing symbolically the common properties of sets of similar experimental data in terms of differential equations? The associated task is a solution to the so-called *general identification problem.*

A generalization of the description of natural laws by excluding some parameters from a family of functions corresponds, in fact, to a reduction of information. In order to describe a particular example from a family, one must re-introduce the corresponding information or specific parameters back into the description. This task, which is the inverse of a generalization, corresponds to the solution or the integration of a differential equation and a specification of the corresponding integration constants. Complementary to the general identification problem, we may ask the question how one might build a system which is capable of symbolically describing the solutions of differential equations corresponding to given conditions. Both problems correspond to the situation in which a family of typical records of functions describing particular examples of certain phenomena exist and we wish to find a similar function which corresponds to some given conditions. A strategy for solving this problem can be separated into two steps. One is the specification of the parameters appearing in the differential equation and this is generally referred to as *identification*. The second step is the solution of this equation which is usually called *integration*. Both of these steps generally represent a formidable mathematical task. Therefore we briefly mention another capability which we shall later hope to realize.

Humans and other animals perceive and remember various examples of natural phenomena. When a person observes a phenomenon in a particular environment under specific conditions, then a comparison of the perceived state of the surroundings with the memorized samples enables the person to imagine the development of the phenomenon in other surroundings. This, in fact, corresponds to finding a particular solution for the development of states in time from the given conditions. It is characteristic that this capability completely relies on an association of the current state of the system with memorized examples and it does not require a description of the natural phenomena in terms of differential equations. This is a very powerful capability in practical applications. It is for this reason that the development of an information processing system possessing the same capabilities, represents one of the main goals of research in the field of automatic modeling of natural phenomena. We proceed to the description of this possibility by reviewing some fundamentals of empirical modeling.

2.7 Methods of Empirical Modeling

2.7.1 The Role of Models

Let us return again to the example of the relation between the volume and the mass of an object. We define a coordinate system in terms of the axes describing the volume and the mass and use it to graph the measured data points. To each measurement there corresponds a point in the plane so that all the sample values correspond to the set of points in a graph such as shown in Fig. 2.8. The corresponding natural law is represented by the straight line through the data points.

Knowledge of the natural law permits us to process a measured value of some selected variable and to recover the value of an unknown, related vari-

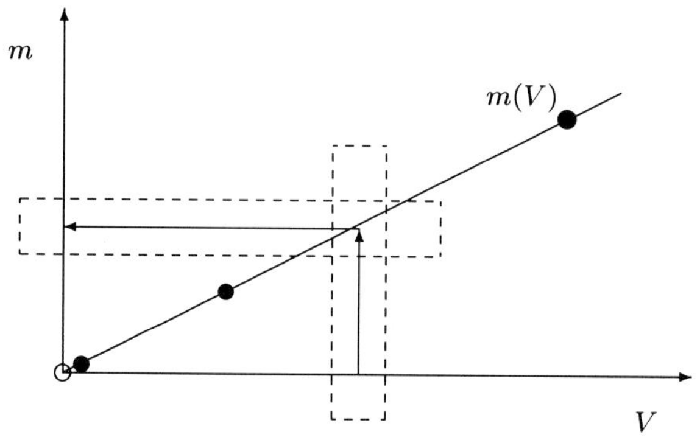

Fig. 2.8. Graph of the relation between the volume and the mass of an object

able. For example, knowing the natural law relating material mass and volume for a particular material, it is a straightforward procedure to determine the mass of a particular specimen from measurement of its volume. The graphical representation of the natural law permits us to determine from a given value of specimen volume, the corresponding mass using only the construction of a parallel projection from the abscissa to the line and then a second projection from the point of intersection on the line to the ordinate axis. A similar procedure permits finding the specimen volume which corresponds to a particular specimen mass. The complete procedure can be carried out either on paper or with an appropriately constructed system of sliding rulers. Regardless of the construction, such a system is generally called a geometrical model of a natural law. Various technical museums have in their collections sophisticated models that have been designed and constructed for solving a diverse variety of problems in this way. A common characteristic of all such models is that an observed state of Nature is connected by a certain reversible relation to a state of the model. The development of such a connection corresponds to a symbolic description of Nature and this can generally be treated as a mapping. Using such a mapping, a relationship of the properties of Nature can be mapped into a relationship among the corresponding variables of the model. Such a relationship enables prediction by means of a reverse mapping from the properties of the model and back to the properties of the system in Nature. A model can also exhibit the properties of a quantitative physical description of the applied variables provided that the corresponding scales of the physical units are mapped to the scales of the model. However, in some applications, knowledge of the scales of the variables in the model are not required.

It is of great technical importance that a mapping between Nature and a model can often be performed autonomously by an instrument, that is, without intervention by the user of the model. The relationship between Nature

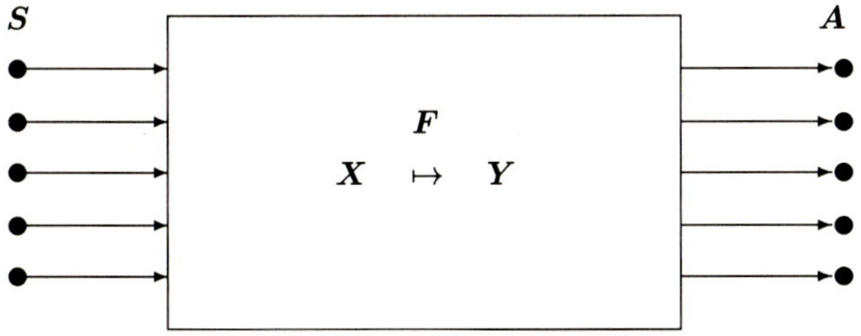

Fig. 2.9. Block diagram of a modeler

and a model can be established by a system comprised of an array of sensors, a transformation unit and an array of actuators which is shown in Fig. 2.9. The function of the transformation unit is to map signals of the sensors into an output of the actuators. In the biological world, neural networks generally perform this mapping while in a technical environment, programmable electronic computers are most generally used. In this application, a program must provide the proper characteristics for the operation of a complete system. In practical applications one often knows the general form of the natural law so that one adapts it to a specific example by fitting a set of parameters that appear in the expression. This corresponds to a global analytical modeling. But, in many cases we do not know in advance the proper form of the natural law which might be used to provide a description of the phenomenon under consideration. This is especially important when we deal with complex and stochastic phenomena. In such cases, we need a complementary method by which the information provided by measurements can be cast into a proper analytical relation without assuming in advance the form of the appropriate natural law. For this reason and in order to minimize repetitions in the programming that is needed for the modeling of specific phenomena, it is reasonable to ask what the general characteristics of a program should be that make a system comprised of sensors, a processor and actuators into a general and automatic modeler of natural phenomena. With this as our goal, we present in the next subsection a method by which a piecewise linear function can be adapted to an arbitrary set of measured data points that represent an arbitrary non-linear natural law in two dimensions. In subsequent chapters, this method will be generalized by considering the elements of a data acquisition system, the random character of physical variables, a self-organized storage of data and an optimal extraction of information from stored information.

2.7.2 Piecewise Linear Models of Empirical Natural Laws

Let us consider a phenomenon that can be characterized by a pair of variables x, y and assume that measurements provide a set of N sample data points $\{x_1, y_1; x_2, y_2; \ldots; x_N, y_N\}$ with a constant spacing $\Delta x_i = x_{i+1} - x_i$ for $i = 1, \ldots, N-1$. Our goal is to analytically describe a smooth interpolating function $y(x)$ which passes through these data points. If we have no additional information other than the sample points then this problem is ill-posed and has no unique solution. In order to regularize it we require that the function $y(x)$ follows the shortest path between adjacent sample points which is determined by straight line segments as shown in Fig. 2.10. In order to obtain an analytical description of such a function we introduce a normalized piecewise linear sigmoidal basis function

$$\Psi(x) = \begin{cases} 0 & \ldots \; x < 0 \\ x & \ldots \; 0 \le x \le 1 \\ 1 & \ldots \; x > 1 \; . \end{cases} \tag{2.8}$$

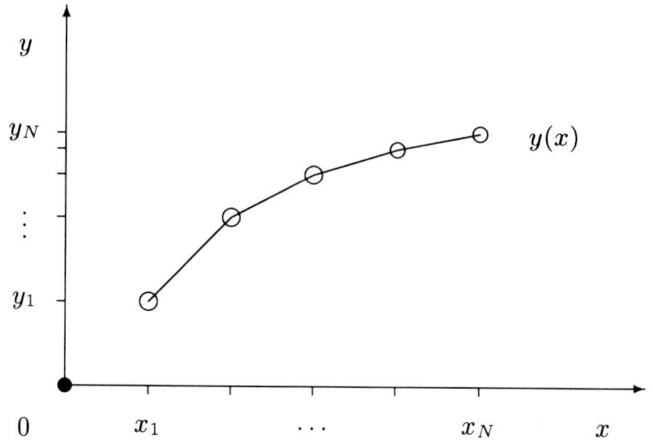

Fig. 2.10. An example of a linear interpolating function

Using the parameters $\Delta y_i \equiv y_{i+1} - y_i$, $c_i \equiv 1/\Delta x_i$, and $\theta_i \equiv x_i/\Delta x_i$ we can then express the piecewise linear interpolating function as a series

$$y(x) = y_1 + \sum_{i=1}^{N-1} \Delta y_i \, \Psi(c_i x - \theta_i) \; . \tag{2.9}$$

This interpolating function represents one possible analytical model of a continuous function satisfying the empirical data. The sigmoidal function $\Psi(c_i x - \theta_i)$ can also be called a *spline* and it represents a unit jump occurring at the transition from the point x_i to the next point x_{i+1} as shown in Fig. 2.11. We denote such a transition by the index i and define the i-th sigmoidal basis function as

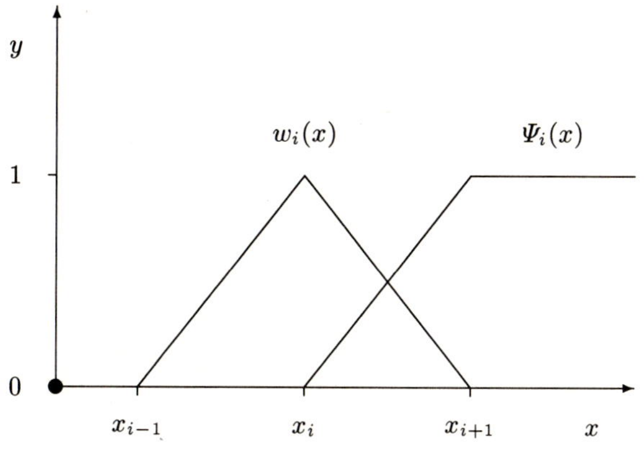

Fig. 2.11. Examples of a piecewise linear sigmoidal and triangular basis function

$$\Psi_i(x) \equiv \Psi(c_i x - \theta_i) \equiv \begin{cases} 0 & \dots\ x < x_i \\ (x - x_i)/\Delta x & \dots\ x_i \leq x \leq x_{i+1} \\ 1 & \dots\ x > x_{i+1} \quad . \end{cases} \tag{2.10}$$

Using these basis functions in Eq. (2.8) we obtain for the linear interpolation function, Eq. 2.9

$$y(x) = y_1 + \sum_{i=1}^{N-1} (y_{i+1} - y_i)\, \Psi_i(x) \ . \tag{2.11}$$

This expression indicates that we can introduce the triangular symmetric basis functions

$$w_{i+1}(x) = \Psi_i(x) - \Psi_{i+1}(x) \tag{2.12}$$

for $1 < i < N$. From the properties of the sigmoidal function we obtain the expression

$$w_i(x) = \begin{cases} 1 - \frac{|x - x_i|}{\Delta x} & \dots\ |x - x_i| \leq \Delta x \\ 0 & \dots\ |x - x_i| > \Delta x \end{cases} \tag{2.13}$$

which is also applicable at the end points x_1 and x_N. An example of such a function is shown in Fig. 2.11. By taking into account the properties of the triangular functions it turns out that we can express the sigmoidal basis functions as

$$\Psi_i(x) = \frac{w_{i+1}(x)}{w_i(x) + w_{i+1}(x)} \ . \tag{2.14}$$

This expression yields undetermined values for $x < x_{i-1}$ and $x > x_{i+2}$, but they can be defined by extending the function Ψ_i beyond the interval $x_{i-1} < x < x_{i+2}$ which yields the values 0 and 1 respectively. By inserting Eq. (2.14) into Eq. (2.9) and taking into account the properties of the overlapping triangular basis functions we can proceed as follows:

$$
\begin{aligned}
y(x) &= \frac{y_1 w_1(x)}{w_1(x) + w_2(x)} + y_2 w_2(x) + \ldots \\
&\quad \ldots + y_{N-1} w_{N-1}(x) + \frac{y_N w_N(x)}{w_{N-1}(x) + w_N(x)} \\
&= \frac{y_1 w_1(x)}{w_1(x) + \ldots + w_N(x)} + \ldots + \frac{y_i w_i(x)}{w_1(x) + \ldots + w_N(x)} + \ldots \\
&= \frac{y_1 w_1(x) + \ldots + y_N w_N(x)}{w_1(x) + \ldots + w_N(x)} \; . \qquad (2.15)
\end{aligned}
$$

In going from the first to the second line of the above equation we have taken into account, first, that the value of the fraction is not changed by adding to the denominator any function that equals 0 on the interval in which the numerator differs from zero. And second, that the sum of the triangular basis functions between the end points is equal to 1, as shown in Fig. 2.12.

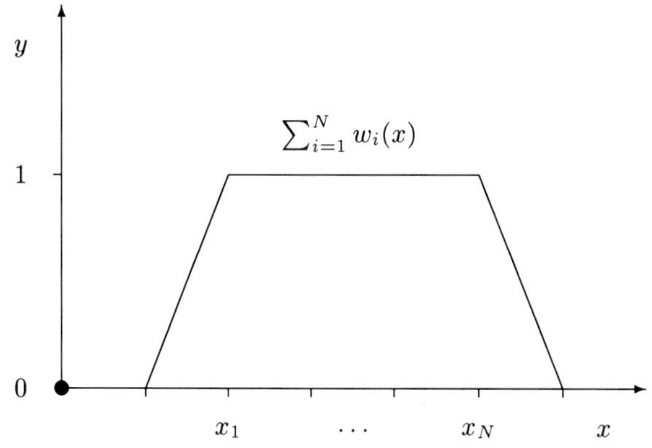

Fig. 2.12. The sum of the triangular basis functions

The last equation can then be rewritten in the simple form

$$y(x) = \sum_{i=1}^{N} C_i(x)\, y_i \qquad (2.16)$$

which statistically can be simply interpreted as a conditional average of y at a given x. [6] Here the new basis function

$$C_i(x) = \frac{w_i(x)}{\sum_{j=1}^{N} w_j(x)} \qquad (2.17)$$

describes a normalized measure of similarity between the given value of x and the sample value x_i. The normalization is achieved by the sum in the denominator, which equals 1 inside the interval between the end points. Therefore, the contribution of points inside this interval to the sum of Eq. (2.16) is simply determined by the triangular basis functions $w_i(x)$. In contrast to this the normalized basis functions at the end points are determined by the sigmoidal functions: $C_1(x) = 1 - \Psi_1(x)$ and $C_N(x) = \Psi_{N-1}(x)$ as is depicted in Fig. 2.13.

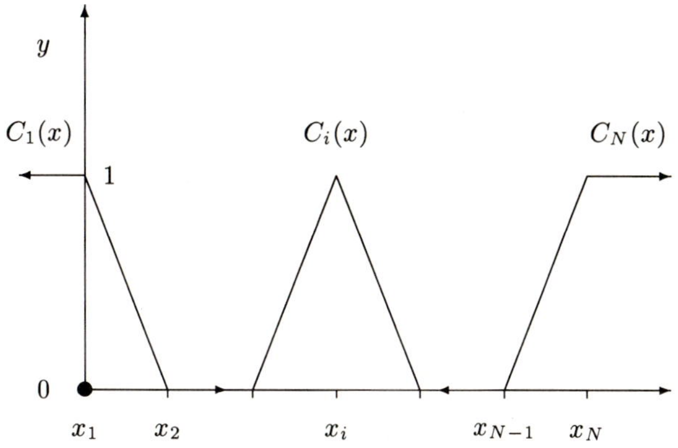

Fig. 2.13. Normalized basis functions $C_i(x)$

Consequently, we obtain from Eq. (2.16) the following rule:

$$y(x) = \begin{cases} y_1 & \ldots\ x < x_1 \\ y_N & \ldots\ x > x_N\ . \end{cases} \tag{2.18}$$

The same rule also follows from Eq. (2.9) which is based on sigmoidal basis functions. This rule in fact represents an extrapolation of the function $y(x)$ beyond the end points as shown in Fig. 2.14.

The piecewise linear sigmoidal function and the triangular basis function appear to be rather artificially introduced by requiring that the interpolating function must follow the shortest path between the data points. This requirement yields a piecewise linear interpolating function that does not possess smooth derivatives. One might expect that for modeling natural laws it might be more appropriate to have basis functions with similar gross properties but with smooth derivatives. Examples include the function $y = [1+\tanh(x)]/2$ and the Gaussian function. In fact, they could have been selected *ad hoc* at the beginning of our discussion. But in this case, the equivalence of Eqs. (2.11) and (2.16), which is essential for later interpretations, is only approximate, making it more difficult to make the transition from

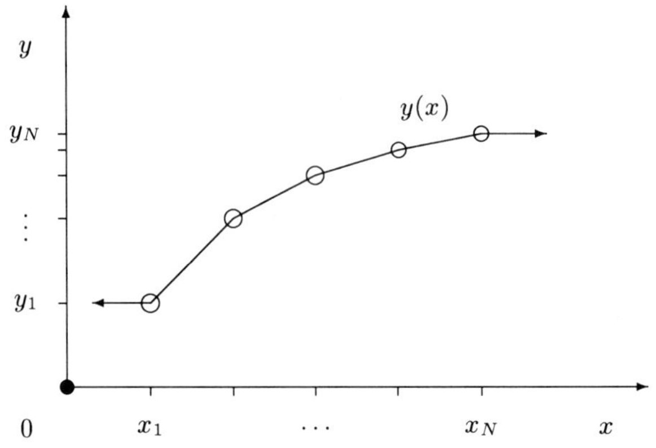

Fig. 2.14. Extrapolated function $y(x)$

sigmoidal to triangular functions. However, an argument in favor of using the Gaussian function is obtained from the following consideration. The triangular basis function $w(x)$ represents an approximation of the Dirac delta function $\delta(x)$. Therefore, one could expect that any approximation of the delta function could be used to represent $w(x)$. When dealing with experimental data, the measured values are usually normally distributed around the corresponding mean values as described by a Gaussian function. Interpolation of the function $y(x)$ can be interpreted as a spreading of a sample point into its surroundings because of experimental uncertainty. [7] As will be shown in Chap. 5, such a description introduces the least information if we use a Gaussian distribution that is characterized by the mean value and the variance of measured data.

Eqs. (2.11) and (2.16) represent two different versions of an identical description of a piecewise linear interpolating function running through a given set of sample data points. Either expression is easily programmed on a computer. But they can also be easily implemented in hardware, which is of advantage for the automatic modeling of natural laws that satisfy given empirical data. It is important that both expressions can be further generalized to the case of multivariate and stochastic data. Therefore, the structure of these expressions is considered to be fundamental for the development of automatic modelers of natural phenomena. [6] An argument in favor of this conclusion stems from an unrelated field of science that is the research of biological neural networks and their modeling. In view of this, we describe in the next section some fundamental properties of neurons and their networks while a more detailed description of the related topics is in Chap. 12.

2.8 Introduction to Modeling by Neural Networks

In asking what concepts are needed for the development of devices capable of automatically modeling natural phenomena, it might be useful to consider how Nature has solved this same task with sensory-neural networks. It is this approach that is followed by many researchers of this topic. In fact, research on the transition from biological observations to the electronic simulations of neurons and their networks is extraordinarily broad and so we shall restrict ourselves here to discuss only those fundamental properties which are of importance for the modeling of natural laws. [14, 16]

It is intuitively clear that thinking proceeds in brains and that thoughts are related to representations of natural phenomena in the brain. The question how the brain operates has been asked since the beginnings of scientific research and thought. Partial answers to this question can be traced to Egyptian notes written on papyrus about four millennia ago, to the works of Plato and Aristotle and to the medieval and modern philosophers. But the era of a physically-based description of the properties of the brain really began only half a century ago with the work of McCulloch and Pitts who carried out quantitative experimental research on this question. [1, 21] Devices intended for the simulation of intelligent operation have evolved similarly. The first were developed about two millennia ago in Egypt where Heron the Alexandrian built the first hydraulic automata. [19] Several centuries ago, the first attempts to build devices for the execution of arithmetic operations appeared, while in the past century various mechanisms were developed which were capable of performing logical operations. [27] An essential development that measurably contributed to the growth of this field was the invention of various digital electronic circuits in the middle of this century. Many credit von Neuman with playing a key role in the development of electronic computers based on digital logic circuits. [1, 22] In his approach, he followed the logical operation of the mind rather than the actual structure of the brain. For instance, he utilized a sequential processing of data by the digital elements which are controlled by a common clock. This does not resemble the parallel operation of the brain which is not synchronized by a clock.

A significant step in constructing a simulated neural network was the development of the *perceptron* which was based on the structure of a real biological neural network. [1, 24] In the perceptron, analog electronic elements resembling the functional properties of operating neurons were connected randomly in parallel and operated without the synchronization of a common clock. However, much of the research related to computing, ranging from algorithms to the development of new hardware technologies has focused on digital computers in which the information is processed sequentially. In the shadow of digital computation, research on analog neural networks, capable of processing information in parallel stagnated. However, because of inherent limitations of digital computers, which result from their sequential operation, the field of analog computation has again drawn considerable interest. The transition from a sequentially operating, digital network to an analog, parallel network might appear to be only of quantitative or operational character. But it should be also qualitative, because we expect that it will open the way for an effective solution of a number of problems which are difficult to solve using sequential digital computers. Included among these are: pattern recognition, control and decision based on learning from given examples, the self-programming of computers, the automatic creation of logical concepts on the basis of previous given examples, and finally, the self-creation of logical concepts and autonomous scientifically oriented operation in particular environments. These problems and the related tasks are, in fact, not only technological problems, but are all related to the optimal processing of information. The solution of each problem requires significant progress in a number of fundamental scientific fields.

Since the advent of digital computers in the 1940's, a number of scientists have attempted to reproduce some of the properties of biological neural networks by simulated networks. The main goal, however, has been to develop new generations of inexpensive parallel computers. It seems that the most promising approach will be simulated analog neural networks which consist of relatively simple analog elements resembling in their functional aspects, biological neurons. These are consequently named *formal neurons.*

An imitation of biological sensory-neural networks, however, has its limitations. Networks which can perform complicated tasks are extraordinarily complex and we possess insufficient knowledge of all the evolutionary influences that have contributed to the development of biological neural networks. It is therefore difficult to translate the experimental discoveries made on biological networks to artificial ones. Consequently, the research and development of artificial neural networks has proceeded along its own directions, supported by numerical computer experiments. In this effort, the biological system and a specific functional property often only represent a particular idea or the basis for the development of the corresponding analytical model which is then further examined and perfected by simulations carried out on digital computers. Such models are generally called *artificial neural networks.* Their study often leads to a better understanding of the corresponding biological networks, as well as opening new possibilities for theoretical formulations and applications. [1]

In the following paragraphs, we shall not try to answer the question why or how the observed properties of sensors and neurons or networks of them have developed in biology, rather we try to describe only their functional characteristics. It seems intuitive that these characteristics can be determined by studying animals in their environments. We presume that an animal's ability to recognize objects and events in its environment and to control its behavior has played a decisive role in the process of selection during evolution. Further, as a consequence of natural selection, the sensory neural network comprised of sensors, neurons and actuators or muscles developed in the biological organisms.

The fundamental characteristics of a sensory neural network is its ability to acquire and to memorize the states of its environments and the influences from its surroundings and to control the behavior of its host, according to learned, previous experiences. In this way, the sensory neural network can improve the operating efficiency of the living being in its surroundings. It is interesting that adaptive properties, similar to those achieved by one organism, can also be developed in an entire species. The evolved properties may be consequences of very simple rules of interaction or communication between the members of the species. [3] Because of the adaptability of the controlling system, it cannot be physically described as a technical system with constant parameters, but rather we must permit an adaptability of the system as we are formulating its operating characteristics. Adaptability can be expressed by the changes of the system at a transition from one generation to another or by changes in the behavior of particular individuals in one generation as a result of learning from past experiences. In a study of neural networks, both types of adaptability must be considered, although the learning of individuals is most often the focus of research.

Let us differentiate between *supervised* and *unsupervised* learning. In supervised learning, a living being is learning from presented examples until it achieves that behavior which corresponds to a predetermined goal. In contrast, in unsupervised learning, the goal is not predetermined. Instead, the being collects experiences from its environment which results in the development of an internal representation of all possible situations that might be encountered by the being. In the memory of highly intelligent beings such as humans, a set of prototype notions is formed that represents various characteristic properties of the objects or processes in Nature. Learning of this kind is governed by the frequency of appearance of particular objects or phenomena in Nature.

In addition to their adaptability, the most characteristic property of biological sensory-neural networks is their ability to recognize objects and events in Nature. This is a consequence of their associative operation. A being that has learned to recognize its surroundings can estimate or forecast under certain situations some hidden properties of Nature even when only partial information is contained in the sensory signals. For instance, the taste of an apple can often be predicted from the color of its skin. The capability of transforming one datum into another is often the principal goal in the development of a particular artificial, sensory-neural network. In terms of physical variables, this property can be expressed as a mapping of one set of variables into another and it generally corresponds to the modeling of some natural law. From this perspective, neural networks can be treated as modelers of natural phenomena. Instead of following an analytical approach in developing modelers of natural phenomena whose basis is the statistical description of relations between physical variables, we follow in this section a complementary, connectionistic approach to the solution of a given problem. In the connectionistic approach, we select some simple adaptive elements *a priori* and connect them into a network of a given architecture and then examine its operation to see whether it performs as expected. However, the idea for the architecture as well as its application often stems from the study of existing biological neural networks.

In the remainder of this chapter, we review the fundamental properties of neurons which are needed for the development of artificial sensory-neural networks while some fundamental properties of sensors will be given in Chap. 3. We begin with the description of neurons and the simplest version of the fundamental structure of a neural network called a *perceptron*. Their adaptation will be described more extensively in Chap. 12, but before reaching this, we will need to describe the fundamental properties of adaptation, self-organization and estimation in information processing systems.

2.8.1 Functional Properties of a Neuron

Next to the sensing element, the most important element of a biological information processing system is a neuron. A common feature of neurons is a cell body, called the *soma*. Attached to the soma are many thin branch-like parts called *dendrites*. There are generally two types of neurons: Local or interneuron cells and transmitting or output cells. These are illustrated in Fig. 2.15. The transmitting neurons connect either different regions of a brain to each other, or they link sensory organs or muscles to the brain. The connection is made over a thicker, usually longer fibril, called the *axon*. In

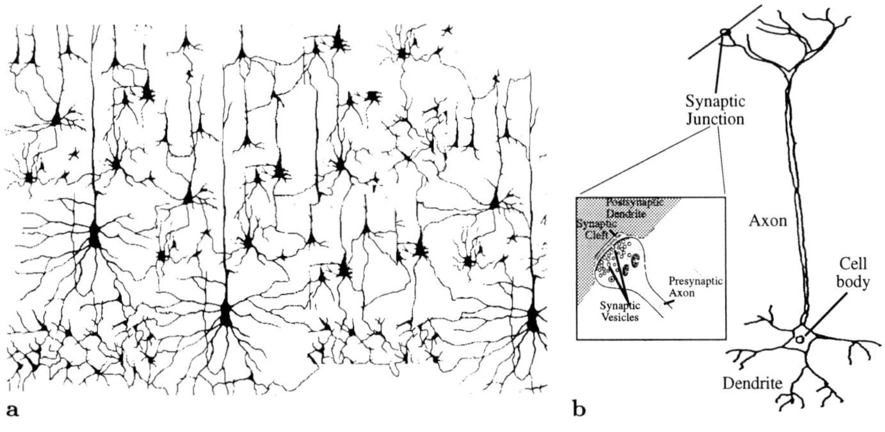

Fig. 2.15. **a** Schematic drawing of an array of neural cells; **b** Individual neuron showing a synapse between an axon and a dendrite

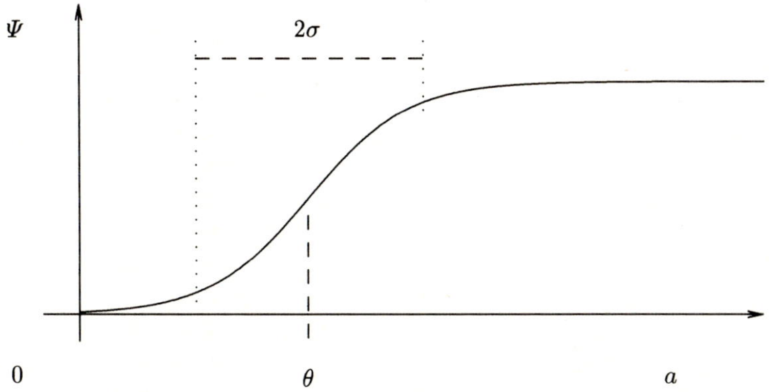

Fig. 2.16. Sigmoidal response function of a neuron Ψ – Output; a – activation or input; θ – threshold level; σ – width of the transition region

contrast, the interneurons are connected mainly by dendrites. The contact regions between neurons are bulb-like *synapses*. The length of the dendrites is of the order of 100 μm while the axons can exceed 1 m in length. The human brain contains about 10^{11} neurons and about 10^{14} synapses. [28]

A neuron is an electrically active cell. Its state can be described by the difference of the electric potential $U - U_0$ between the inside and outside of the membrane surrounding a neuron, respectively. It is convenient to express the state of activation by the relative variable $a \equiv (U - U_0)/\Delta U$ where the reference ΔU is a *threshold* potential difference between a threshold potential U_θ above which the neuron is activated and the potential U_0 which exists in the region exterior to the neuron. This potential difference is written as: $\Delta U = U_\theta - U_0$. Typical values of constants are $U_0 \approx -70$ mV and $\Delta U \approx +10$ mV. [28] The influences from the surroundings are added in the soma and result in a corresponding change in the membrane potential called *activation*. The membrane potential further influences the behavior of the neuron described as its excitation state. Most commonly the neuron is not excited. This is its quiet state. When enough inputs from the surroundings simultaneously become active, then the neuron responds by making the transition into a state of high excitation. In this state a neuron generates voltage pulses called *action potentials* that propagate along the axons to other elements of the network. This occurs when the action potential exceeds the threshold value $U_\theta = U_0 + \Delta U$. The action potential is a short voltage spike of approximately 100 mV and $\sim$ 1 ms in duration. The repetition rate of action potentials depends on the level of activation of the neuron, with higher activations resulting in higher repetition rates. In modeling neural networks, we usually do not describe particular action potentials but rather the *activity* of the neuron which is expressed by the repetition rate of action potentials. This represents the excitation of the neuron or its response and it is generally time-dependent. We will ascribe the variable y to correspond to the rate of action potentials and it is the output signal supplied by a neuron to the network. The output of a neuron in a quasi-stationary state is a nonlinear S shaped or sigmoidal function of the activation a as shown by Fig. 2.16. It represents a transition from the quiescent $(y = y_{\min})$ to the excited state $(y = y_{\max})$ of a neuron that occurs when the activation exceeds some interval around the characteristic relative threshold level, that is, $a_{\text{exc}} = \theta = 1$.

In addition to this parameter, there are two other parameters that describe the properties of the sigmoidal function. These are the span of the output and the width of the transition region. The first is determined by the difference between the output in the quiet and most excited state. That is, $\Delta y = y|_{a\to+\infty} - y|_{a\to-\infty} = y_{\max} - y_{\min}$. Without loss of generality, we can describe the operation of the neuron in normalized units in which the span becomes non-dimensional. The second important parameter of the sigmoidal function is described by the slope of the function in the middle of its transition region. This we denote by $d\Psi/da|_\theta = 1/\sigma$ where σ corresponds to the width of the transition region observed when the graph of neural response is approximated by the piecewise linear sigmoidal function $\Psi = [1+(a-\theta)/\sigma]/2$ for $\theta - \sigma < a < \theta + \sigma$. In addition to this approximation, other representations of the sigmoidal function use the logistic signal function

$$\Psi(a) = \frac{1}{1+\exp[-(a-\theta)/\sigma]} \tag{2.19}$$

or the modified hyperbolic tangent function,

$$\Psi(a) = \{1+\tanh[(a-\theta)/\sigma]\}/2 \; . \tag{2.20}$$

In the modeling of neural networks, it is sometimes assumed that the excitation ranges between $(-1, +1)$ and that the threshold level θ is zero. Further, the neurons are approximated as two-state elements being either in the quiet or the excited state. That is, neurons whose transition width σ is zero. The corresponding response function of such neurons is a discontinuous unit-step function. However such a model cannot be treated as a general one.

The action potentials are transmitted from one neuron to another in the network by releasing chemicals called *neuro-transmitters* in the synapses. These chemicals alter the flow of ions across the dendrites and the membrane of a neuron that further results in a change of its action potential and state of excitation. The effect of the neuro-transmitter may be either an increase of the action potential, called *activation*, or its decrease, called *inhibition*; but a particular synapse can act only in one mode. A neuron can possess thousands of synapses at its inputs at the dendrites as well as at its outputs at the end of the axon. The result is a massive interconnectivity of neurons in the brain. By changing this interconnectivity various transformations of input signals are obtained.

One of the most important properties of the synapses is that the effectiveness of the transmission of signals can be modified by the signals transmitted across them and hence the excitation of the neurons is altered. It is generally believed that the memory of the brain is a consequence of the changeable properties of the synapses in the neural network. [1]

In order to describe the effect of a synapse on the transmitted signal, we introduce the *conductivity* or *efficacy* of the synapse joining the neurons with indexes i and j by the weight m_{ij}. All the synaptic joints of the i-th neuron can then be represented by the weight vector $\boldsymbol{m}_i^T = (m_{i1},\, m_{i2},\, \ldots,\, m_{ij},\, \ldots)$. If a neural network is comprised of N completely interconnected equal neurons then the corresponding memory can be represented by the weight matrix

$$\boldsymbol{M} \;=\; [m_{ij}]\;. \tag{2.21}$$

The study of this matrix and its adaptation is one of the fundamental areas of research in the field of the artificial neural networks. For understanding and modeling neural networks, Hebb's hypothesis about the adaptive formation of synaptic joints is essential. Hebb's hypothesis postulates [1, 15]

The efficacy of signal transmission of a neuron is enhanced if its pre- and post-synaptic parts are simultaneously excited.

We shall utilize this hypothesis later when developing a dynamical description of the adaptation and learning in neural networks.

A signal x_j transmitted from the j-th neuron over the synapse with weight m_{ij} contributes to the activation of the i-th neuron the amount $\Delta a_i = m_{ij}\, x_j$. The influence of all transmitted signals on the activation state of the i-th neuron is then described by the sum of the contributions from all the synapses. That is,

$$a_i \;=\; \sum_{j=1}^{N} m_{ij}\, x_j\;. \tag{2.22}$$

Here N denotes the number of synaptic inputs to the i-th neuron which can either be transmitted from the sensors or other neurons in the network. The corresponding simplest, quasi-static physical model of a neuron then relates the response of the neuron to the excitations from the surrounding by some sigmoidal function [21]

$$y_i \;=\; \Psi\left(\sum_{j=1}^{N} m_{ij}\, x_j - \theta_i\right)\;. \tag{2.23}$$

Similar to the synaptic conductance, the threshold θ_i is generally treated as a variable parameter of a neuron. By introducing a fixed input $x_0 = 1$ and a synaptic weight $m_{i0} = -\theta_i$ the last equation can be written in the simple form

$$y_i \;=\; \Psi\left(\sum_{j=0}^{N} m_{ij}\, x_j\right)\;. \tag{2.24}$$

In the past, the sigmoidal step-like response function of a neuron has led to the erroneous assumption by some researchers that neurons can simply be replaced by digital electronic elements whose transduction characteristics were similar to a neuron and that the operation of the digital electronic circuits would resemble that of a neural network. Because the response of a neuron depends on the weighted sum of the input signals, the operation of a neuron more closely resembles that of analog nonlinear electronic elements. This fact leads to essential differences between the properties of digital circuits and neural networks even though some functional properties might be similar. More to the point, a biological neuron is, in fact, not a stationary element, as represented by the simplified description above, but rather a living cell possessing an intricate dynamical behavior. We shall return to its description when treating the adaptation of neural networks in Chap. 12.

2.8.2 Empirical Modeling by a Perceptron

Biological neural networks have evolved to control biological organisms. To achieve this, a proper transformation of sensory data into muscle excitations must be performed. A transformation which maps input sensory signals into output excitation signals can generally be described by some function. In the past several decades much research has focused on learning how neural networks can execute the corresponding mappings. [2, 13, 14, 16] It was found that one can consider a network comprised of three layers of neurons as an empirical modeler. The physical model of such a neural network is called a *three-layer perceptron.* [2, 14, 16] In order to demonstrate its mapping capability, we consider here the simple mapping of the one-dimensional variable x to the one-dimensional variable y. The multi-dimensional case is treated in Chap. 12. We assume that the mapping can be described by some continuous function $y(x)$ running through the set of sample points $\{x_i, y_i\ ;\ i = 1, \ldots, N\}$. Using the piecewise linear interpolation described by Eq. (2.9) we obtained earlier in this chapter the interpolation function

$$y(x) = y_1 + \sum_{i=1}^{N-1} \Delta y_i \Psi(c_i x - \theta_i) \ . \tag{2.25}$$

This function can be interpreted as an input–output relation of a three-layer perceptron. The structure of the perceptron which is capable of performing the corresponding mapping is depicted by the diagram shown in Fig. 2.17.

This simplified perceptron has just one input and one output neuron and a middle layer that is comprised of several neurons. The neurons of the perceptron are interconnected by synaptic connections which are characterized by their conductivities c_i. The input neuron distributes the signal x equally to all neurons of the second layer which consists of N neurons. The i-th neuron has only one input synapse with the conductivity c_i and the excitation threshold value θ_i. The signal generated by the i-th neuron is

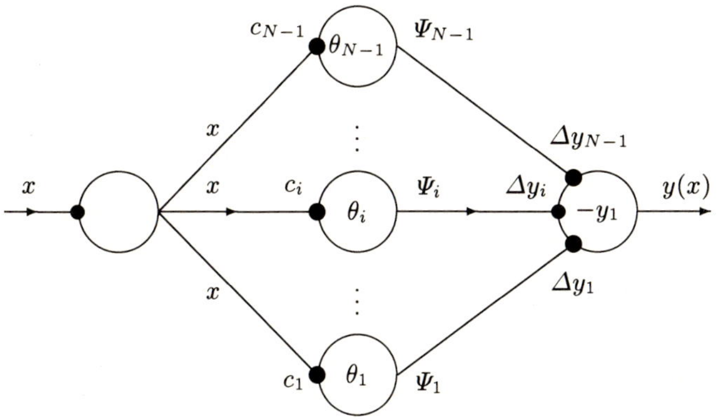

Fig. 2.17. Scheme of a three-layer perceptron capable of modeling a function $y(x)$ extending over the given set of data points $\{x_i, y_i \; ; \; i = 1, \ldots, N\}$

$\Psi_i = \Psi(c_i x - \theta_i)$ and this is transmitted over the synapse with the second layer conductivity Δy_i to the neuron in the final layer whose threshold is $-y_1$ and possesses a linear response function. The transformation of the input signal in such a three-layer perceptron is seen to be equivalent to the operation described by Eq. (2.25). We thus conclude that a three-layer perceptron is capable of modeling a piecewise linear interpolating function running over the given set of representative data points $\{x_i, y_i \; ; \; i = 1, \ldots, N\}$ provided that its parameters are set in accordance with the Eq. (2.9). However, how this setting can be achieved remains to be explained.

The fact that the perceptron represents the basic structure of biological neural networks, which, up to now are still the best modelers of natural phenomena, has led us to consider the piecewise linear interpolating function or its approximation as one of the basic paradigms which might be applicable for the development of automatic modelers of natural phenomena. However, in order to proceed with this development, we must first generalize the description presented above to multivariate and stochastic signals. For this purpose, we must include into our treatment a statistical description of natural phenomena. For this, the description of an interpolating function based on symmetrical basis functions appears analytically to be more appropriate, because it can simply be interpreted as a conditional average estimator. Such an estimator generally corresponds to a statistically optimal estimator of the functional relationships between multivariate stochastic variables. [6, 7] However, for an implementation, the representation of the interpolating function by sigmoidal functions might be more appropriate. At least, that might be presumed because of their existence in biological neural networks. [2]

One of the most difficult challenges related to the modeling of natural phenomena by neural networks is how one can achieve an automatic setting of the characteristic parameters that appear in the model. It is also not clear from the above simplified review of the perceptron how a neural network can achieve a proper ordering of parameters that is implicit in the expression of the interpolating function. These problems will lead us in subsequent chapters to consider the fundamentals of statistical description of natural phenomena and their adaptive modeling. We will see that this approach will later help us in explaining some basic properties of neural networks and developing an applicable empirical modeler of natural phenomena.

3. Transducers

3.1 The Role of Sensors and Actuators

The ability of a living organism to accommodate to a changing environment is an essential element for the existence of life itself. How can a living being achieve this? It must have an ability to sense its environment and to respond. What is *sensing* and what is *responding*? If we observe an animal, we quickly recognize that it possesses special organs composed of cells which are selectively sensitive to influences of their environment. We call such cells *sensors*. The sensors of living beings respond to light, sound, to mechanical stimuli or to a variety of chemicals. There are also other characteristic cells called *actuators* which respond to stimuli to generate motions. In order to function properly, the sensors and actuators of a living being require a *link* which is provided by an amazing network of neurons and the brain. This link is an information processing system because it transforms the information of a stimulus into an output response of the system.

The combination of sensors and actuators with a neural network that includes a memory, results in a truly remarkable system. The relative sliding of actin and myosin filaments when millions are stimulated in concert, results in muscle contractions which can accelerate animals weighting several tons or generate forces in the jaws of dogs of several hundred kiloponds.

When referring to sensors and actuators collectively, we shall use the term *transducer*. A transducer (from the Latin, **transduco**: to lead through) is a device which converts one physical variable to another or a signal from one form of energy to another. Of interest here are the signals of six different kinds of physical variables: mechanical, thermal, magnetic, electrical, optical and physical. In biological systems and in most measurement systems, either the input or the output variable is an electrical signal. The sensory system, in effect, provides the connection between the environment and the processing unit. The muscular system in an animal is a complex system of actuators acting on the skeletal system that connects the processor to the environment. The process of natural selection over the millennia has resulted in the evolution of living beings possessing a diversity of *sensors* and *actuators* which are well adapted to survival.

The general form of a measurement system consisting of the functional elements called the *input* stage, *processing* unit and the *output* stage is shown

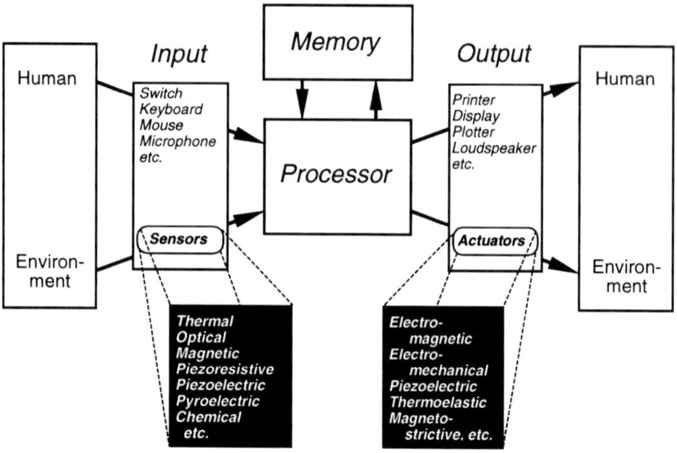

Fig. 3.1. Schematic diagram of a general measurement system

schematically in Fig. 3.1, where the *input* and *output* stages, which are the focus of our interest here, are shown in additional detail. The input or sensing stage is the system's interface to either a living being or to the environment of the *measured* system. In the former, the sensing stage may be a switch which may be simply of *on-off* type or of variable control, a keyboard, a 'mouse', a light pen or any other means for entering information into the processor. When interfaced with the environment, a variety of sensors can be used, whose principle of operation may be any of a number of physical phenomena, with the choice dependent on the quantity to be detected and measured, that is, the *measurand*, as well as the output signal sought.

The information from the sensors is manipulated according to a particular procedure or algorithm in the *processor* in order to extract specific information or to generate a particular output signal. There is usually also provision for storing information in a *memory*. The information stored in the memory may be identical to the input or the output data or it may be kept in some coded, intermediate form.

The characteristics of the output or actuating stage are also dependent on whether this stage interfaces the processor to a living being or to the environment. For the former, the output might be a printer or a similar display device, electrodes, a loudspeaker, or others. When the interface is to a non-biological system, the output may be one of a number of actuators. A listing of some of the physical phenomena which can be used as the basis of an actuator is given in the enlargement in Fig. 3.1. It is seen that several of these are identical to those used as the basis of a sensor. Devices which utilize such phenomena are said to be *reversible* because they can operate either as a sensor or as an actuator.

It is noteworthy that the measurement systems developed by human beings possess the same principal structure and components that are present in all biological systems. Developments over the last twenty years have led to the design and fabrication of "smart" or "intelligent" transducers that in addition to their sensing or actuating element, also incorporate electronic elements which condition the signal. Sensors of this type are known as *integrated sensors*. There may also be components for providing self-calibration and transducer diagnostics capabilities and possibly also a processing unit that permits an analysis and possibly even a rudimentary interpretation of the signals [1]. Additional details of such transduction systems will be reviewed in Sect. 3.4.1.

Our aim in the following sections is to consider the operational characteristics of transducers and how they affect the operation of a measurement system. We briefly survey the remarkable transducers and their operating characteristics that have evolved in biological systems. We then consider aspects of man-made transducers which form part of most measurement systems. It is not our intent here to review all the physical phenomena that can be used as the basis of a transducer, but rather to provide a foundation by which we can understand the transmission of information through such devices.

3.2 Sensors and Actuators of Biological Systems

In order to survive in a changing environment, an animal must be able to quickly and reliably detect the state of its surroundings and especially changes therein. For this purpose, an animal's sensors must be capable of responding quickly and require little input energy for their excitation and yet, at the same time, be capable of reliably responding under adverse conditions. The input element to all biological information processing and control systems is a sensory receptor. Arrays of sensory receptors, in which each cell exhibits the required response characteristics have evolved over the millennia. A partial listing of sensory receptors possessed by living beings is given in Table 3.1 which has been adapted from Ref. [2]. Included are the well-known receptors associated with the five senses – hearing, seeing, tasting, smelling and touching. The list is supplemented to include the two kinds of sensory receptors in the skin which sense warmth and cold and the pain receptors which are stimulated by specific chemicals that are released as body tissue becomes damaged. Also included are all those receptors which are *visceral*, that is, those which operate as part of the autonomous nervous system of a living being. Examples include the temperature sensors in the hypothalmus region of the brain which are essential to maintaining the temperature of the body, the vestibular pressure sensors in the corotid arteries that are critical to maintaining a nearly constant arterial pressure in the brain, sensors in the

Table 3.1. Sensory receptors of living beings (after Ref. [2])

Receptor Type	*Stimulus*
Auditory	Dynamic Pressure Changes
Visual	Light
Taste	Chemical
Olfactory	Chemical
Touch	Static Pressure
Cold	Temperature
Warmth	Temperature
Pain	Excessive Stress or Damage
Visceral	Various

inner ear for maintaining balance and the chemoreceptors located in arteries to monitor blood chemistry.

3.2.1 Performance Characteristics of Biological Sensors

To illustrate how biological evolution has led to extremely well adapted sensors, we need to look no farther than those possessed by most persons. The sensory system of a human being is, simply put, extraordinary. Each of the five senses exhibits a remarkable *threshold of detection* as well as a *dynamic range* which is only rarely equaled by man-made sensors. The human ear has a *threshold of hearing* or *threshold of audibility* of approximately $0.0002\,\mu$bar at frequencies around 3 kHz. This remarkable threshold pressure corresponds to the change in atmospheric pressure corresponding to just 0.7 mm change in altitude at sea level. Because the changes in pressure resulting from the thermal agitation of air molecules are of the same order as the threshold of hearing, the human ear is, in fact, optimally sensitive to sounds whose frequencies are in the low kHz range. Yet, the dynamic range of the human ear extends to the *threshold of feeling* which corresponds to pressures more than twelve orders of magnitude greater and which is possible only because of the non-linear response characteristics of the auditory system. At higher sound intensities, pain and permanent hearing damage result. The audible range of frequencies of sound spans three decades of frequency from 20 Hz to 20 kHz in a healthy ear.

The eye serves as the sensor of our vision system. It is sensitive to light whose wavelengths extend from approximately 400 to 700 nm. The minimum *threshold energy*, that is, the minimum energy of light incident on the eye which will likely produce a visual response is approximately 4×10^{-17} J at 510 nm. This energy corresponds to approximately 10^2 photons incident on the cornea. The best man-made sensing system is only a little more sensitive, being able to detect a few photons per second. The dynamic range of our sense of vision is typically 5×10^5 which can only be realized by a sensing system possessing non-linear response characteristics.

Not unexpectedly, the threshold of man's sense of taste strongly depends on the substance tasted – sour, bitter, sweet or salty. The tasting threshold for examples of each of these substances is 0.0009 N (Normal) for *HCl* (sour); 0.8×10^{-5} M (Molar) for quinine (bitter); 0.01 M for sucrose (sweet) and 0.01 M for *NaCl* (salty).

The olfactory sense of humans is also extraordinary. For example, a concentration as small as 1 part in 2.5×10^{10} mg of the substance methyl mercaptan can be detected by a human being. Some animals have far lower threshold sensitivities. However, the dynamic range of our olfactory system is somewhat limited, being only a factor of 10 to 50 of the threshold value. It is presumed that during evolution, man's sense of smell appears to have adapted only sufficiently to act as a warning system, capable of detecting the presence or absence of an odor but not quantifying its amount.

Quite different from the foregoing is our sense of touch which is a consequence of at least six different tactile receptors. Those in the skin or near its surface are important to touch and pressure and respond for only a finite time, adaptively becoming less sensitive. Sensors deeper in the body can detect vibrations as well as pressure signals.

3.2.2 Structure of Biological Sensors

In the previous subsection we briefly summarized some of the characteristics of the remarkable sensors and actuators possessed by living beings. In order to permit the operation of biological sensors over a broad dynamic range, the information processing in biological systems often relies on a special encoding (kind of presentation) of the detected stimuli. The magnitude of the generated local potential from a sensory cell, the so-called *receptor potential*, is a direct and usually non-linear function of the input stimulus. The sensory cells are adapted for directly communicating with neurons or axons, which communicate with other neurons in a similar way. The characteristics of neurons have been summarized in Chap. 2.

There are strong similarities between the response characteristics of neurons and sensory receptors. Low-level signals from the sensory cells corresponding to the receptor potentials attenuate and distort as they propagate over even a short length of nerve fiber or axon. But when the stimulus and corresponding signal is large enough, that is, when it exceeds an *adequate stimulus*, the result is a depolarization of the potential in the interior of the nerve fiber making it less negative than the exterior. On reaching a critical threshold level, short, impulse-like *action potentials* are triggered in the nerve. Such action potentials can propagate unattenuated for long distances along the nerve at speeds related to the thickness dimension of the axon.

Stronger stimuli result in higher impulse firing frequencies, resembling the pulse position modulation (PPM) data encoding scheme for digitally representing analog data. Such frequency-modulated encoding of sensory information has many advantages, the most important among them is that it results in a noise-insensitive transmission of signals spanning a wide dynamic range.

The functionality of the sensory receptor of a biological system can be represented using the block diagram shown in Fig. 3.2. The transformation element comprising the first stage may focus, amplify, filter or modify the stimulus so that it matches the properties of the *sensory receptor* which is the specialized ending of a sensory nerve that responds to an external stimulus. The sensory receptor defines the limits of sensitivity and determines the range of stimuli which can be detected. In most cases, the exact mechanisms which form the basis of the sensory receptor are still a mystery and are a topic of current research.

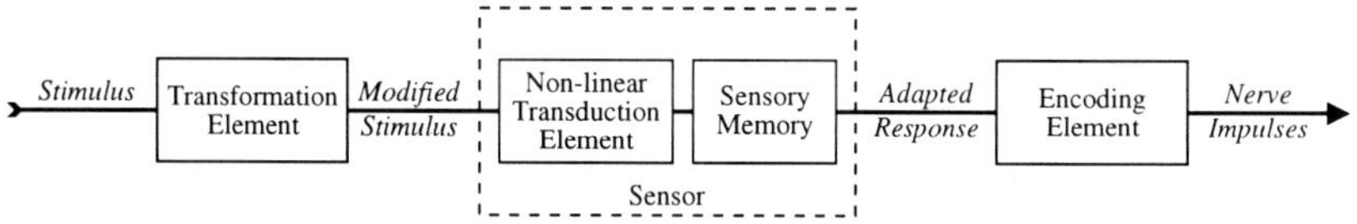

Fig. 3.2. Block diagram of a biological sensory system

In the retina there are short receptors identified as *rods* and *cones* according to their shape. The rods are used to see in dim light while the cones are less sensitive but are of three types according to their color sensitivity to red, green or blue light. In each receptor, a stimulus of light results in a modulation of the membrane potential which then passively spreads from the sensory region to the synaptic region of the sensory cells where they join other neurons. The receptor potential corresponding to the original signal may spread in this way for up to several mm in the mechano-receptors of some crustaceans [4] and leech [5] or the photo-receptors found in the barnacle eye[6]. There are also a number of short receptors such as the photo-receptors of the vertebrate retina which respond to input signals exceeding a particular threshold with a hyperpolarizing change in their membrane potential. Such potentials correspond to an increase in the (negative) magnitude of the potential in the interior of the nerve fiber, whereby it becomes more negative than the region exterior to the fiber. The result is a temporary reduction in sensitivity of the sensory receptor.

Thousands of hair cells attached to the basilar membrane which extends along the length of the cochlea serve as the mechano-receptors of sound. Bundles of variable length hairlike fibers extend from each hair cell with the longest of these making contact with the overlying *tectoral membrane*. Sound-induced motions of the basilar membrane result in shearing motions of the cilia which are in contact with the membrane. These motions open or close a channel that controls the flow of potassium-rich current which in turn results in a depolarization or hyperpolarization of the hair cell. This mechanism is capable of detecting sub-Angstrom displacements of the basilar membrane. And yet, because of various cilia lengths, the transduction mechanism is highly non-linear which can operate over a large dynamic range. By being sensitive to sounds of different frequencies at points along the basilar membrane, the cochlea acts as an effective frequency analyzer. Further, since the bending is directional, the auditory sensory cells have membrane potentials which are either depolarizing or hyperpolarizing. These membrane potentials result in changes in the shape of the hair cells, leading to a positive feedback mechanism that sharpens the frequency selectivity of the basilar membrane.

Long receptors, so called, because of the distance between the sensory receptor and the synaptic region of the sensory cell, are found in the skin and muscles of animals. In these cells, the depolarization potential also gives rise to a train of action potentials. In both types of sensors, when the series of impulses from the sensory axon reaches the synapses at specific locations in the central nervous system, that is, the spinal chord or the brain of the animal, it results in a depolarizing potential there, which, in turn, generates a train of action potentials that propagate to the muscles, resulting in their contraction. The entire process from sensing to action response usually occurs quickly. For example, in a reflex reaction in humans, the entire process is completed in less than 50 ms.

It is interesting to note that the electrical signals in the neurons corresponding to particular sensory inputs are indistinguishable as regards to the measurand. Further, the signals are similar in different animals. For these reasons, the cognition of sensory information in a living system depends not only on the characteristics of the electrical signal, but also on the origins and connections of the nerve fibers. The different sensory neurons are connected to different parts of the spinal chord or brain of an animal.

3.2.3 Transduction Characteristics of Biological Sensors

Living beings which are more advanced than others have become so because they possess highly developed sensing and actuating systems coupled to an efficient information processing system. Although we have developed electronic information processing systems, we are only now beginning to understand how biological information processing systems operate. We can undoubtedly learn much from these optimized systems for the future development of our own measurement and control systems. Thus we may ask, what are the essential characteristics of sensors or actuators if they are to mimic their counterparts in biological systems? We consider those characteristics as to their temporal (including frequency), spatial and system characteristics. The latter refer to the operation of the sensor or actuator as an integral part of the biological system.

The temporal response of biological sensors is usually described as being fast or slow-responding. Sensors may also possess the capability of adapting their response to a continuing input stimulus. That is, the response of a sensory receptor depends on the transient as well as long-term characteristics of the stimulus. The steady-state or DC response is the response of a sensor that has fully adapted to a stimulus. In contrast, the transient, or AC response corresponds to the response of the sensor immediately following a sudden change of the input stimulus. To model these characteristics of a sensor, Deutsch, et al. [2, 3] have proposed a two-component model for the sensory element which is shown in Fig. 3.3.

In human beings, the steady-state response of both the auditory and touch receptors is essentially negligible. The adaptation of the other sensory receptors is a consequence of the time-dependent electrical, mechanical or chemical effects which are present. The adaptation of the eye to dim illumination is an example of chemical effects adjusting the response of the photo-receptors in the retina. The rate of adaptation of receptors to an input stimulus varies considerably. In mechano-receptors the adaptation may be the result of the viscoelastic properties of the attachments between the nerves and the muscle fibers, changes in the positive Na^+ ion concentration or an increase in the Potassium conductance in the axon, both of which serve to reduce the receptor potential and thus the amplitude and frequency of the action potential impulses. Hence, in modeling the response of sensors which adapt to the input stimulus, a "memory" element is required. In the simple model shown in Fig. 3.3, this is obtained using the linear RLC network in which there is a conductance G which is in parallel with a capacitance C and these elements are connected to other resistive and inductive elements as shown.

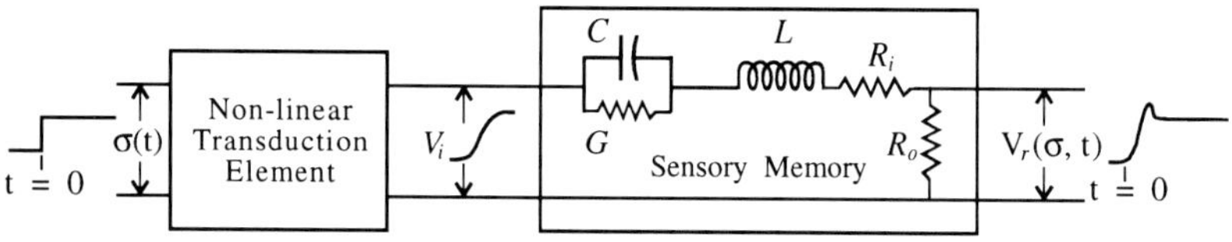

Fig. 3.3. Model for a sensory element (after Ref. [3])

The steady-state transfer characteristics of the network shown is given by

$$H(t \to \infty) = \frac{R_o}{R_i + 1/G} . \tag{3.1}$$

While the transfer function of the memory element is in reality very likely non-linear, in this model, such non-linear effects are imbedded in the first component of the sensor element. The steady-state response of various sensory receptors is listed in Table 3.2 and these are shown graphically in Fig. 3.4.

Table 3.2. Response after complete adaptation of various sensory receptors (after Ref. [3])

Sensory Receptor	*Behavior*	*Input/Output Characteristics*
Touch Auditory	No Response	None
Olfactory Taste Muscle Stretch	Exponential	$1 - e^{\sigma}$
Visual	Logarithm squared	$[\ln(1+\sigma)]^2$
Warm Cold Pain Visceral	Quadratic Power Law	σ^2

By having both excitory as well as inhibitory axons, the biological sensory receptors exhibit a capability of controlling their sensitivity. Further, the central nervous system of animals is also capable of editing sensory information as well as exercising feedback control of the information it receives. Known as *centrifugal control*, it is an important element in controlling the contraction of muscles which is of particular importance for muscles which are capable of fine control, such as those in the hand. It also plays a role in adapting and optimizing the response characteristics of certain sensors, such as certain hair cells in the auditory system.

In many cases, the dimensions of the sensory receptor that is exposed to the physical variable being sensed determines the spatial response or receptive

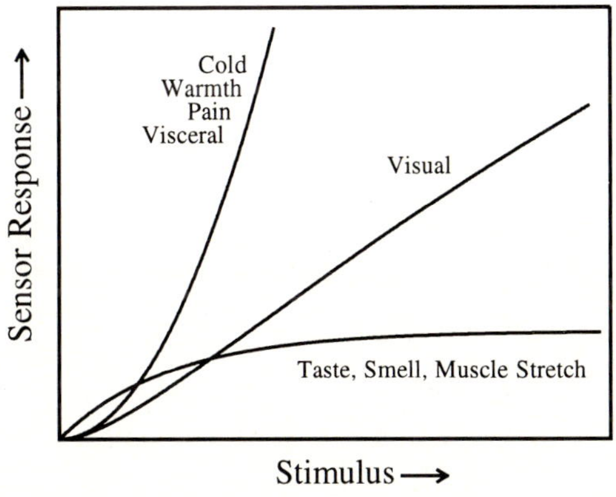

Fig. 3.4. Response curves of sensors which are fully adapted (after Ref. [3])

field of the biological sensor. The spatial response of a sensor is of importance when the sensor collects dynamic measurements or when it is used to localize the sensory information. For example, the resolving power of the healthy eye is determined by the dimensions of the rods and cones in the retina. In the *foveola* of human beings, which is the most sensitive portion of the retina, the rods are about 3 μm in diameter. This leads to a resolving power of about 0.011 degrees or an angle of 1 minute. The opening of the auditory canal and the diameter of the ear drum are less than the shortest wavelength the auditory system is capable of hearing. Hence to a good approximation, the ear acts as a point detector.

The mechano-receptors in the skin, known as the somatic sensory cells, are of several types depending on their location in the skin and their rate of adaptability to a touch stimulus. In non-hairy skin, there are rapidly responding Meissner's corpuscles and the more slowly adapting Merkel's disks. Deeper situated are the rapidly responding Pacinian corpuscles and the slowly adapting Ruffini's corpuscles. In addition, there are free nerve endings, so-called because they lack a distinct morphological structure, and they sense pain and temperature. In hairy skin, nerve terminals surrounding the hair follicles give a rapid response to a bending of the hairs.

In living beings, arrays of individual transduction elements may also operate in concert. For example the skin of the human hand, which is densely innervated, has 15 to 20 thousand mechano-receptors reaching a density of about 100/cm^2 at the fingertip which is about three or four times greater than their density in the palm of the hand and many times greater than in other parts of the body such as the back. Consequently, the hand possesses an incredible sense of touch, while also providing rapid warning of potentially harmful object temperatures and an adequate sense of object temperature recognition.

3.3 Operational Characteristics of Transducers

3.3.1 Transducer Classification

There are a number of different approaches for classifying transducers. White [7] has proposed a general scheme for categorizing sensors that considers the various sensor characteristics, grouped according to their measurand, their operational characteristics, their means of detection, their operative conversion phenomena, materials and their field of application, that is, transducers can be grouped according to their operating principle or phenomenon, by the property or physical variable that is sensed or manipulated, as to their application or a variety of other distinguishing features. The most general description, however, will be a combination of all these approaches. This corresponds to a listing of the responses to a number of questions related to the various aspects related to the operational characteristics of a transducer, including:

Concept - What physical quantity or quantities does the sensor detect, or the actuator generate?
It is usually assumed and it is often desired that a transducer measures just one variable or one component of a vector variable. The *primary* quantities include such variables as linear and angular displacements; linear and angular velocities; linear and angular accelerations; force; temperature; light and time. It is possible to compute from these primary measured quantities several derived quantities. For example, the output signal of a displacement sensor may be used to derive quantities such as length, width, thickness, position, level, material erosion, wear, surface quality, strain, vibration amplitude, etc. Measurements of temperature may be used to monitor heat flow, fluid flow, gas pressure, gas velocity, angle of flow, turbulence and sound amplitude. [13]

Model - In practice a large number of actuators rely on an electrical input signal to generate the signal of the desired variable while most sensors generate an electrical output signal in response to their input variable. The elementary picture of a transducer as an energy converter is too broad to be very useful for interpreting and quantifying a transducer signal. Instead one needs to use a model that describes reasonably well the complete operation of the transducer. To develop this model, it is customary to critically look at the physical transduction principle which forms the basis of the operation of the transducer? But one also needs to address the question whether there is more than one transduction principle operative. It is recognized that if the mathematical model is to accurately describe the operation of a transducer, it will be, of necessity, complicated. This difficulty will be addressed later in this chapter.

Realization - What are the transducer's design features? And what are the transducer's operational requirements and limitations? The response to these questions is developed by the transducer designer and fabricator.
In particular, one may ask what is the choice of material comprising the transducer element? What are its material properties? What is its geometry? What additional features does the transducer possess? (e.g. Coatings, wear plate, attached lens, fins, tuning coils and other impedance matching devices, etc.)
It is equally important that the designer of the instrument or measurement system in which the transducer is imbedded know what the operational features of the transducer of a transducer are. That is, what is its interfaceability with conditioning elements or a processor; and with the environment? What are its environmental characteristics? What is the transducer's reliability and ruggedness? What is the transducer excitation? Is it self-powered, or *active*, such that the output signal is derived or

generated from the input signal, or does it use an external power source, or *passive*, in which the input signal controls or modulates the output.

Application and Operation - The user of a transducer and its associated measurement system needs know and be able to determine a device's measurement characteristics. These may include: Its spatial characteristics; frequency characteristics; dynamic range; time response; the presence or absence of "ill" or undesirable characteristics; the transducer's signal-to-noise ratio/threshold; its sensitivity and the sensitivity shift with time as well as its dependence on signal amplitude and duration? What is the effect of creep; and the shift of the zero with time and with signal character?

We note that of the above questions, a user will generally only focus on those in the first and last area. While the transducer designer and instrument maker will generally only focus on the second and third areas.

What is an "ideal" transducer? An ideal transducer is one that efficiently transforms the signal of a physical variable from one form of energy to another, with negligible modification of the input signal from any point or region in a test medium and with negligible influence on either the test medium or the instrumentation with which the transducer interacts. Most transducer designs represent a compromise between the various characteristics of the ideal transducer. Techniques to minimize any deficiencies can be applied at the transducer element or its mounting directly, or implemented as modifications of the excitation or detection circuitry.

3.3.2 Transduction Characteristics

There are many factors that affect the operational characteristics of a transducer. For a transducer in which either its input or its output is an electrical signal, these characteristics are determined by the complex interaction of the transducer's electrical, transduction and field characteristics. The *transduction characteristics* of a transducer refer to the relationship between the input and output signals across the transducer. The functionality of a transducer can be schematically represented as a box-like element with *input* and *output* signals $q_i(t)$ and $q_o(t)$. The case of an actuator is shown in Fig. 3.5. While some transduction devices can be designed to minimize the influence of the electronic instrumentation and the test medium on the transduction process, there is no transducer that eliminates all of them completely. It is recognized that the precise nature of the transduction process must be understood if truly quantitative information is to be extracted from a measurement.

Let us consider the operation of a sensor in more detail. We denote the time evolution of the input variable to the sensor situated at position $\boldsymbol{r} \equiv (x, y, z)$ of the test medium as $q_i(\boldsymbol{r}, t)$. Since we are principally interested in sensors which produce an electrical signal at their output, the output signal

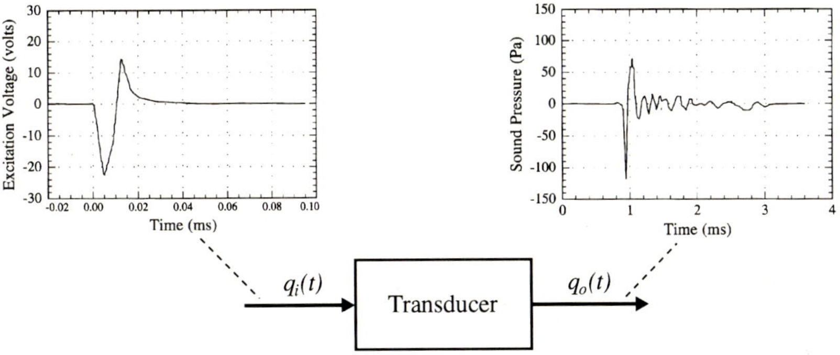

Fig. 3.5. Schematic representation of the operation of a transducer. The signals shown correspond to the input electrical voltage signal and the output acoustic pressure signal from a simple loudspeaker. (Note the expanded time scale of the signal $q_i(t)$)

$q_o(t)$ may represent the voltage $V_o(t)$ or the current $I_o(t)$. The output signal of a transducer is time-shifted relative to the input signal and so we may write for the output signal, the complex time functions $\boldsymbol{V}_o(t)$ or $\boldsymbol{I}_o(t)$ whose magnitudes are respectively, $V_o(t)$ and $I_o(t)$, and whose phases are given by ϕ_V and ϕ_I . An equivalent description can be given in the frequency-domain as $\boldsymbol{Q}_o(\omega)$ whose magnitude is $Q_o(\omega)$ and phase is $\phi_o(\omega)$. The quantity ω denotes the circular frequency ($= 2\pi f$, where f is the frequency in [Hz]) of the signal. When vector quantities are detected, we will write for the input signal, $\boldsymbol{q}_i(\boldsymbol{r},t)$. An example is an electro-mechanical sensor which is sensitive to input forces $\boldsymbol{F}_i(\boldsymbol{r},t)$ and displacements $\boldsymbol{U}_i(\boldsymbol{r},t)$. A similar description can be given for an actuator that takes an electrical input signal $q_i(t)$ corresponding to $\boldsymbol{V}_i(t)$ or $\boldsymbol{I}_i(t)$ and generates the vector output signal $\boldsymbol{q}_o(\boldsymbol{r}',t)$ at the location $\boldsymbol{r}'$ of the test medium. For an electro-mechanical actuator, the output signal may correspond to forces or displacements in the medium. These are summarized in Fig. 3.6 which is taken from Ref. [15].

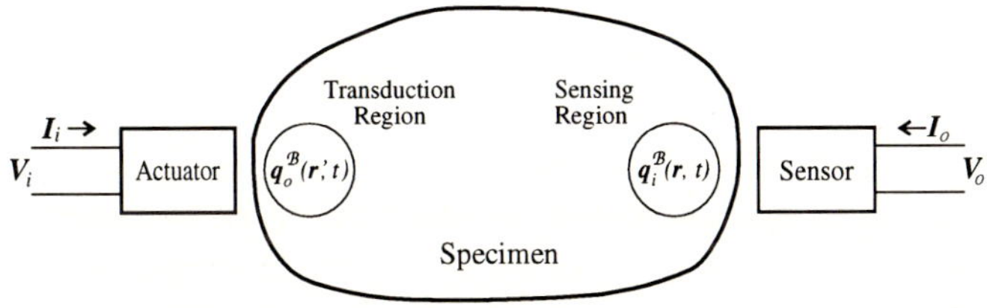

$$\text{Actuator:} \quad \left.\begin{matrix} \boldsymbol{V}_i(t) \\ \boldsymbol{I}_i(t) \end{matrix}\right\} \Rightarrow \left\{\begin{matrix} \boldsymbol{F}_o(\boldsymbol{r}',t) \\ \boldsymbol{U}_o(\boldsymbol{r}',t) \end{matrix}\right. \qquad \text{Sensor:} \quad \left.\begin{matrix} \boldsymbol{F}_i(\boldsymbol{r},t) \\ \boldsymbol{U}_i(\boldsymbol{r},t) \end{matrix}\right\} \Rightarrow \left\{\begin{matrix} \boldsymbol{V}_o(t) \\ \boldsymbol{I}_o(t) \end{matrix}\right.$$

Fig. 3.6. Conceptual model of the transduction process for an electro-mechanical actuator and a sensor

The transduction characteristics of a transducer need to be known in order to predict the transducer's output for a given input signal. This is the *forward problem*. Usually more difficult is the *inverse problem*, that is, of determining the input signal from a transducer's output signal. Knowledge of a sensor's transduction characteristics permits one to process an output signal to determine the corresponding input signal of the sensed physical variable. Similarly, one can compute the signal that should be input to an actuator in order to generate a particular excitation signal in the test medium.

In describing the transduction characteristics of a transducer, one needs to know the specific variables to which the input signal to a sensor $\boldsymbol{q}_i(\boldsymbol{r},t)$ or the output signal of an actuator $\boldsymbol{q}_o(\boldsymbol{r}',t)$ correspond. All sensors and actuators generate an output to varying degree to more than one input variable, called the transducer's *secondary* inputs. Once the variables have been specified, it would appear that for static signals, this operation is straightforward, provided one knows the transducer's sensitivity. For dynamic signals, the transducer's *transfer function* is needed. It will become clear in the following paragraphs that this is the case only under special circumstances and even then, it can only provide a reasonable approximation between input and output characteristics for most transducers.

In order to consider this point further, let us review the operation of a linear, finite-aperture, electro-mechanical sensor in more detail. The description presented here follows that given in the review by Sachse and Hsu.[15] We restrict the discussion to a sensor whose output is an electrical signal. As shown in Fig. 3.6, this output signal is characterized by the voltage $\boldsymbol{V}_o(t)$ across the transducer and the current $\boldsymbol{I}_o(t)$ flowing through it. The input to the sensor can be characterized by the distributed field quantities of the physical variables being sensed over the sensor's aperture region $\mathcal{B}$. The physical input variable of a finite-aperture sensor whose centroid is located at $\boldsymbol{r}$ can be written as $\boldsymbol{q}_i^{\mathcal{B}}(\boldsymbol{r},t)$ which can be defined by

$$\boldsymbol{q}_i^{\mathcal{B}}(\boldsymbol{r},t) \equiv \frac{1}{S_{\mathcal{B}}} \int_{\mathcal{B}} \boldsymbol{q}_i(\boldsymbol{r}, \boldsymbol{s}-\boldsymbol{r};\, t)\, dS \tag{3.2}$$

where $S_{\mathcal{B}}$ corresponds to the effective area of the transducer. It is noted that most often this is less than the physical area of the transducer element. The term $\boldsymbol{s}-\boldsymbol{r}$ is defined as the distance from the area dS of the transducer element to the centroid of the transducer. Examples of transducer input quantities include vector quantities such as tractions (force per unit area) $\boldsymbol{F}_i^{\mathcal{B}}(\boldsymbol{r},t)$, the particle velocities $\boldsymbol{U}_i^{\mathcal{B}}(\boldsymbol{r},t)$ or scalar quantities, such as pressure $p_i^{\mathcal{B}}(\boldsymbol{r},t)$ or the temperature $T_i^{\mathcal{B}}(\boldsymbol{r},t)$. By writing the physical input variables explicitly, we emphasize their vectorial characteristics as well as their spatial and temporal dependencies. The complexity of the transduction process rests upon the following situation: (1) The input field quantities, and specifically their extent and their distribution are often difficult to measure or to predict; and (2) Because of loading effects by the transducer on the input variable $\boldsymbol{q}_i$, the relationship between generated voltage and current, the electrical *output impedance* of the sensor, is, in general, dependent on the medium to which the sensor is coupled; and (3) The transduction process involves many indirect physical interactions for which the mechanisms may not be well known. For example, a piezoelectric transducer requires a direct mechanical coupling between the transducer element and the test medium which is often obtained by using a coupling fluid. This coupling is often poorly characterized and its effects on the transmissivity of the mechanical signal through it and into the transducer element are difficult to determine.

An electromagnetic transducer utilizes an induced eddy current field which depends not only on the characteristics of the signals, but also on other factors such as, for example, the local material properties in the field as well as the gap between the material and sensor. The development of optical-based sensing systems have resulted in non-contact measurement techniques and for these, provided that they are non-damaging to the test specimen, permit neglecting, to a good approximation, the mechanical loading effect of the sensor on the test medium. Nevertheless, the measured response to temporally- and spatially-varying physical variables may only be an approximation of the actual field.

The transduction process associated with an actuator can be formulated similarly. The excitation signal generated in the medium is related in a complicated way to the input electrical signal specified in terms of the voltage and current inputs to the actuator. The observations made regarding sensors which are directly coupled to a medium being sensed are also applicable to directly coupled actuators. Much recent work has focused on the development of laser-generated mechanical excitations in fluids and solids in which the excitation mechanism is based on an absorption of thermal radiation energy and the resulting local thermal expansion of the medium (c. f. Scruby[16]). While the qualitative nature of the mechanisms is understood and a quantitative analysis of the process has been made, the force and velocity vector fields as well as those of other quantities which are generated in an actual test medium, especially if the medium is a solid, are dependent on the input laser power in a very complicated way.

Simplifying Assumptions and the Transduction Matrix. The above description of a transducer's input and output characteristics, while complete, is not readily amenable to analysis; assumptions are required to permit obtaining a practical solution. At the outset, it is assumed that the physical variable detected by a sensor or generated by an actuator is a scalar quantity. In other words, when a vector quantity is the physical variable, it is assumed that the components of the vector are uncoupled and that they can be separately analyzed. That is, we assume that: $\boldsymbol{q}^{\mathcal{B}}(\boldsymbol{r},t) \rightarrow q^{\mathcal{B}}(\boldsymbol{r},t)$. With the exception of the sub-miniature *micro-sensors*, the primary sensing element in most real transducers is of finite size, ranging most commonly from mm to several cm in extent. The second assumption made is that the interaction region $\mathcal{B}$ of the transducer and the test medium is small and localized. This assumption is equivalent to assuming that the detected or generated physical variable fields are uniform and hence, independent of the spatial variable $\boldsymbol{r}$ in the region $\mathcal{B}$. That is, we make the simplification: $q^{\mathcal{B}}(\boldsymbol{r},t) \to \bar{q}(\boldsymbol{r}_0,t) \to \bar{q}_{\boldsymbol{r}_0}(t) \equiv q(t)$, the value of the variable at the position of the centroid of the transducer, $\boldsymbol{r}_0$. This is equivalent to evaluating the following integral

$$\bar{q}_{\boldsymbol{r}_0}(t) \;\equiv\; q(t) \;=\; \frac{1}{S_{\mathcal{B}}} \int_{\mathcal{B}} q(\boldsymbol{r}_0, \boldsymbol{s}-\boldsymbol{r}_0; t)\, dS \;. \tag{3.3}$$

The above two assumptions reduce the transduction process to a one-dimensional model.

The third assumption made is that the transduction process is linear. That is, if there are two input signals $q_i^{(1)}(t)$ and $q_i^{(2)}(t)$, then we assume the following linear relationship holds

$$H[aq_i^{(1)}(t) + bq_i^{(2)}(t)] \;=\; a\,H[q_i^{(1)}(t)] + b\,H[q_i^{(2)}(t)] \tag{3.4}$$

where the operator H denotes the response of the system to a time-dependent excitation and the a and b denote the input amplitude multipliers. It is then convenient to express any arbitrary input function as the linear superposition of time-shifted impulses. That is,

$$q_i(t) \;=\; \int_{-\infty}^{\infty} q_i(t')\,\delta(t-t')\,dt' \;. \tag{3.5}$$

In this case we obtain for the output or response of the system to the excitation $q_i(t)$

$$q_o(t) \;=\; H[q_i(t)] \;=\; H\left[\int_{-\infty}^{\infty} q_i(t')\,\delta(t-t')\,dt'\right] \;. \tag{3.6}$$

By defining $H\,[\delta(t-t')] \equiv h(t-t')$ which describes the response of the transducer to a perfect impulse excitation, we can express the output of the transducer as the convolution integral

$$q_o(t) \;=\; \int_{-\infty}^{\infty} q_i(t')\,h(t-t')\,dt' \;. \tag{3.7}$$

When the transducer signal is a harmonic input signal which is written as

$$q_i^{(\omega)}(t) \;=\; Q_i(\omega)\cos(\omega t)\,, \tag{3.8}$$

it is convenient to consider the response of the system to the real component of the generalized complex harmonic signal

$$q_i^{(\omega)}(t) \;=\; Q_i(\omega)\,\exp(\jmath\omega t)\;. \tag{3.9}$$

The response to such an input signal is the real component of the complex output response, written as

$$\Re\{\mathbf{q}_o\} \equiv q_o^{(\omega)}(t) = |\boldsymbol{H}(\omega)| \cdot Q_i(\omega) \cdot \left|e^{\jmath\omega t+\phi(\omega)}\right| \tag{3.10}$$

where we have introduced the complex *frequency response function* by the expression

$$\boldsymbol{H}(\omega) = \int_{-\infty}^{\infty} h(t')\, e^{\jmath\omega t'}\, dt' = |\boldsymbol{H}(\omega)|\, e^{\jmath\phi(\omega)} = H(\omega)\, e^{\jmath\phi(\omega)} \tag{3.11}$$

where $|\boldsymbol{H}(\omega)|$ $(\equiv H(\omega))$ and $\phi(\omega)$ represent the magnitude and phase of the response function.

In operation, a transducer can transduct a number of mechanical or electrical variables, such as force, velocity, displacement, voltage, current, etc. For a linear transducer which is sensitive to input forces $f(t)$ and input velocities $u(t)$, the output voltage response can be written as the linear superposition of the response of the transducer to each of these variables. That is,

$$V(t) = \int_{-\infty}^{\infty} h(t)\, f(t-t')\, dt' + \int_{-\infty}^{\infty} g(t)\, u(t-t')\, dt' \,. \tag{3.12}$$

For the case of complex harmonic signals this is equivalent to

$$V(t) = H_{11}F(\omega) + H_{12}U(\omega) \tag{3.13}$$

where

$$H_{11}(\omega) \equiv \int_{-\infty}^{\infty} h(t') \cdot e^{-\imath\omega t'}\, dt' \quad \text{and} \quad H_{12}(\omega) \equiv \int_{-\infty}^{\infty} g(t') \cdot e^{-\imath\omega t'}\, dt' \,. \tag{3.14}$$

Analogous relations exist for the output current signal from the transducer. Thus the transduction process can be characterized by a complex matrix which relates the electrical input parameters to the mechanical output parameters in the frequency-domain. This matrix, relating any transducer's input and output parameters, can be denoted as the transduction matrix $[\mathbf{H}]$ of the transducer. That is,

$$\begin{bmatrix} \boldsymbol{V}_o \\ \boldsymbol{I}_o \end{bmatrix} = [\mathbf{H}] \begin{bmatrix} \boldsymbol{F}_i \\ \boldsymbol{U}_i \end{bmatrix} = \begin{bmatrix} \mathbf{Z}_{11} & \mathbf{Z}_{12} \\ \mathbf{Z}_{21} & \mathbf{Z}_{22} \end{bmatrix} \begin{bmatrix} \boldsymbol{F}_i \\ \boldsymbol{U}_i \end{bmatrix} \tag{3.15}$$

at every frequency ω. The matrix elements $\mathbf{Z}_{ij}$ correspond to blocked and free electrical impedances and open- and short-circuit mechanical impedances, according to the subscripts.

In the discussion up to now, emphasis has been placed on the operational aspects of the transduction process and the assumptions that are needed to permit its characterization with experimental techniques. The discussion has not been intended to be a rigorous derivation of the transduction matrix. However, for any electro-transduction device, the derived results are expected to be applicable. Furthermore, it has been shown that for a particular transducer design, the matrix elements can be computed directly from the electro-transduction properties of the elements constituting the transducer. [17] For example, the case of piezoelectric transducer possessing an arbitrarily layered structure was computed by Sittig. [18].

By applying the same assumptions delineated previously (mode uncoupling, spatial field independence, and linearity), the operation of a mechanical actuator can similarly be characterized by a transduction matrix relating the mechanical output parameters to the electrical input parameters; hence,

$$\begin{bmatrix} \boldsymbol{F}_o \\ \boldsymbol{U}_o \end{bmatrix} = [\mathbf{H}'] \begin{bmatrix} \boldsymbol{V}_i \\ \boldsymbol{I}_i \end{bmatrix} = \begin{bmatrix} \mathbf{Z}'_{11} & \mathbf{Z}'_{12} \\ \mathbf{Z}'_{21} & \mathbf{Z}'_{22} \end{bmatrix} \begin{bmatrix} \boldsymbol{V}_i \\ \boldsymbol{I}_i \end{bmatrix} \tag{3.16}$$

at every frequency ω. The transduction matrices given by Eqs. (3.15) and (3.16) are inverses of each other.

Further simplifications of the linear matrix model are also possible. For instance, if the coupling medium has a definite and known mechanical impedance, $\boldsymbol{Z}_m$, then the electrical input impedance, $\boldsymbol{Z}_E$, of a transducer operating as an actuator, can be computed from the matrix elements. That is,

$$\boldsymbol{Z}_E = \frac{\boldsymbol{V}_i}{\boldsymbol{I}_i} = \frac{\mathbf{Z}_{12} - \mathbf{Z}_{22}\mathbf{Z}_m}{\mathbf{Z}_{21}\mathbf{Z}_m - \boldsymbol{Z}_{11}} . \tag{3.17}$$

In this equation, the input voltage and current signals are written as $\boldsymbol{V}_i$ and $\boldsymbol{I}_i$ to emphasize that they are complex quantities, possessing both magnitude and phase components.

If, in addition, both the loading of the physical variable on the transducer and the electronic instrumentation to which the transducer is connected are fixed, that is, the transducer is coupled to a particular test medium and connected to a specific electronic instrument whose input impedance is known, then the input physical variable is linearly related to the output signal voltage. The transduction relation is thus reduced to a simple linear transfer function equation which can be written either in the time- or frequency-domains. The output voltage of a sensor detecting the physical variable $q_i(t)$ whose complex Fourier transform is written as $\mathbf{q}_i(\omega)$, is thus given by

$$\boldsymbol{V}(\omega) = \boldsymbol{H}(\omega) \cdot \mathbf{q}_i(\omega) \tag{3.18}$$

where $\boldsymbol{V}(\omega)$ denotes the complex Fourier transform of the generated sensor voltage and $\boldsymbol{H}(\omega)$ denotes the transfer function of the transducer. Normally, one treats the transducer as a system of constant parameters and under these conditions, the transfer function of a transducer is time-invariant. This may not be valid when the transducer is subject to environmental stress or to damage.

The transduction equation, expressed by Eq. (3.18) can also be expressed as a time-domain convolution, and it is then written in a form similar to Eq. (3.7). When the input signal is of finite duration $0 \le t \le \mathrm{T}$, then one writes for the sensor output voltage signal

$$V(t) = \int_0^{\mathrm{T}} q_i(t')\, h(t-t')\, dt' \tag{3.19}$$

which is usually written more compactly as:

$$V(t) = h(t) * q_i(t) \ . \tag{3.20}$$

The foregoing is perfectly general and is not restricted to sensors sensitive to any particular mechanical or other variables. Similar linear transfer functions can also be written for a transducer operating as an actuator. In this case, the generated physical variable in the medium is formally written as a convolution of the transfer function of the actuator and the excitation voltage (or current). Eqs. (3.18) and (3.20) represent the solution to the forward problem of the transducer, that is, from an input signal to a transducer, one can predict its output response.

In an ideal transducer, all the assumptions that have been made above are realized perfectly and the transduction matrix corresponds to the identity matrix multiplied by a scalar. In this case, the electrical output signal from the sensor can be interpreted as corresponding directly to the physical quantity of interest. Similarly, for an actuator, the desired physical variable can be obtained via *open loop control* by applying the appropriate electrical excitation signal.

Transfer Function of a System. The transduction element may be just one of the components in a chain comprising a complex sensing or actuating system. If one wishes to process the output signal of such a system, the transfer functions of each of the components comprising the system must be known. If N elements comprise the entire chain, then the corresponding frequency response function of the total system is given by

$$\boldsymbol{H}_{\mathrm{total}}(\omega) = \prod_{n=1}^{N} \boldsymbol{H}_n(\omega) \ . \tag{3.21}$$

Hence the magnitude and phase of the transfer function of the complete N-component system are given by

$$\begin{aligned} \text{Magnitude}: \quad H_{\mathrm{total}}(\omega) &= \prod_{n=1}^{N} H_n(\omega) \\ \text{Phase}: \quad \phi_{\mathrm{total}}(\omega) &= \sum_{n=1}^{N} \phi_n(\omega) \ . \end{aligned} \tag{3.22}$$

A corresponding equation can be written as a time-domain convolution. In either case, the performance of the complete transduction system is a convolution of the transfer characteristics of the individual elements comprising the system.

Determining the Transduction Characteristics of a Transducer. Experimental procedures for determining the transduction characteristics of a transducer in many ways resemble those followed to determine the functional characteristics of any system. That is, one measures the output response of the transducer to a known input signal. Details of the measurement techniques will obviously vary widely according to the physical variable the transducer detects or generates and whether the signals of interest are constant, slowly or rapidly varying functions of time, that is, static, quasi-static or dynamic.

Static Characteristics. The *static characteristics* of a transducer include its *sensitivity* which is the slope of the input/output characteristics of the transducer, its *range* or the maximum and minimum values of the input or output variables, its *span*, the maximum variation of input or output variables, its *non-linearity* which is the difference between a linear input/output characteristic and that of the actual transducer. Additionally, there may be *hysteresis* and *environmental effects*. The former is evidenced when the response of the transducer is affected as to whether the input/output variables are increasing or decreasing and the latter can appear as a modification or an interference in the operating characteristics of the transducer.

Dynamic Characteristics. A transducer's *dynamic characteristics* are expressed by the transducer's transfer function in which the static characteristics described above represent the zero frequency value. In addition to the sensitivity, the other parameters are also of importance when describing the dynamic characteristics.

There are a number of techniques for experimentally determining the transfer function of a transducer, expressed in terms of the transducer's *gain* and *phase factor* as a function of frequency or its *impulse response* in the time-domain. All of the techniques require a known and well-characterized excitation signal in which the characteristics of the physical variable are either known or can be determined in a separate calibration test and the transducer output signals are measured. The input signal must be of sufficiently broad bandwidth, at least such that it exceeds that of the transducer being characterized. While these characteristics may be readily realized for electrical and optical signals, that is usually not the case for signals of other types and, in particular, mechanical signals. The excitation signals may be a series of discrete frequency signals, generated either continuously or in finite-duration bursts, by a continuously-swept frequency excitation, a short-duration transient, or a broadband noise source. With the advent of computer-based data acquisition and processing techniques, the last three of the aforementioned approaches are most commonly used today.

There are advantages to using a transient which are related to its well-defined temporal characteristics. Ideally, this would be the impulse excitation $A \cdot \delta(t)$, but such an excitation is very difficult to realize in practice. Instead, as the example shown in Fig. 3.5 illustrates, one is usually faced with using an actual impulse which is of finite duration requiring that the output signal be deconvolved with the input signal in either the time– or frequency-domains in order to recover the transfer function of the transducer. That is,

$$h(t) = q_o^{\mathrm{cal}}(t) * [q_i^{\mathrm{cal}}(t)]^{-1} \tag{3.23}$$

where $q_i^{\mathrm{cal}}(t)$ and $q_o^{\mathrm{cal}}(t)$ refer to the input and output calibration signals of the transducer, respectively. Most often, this deconvolution operation is carried out in the frequency-domain for which we write $q_i^{\mathrm{cal}}(t) \to \boldsymbol{Q}_i^{\mathrm{cal}}(\omega)$ and $q_o^{\mathrm{cal}}(t) \to \boldsymbol{Q}_o^{\mathrm{cal}}(\omega)$. Then the transfer function of the transducer $\boldsymbol{H}(\omega)$ is obtained from

$$\boldsymbol{H}(\omega) = \frac{\boldsymbol{Q}_o^{\mathrm{cal}}(\omega)}{\boldsymbol{Q}_i^{\mathrm{cal}}(\omega)} . \tag{3.24}$$

Since the operation described by this equation is a calibration procedure for a transducer, we call this equation the transducer's *calibration equation*. The procedure specified by the calibration equation, Eq. (3.24), is more robust if calibration signals whose bandwidth is as wide as possible are used to excite the transducer under test.

Even when it is possible to generate impulsive excitation signals, the foregoing procedure may lead to difficulties because of the high peak powers which may be generated in the transducer and its concomitant non-linearities. For this reason *pseudo-random* or *periodic random* excitation signals are often used in practice. The frequency spectrum of such random excitations spans the frequency range of the sinusoidal signals of arbitrary frequencies and phases that are used to synthesize such excitations. Because of temporal variations of the power spectral density of such random excitations, signal averaging techniques are often used. To obtain a better estimation of the power level of the transfer function plus noise as a function of frequency, one can use *root mean square* (*rms*) averaging. By using a *linear averaging* of the signals, it is possible to improve their *signal-to-noise* ratio (*SNR*) in the determination of the transfer function. This technique, however, requires an appropriate trigger or synchronization signal which is synchronous with the periodic component of the excitation signal.

Absolute Transducer Measurements. The solution to the inverse problem permits us to determine the input signal to a sensor by deconvolving the transducer's transfer function and the sensor-generated output signal. The procedure is similar to that used for determining the transfer function of a transducer described in the previous section. In the frequency-domain, this deconvolution operation is written in a form analogous to Eq. (3.24), that is

$$\boldsymbol{Q}_i(\omega) = \frac{\boldsymbol{Q}_o(\omega)}{\boldsymbol{H}(\omega)} . \tag{3.25}$$

Instabilities arise in this operation when the bandwidth of the sensor is too narrow but procedures for stabilizing the deconvolution have been developed. By inserting the calibration equation given by Eq. (3.24) into Eq. (3.25), we obtain

$$\boldsymbol{Q}_i(\omega) = \left[\frac{\boldsymbol{Q}_o(\omega)}{\boldsymbol{Q}_o^{\text{cal}}(\omega)}\right] \cdot \boldsymbol{Q}_i^{\text{cal}}(\omega) \tag{3.26}$$

which shows that any signal can be expressed in terms of a calibration signal $\boldsymbol{Q}^{\text{cal}}$. For example, the proportionality between $\boldsymbol{Q}_i(\omega)$ and $\boldsymbol{Q}_i^{\text{cal}}(\omega)$ is given by $\boldsymbol{Q}_o(\omega)/\boldsymbol{Q}_o^{\text{cal}}(\omega)$. This factor of proportionality can be interpreted as being a *measure of similarity* between the measured and calibrated output signals. This interpretation is similar to the recognition in terms of "prototype objects" which is described in later chapters.

In those cases in which the signals are not of short-duration and in which the information of greatest interest lies in a particular portion of the detected signal, it may be advantageous to use time-domain signal deconvolution procedures to find the $h(t)$ such that when it is convolved with a known input signal, $q_i(t)$, one obtains the measured output $q_o(t)$. The deconvolution process involves finding that $[h(t)]^{-1}$ which, when convolved with the sensor output signal $q_o(t)$, yields the corresponding input signal $q_i(t)$

$$q_i(t) = [h(t)]^{-1} * q_o(t) . \tag{3.27}$$

The perfect inverse of the impulse response $[h(t)]^{-1}$ has the property that

$$h(t) * [h(t)]^{-1} = \delta(t) . \tag{3.28}$$

The inverse impulse response is that filter whose output is a delta function when the input is $h(t)$. Once the impulse response of a transducer is known, its inverse function can be found, at least in principle, although for most transducers this is not straightforward. In some practical applications, it may be advantageous to find that inverse function $[h_G]^{-1}$ corresponding to the filter whose output is a Gaussian pulse of pre-selected width σ and time delay, t_0, that is,

$$h(t) * [h_G(t)]^{-1} = \exp[-(t - t_0)^2/2\sigma^2] . \tag{3.29}$$

Deconvolution processing methods with applications in seismology, electrical devices, communications and measurement science have been studied for many years and a variety of algorithms have been developed.[19, 20, 21, 23] In most cases involving real data, such processing is inherently unstable, that is, the output increases without limit with increasing time. The choice of algorithm depends in large measure on the characteristics of the signal to be deconvolved and its length. *Optimum* or *least-squares-filters* (c.f. Robinson and Treitel [21]), which are discussed in more detail in Chap. 10, are based on discrete time-domain methods that have have been successfully used in a number of applications.[20, 22] In these, the mean square error, corresponding to the energy or power between the actual and the desired filter outputs is minimized. The algorithm, which yields an optimum result for the deconvolved, or inverse signal, requires the representation of the continuous transducer input signal by a finite-length discrete approximation. The input signal and the deconvolution output must be of sufficient length and the arrival of the energy maximum in the output must be so adjusted that the processing yields an optimum result. The algorithm requires the evaluation of the autocorrelation function of the input signal, $\mathcal{Q}$, and the cross-correlation of the input with the output signals $\mathcal{G}$. Then the solution of the equation $\mathcal{Q} \cdot X = \mathcal{G}$ for the elements of X corresponds to the least-squares inverse solution. Details of the processing algorithm have been given by Robinson.[21]

An alternative formulation can be carried out in the frequency-domain, provided that the transducer signal amplitude decays rapidly enough. The processing is then carried out using a discrete Fourier transform deconvolution procedure. The result that is obtained is not exactly identical to that obtained via the time-domain least-squares solution outlined above because of the circular nature of the digital Fourier transform that is used to perform the computation. The results are also affected by the windowing that is used to truncate the signal. However, with care, the results obtained from either of the two processing procedures are essentially equivalent.[22]

Transducer Performance Enhancements. The transducers used in typical man-made measurement systems are often far more limited in bandwidth than many of those found in nature. And in applications that require a maximum of transduction efficiency, the transducer's operating bandwidth is sometimes intentionally reduced. Here we describe a procedure by which a transducer's performance can be enhanced by appropriate signal processing of its output. By performance enhancing, we are referring to an actual device or a signal processing algorithm by which particular features in a signal can be more easily and unambiguously identified. For example, specific signal arrivals or amplitudes are features that might be sought.

As a specific example consider a measurement system in which a source transducer generates an excitation signal in a specimen and this signal is subsequently detected by a sensor. The latter could be identical to the source transducer if reversible devices are used. The input of the source transducer is connected to a source generator and the output of the receiving transducer is connected to the processing and display instruments. In this example, the spatial variable of the measurement is fixed, so that the displayed signal corresponds to the multiple convolution of the temporal responses of the individual components. In the time-domain, this can be written as

$$V(t) = V_0(t) * h_s(t) * G(t) * h_r(t) * D(t) \quad (3.30)$$

where:

$V(t)$	-	Displayed signal amplitude
$V_0(t)$	-	Source signal
$h_s(t)$	-	Source transducer transfer function
$G(t)$	-	Specimen response
$h_r(t)$	-	Detecting transducer transfer function
$D(t)$	-	Display instrument response .

By combining the responses of all of the components comprising the measurement system into one response, defined as $M(t)$, that is, the transfer function of the measurement system, Eq. (3.30) becomes

$$V(t) = G(t) * M(t) \quad \text{where: } M(t) \equiv V_0(t) * h_s(t) * h_r(t) * D(t) . \quad (3.31)$$

In order to realize an enhancement of measurement system performance, we need to find an element whose transfer function characteristics are described by $p(t)$ and which, when it is incorporated into the measurement system, will result in the system's enhanced performance. If the enhanced system performance that is desired is specified by $M^+(t)$, then we express the enhancement as

$$M^+(t) = M(t) * p(t) . \quad (3.32)$$

The characteristics of the enhancing element can be found by deconvolving this equation. That is,

$$p(t) = M^+(t) * [M(t)]^{-1} \quad (3.33)$$

which can be written in the complex frequency-domain as

$$\boldsymbol{P}(\omega) = \frac{\boldsymbol{M}^+(\omega)}{\boldsymbol{M}(\omega)} . \quad (3.34)$$

Before this can be carried out, however, one must first determine the impulse response of the measurement system. To find this, one makes use one of the approaches outlined in Sect. 3.3.2 using a calibration or a test specimen. Such a specimen possesses a known transfer function, $G^\circ(t)$. For the special case in which the specimen exhibits a perfect impulse response, that is, when, $G^\circ(t) = \delta(t - t_0)$ the response of the measurement system is obtained directly,

$$\begin{aligned} V(t) = G^\circ(t) * M(t) &= \delta(t - t_0) * M(t) \\ &= M(t - t_0) . \end{aligned} \quad (3.35)$$

In the more more general case, then

$$M(t) = V(t) * [G^\circ(t)]^{-1} . \quad (3.36)$$

Once $M(t)$ is known, then the characteristics of the enhancing element can be found from Eqs. (3.33) or (3.34). The enhanced signal output from the measurement system, V^+, is then given by an equation analogous to Eq. (3.32),

$$V^+ = V(t) * p(t) = G(t) * M(t) * p(t) = G(t) * M^+(t) . \tag{3.37}$$

The application of the foregoing procedure is dependent, of course, on the frequency bandwidth of the transducer's response and also the level and bandwidth of the noise that is present. If noise is present and if the bandwidth of the measurement system is finite, then the deconvolution operation according to Eqs. (3.33) or (3.18) is unstable at high frequencies, where $\omega \to \infty$, as well as at low frequencies (i.e. $\omega \to 0$). The addition of a small constant n_0 in the denominator of Eq. (3.34) can be used to stabilize the deconvolution calculation. In the frequency-domain, this is

$$\boldsymbol{P}(\omega) = \frac{\boldsymbol{M}^+(\omega) \cdot \boldsymbol{M}^*(\omega)}{\boldsymbol{M}(\omega) \cdot \boldsymbol{M}^*(\omega) + n_0(\omega)} . \tag{3.38}$$

Example: *High-resolution Ultrasonic Measurements*: Transducers that are used to generate and detect ultrasonic waves in solids, do so with finite bandwidth. As a result, the precise determination of signal arrivals is difficult when signals arrive a short interval after earlier arrivals. This situation occurs when high-resolution measurements are sought from low-frequency or narrowband transducers in samples whose dimensional features are small.

As an example of the foregoing, we show how measurements made with narrowband ultrasonic transducers can be enhanced to permit resolving features in a waveform which are at the limit of the bandwidth of the transducers. This example is taken from the work of Michaels. [22]

The specimen consisted of a 1/8 in. (3.2 mm) thick aluminum layer that was in intimate contact with a 1 in. (25.4 mm) thick Plexiglas (lucite) plate-like specimen. A piezoelectric, ultrasonic transducer, 12.7 mm in diameter, was operated reversibly as a source and a receiver, that is $h_s(t) = h_r(t)$. It was coupled to the Plexiglas side of the specimen as shown in Fig. 3.7a. One set of measurements was carried out with a broadband transducer whose frequency of optimum response, f_0, that was near 3 MHz and it had a bandwidth BW of about 3 MHz. Thus, its *fractional bandwidth*, given by $(BW/f_0) \times 100\%$, was about 100%. By adjusting the electrical parameters of the electrical circuit which controls the excitation voltage, $V_0(t)$, the transducer's optimum response could be reduced to 2.4 MHz where it exhibited a reduced bandwidth of 0.7 MHz, corresponding to a fractional bandwidth of only about 30%. In these measurements, the performance of the transducer was the performance controlling component in the ultrasonic measurement system.

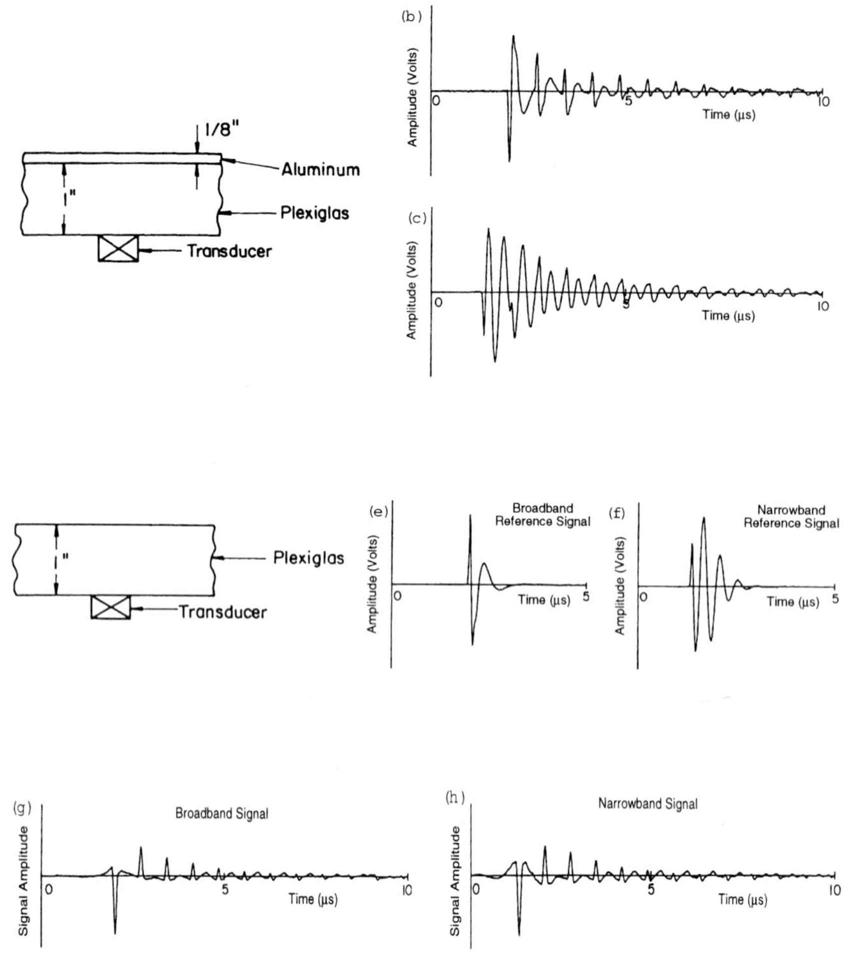

Fig. 3.7. Pulsed ultrasonic measurements of a thin layer: **a** Two-layer aluminum/Plexiglas test specimen; **b** Signal recorded with a broadband transducer (100% fractional bandwidth) with the echoes from the thin layer identified; **c** Signal recorded with a narrowband transducer (about 30% fractional bandwidth) in which the individual echoes cannot be delineated; Measurement of system impulse response: **d** Ideal reflector test specimen; **e** Reference signal recorded with the broadband transducer; **f** Reference signal recorded with the narrowband transducer; Result of the processing: **g** Enhanced broadband signal; **h** Enhanced narrowband signal. Because of the change of impedance across the reflecting surface in measurement situations **a** and **d**, the polarity of the reference signals is opposite that of the original signals. (In the waveforms, the time origin $t = 0\,\mu s$ is arbitrary.)
(from Michaels, Ref. [22])

The signal that is displayed when the measurement system is operating as a broadband system is shown in Fig. 3.7b. The first, large signal corresponds to the echo signal from the lucite/aluminum interface while the sequence of subsequent signals are the reverberations in the thin aluminum layer. The arrival times of these reverberations correspond to the thickness-to-wavespeed ratio of the thin layer. In the detected narrowband signal that is shown in Fig. 3.7c, only the first interface reflection signal can be identified because the ringing of the narrowband transducer dominates remainder of the detected signal to such an extent that no reverberation echoes in the thin layer can be identified.

The measurement system response $M(t)$ is determined using the "calibration" test with the ideal reflector test specimen depicted schematically in Fig. 3.7d. The detected voltage waveforms corresponding to the reference signals obtained with the broadband and narrowband transducers are shown in Figs. 3.7e and f, respectively. The enhanced signal should possess sharp arrivals corresponding to the echoes in the thin layer. Using the processing steps outlined in the previous sub-section, we find the enhanced system performance M^+ and from it, the enhanced signal output V^+ for the broadband as well as narrowband waveforms. These are shown respectively in Figs. 3.7g and h. The desired signal enhancement has been realized for both signals.

3.3.3 Sensor Loading Effects

In its operation, it is unavoidable that a transducer, which converts one physical variable to another, extracts energy from the system or medium from which it obtains its input signal and adds energy to the system on which it acts. The effect of a sensor is to change the value of the *measurand* from what it would be if the sensor were not there. Similarly, the signal generated by an actuator is influenced by the medium on which it acts. Further, the operation of a transducer affects the operation of the components of the measurement system with which it is interfaced. This is known as the *inter-element loading effect* of a transducer. Such loading effects may completely dominate the output signal of the transducer. We consider here these *loading effects* and how one might compensate a measurement or an action for their presence.

The important question is how much energy or power does a transducer take from the input device or system? The problem is illustrated in Fig. 3.8. In most applications it is hoped that this will be small or even negligible, even if it is always non-zero. As might be expected, it is usually also a function of the active area of the transducer. It is important to remember, however, that those transducers which exhibit a minimal loading effect, often also generate a small output signal or worse, they generate an output signal whose *SNR* is small. We recognize that a signal low in amplitude but with sufficient *SNR* can be amplified to increase its magnitude which is not possible for signals with a low *SNR*. The search for new phenomena which can form the basis

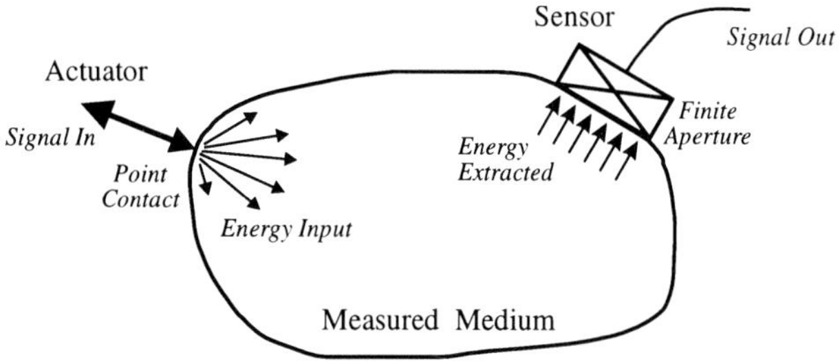

Fig. 3.8. Transducer loading effects on a medium

of a new sensor for a particular application and whose loading effect will be small is always of engineering interest. A number of new sensors based on electromagnetic and optical phenomena have been developed over the last several decades. The compact disc player with its laser diode-based sensor to read the pits and lands of a spinning disc is a well-known example.

To estimate the loading effect of the transducer on the environment or phenomenon being measured, one can model the sensor-material interactions. One common approach relies on a two-port network equivalent circuit of the sensor-based measurement system that is shown in Fig. 3.9. This approach is applicable for static or low-frequency signals for which a *lumped parameter* representation of the transduction system is valid. When focusing on the sensor input, and hence the interactions between the sensor and the measured phenomenon in this approach, one describes the loading effects of the sensor in terms of the output impedance of the sensor environment, $\mathbf{Z}_{\mathrm{out}}$, and the input impedance of the sensor, $\mathbf{Z}_{\mathrm{in}}$. For mechanical elements, these impedances correspond to the ratio of *force/velocity* over the active region of the transducer, $\mathcal{B}$. For electrical elements, these impedances are expressed in terms of the voltage to current ratio.

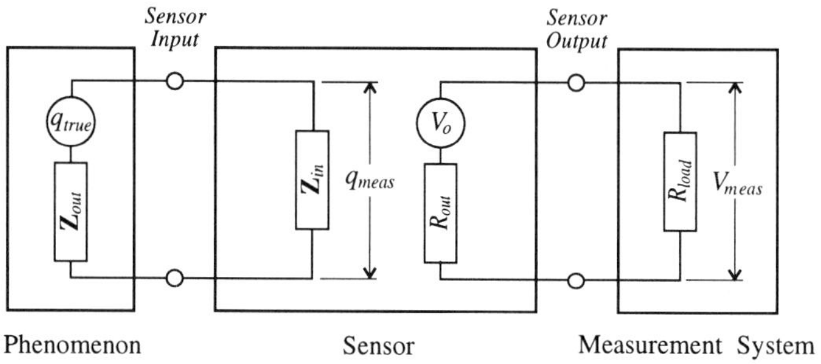

Fig. 3.9. Equivalent circuit for a sensor represented as a two-port network

Let us consider the operation of a sensor. An analysis of the conservation of energy per unit time across the sensor/environment interface permits calculation of the transient response of the sensor to a particular input. The steady-state solution is a statement of the equilibrium between a sensor and its environment. The actual, measured quantity which the sensor transducts into an output voltage V_o is denoted by q_{meas}. It can be shown that the measured sensor output signal is expressed as a function of the *true* or *actual* signal by

$$q_{\text{meas}} \equiv q_i = \left(\frac{\mathbf{Z}_{\text{in}}}{\mathbf{Z}_{\text{in}} + \mathbf{Z}_{\text{out}}}\right) q_{\text{true}} = \left(\frac{1}{1 + \mathbf{Z}_{\text{out}}/\mathbf{Z}_{\text{in}}}\right) q_{\text{true}} . \tag{3.39}$$

The analysis of the equivalent circuit leads to the prediction that $q_{\text{meas}} \approx q_{\text{true}}$ when sensor loading effects are minimized. This is realized when the input impedance of the measuring transducer is much higher than the output impedance of the measured system. The application of this approach to determine the loading effect of sensors which are used for making static mechanical measurements requires replacing the impedances by the static stiffnesses. Alternate formulations are expressed in terms of system *admittances* and *static compliances*.[8]

Based on the equivalent circuit, the amount of energy extracted from the measured medium is given by

$$P = \frac{q_{\text{meas}}^2}{Z_{\text{out}}} = \left|\frac{\mathbf{Z}_{\text{out}}}{(\mathbf{Z}_{\text{out}} + \mathbf{Z}_{\text{in}})^2}\right| q_{\text{true}}^2 . \tag{3.40}$$

Based on this equation, in order to minimize the extraction of energy from the measured system, the input impedance of the measuring transducer should be much higher than the output impedance of the measured system. For example, in order to minimize the mechanical power extracted from the measured system, one should use a sensor or sensing principle which exerts a small force on the system or one which possesses an extremely high or very low mechanical input impedance. This usually means that transducers operating in series with the measured system should possess a high stiffness, while those operating in parallel should be flexible or be of low stiffness. The trend for many displacement measurement sensors is to optical-based sensing systems which exert forces that are essentially zero. For this reason, sensors based on optical techniques closely approach the ideal.

Example: *Loading Effects of a Thermal Sensor*: The loading effects associated with a thermal sensor is discussed in a number of textbooks.[8, 14] The conservation of thermal power in the sensor leads to the equation for the actual temperature of the sensor T_{meas} that has been placed in an environment whose temperature is to be measured and is denoted by T_{true}

$$\left(\frac{mC_p}{UA}\right)\frac{dT_{\text{meas}}}{dt} + T_{\text{meas}} = T_{\text{true}} \tag{3.41}$$

where m is the mass of the sensor element and C_p [J kg^{-1} °C^{-1}] is the specific heat of the sensor material. The quantities U and A refer respectively to the heat transfer coefficient between the sensor material and the environment and the area of the sensor over which the heat transfer takes place.[8] For time-varying, harmonic temperature changes, we obtain

$$\left(\frac{mC_p}{UA}\right) \jmath\omega T_{\text{meas}} + T_{\text{meas}} = T_{\text{true}} . \tag{3.42}$$

Eqs. (3.41) and (3.42) are analogous to the equation which describes the voltage at the output of a *series-RC* circuit, resembling the sensor input portion of the equivalent circuit shown in Fig. 3.9 in which $\mathbf{Z}_{\text{out}} \leftrightarrow 1/UA$ and $\mathbf{Z}_{\text{in}} \leftrightarrow mC_p$. This leads to the following associations

$$\mathbf{Z}_{\text{out}}(\omega) = \left.\frac{1}{UA}\right|_{\text{sensor}} \equiv \frac{1}{U_s A_s} \quad \text{and} \quad \mathbf{Z}_{\text{in}}(\omega) = \frac{1}{\jmath\omega m_s C_p^s} . \tag{3.43}$$

Thus the temperature measured by the thermal sensor is related to the actual value by

$$\boldsymbol{T}_{\text{meas}}(\omega) = \left(\frac{1}{1 + \jmath\omega m_s C_p^s / U_s A_s}\right) \boldsymbol{T}_{\text{true}}(\omega) \tag{3.44}$$

which is analogous to Eq. (3.39). This result is not unexpected. When the mass of the sensor or its specific heat is small or if the heat transfer coefficient and sensor area are large, then the sensor will accurately respond to the true value of the temperature of the medium provided that the frequencies associated with the temperature change are not too high.

Analyses as these, point to engineering solutions for designing a sensor to respond more accurately to the actual value of the measurand. In a thermal sensor, one can reduce the mass of the sensor or change its material to one possessing a lower specific heat value. Alternatively one can increase the heat transfer coefficient of the sensor by, for example, adding fins to the sensor. We note that it is difficult to reduce the mass of the sensor by making it smaller and at the same time increasing its area by making it larger. Furthermore, a reduction in a sensor's mass is often accompanied by a reduction of the sensor's sensitivity. This example illustrates the challenge one faces in selecting a sensor possessing good sensitivity and faithful response because of minimized loading.

The complete measurement system includes the sensor followed by one or more components whose function is to process the sensor output signal, display and record it or possibly convert it into digital form for subsequent detailed analysis. The equivalent circuit modeling approach described above is equally applicable in predicting the performance of this portion of the measurement system. That is, to estimate the inter-element loading effects which each of the components comprising the measurement system has on the original sensor signal. In the example depicted in Fig. 3.9, the sensor possesses an output resistance R_{out} and the subsequent measurement system has an input resistance R_{in}. The measured voltage signal in this measurement system V_{meas} corresponding to the voltage signal V_o that is generated in the sensor will be given by an equation analogous to Eq. (3.39). That is

$$V_{\text{meas}} = \left(\frac{1}{1 + R_{\text{out}}/R_{\text{in}}}\right) V_o \,. \tag{3.45}$$

3.3.4 Transducer Field Characteristics

Up to now we have focused exclusively on the transduction characteristics of a transducer. Transducers which generate or detect time-varying signals in a medium, do so under certain operating conditions with unequal effectiveness in every direction. These characteristics of a transducer are known as its *field* or *radiation characteristics*. A well-known example is the radiation of sound from a loudspeaker which is shown in Fig. 3.10. Knowledge of these characteristics is essential if a correct analysis of the detected signals is to be made. For this reason, the determination of the field characteristics of a transducer forms an essential component of any complete transducer characterization procedure.

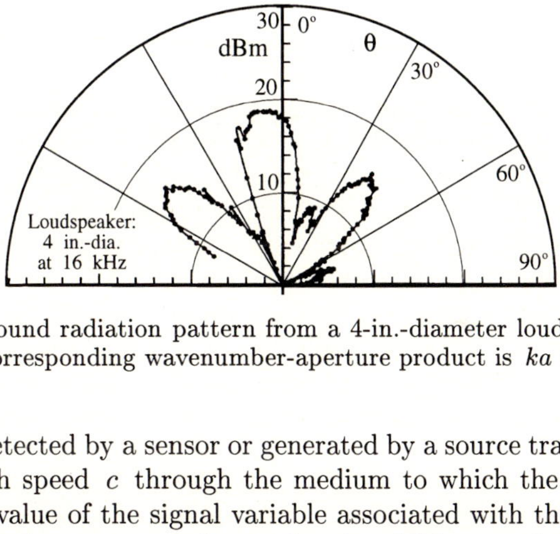

Fig. 3.10. The sound radiation pattern from a 4-in.-diameter loudspeaker driven at 16 kHz. The corresponding wavenumber-aperture product is $ka \approx 30$

The signal detected by a sensor or generated by a source travels as a wave propagating with speed c through the medium to which the transducer is interfaced. The value of the signal variable associated with the wave amplitude is not everywhere constant, but rather, it exhibits a periodicity in space which is equal to the wavelength of the wave. When the frequency of the wave or one of its components is f, then the corresponding wavelength, λ, is given by the well-known relation $\lambda = c/f$ and the wavenumber k is defined as $2\pi/\lambda$. That is, waves of low frequency and long wavelengths, possess a correspondingly low spatial periodicity. For such waves, the signal variable does not vary significantly over large distances and thus, the assumption implied by Eq. (3.3) regarding the spatial independence of the signal is justified. If a is the aperture of the transducer, then in that case, $ka \leq 1$. In contrast, with high-frequency waves, such that $ka \gg 1$, the signal varies significantly over short distances and the assumption expressed by Eq. (3.3) is not met. A finite-aperture transducer averages the total response across its active element. The output signal of the transducer will depend on the wavelength such that its amplitude is no longer a single-valued function of the input signal. Rather, it depends on ka and the angle between the direction of wave propagation and the normal to the transducer aperture. The most common consequence is a smoothing or a loss of high-frequency information in a detected signal. The faithful detection or generation of high-frequency signals thus requires correspondingly smaller aperture sensors and actuators, respectively.

Ideally, point-like transduction devices, that is, devices whose aperture is less than the wavelength of the signal being generated or detected, could be used to solve this problem for they possess the important property that they generate and detect signals over a broad angular spectrum in a material. But this lack of directionality has both beneficial as well as detrimental features. They permit simultaneous measurements along different directions in a material. This is particularly useful for elucidating the properties of anisotropic materials. But this is of disadvantage in those measurement situations benefiting or requiring directional measurements. Coupling point-like transducers to solids is made more reproducible when the region of required surface preparation is small. But such transducers have the disadvantage that their *SNR* is generally lower than that obtained for larger aperture devices. It is for this reason that most point-like transducers are fabricated with a small, yet non-negligible aperture.

In view of the foregoing, it is appropriate to consider in more detail what effect a transducer's aperture has on the detected or generated signals and how one might process the signals to compensate for these effects. It is shown in Sect. 3.3.2, that one can represent the operation of a transducer as a one-dimensional model, according to Eq. (3.3), provided that the value of the transduced variable is replaced by its value at the centroid of the transducer. We consider now the case of a finite-aperture sensor which forms part of a dynamic measurement system in which the excitation is a point source whose strength is specified by $f(t)$. The signal detected and integrated by the finite-aperture sensor whose centroid is located at position $\boldsymbol{r}_0$, can be written as

$$\bar{q}(\boldsymbol{r}_0, t) \;\equiv\; q(t) \;=\; \int_{-\infty}^{\infty} f(t')dt' \int\!\!\int_{\text{Area}} \varrho(\boldsymbol{r}')\, G(\boldsymbol{r}|\boldsymbol{r}', t-t')\, dS. \tag{3.46}$$

In this equation, the term $\varrho(\boldsymbol{r}')$ is the spatial sensitivity function across the active area of the transducer. The dynamic Green's function of the structure for the source located at $\boldsymbol{r}$ and one of the receiver points denoted by $\boldsymbol{r}'$ in area element dS is denoted by $G(\boldsymbol{r}|\boldsymbol{r}', t-t')$. In this equation, we assume that the transduction characteristics of the sensor are known. Chang [10] has demonstrated a procedure for separating the spatial and temporal effects from the measured signals $\bar{q}(\boldsymbol{r}_0, t)$ for such a sensor. We summarize this procedure in the following paragraphs.

One approach for performing the areal integration appearing in Eq. (3.46) is to spatially digitize the problem. That is, one divides the transducer into a number of segments and sums the response over all the segments. For this, we discretize the terms appearing in Eq. (3.46). That is, for the signal $\bar{q}(\boldsymbol{r}_0, t) \equiv q(t)$ at the specific instant $t = t_n$ we obtain the discrete value

$$q(t_n) \;\rightarrow\; q(n\Delta t) \;\rightarrow\; \mathsf{q} \;. \tag{3.47}$$

The set of all discretized samples of the original signal is written as $[\,\mathsf{q}\,]$. The other terms appearing in Eq. (3.46) can be written similarly, with $[\,\mathsf{f}\,]$ corresponding to the discretized force and $[\,\mathsf{G}\,]\cdot[\,\mathsf{W}\,]$ the dynamic response of the specimen summed over the transducer area. The matrix $[\,\mathsf{W}\,]$ denotes a matrix representing the *sensor weighting function*. Thus, the complete equation can be re-written in compact form as a *bi-linear equation*

$$[\,\mathsf{q}\,] \;=\; [\,\mathsf{f}\,] * [\,\mathsf{G}\,]\cdot[\,\mathsf{W}\,]\,. \tag{3.48}$$

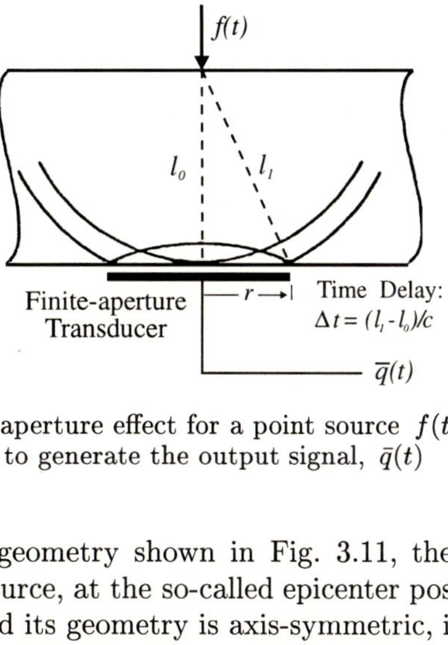

Fig. 3.11. Origin of the aperture effect for a point source $f(t)$ acting on the axis of the sensor of radius r to generate the output signal, $\bar{q}(t)$

For the measurement geometry shown in Fig. 3.11, the sensor is located directly opposite the source, at the so-called epicenter position. If the sensor is operating linearly and its geometry is axis-symmetric, its spatial response can be expressed in terms of a linear superposition of the responses of discrete rings where A_i is the area of the i-th ring of the discretized active area of the sensor and ϱ_i is the average sensitivity of the sensor over this area. The value of the weighting function of the i-th ring can then be written as

$$W_i \;\equiv\; A_i \cdot \bar{\varrho}_i \;. \tag{3.49}$$

By this procedure, the sensor is, in effect, approximated by a line receiver having point detectors with differing sensitivities along its length x which spatially sample the signal. The spatial sampling period, which is not uniform, is so selected that the corresponding arrivals of the signal at each point receiver are delayed by exactly one sampling time period. With this representation, the spatial weighting function $W(x)$ becomes equivalent to a time function $W(t)$. Thus, the equivalent spectrum of a transducer's aperture equals the Fourier transform of $W(t)$.

The spatial and temporal effects of a transducer can be separated, provided that the aperture of the transducer is not too great such that the Green's functions corresponding to each transducer point are nearly identical and equal to an equivalent Green's function G_e. The original convolution equation, expressed by Eq. (3.46), can then be rewritten as

$$q(t) = [f(t) \cdot W(t)] * G_e(\boldsymbol{r}|\boldsymbol{r}'_0, t) . \tag{3.50}$$

Thus, only the product of the source function and the weighting function, i. e. $f(t) \cdot W(t)$, can be recovered by deconvolving the measured data $q(t)$ with knowledge of the equivalent dynamic Green's function of the measurement configuration.

When the Green's functions are dependent on the position of the receiver element, no equivalent Green's function can be used. Instead, the Green's function appropriate to the position of each receiver element must be substituted into the bi-linear equation and the source function $f(t)$ and the weighting functions W_i of each sensor element found from the measured data $q(t)$ using a double iterative deconvolution procedure. [9] Once the weighting function is known, the sensitivity of the sensor across its aperture, expressed as its diameter, $\varrho(x)$, can be found, provided that the discretized area associated with each time-step is known.

Example: *Separation of Spatial and Temporal Effects in a Transducer*: To illustrate the foregoing, we show how measurements made with a finite-aperture ultrasonic capacitive sensor can be processed to permit recovery of the true characteristics of the signal source as well as the weighting function of the transducer. This example is taken from the work of Chang [10] who used capacitive sensors to measure the displacements normal to the specimen surface. Under proper operating conditions, the displacements correspond to changes in the gap dimension between the specimen surface and the sensor electrode and the associated changes of electric charge on the electrode are subsequently converted into a voltage signal. Over a reasonable range of operating frequencies, the output voltage is directly proportional to the specimen displacements. [11, 12]

Measurements were made on glass plate specimens of two thicknesses 0.25 in. (6.35 mm) and 0.75 in. (19.05 mm). Step-like point source excitations were obtained from the fracture of a glass capillary on the surface of the plates. Capacitive sensors with two apertures were used: 0.125 in. (3.175 mm) and 0.25 in. (6.35 mm) in diameter. Each sensor was mounted to detect the signals at the epicentral receiver position on the side of the plate opposite to the location of the source. The displacement signals were differentiated to generate the normal velocity signals. The signals detected on the thinner plate with each of the two sensor apertures are shown in Fig. 3.12. The general features of wave arrivals in both waveforms are similar. But there are significant differences. The larger aperture sensor generates a higher amplitude output signal and the duration of the wave arrivals is longer than those detected using the smaller aperture sensor. This demonstrates the loss of high-frequency information when using a larger aperture sensor. Also, the second arrival, denoted by 'S' and corresponding to the shear wave signal is distinctly different in the two measurement situations.

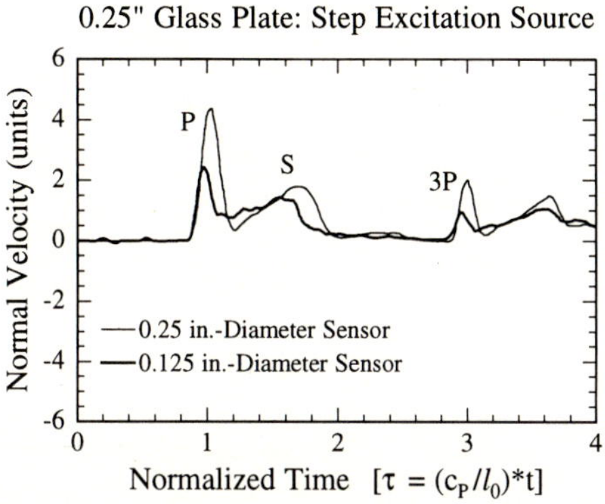

Fig. 3.12. Normal surface velocity signals detected using two different aperture sensors on a 6.35 mm glass plate from a step vertical force source at epicenter. 'P' and 'S' denote the arrivals of the longitudinal and shear wave signals, respectively, and '3P' is the multiply-reflected signal in the plate

If one assumes that the signal has been detected by a point sensor for which the weighting function $W(x) = \delta(x - x_0)$ and performs the deconvolution using the assumption that there is only one Green's function corresponding to the source and receiver positions, one finds a time function for the source that exhibits a number of artifacts. The results are shown in Fig. 3.13a. The artifacts include a slow risetime and a periodicity in the time function. The risetime is the consequence of averaging the signal across the aperture of the transducer and the periodicity results from using a Green's function that does not reflect reality. By using a double-iterative deconvolution algorithm to determine both the correct source-time function and the associated sensor weighting function from Eq. (3.48), one finds the correct source-time function that is also shown in Fig. 3.13a.

The corresponding weighting function is shown in Fig. 3.13b. It is seen that the weighting function decreases to zero when the radius is near $0.6\, l_0$, where l_0 is the thickness of the specimen. The apparent radius exceeds the

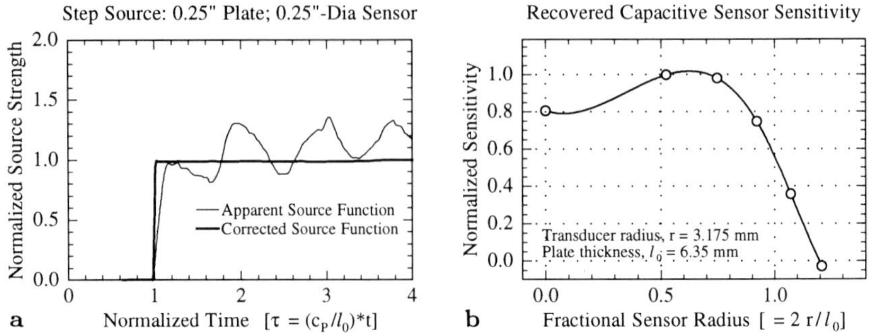

Fig. 3.13. **a** Recovered source-time functions. **b** Recovered sensor weighting function

actual radius of the transducer. This is likely related to the fringing field of the capacitive sensor.

3.4 Fabricated Transducers

Because biological transducers have evolved over the millennia to adapt to their operating environments, the performance capabilities of man-made sensors in many respects pale in comparison to those of the biological sensors. While the performance of most fabricated transducers is more limited than their biological counterparts, transducers have been developed for transforming the signals of physical variables for which biological transducers are not well-adapted. But transducers have also been designed and fabricated to transduct signals of the same physical variables as those of importance to biological systems. These, however, may be designed to operate more reproducibly, at frequencies unattainable by their biological counterpart, with higher precision and in extreme environmental conditions which are far more variable than those that biological transducers can function in.

A wide variety of mechanical, thermal, magnetic, optical, electrical, chemical and other phenomena which are combinations of these have been used as the basis of transduction devices that can be used to transform the signals of a diversity of physical variables. Transducers may operate reversibly, that is, they can be used as sensors as well as actuators. Examples of the latter include devices based on piezoelectric, electromagnetic, electrostatic, and magnetostrictive effects.

Examples of sensors which are used to detect specific signals include *pressure transmitters, detectors* of various types, *load cells, strain gages, vibration pick-ups, accelerometers, IR sensors* and a diversity of optical, thermal and chemical sensors. The list of actuators is not as extensive. Examples of mechanical actuators include *vibrators, loudspeakers, shakers, piezoelectric elements*, among others. Thermal actuators range from simple heaters to pulsed beams of photo-thermal radiation such as those derived from lasers, or other pulsed electromagnetic sources. Pulsed thermal sources have come into routine use as actuators of mechanical stress waves in solids and fluids via the transient thermoelastic expansion of the target region of a specimen. Such techniques permit the non-contact generation of mechanical stimuli whose area of application is easily varied and whose site of application can be scanned.

3.4.1 Microsensors and Integrated Sensors

The last decade has seen the explosive growth in the field of *micro-sensors* and *micro-actuators*. As their names imply, these micro-mechanical devices, sometimes called *MEMS* for *micro-electro-mechanical structures*, are small, ranging in size from about 10^{-1} μm to 1 mm which makes them suitable for a multitude of applications. Because they are usually fabricated using techniques similar to those used to fabricate micro-electronic devices, their unit cost can be low. The development of such devices has required the solution to a number of multi-disciplinary problems spanning the fields of materials science, mechanics, electronics, chemistry and even biology and for this reason, their development is today an important area of interdisciplinary research.

Today, the material of choice for fabricating many transducer structures is silicon. This material has been well characterized, it is strong and it lacks mechanical hysteresis. In fact, silicon possesses remarkable electronic and mechanical properties with good thermal characteristics of high thermal conductivity and low thermal expansion. The material can be micro-machined into a vast variety of intricate shapes whose dimensions are on the micrometer scale. The fabrication can be carried out in mass production using wafer fabrication lines. The structures that have been fabricated are electro-mechanical elements that may act either as the basis of either sensors or actuators. A compilation of articles reviewing the current state-of-the-art of semi-conductor sensors forms the basis of a recently published textbook. [24]

There is a special group of sensors, called *biosensors* in which the stimulus is recognized by a *bio-receptor* analogous to those used in living systems. This element forms the sensitive component of the sensor. [25] Because their operation is based on molecular recognition, the characteristics of bio-sensors include their specificity to recognize an individual component among a number of other substances in a sample. Examples of bio-receptors include various enzymes, micro-organisms, immunoagents, chemo-receptors and certain tissues. In contact with the sensed medium, the bio-receptor becomes modified which is converted into an electrical output signal by a conventional transduction element, such as an electrode, a transistor, thermistor, optical fiber or a vibrating piezoelectric crystal.

The fabrication techniques that are employed to make semi-conducting sensors include bulk micro-machining that relies on selectively etching away wafer material to obtain a particular mechanical structure. The etching processing operation results in device geometries possessing particular aspect ratios but this limitation can be overcome by modifying the bulk micromachining process or by using surface micro-machining techniques. In these, specific, thin-film materials are selectively added and removed from the substrate wafer. This procedure is capable of fabricating large arrays of transduction elements for which the substrate can be used to mount the interface circuitry. Examples of surface micro-mechanical devices include 2048×1152 addressable arrays of digital micro-mirrors devices whose motions can be controlled to form the basis of high-definition television displays. Based on similar fabrication technologies, there are micro-relays, micro-resonators, micro-accelerometers, room-temperature infrared temperature arrays, electrostatically-operated micro-tweezers and even miniature motors and turbines. The surface machining technique has even been used to fabricate structures which contain hinges that permit a portion of the transduction element to move out of the plane of the wafer thus serving as a three-dimensional structure. Fabricated have been a 16-joint anthropomorphic skeleton of a hand, less than 0.55 mm in dimension, that also comprises a micro-gripper actuator. Transduction elements based on electrostatic, piezoresistive, piezoelectric and pyrolectric phenomena have been reported.

More elaborate transduction structures consisting of layers of bulk machined elements are obtained by diffusion bonding individual layers together. This procedure permits the fabrication of structures possessing thickness an order of magnitude greater than is possible with surface micro-machining fabrication. Further, with such procedures it is possible to realize enclosed cavities which are a useful feature for obtaining damping or vibrating elements and hence in realizing transducers with wider bandwidth.

In order to fabricate devices possessing large vertical aspect ratios, that is structures with lateral dimensions of the order of μm's and whose vertical dimensions may be up to 1 mm high, X-ray lithography techniques are combined with micro-electroplating and micro-molding procedures. Called LIGA (LIthographie, Galvanoformung, Abformung), a photoresist pattern is generated on a conductive substrate using X-ray lithography. Features in the resist pattern are preferentially electroplated to give a highly accurate negative replica of the original resist pattern. The metal electroplating may be conductive as well as magnetic and it may subsequently be used on a mold for a variety of polymeric resins or ceramic slurries. This technology has been used to fabricate micro-gears and linkages as well as complete micro-motors, turbines and magnetic actuators.

Today, about 25 million micro-sensors are fabricated annually as automotive pressure sensors for metering the fuel/air mixture. The application of micro-sensors as accelerometers for airbag deployment applications is also growing rapidly. A commercially-available accelerometer for this application is shown in Fig. 3.14. The operation of this sensor relies on a capacitive mea-

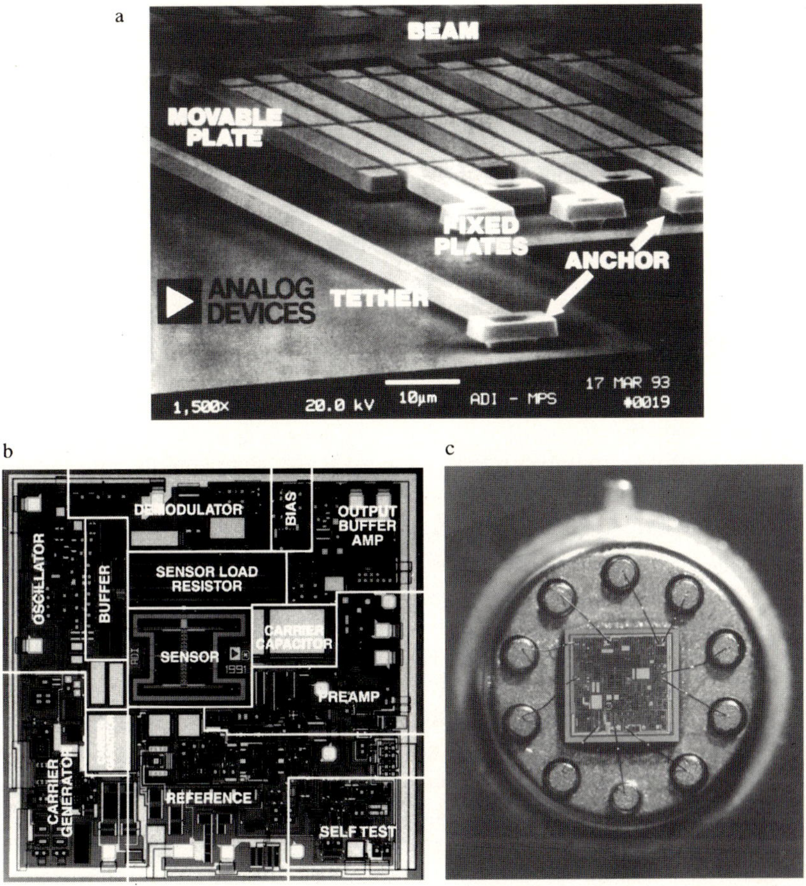

Fig. 3.14. A commercially-available micro-accelerometer: **a** Surface micromachined sensing elements. **b** Layout of the acceleration sensing system including sensor, signal processing and sensor calibration circuits; **c** Top view of the encapsulated sensor (with cover removed) (Photos courtesy of Analog Devices, Inc.)

surement of the spacing between movable and fixed plates shown in Fig. 3.14a. Because this sensor is sensitive to accelerations lateral to the movable and fixed plates it is important that the sensor be properly aligned with the direction of acceleration to be measured. Fig. 3.14b shows that the sensing element forms part of a complete sensing measurement system. Included on the monolithic integrated circuit are components for self-testing the accelerometer, electronic elements for setting the zero-g level and elements for adjusting the output sensitivity. The entire unit is hermetically sealed in a miniature can which is less than 10 mm in diameter. The entire assembly is fabricated into a robust package, shown in cut-away view in Fig. 3.14c, that is capable of operating reliably in harsh environments.

It is predicted that in the future the output of accelerometers will be combined with appropriate control circuits to create devices capable of generating control signals that can be used to adjust the operation of the car's engine and suspension in response to chassis acceleration data. Such sensors are called *smart sensors* because these devices are capable of sensing the system and they also possess the necessary circuitry and algorithm that will permit them to monitor and control the system to which the sensor is interfaced.

The second largest application of micro-sensors is to medical applications. In these, the largest number of sensors are used as sensors in blood pressure measurement applications. Other microsensors operate as pressure sensors in intrauterine pressure sensors; angioplasty pressure sensors, catheter tips, respirators, lung capacity meters and intra-cranial pressure sensors. Other micro-sensors find application in monitoring medical procedures and kidney dialysis equipment.

Micro-sensors capable of measuring pressure, temperature and acceleration are the integral component in a rapidly growing number of consumer products and industrial applications. In a number of cases for the latter, the sensor is integrated with the requisite processing electronics on the chip to transform the sensor output into a signal that is suitable for long-distance transmission. Work is currently underway for integrating accelerometers to monitor the vibrations of machine components and miniature force sensors to monitor the multi-component forces acting on a machining tool when in use. The goal is to characterize and ultimately to control the machining operation.

Much of the recent work has focused on the integration of the micro-sensor with the signal processing electronics on the same chip. These components provide not only sensor analog signal amplification, filtering and conditioning functions but they also permit implementing functions such as correcting the signal offset and adjusting the sensor sensitivity. Further, in a number of cases, there are on-chip components that provide a suitable interface that facilitate the communication of a sensor with external digital processing equipment. On the horizon are the development of standard digital sensor signal formats and a smart transducer network that will establish the communication between any sensor and the central network to transmit information about a sensor's measured variable, its range, performance, address and communication parameters. The result will be a sensor that could operate like a computer peripheral device.

The micro-sensor fabrication technology has already been used to form a thin diaphragm with small holes in it in order to fabricate a neural interface chip which can be used to connect the nerves of living systems with external electronic and computing circuits. The chip is implanted in living systems so that nerve cells can grow through the holes to establish the required electrical connections.

Microfabrication techniques can also be used to fabricate a variety of micro-actuators. Already demonstrated have been individual as well as ar-

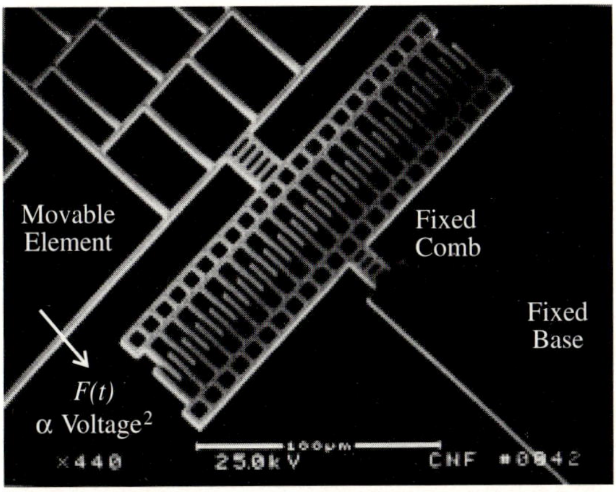

Fig. 3.15. Comb-drive, electrostatic actuator. The upper left element is the movable structure while the lower right component is the fixed attachment. The 22-finger comb structure generates forces which are proportional to the square of the applied voltage and independent of the displacement. For the actuator shown, a 2-Volt excitation results in a force of about $0.1\,\mu$N, independent of actuator movements of about $4\,\mu$m (Photo courtesy of S. G. Adams and N. C. MacDonald, Cornell University)

rays of actuators ranging from simple electromechanical tweezers, movable platforms, to turbines and micro-motors. For example, micro-tweezers based on electrostatic forces have been fabricated [26] but their application has been limited because of problems related to packaging and the inverse dependence of the weak electrostatic force on tweezer opening. One solution to the weak interaction problem is to design actuators which are based on arrays. For example, in Fig. 3.15 is shown the small array, a comb-drive, electrostatic actuator which has been fabricated using the single crystal reactive etching and metallization (SCREAM) process. [27] This device produces forces which are proportional to the square of the voltage applied across the actuator and independent of actuator deflection. Actuators such as these may be used as the excitors of resonant sensors.

Larger arrays comprised of as many as 5-10$\times 10^3$ elements are capable of generating forces on the order of 1 mN. [28] Such actuators can form the basis of a micro-mechanical, materials test system which is capable of subjecting micro-specimens to buckling and fracture. [28] Using similar actuator structures, displacement ranges of 1-30 μm are possible. The modification of the response of mechanical structures using micro-actuators has recently been explored.[29] In these applications, the goal is to dynamically modify the linear and non-linear stiffness coefficients of a mechanical system by appropriate micro-actuators which thus control the dynamics of the mechanical system.

An important focus of current actuator research is the design and fabrication of self-propelled micro-machines, sometimes called micro-robots. Included are devices which walk [30], swim [31] and fly. [32] In addition to the design and fabrication of such devices, the critical problems related to device friction, packaging and the supply of power for autonomous structures topics of current research.

3.4.2 Synthetic Bio-sensors and Neurobiology

Mead has pioneered an innovative approach for using microelectronic VLSI design and fabrication techniques to obtain electronic systems that mimic the signal transmission and processing features of living neurons, including circuits possessing the functionality of the retina and cochlea.[33] Most commonly today, the execution of neural-like processing algorithms is on serial or parallel digital computers. Mead's revolutionary approach instead is based on the capabilities of silicon-based VLSI circuits to produce low-power, analog and highly non-linear circuits which exhibit a broad dynamic range and are capable of executing computing functions involving many input variables. This approach has been termed *synthetic neurobiology*.

Mead and his associates have applied VLSI technology to produce a number of extraordinary processing sensor systems.[34, 35] The *silicon retina* is an electronic chip that generates, in real time, output signals which are similar to those obtained from the retina of a biological system. The silicon retina possesses a direct structural relationship with the retina of vertebrates and shares a number of essential, operational features with it, in particular, its ability to detect optical signals over several orders of magnitude of illumination and a broad range of contrast. Further, it provides gain control and image enhancement and by selectively rejecting irrelevant information, it permits a vast reduction of information that needs to be transmitted to a processor.

The input stage of the silicon retina consists of a 48×48 pixel array of photodetectors. The incident light is transformed in the photodetectors and their associated metal oxide semiconductor (MOS) transistors into an electrical stimulus with a response that, just as in the retina of living beings, is proportional to the logarithm of the incident intensity. This leads to a large dynamic range for the sensor and an output between two points in the visual field which is proportional to the contrast ratio between the two points, independent of illumination level. Directly under the photoreceptors in the retina of vertebrates is an array of horizontal cells that forms a resistive network which leads to a spatial as well as temporal average of the photoreceptor output. Also in this horizontal layer are bi-polar cells which generate an output that is proportional to the photoreceptor signal and the horizontal cell signal. The silicon retina incorporates a similar resistive network configured in a hexagonal array in which at each node of the network is an electronic pixel. Here the photodetector signal is modified to drive the resistive network and the strength of the six resistive connections can be adjusted using a bias circuit. The result is a level normalization which corresponds to a local average of light intensity. This is followed by a "triad synapse" such that at each point of the retina there will be generated a signal corresponding to the difference between the intensity at the pixel and the weighted average of the intensities of the surrounding pixels. The synthetic retina is completed with a scanning circuit that permits accessing any pixel in the array or access to all of the pixels in a serial mode to generate an image.

The performance of the silicon retina is found to be remarkably similar to the biological one. Both retinas exhibit sensitivity curves (the response as a function of incident light intensity) which resemble a $tanh$-characteristic, generating a constant output at very low and at very high intensities and being most sensitive to intensity changes over the middle of its operating range which is centered according to the light intensity incident on a pixel over that which is incident on the pixels in a local neighborhood. The time response to a step as well as a ramp in light intensity and the response to smooth edges in an image by synthetic and biological retinas is also very similar.

Mead and his associates have shown how one can combine the logarithmic photodetector of the silicon retina with powerful analog computational elements to determine the 2-dimensional velocity vector of a moving object.[36] The system serves as an integrated motion detector. It uses a 8×8 pixel array of photo-detectors coupled with analog VLSI circuit elements to implement a gradient descent solution for estimating an object's two-dimensional velocity components in a least squares sense from the measured local intensity information and its variation with position and time. The motion detector is robust and requires no pre-processing of the image, such as edge detection or recognition of the object.

Mead and Lyon have described the elements of a silicon cochlea that permits the detection and analysis of sound.[37, 38] It utilizes a synthetic basilar-membrane fabricated out of silicon that is a transmission line consisting of 480 amplifiers connected serially through which the velocity of signal propagation can be tuned electronically using positive feedback. The result is similar to the action of an outer hair cell in the biological cochlea in which the signal speed decreases exponentially with distance along the length of the transmission line. Both the biological and synthetic cochleas convert time-domain pressure variations into variations of the signal in space. The "membrane" has output taps to permit extracting the signal along the length of the cochlea. The characterization of the operation of the silicon cochlea is made in terms of its frequency response, transient response and gain control. The first two characteristics can be varied by design of the nonuniform wave medium and the associated VLSI circuits. A comparison of the performance between the synthetic and a biological cochlea is made using iso-output curves which correspond to the sound pressure input required to obtain a fixed displacement or velocity of the membrane or a particular firing rate of an auditory nerve fiber as a function of frequency. There is remarkable agreement between the actual, measured mechanical iso-output curves on a biological cochlea and those synthesized using typical design parameters of the synthetic one.

One application of the neural-like VLSI circuits is to augment an impaired sensory system of a handicapped person. In the *SeeHear* device [33], signals corresponding to the intensity and position of a light source are mapped into synthesized binaural acoustic signals whose temporal characteristics possess those psychophysiological cues that permit blind persons to imagine their visual surroundings.

In the *SeeHear* system a lens is used to project a scene onto a 32×36 two-dimensional array of photo-sensors. To preserve the dynamic range of the visual scene and to detect motion, the analog signals are logarithmically amplified and differentiated. The sound synthesis consists of an array of signal sources and analog delay lines so as to synthesize horizontal characteristics corresponding to the effects of interaural time delays and acoustic headshadow. The vertical characteristics are synthesized by modeling the pinna-tragus characteristics of the ear. The system has demonstrated capabilities of generating audio signals by which a person can localize horizontal variations in light intensity. In the vertical direction, it does provide the proper cues of the visual scene but the results are not as striking as for the horizontal characteristics unless the light source was moving.

3.5 Transducers in Intelligent Measurement Systems

The approaches pioneered by Mead and his co-workers in coupling analog VLSI and neural-like systems clearly provide a starting point for developing autonomous systems which can hear and see. The general problem of sensors and actuators that form part of an intelligent measurement system which is capable of learning from past experiences and adapting appropriately, will not require transducers that are well-characterized or "calibrated". Such transducers will only need to possess sufficient bandwidth and spatial response that is required to properly interact with the environment. As will be shown in later chapters, such systems will learn from past experiences and store in their memory, reference signals of the phenomena. Provided that the operation and characteristics of the transducers are time-invariant, the relationships between the reference and observed phenomena can be described on the basis of the memorized signals. In view of this, a transducer characterization and "calibrations" are not needed because the signals from the reference phenomenon play that role.

Because of the complexity of natural phenomena we can expect that a multitude of sensors will often be required to characterize the phenomenon under observation. It is noted that the natural phenomena need not always be deterministic and therefore multi-variate random variables and corresponding sensor arrays will be needed for their observation. In order to make a first step in this direction, we will need to review the fundamentals of probability, information and adaptive modeling of natural phenomena. These topics form the basis of the following chapters.

3.6 Future Directions in Transducer Evolution

In this chapter we have summarized the operational characteristics of transducers and, in particular, the characteristics of transducers that are typically used in today's measurement systems. While such transducers form the basis of countless, effective measurement applications, they often are the limiting element in a measurement system. We consider in this section the possible directions in the evolution of transduction devices which we might expect to see in the future. We begin by making a number of observations.

Future development of the sciences will require actuators and sensors that are capable of generating and detecting signals which are weaker, stronger, in new frequency regimes and possibly even new variables. In the fields of chemistry and materials science, the synthesis of ever more complex materials will depend on measurements in new environments on a variety of materials that will require sensors to provide current information about the material and its state as well as its interaction with its environment.

But the needs of technology are also great. There will likely be adaptive and controlled measurement systems that will require new sensors and actuators which are capable of providing the information about the measured system and which can also properly control it, to achieve a desired operating condition.

Not to be neglected are the needs of the military systems which are often the most advanced and which must properly function in a robust manner under the most adverse conditions. Often, an essential requirement is the detection and processing of various field variables which requires the use of multi-sensor arrays. Military systems also depend on a variety of sensors for detecting a multitude of variables; military operations are blind without them. Well-known examples include night vision devices, radar and infrared imaging systems some of which may be connected to pattern recognition processing for the detection and location of objects. History has shown that the development of sensors and sensor signal processing systems can determine the evolution of military equipment and thus even strategies.

The developments in transduction will certainly be evolutionary and possibly even revolutionary. We expect that the evolutionary developments will follow along the following directions: (1) Performance optimization; (2) Miniaturization; (3) Integration; (4) Multiplication; and (5) Extremization. For advances in some of these areas, the transducer designer will look to nature to see how it has solved similar problems. That is, the new developments in transduction will likely mimic biological systems. They will usually not depend on biological phenomena, but rather, on the phenomena which already form the basis of transduction devices today. We consider now each of the directions.

Performance Optimization: We can expect that sensors and actuators will be designed and built that will operate over greater bandwidth, with higher fidelity, and possess better signal-to-noise ratio, and with negligible loading on the system. We also expect to see the development of new sensors which are capable of making absolute measurements of scalar, vector and possibly even tensor variables.

Miniaturization: The developments of micro-sensors and micro-actuators over the past decade will likely continue. The possibility of generating and detecting signals with point-like transducers would be of great utility in a number of measurement situations. Further, micro-transducers that permit measurements on miniature structures, in nearly inaccessible locations, with minimum loading and do it cost-effectively will drive this evolution.

Integration: We can expect the further development of integrated sensors. The capabilities of micro-electronic fabrication techniques will permit a cost-effective fabrication of electronic components and transduction elements on the same substrate. For actuators, this means that signal generation, D/A converters or other signal encoding elements and the inclusion of actuator drive electronics. For sensors, the integration will include components on the chip that are required for signal conditioning, pre-processing, A/D conversion or signal decoding and possibly even the processing electronics itself.

Multiplication: The measurement of field variables necessitates the use of sensor arrays. Today's transducer arrays represent only the first step in the development of one-, two- and possibly even three-dimensional transducer arrays. Their fabrication will depend on nano-fabrication techniques and new and possibly novel integration techniques that will result in a dramatic reduction of un-necessary information that large arrays provide.

Extremization: By *extremization* we mean the development of transduction devices which are capable of performing under extreme conditions of temperature, pressure, radiation and environment. Transduction devices will become available that can transduct scalar, vector and possibly even tensor quantities, but there will also be developed sensors capable of sensing the performance of a system, including its wear, lubrication, sharpness, etc.

But further, we include here the development of transduction devices whose performance mimics as closely as possible the performance of the transducers in healthy living systems, and, in particular, in human beings. Such transduction devices would interface directly with the living being and will result in its performance which is undiminished from its healthy condition.

Work along this direction has already begun. Examples include the development of the so-called *silicon retina* or cochlea which were described in Sect. 3.4.2 and whose outputs can be directly connected to visual or auditory neurons.[33] Today, the connections may be small in number, but in the future, they will surely increase. An example of a transduction device includes prosthetic devices such as artificial limbs whose motions are controlled by signals from the brain and which are capable of providing feedback signals to the brain regarding their interaction with the environment.

Sensors for measuring and monitoring the state or condition of a living system can also be expected. For example, available today are sensors for body temperature, blood pressure, electrocardiograph, blood sugar and blood composition. The future will likely see the development of far more capable sensors for these applications. Sensors resembling the immune system of the body will likely be developed for the early detection of disease and possibly even prediction of its course.

For predicting *revolutionary* changes in transduction, we look to the past. At the beginnings of the development of tools and machines, a human being relied only on himself to survive. Civilization has progressed in relation to tools and processes required for survival and its related activities. It has led from simple tools and devices to ever sophisticated ones which include also the processing of information. The simple, single-channel electronic feedback controller which usually is based on just one sensor and one actuator has led to today's large multi-channel systems.

The most advanced system for processing information is usually considered to be the human brain which is based on neurons whose characteristic response time is about 1 ms. It is possible today to fabricate transduction devices that resemble the operational characteristics of neurons but which possess response times of order of 'ns'. Such a remarkable increase in throughput when properly included in a network can be expected to yield ultrafast, adaptive systems which could further lead to a revolutionary advancement in capability of information-based processing devices. Such a revolutionary step will require transduction devices capable of high-speed performance, possessing either a broad frequency or broad spatial frequency response and thus concomitant increase in sensory information or action on their environment. This will require the development of appropriate high-speed information transfer systems. Such systems may, in fact, also lead to the development of systems capable of training the brain of living systems, an exciting and possibly dangerous prospect.

4. Probability Densities

4.1 Estimation of Probability Density

4.1.1 Parzen Window Approach

It was mentioned in Chap. 2 that a proper analytical description of empirical data is one of the primary tasks of empirical science. A basic related problem is the estimation of the probability density function (PDF) $f(x)$ of a continuous, random variable X from a set of N independent, identically distributed sample values $\{x_1, \ldots, x_N\}$. Hereafter we assume that only continuity of the random variable and no other *a priori* information about its properties is known, therefore a non-parametric estimation is considered. Among various non-parametric techniques the presentation of the PDF by a kernel estimator is most common [4, 3, 6, 5] and this is also the focus of our treatment.

Using only the fundamentals of statistics which are summarized in Appendix A, we assume that the probability density of a random variable can be estimated by a function ψ which is the quotient of relative frequency of the variable corresponding to a narrow interval Δx and the width of the interval. That is,

$$\psi(x; \Delta x) = \frac{\Phi_{\Delta x}}{\Delta x} \,. \tag{4.1}$$

Here, $\Phi_{\Delta x} = N_{\Delta x}/N$ denotes the relative number of experiments yielding the value of X in the interval Δx.

Since the estimation of the probability density function represents one of the central problems related to the modeling of natural phenomena, the last estimator deserves additional attention. Let us first recall that the probability density is determined by the derivative $f(x) = dP/dx$. Estimation of this function by the estimator $\psi(x; \Delta x)$ includes two different sources of error. The first is related to the statistical estimation of probability. The second source of error is inherent in the selection of the width of the interval Δx. That is, the interval cannot be taken arbitrarily small if at least one sample out of a finite total number of samples N should occur in it. It is therefore not immediately clear how the limit process $\Delta x \to 0$ should be performed. There are two limiting possibilities. In the first one, the interval Δx is kept constant and a number of samples is selected such that $\Phi_{\Delta x}$ is determined with the required accuracy. This procedure is then repeated with decreasing interval width in order to estimate the limiting value of estimator ψ. In the second possibility, the relative frequency and with it the number $N(\Delta x)$ is selected *a priori* and then the corresponding width Δx is determined for ever greater total number of samples N. In any case, the quantities appearing in the definition of the estimator ψ cannot be taken independently. If ψ is to converge to $f(x)$, three conditions must be simultaneously fulfilled:

$$\text{I.} \lim_{N\to\infty} \Delta x = 0 \tag{4.2}$$

$$\text{II.} \lim_{N\to\infty} N(\Delta x) = \infty \tag{4.3}$$

$$\text{III.} \lim_{N\to\infty} N(\Delta x)/N = 0 \, . \tag{4.4}$$

By changing the interval width Δx into a volume ΔV the above conditions can be readily generalized to a multivariate case. In the first mentioned possibility, these conditions can be fulfilled by specifying the interval width Δx as a function of N, for example $\Delta x \propto N^{1/2}$ such that $N(\Delta x)$ is empirically determined. In the second possibility, the number $N(\Delta x)$ is specified as a function of N, for example $N(\Delta x) \propto N^{-1/2}$, and the interval width is increased until it covers $N(\Delta x)$ neighbors of x. Both possibilities yield convergent, practically applicable procedures which are known as the *Parzen-window* and the *nearest-neighbor approach*, respectively, and these will be discussed later. In the following section, we first discuss the Parzen-window approach.

A formal expression for the estimator of the probability density can be obtained from the following treatment that was developed by Parzen. [4, 5] Let us express the sample mean of an arbitrary function $y(x)$ in a similar way as the expected value:

$$\langle y(x) \rangle = \frac{1}{N} \sum_{i=1}^{N} y(x_i) = \int_{-\infty}^{\infty} y(x)\, f_e(x)\, dx \, . \tag{4.5}$$

Here $f_e(x)$ denotes the empirical density function. Comparison of the sum and integration of the last expression reveals that this function can be expressed as

$$f_e(x) = \frac{1}{N} \sum_{i=1}^{N} \delta(x - x_i) \, . \tag{4.6}$$

This corresponds to a density of a discrete random variable with the sample space determined by the sample values x_i and equal probabilities $p_i = 1/N$ for $i = 1$ to N. If x_i is treated as a realization of a random variable X_i, then $f_e(x)$ can be treated as a random variable associated with the sample vector $\{x_1, x_2, \ldots, x_N\}$. The expected value of this random variable is

$$\mathrm{E}\left[f_e(x)\right] = \int_{-\infty}^{\infty} \frac{1}{N} \sum_{i=1}^{N} \delta(x - x_i) f_{X_i}(x_i)\, dx_i$$

$$= \frac{1}{N} \sum_{i=1}^{N} f_{X_i}(x) = f_X(x) \ . \tag{4.7}$$

Here we have taken into account that all components are identically distributed. The last equality shows that the empirical distribution function is an unbiased estimator of the probability density. Its variance is

$$\mathrm{var}(f_e) = \mathrm{E}\left[\left(f_e(x) - f_X(x)\right)^2\right]$$

$$= \frac{1}{N^2}\mathrm{E}\left[\sum_{i=1}^{N}\sum_{j=1}^{N}\{\delta(x - X_i) - f_X(x)\}\{\delta(x - X_j) - f_X(x)\}\right] . \tag{4.8}$$

By assuming that the sample components are mutually independent and identically distributed we obtain the result

$$\mathrm{var}(f_e) = \frac{1}{N}\left[\delta(0)\, f_X(x) - f_X(x)^2\right] \ . \tag{4.9}$$

This indicates that the variance diverges because of the singular delta function that is used in its definition according to Eq. (4.6). The empirical density is thus a non-consistent estimator. The divergence can be avoided by substituting the delta function through a regular, non-singular approximation δ_a. The regularization of the delta function corresponds to its stretching into the region surrounding its point of singularity. The stretching can generally be described by the convolution integral

$$\delta_a(x) \ = \ w(x) \ = \ \int w(x - x')\delta(x')\, dx' \ = \ w * \delta \tag{4.10}$$

where $w(x)$ is an appropriate kernel, or Parzen window, peaked at $x = 0$ which satisfies the following conditions:

$$w(x) \ \geq \ 0 \tag{4.11}$$

and

$$\int w(x)\, dx \ = \ 1 \ . \tag{4.12}$$

These conditions indicate that the kernel must itself represent a possible probability density function, such, as for example, a Gaussian

$$w(x) = \frac{1}{\sqrt{2\pi}\,\sigma} \exp\left[-\frac{1}{2}\left(\frac{X}{\sigma}\right)^2\right] \ . \tag{4.13}$$

The filtered probability density estimator is described by the convolution equation

$$\phi(x) = w * f_e = \int w(x-x')f_e(x')\,dx' = \frac{1}{N}\sum_{i=1}^{N} w(x-x_i)\,. \qquad (4.14)$$

The statistical properties of the estimator are characterized by the hypothetical average and variance

$$\begin{aligned}\mathrm{E}\,[\phi(x)] &= \int w(x-x')\mathrm{E}\,[f_e(x')]\,dx' = \int w(x-x')f_X(x')\,dx' \\ &= m_\phi(x) \neq f_X(x) \quad \text{for } w \neq \delta\,,\end{aligned} \qquad (4.15)$$

$$\begin{aligned}\mathrm{var}[\phi(x)] &= \mathrm{E}\,[\{\phi(x)-m_\phi\}^2] \\ &= \frac{1}{N^2}\mathrm{E}\left[\sum_{i=1}^{N}\sum_{j=1}^{N}\{w(x-X_i)-m_\phi\}\{w(x-X_j)-m_\phi\}\right].\end{aligned} \qquad (4.16)$$

By considering an identical distribution of all samples and their mutual statistical independence we obtain after some arithmetical steps the result

$$\mathrm{var}[\phi(x)] = \frac{1}{N}\left\{\int w(x-x')^2 f_X(x')\,dx' - m_\phi^2\right\}. \qquad (4.17)$$

This equation can be used to estimate the bound of $\mathrm{var}(\phi)$. For this purpose, we first drop the second term and then estimate the first term as follows:

$$\begin{aligned}\mathrm{var}[\phi(x)] &\leq \frac{1}{N}\int w(x-x')^2 f_X(x')\,dx' \\ &\leq \frac{\sup(w)}{N}\int w(x-x')f_X(x')\,dx'\end{aligned} \qquad (4.18)$$

$$\mathrm{var}[\phi(x)] \leq \frac{\sup(w)\,m_\phi}{N}\,. \qquad (4.19)$$

The window function can generally be expressed by its width σ_w as $w(x) = g(x)/\sigma_w$, where $g(x)$ is a non-dimensional function. The last inequality then becomes

$$\mathrm{var}[\phi(x)] \leq \frac{\sup(g)\,m_\phi}{N\sigma_w}\,. \qquad (4.20)$$

The numerator remains finite when $N \to \infty$ and therefore σ_w may approach 0 and we can still have $\mathrm{var}(\phi) \overset{N}{\to} 0$ provided that $N\sigma_w \to \infty$. By letting

$$\sigma_w = \sigma_1/\sqrt{N}\,, \qquad (4.21)$$

or

$$\sigma_w = \sigma_1/\log(N+1)\,, \qquad (4.22)$$

or any similar function that satisfies conditions I-III in Eq. (4.4) we can thus assure the convergence of the variance and consistency of the estimator

ϕ. These conclusions can easily be generalized to multi-dimensional variables as well.

Unfortunately, filtering with $w \neq \delta$ generally yields a non-zero estimator bias

$$\eta = |m_\phi(x) - f_X(x)| \neq 0 \tag{4.23}$$

that increases if the window width at given N is increased. The question therefore appears how to select the window function, or for a given window function, its width, in order to achieve an acceptable balance between the bias error, introduced by filtering and the statistical fluctuations of the estimator. This problem is addressed in the next section.

4.1.2 An Optimal Selection of the Window Function

In order to avoid any arbitrariness in the selection of the window function mentioned above, we employ an optimization principle. We first presume that the hypothetical probability density function $f(x)$ is given. This presumption must be critically evaluated later because we, in fact, do not need any estimator if $f(x)$ is given. Next, we define the square discrepancy between this function and its estimator $\phi(x)$ by the integral

$$D(\phi, f) = + \int_{-\infty}^{+\infty} \{\phi(x) - f(x)\}^2 \, dx \, . \tag{4.24}$$

This is a global measure of the difference between both functions because it includes integration over the complete span of the variable X. The estimator is further treated as a random variable defined on the sample vector. Therefore we define the mean square discrepancy as

$$\mathcal{F}(w) = \mathrm{E}\,[D(\phi, f)] = \mathrm{E}\left[\int \{\phi(x) - f(x)\}^2 \, dx\right] . \tag{4.25}$$

Let us further express the mean square discrepancy by the variance and bias of ϕ as follows

$$\begin{aligned} \mathcal{F} &= \mathrm{E}\left[\int \{\phi(x) - f(x)\}^2 \, dx\right] = \mathrm{E}\left[\int \{\phi(x) - m_\phi + m_\phi - f(x)\}^2 \, dx\right] \\ &= \int (\mathrm{E}\,[\{(x) - m_\phi\}^2] + \{m_\phi - f(x)\}^2) \, dx \\ &= \int \{\mathrm{var}[\phi(x)] + \eta(x)^2\} \, dx \, . \end{aligned} \tag{4.26}$$

where $\int \eta^2(x)\,dx = \int \{m_\phi - f(x)\}^2 \, dx$ describes the bias of the estimator ϕ. The first term of this integral generally decreases, while the second one increases with increasing window width. We therefore conjecture that there exists a minimum of the mean square discrepancy with respect to the applied window function.

The estimator depends on the window function and therefore the mean square discrepancy represents a functional $\mathcal{F}(w)$. Our goal is to determine among all the possible window functions, an optimal one w_o that yields the minimum of this functional. This is a variational problem

$$\min[\mathcal{F}(w)] \Rightarrow w_o \ . \tag{4.27}$$

Before we proceed to the solution of this problem, let us explicitly express the mean square discrepancy by the window function. For this purpose we utilize the previously derived expressions for $\mathrm{var}(\phi)$ and η and obtain after several arithmetic steps the following result

$$\begin{aligned}\mathcal{F}(w) &= \int \left\{ \frac{1}{N} \left(\int w(x')^2 f(x-x')\,dx' - \left[\int w(x')^2 f(x-x')\,dx' \right]^2 \right) \right. \\ &\quad \left. + \left[\int w(x') f(x-x')\,dx' - f(x) \right]^2 \right\} dx \ . \end{aligned} \tag{4.28}$$

The variation $w \to w + \delta w$ yields the following variation of $\mathcal{F}$:

$$\begin{aligned}\delta\mathcal{F}(w) = \frac{2}{N} &\left\{ \int \int w(x') f(x-x') \delta w(x')\,dx'\,dx \right. \\ &- \left. \int \left[\int w(x'') f(x-x'')\,dx'' \right] \left[\int \delta w(x') f(x-x')\,dx' \right] dx \right\} \\ &+ 2 \int \left[\int w(x'') f(x-x'')\,dx'' - f(x) \right] \left[\int \delta w(x') f(x-x')\,dx' \right] dx \ . \end{aligned} \tag{4.29}$$

This variation must equal 0 for arbitrary $\delta w(x')$ which occurs when the integrand with respect to x' equals 0. This condition yields the following integral equation for the optimal window function w_o

$$\begin{aligned}\frac{1}{N} &\left\{ w(x') \int f(x-x')\,dx - \int \int w(x'') f(x-x'')\,dx'' f(x-x')\,dx \right\} + \\ &+ \int \left[\int w(x'') f(x-x'')\,dx'' - f(x) \right] f(x-x')\,dx = 0 \ . \end{aligned} \tag{4.30}$$

This expression can be simplified by taking into account that $\int f(x-x')\,dx = 1$ and introducing the correlation function of the probability density

$$R(x', x'') = \int f(x-x') f(x-x'')\,dx = \int f(s) f(s + x'' - x')ds \ . \tag{4.31}$$

It is evident from the last integral that the correlation function depends on the difference $\Delta x = x'' - x'$ and it is an even function. We therefore write $R(x', x'') = R(\Delta x)$ and when this is substituted into Eq. (4.30), the result is the linear integral equation

$$\frac{1}{N}w(x) - (1 - \frac{1}{N}) \int w(x')R(x - x')\,dx' - R(x) \;=\; 0\,. \tag{4.32}$$

The trivial solutions of this equation can immediately be determined for $N = 1$ and $N \to \infty$. When $N = 1$, it follows that $w(x) = R(x)$, while for $N \to \infty$ we obtain the equation

$$\int w(x')R(x' - x)\,dx' - R(x) \;=\; 0 \tag{4.33}$$

which has the solution $w(x) = \delta(x)$. As N increases from 1 to ∞, we expect that the resulting w, as it evolves from $R(x)$ to $\delta(x)$, generally represents an approximation of a delta function with ever decreasing width.

Because of the linearity of the integral equation for w we can find its general solution by Fourier transform methods. This yields the equation for the Fourier transform $W(\omega)$ of $w(x)$

$$\frac{1}{N}W(\omega) - \left(1 - \frac{1}{N}\right) W(\omega)S(\omega) - S(\omega) \;=\; 0\,. \tag{4.34}$$

Here we have introduced the spectral density $S(\omega)$ which is the Fourier transform of the correlation function $R(x)$ and recognized that the Fourier transform of the convolution integral yields the product of corresponding transforms. The solution of the last equation can be explicitly written as

$$W_N(\omega) \;=\; \frac{S(\omega)}{\frac{1}{N} + (1 - \frac{1}{N})S(\omega)} \;=\; \frac{N\,S(\omega)}{1 + (N-1)S(\omega)}\,. \tag{4.35}$$

The previously mentioned trivial solutions $w(x) = R(x)$ and $w(x) = \delta(x)$ correspond to $W_1(\omega) = S(\omega)$ and $W_\infty(\omega) = 1$, respectively.

The general solution expressed by Eq. (4.35) indicates that the optimal window function cannot generally be determined because it is task-dependent. For each phenomenon being investigated there generally exist a specific spectrum of the probability function $S(\omega)$ and hence a corresponding specific window function. However, for $N \to \infty$ the asymptotic window function is task independent and converges in all cases to the delta function $\delta(x)$. This analysis explains why so many attempts to specify a common window function, published in the literature, have not been successful. Relative to this conclusion there appears an interesting question. At the start of any experimental examination of a given phenomenon, one generally knows nothing about the probability density or its spectrum. How can one select an appropriate window function for estimating the probability density from the empirical data? An answer to this question can be obtained from an examination of two typical examples, one corresponding to a normal and the other, to an uniform probability density function.

Let us first consider the case of normal variable with $m = 0$ and $\sigma^2 = 1/2$. The corresponding probability density, its correlation function, the spectral density, and optimal window function are as follows

$$f(x) = \frac{1}{\sqrt{\pi}} \exp[-x^2] \tag{4.36}$$

$$R(x) = \frac{1}{\sqrt{2\pi}} \exp[-\frac{x^2}{2}] \tag{4.37}$$

$$S(\omega) = \int_{-\infty}^{+\infty} R(x) \exp[-\imath\omega x]\, dx = \exp[-\frac{\omega^2}{2}] \tag{4.38}$$

$$W_N(\omega) = \frac{N \exp[-\frac{\omega^2}{2}]}{1 + (N-1)\exp[-\frac{\omega^2}{2}]} . \tag{4.39}$$

The window function in the frequency-domain decreases monotonically with increasing ω. The properties of the window function in the x-domain can generally be described by its gross structure on a large scale and its details on a smaller scale. The gross structure is determined by the Fourier transform at low frequencies while the details depend on the high frequency components. Information about the gross structure of the window function can thus be obtained by analyzing the behavior of $W(\omega)$ at $\omega \to 0$. Using the Taylor expansion $\exp[-\frac{\omega^2}{2}] \approx 1 - \frac{\omega^2}{2}$ we obtain from the last equation when $\omega \to 0$, the approximation $W(\omega) \approx 1 - \frac{\omega^2}{2N} \approx \exp(-\frac{\omega^2}{2N})$ which possesses the characteristic width $\omega_o = \sqrt{N}$. Its Fourier transform yields for the large scale form in the x-domain an approximately Gaussian function with the standard deviation $\sigma_w \approx 1/\sqrt{N}$. This result corresponds to the normal density with $\sigma_x = 1/\sqrt{2}$. For another σ_x we must apply the estimate

$$\sigma_w \approx \sigma_x \sqrt{2/N} = \sigma_{w1}/\sqrt{N} . \tag{4.40}$$

The small scale behavior is determined by the high frequency region where $W(\omega)$ effectively falls to 0. From the form of $W(\omega)$ it follows that the corresponding characteristic frequency can be estimated from the equation $W_N(\omega_{1/2}) = 1/2$ as

$$\omega_{1/2} = \sqrt{2\ \log(N+1)} . \tag{4.41}$$

The corresponding period, given by

$$x_{1/2} = 2\pi/\omega_{1/2} \propto 1/\sqrt{2\log(N+1)} \tag{4.42}$$

shows that the window function on the small scales approaches a delta function with increasing N less efficiently than indicated by its gross behavior. A broad window function results in a more efficient smoothing effect in the filtering process over small scales.

A similar procedure performed with the uniform probability density yields:

$$f(x) = \begin{cases} 1 & \text{for } 0 \le x \le 1 \\ 0 & \text{elsewhere} \end{cases} \tag{4.43}$$

$$R(x) = \begin{cases} 1 - |x| & \text{for } -1 \le x \le 1 \\ 0 & \text{elsewhere} \end{cases} \tag{4.44}$$

$$S(\omega) = \left[\frac{\sin(\frac{\omega}{2})}{(\frac{\omega}{2})} \right]^2 = \operatorname{sinc}^2\left(\frac{\omega}{2\pi}\right) \tag{4.45}$$

$$W_N(\omega) = \frac{N \operatorname{sinc}^2\left(\frac{\omega}{2\pi}\right)}{1 + (N-1)\operatorname{sinc}^2\left(\frac{\omega}{2\pi}\right)} \,. \tag{4.46}$$

The long scale behavior is in this case determined by the expansion of function $[\sin(\alpha)/\alpha]^2 \approx 1 - \alpha^2/3$ which yields $W(\omega) \approx 1 - \frac{\omega^2}{3N} \approx \exp(-\frac{\omega^2}{3N})$ with the characteristic width $\omega_o = \sqrt{3N/2}$. The corresponding large scale form in the x-domain is again an approximately Gaussian function with standard deviation $\sigma_w \approx 1/\sqrt{3N/2}$. The small scale behavior is in this case determined by the envelope of the spectral density. It falls with the increasing ω as $1/\omega^2$ which determines the envelope of $W(\omega)$. The half-width of the envelope can be estimated from the equation for $W(\omega)$ by substituting into it the envelope $1/\omega^2$

$$\frac{1}{2} = \frac{N(\frac{1}{\omega})^2}{1 + (N-1)(\frac{1}{\omega})^2} \tag{4.47}$$

yielding the estimate $\omega_{1/2} \approx \sqrt{N+1}$. In this case, the large scale behavior yields approximately the same dependence of the characteristic frequency or period on the number N as the small scale behavior.

These two examples show how one can find, or at least estimate, an optimal window function provided that the information about the corresponding distribution function is specified in terms of the corresponding spectral function. However, at the start of a combined experiment which yields the samples $(x_1, x_2, x_3, \ldots, x_N)$, the probability distribution function is generally not known and one must obtain the corresponding information from the samples. For this purpose we recall that the optimal window function converges in each case with increasing N to a delta function. We thus do not need to specify the structure of the window function in detail, but rather, we require an acceptable estimate of its gross form that is applicable at small values of N which only approximately corresponds to the correlation function $R(x)$ and is a smooth function of a continuous variable. One might try to utilize for this purpose the filtered function ϕ, but this is a logically incorrect step. Recall that one is trying to estimate the window function by which the empirical density is then transformed into ϕ. Therefore, it is the filter function that determines the properties of ϕ and not vice versa. Consequently we may not use ϕ in a specification of w but rather we must specify the properties of the spectral density $S(\omega)$ directly from the empirical data. The gross form of the correlation function is determined principally by the spectral components at low ω and hence the problem is again reduced to an estimation of the behavior of S when $\omega \to 0$. Let us for this purpose consider the correlation function and the spectrum of the empirical probability density $f_e(x)$

$$R_e(x) = \int f_e(s) f_e(s+x) ds = \frac{1}{N^2} \sum_{i=1}^{N} \sum_{j=1}^{N} \delta(x + x_i - x_j) \qquad (4.48)$$

whose spectrum is

$$S_e(\omega) = \frac{1}{N^2} \sum_{i=1}^{N} \sum_{j=1}^{N} \cos[(x_i - x_j)\omega] \ . \qquad (4.49)$$

For $\omega \to 0$ we apply the Taylor expansion $\cos(x\omega) \approx 1 - (\omega x)^2/2$ and obtain

$$\begin{aligned} S_e(\omega) &\approx 1 - \frac{\omega^2}{2} \frac{1}{N^2} \sum_{i=1}^{N} \sum_{j=1}^{N} (x_i - x_j)^2 = 1 - \left(\frac{\omega \sigma_x}{\sqrt{N}}\right)^2 \\ &\approx \exp\left[- \left(\frac{\omega \sigma_x}{\sqrt{N}}\right)^2 \right] \ . \end{aligned} \qquad (4.50)$$

In the last expression σ_x denotes the sample standard deviation of the variable X. Using this function as a first approximation of the spectral density $S(\omega)$, we can then determine the Fourier transform in the frequency-domain $W_X(\omega)$ as in the previously demonstrated example. The corresponding inverse Fourier transform represents the window function in the time-domain. Its starting, smoothest form, corresponding to $N = 1$, is determined by the Gaussian function with the variance $\sigma_{w1}^2 = 2\sigma_x^2$

$$w_1(x) = \frac{1}{\sqrt{2\pi}\,\sigma_{w1}} \exp[-\frac{x^2}{2\sigma_{w1}^2}] \ . \qquad (4.51)$$

This smooth function corresponds to a properly filtered singular correlation function of empirical probability density distribution ϕ. However, the spectrum cannot be estimated from just one sample point x_1.

The expression for $S_e(\omega)$ indicates that the estimation of the spectral density depends critically on the sample average of square mutual distances between sample points which equals $2\,\mathrm{var}(X)$. This fact is also applicable in formulating other estimators of the probability density function.

Relative to the estimation of the probability density by the Parzen window approach, we can also find an answer to the question: How many measurement samples is a reasonable number if one is to obtain an acceptable estimation of the empirical probability density? For this purpose, let us recall that any measurement used in the determination of the value of a continuous variable is related with some unavoidable experimental uncertainty. Without essential loss of generality we assume that it can be described by the normal distribution, or *inaccuracy function,* $g(x - x_o, \sigma_e)$, in which the standard deviation σ_e denotes the experimental or instrument error. This inaccuracy function describes the scatter of experimental output provided that the input is given by the same fixed value x_o. In order to permit the characterization of a random variable X, we generally need to have much less experimental scatter than the statistical scatter of this variable: $\sigma_e \ll \sigma_X$. Rather than using the delta function for the description of the density of the empirical distribution function, we must apply the inaccuracy function. The corresponding experimental density is then determined by the convolution

$$f_{ex}(x) = f_e * g = \frac{1}{N}\sum_{i=1}^{N} g(x - x_i, \sigma_e) \tag{4.52}$$

which can be treated as the response of a linear system with unit response function g to the input f_e. If we compare this expression with the expression for the probability density that has been filtered by an optimal window function $w_N(x - x_i)$

$$\phi(x) = \frac{1}{N}\sum_{i=1}^{N} w_N(x - x_i) , \tag{4.53}$$

we can conclude that it is reasonable to use a number of samples N such that $w_N(x - x_i)$ corresponds approximately to the inaccuracy function $g(x - x_i, \sigma_e)$. If we begin with the most coarse approximation that yields an approximately Gaussian window function then we can apply the result of Eq. (4.40) for the window width described in the previous example with a normal variable

$$\sigma_w \approx \sigma_e \approx \sigma_{w1}/\sqrt{N} = \sigma_x\sqrt{2/N} \tag{4.54}$$

that yields at large N the estimation

$$N \approx 2(\sigma_X/\sigma_e)^2 . \tag{4.55}$$

With an increase of N, the window function becomes narrower which could in principle permit a description of the structural properties of the probability density on the scales finer than σ_e. However, because of the experimental errors, such structural details are indeed blurred and consequently the estimation cannot be significantly improved by an extended sampling. The acquisition of data points in an actual experimental situation can be stopped whenever the width of the window function is reduced to the value of experimental error σ_e. This result is of practical importance because it enables us to construct an approximately optimal window function without knowing the spectrum $S(\omega)$ in detail. The procedure for finding an optimal window function as described here for a scalar variable X can be generalized to multi-dimensional cases without serious obstacles or essential changes in the conclusions.

Considering all the properties of an optimal window function that have been mentioned, we can conclude that there are two essential natural scales related to the empirical characterization of a random phenomenon. These are: the standard deviation of the experimental inaccuracy and the standard deviation of the observed random variable. The first is inherently related with the instrumentation used in formulating an empirical description of the phenomenon, while the second is specifically determined by the nature of the variable. The fact that there exists a finite number N that determines a reasonable extent of sampling is important because it suggests that only a finite number of memory cells is needed to obtain an acceptable characterization of the corresponding data in a computer or any other memory device despite of the continuous nature of the random variable that is used to describe the phenomenon.

4.1.3 Nearest Neighbor and Maximal Self-Consistency Approach

Although the method of optimization of the window function appears to work well, there exists a serious objection regarding its optimality for filtering of multimodal distributions possessing a small number of data samples. If, for example, the probability distribution is comprised of two normal distributions with two different standard deviations σ_1 and σ_2 and centers separated by much more than $\sigma_1 + \sigma_2$, then the standard deviation of such a distribution is principally determined by the separation between centers $\sigma_x \approx |m_1 - m_2|$. If the width of the initial window function σ_{w1} is selected with respect to σ_x, then we cannot expect that it will also be optimal for smoothing of the empirical distributions around both separated centers. This problem is especially acute when dealing with a small number of data samples. At large numbers the window function in any case converges to the delta function and is thus applicable equally well to the filtering of both constituent components. The problem stems from application of just one window function and the application of a global measure of discrepancy between the filtered distribution ϕ and the hypothetical distribution f. For the case of a bimodal distribution function we can simply expect that two different window functions, each of them applied to a proper constituent portion, would be better suited. This means that a window function which depends on the data would be more appropriate for filtering when the number of samples is small. One potential remedy for this problem is to split the sample space into a finite number of connected intervals and to estimate from the samples in each interval a locally optimal window function. The problem then is how to select the intervals. If we wish to estimate the local correlation function, we must be certain that some data lie in the interval that is considered. This corresponds to an inverse problem in which one first specifies an arbitrary number k of N samples and then determines the corresponding interval widths $h_{i,k}$ for $i = 1, 2, ..., I_k = N/k$. A rough estimate of the probability density in the i-th interval is then

$$f_i \approx \frac{k}{N\, h_{i,k}} . \tag{4.56}$$

This procedure is known as a $k-$nearest neighbor histogram estimation and it is a modification of a kernel type definition with variable width

$$\phi(x) = \frac{1}{N} \sum_{i=1}^{N} w(x - x_i, h_k(x)) . \tag{4.57}$$

Here the kernel w represents an arbitrarily selected density function with the $h_k(x)$ determined from the condition that the interval $[x - h_k(x)/2, x + h_k(x)/2]$ contains k sample points. In order to obtain a good estimate of the probability and an asymptotic convergence of this estimator, we must provide for $k \to \infty$ when $N \to \infty$, or k must be an appropriate function $k(N)$. It must grow sufficiently slowly so that the width $h_{k(N)}(x)$ of the interval containing $k(N)$ samples converges to 0 and that an asymptotically unbiased estimation is assured. As in the Parzen window approach, the following conditions

$$\lim_{N\to\infty} \frac{k(N)}{N} = 0 \qquad \text{and} \qquad \lim_{N\to\infty} k(N) = \infty \tag{4.58}$$

are necessary and sufficient to assure a proper convergence of $\phi(x)$ to $f(x)$ in the region where $f(x)$ is continuous. An acceptable function is $k(N) = \sqrt{N}$. From this estimate, we obtain $h_k \approx 1/(f(x)\sqrt{N})$. In this case the window width again falls off as $1/\sqrt{N}$. Its initial value is not determined by a global parameter as is the standard deviation of the complete distribution, but rather, it is determined locally by the probability density at the point of estimation, or specifically, by the samples of the variable X.

Relative to the application of the $k-$nearest neighbor kernel type estimator, there arise two challenges. First, the inverse problem of estimating the window width is difficult in practical situations. The width $h(x)$ must be determined at each point in the interval where the distribution function is described and this is generally a computationally demanding task. The second, less important property is that the function $k(N)$ is arbitrarily selected. We therefore look for a method which is more practical and by which we could avoid these difficulties. Let us for this purpose try to find an optimal window function possessing a variable width, by again minimizing the functional

$$\mathcal{F}(w) = \mathrm{E}\left[\int \{\phi(x) - f(x)\}^2\, dx\right] . \tag{4.59}$$

The kernel-type estimator with variable width is then described by a general form of the window function $w(x, x')$

$$\phi(x) = \frac{1}{N}\sum_{i=1}^{N} w(x - x_i, b(x)) = \int w(x, x') f_e(x')\, dx' \; . \tag{4.60}$$

We might expect that by minimizing the functional $\mathcal{F}(w)$ we would find the form of a general type of window function $w(x, x')$. Unfortunately, this is not the case because the variational procedure leads to the trivial solution $w(x, x') = f(x)$ yielding the estimate $\phi(x) = f(x)$ that is independent of N which is not a useful result because the density $f(x)$ is introduced only as a hypothetical reference. This indicates that the problem is indeed ill-posed and in order to regularize it we must reconsider the previous expression of the window function, $w = w(x - x_i, h_k(x))$. Its structure is described by two functions that are generally not known: the window function $w(x, h)$ itself as well as its width $h.b(x)$. The problem is logically not well-posed until we specify one function. In order to avoid this difficulty, let us recall that our main goal is to find a kernel that represents a smooth approximation of a delta function. One approach is to select the analytical form of the function *a priori*. This is the approach we will follow by selecting a Gaussian function. The remaining task is to find the width of the function $h(x)$. In order to avoid the computationally demanding task related to the calculation of $h(x)$ at each point, we shall assume that it is sufficient to specify the width of the window function only at the sample points. We thus obtain the apparently simple kernel-type estimator

$$\phi(x) = \frac{1}{N}\sum_{i=1}^{N} w(x - x_i, h_i) \tag{4.61}$$

that requires additional interpretation of the nature of widths h_i. A value of x_i, that determines the window center is a realization of the random variable X. Associated with this is the width h_i which we expect will depend on the position of neighboring sample points. It is therefore also a random variable, but it is determined by the complete sample vector $\{x_1, \ldots, x_i, \ldots, x_N\}$ rather than by only one sample point. The window width therefore depends on multipoint distribution functions, which when applied, cause difficulties with their interpretation in the subsequent analysis. We therefore avoid them and instead specify the error functional without a statistical average

$$\mathcal{F}(w) = \int [\phi(x) - f(x)]^2\, dx \; . \tag{4.62}$$

We shall temporarily treat the widths as parameters that can be determined by minimizing this functional. The resulting equations must then be correctly treated as definitions of the random variables h_i.

The minimization of the functional yields the following system of equations

$$\frac{\partial \mathcal{F}}{\partial h_i} = \frac{2}{N}\int [\phi(x) - f(x)]\frac{\partial w_i}{\partial h_i}\,dx = 0, \qquad i = 1,2,\ldots,N \tag{4.63}$$

with $w_i = w(x - x_i, h_i)$. This can be written in the form

$$\frac{1}{N}\sum_{j=1}^{N}\int w_j \frac{\partial w_i}{\partial h_i}\,dx = \int f(x)\frac{\partial w_i}{\partial h_i}\,dx, \qquad i = 1,2,\ldots,N\,. \tag{4.64}$$

In order to find the solution of this system of equations, we must specify $f(x)$ or at least describe its properties. With this goal in mind, let us recall that the principal purpose of filtering is to diminish the 'infinite' contributions of the delta functions by spreading them around the points of singularity to the level of a supposedly smooth distribution function $f(x)$. [12] We therefore expect that around the i-th sample point the principal contribution to the sum in the expression of the window function $\phi(x)$ given by Eq. (4.61) is determined by the term w_i/N which should be approximately equal to the value $f(x_i)$. With this supposition we can then estimate

$$\frac{1}{N}\int w_i\left(\frac{\partial w_i}{\partial h_i}\right)dx \approx \int f(x)\left(\frac{\partial w_i}{\partial h_i}\right)dx, \qquad i = 1,2,\ldots,N \tag{4.65}$$

and obtain from Eq. (4.64), the following equation

$$\sum_{j\neq i}^{N} w_j\left(\frac{\partial w_i}{\partial h_i}\right)dx = 0, \qquad i = 1,2,\ldots,N\,. \tag{4.66}$$

After differentiating each equation with respect to h_i we obtain a system of conditions

$$\sum_{j\neq i}^{N}\frac{h_i}{h_{ij}^3}\left[\frac{(x_j - x_i)^2}{h_{ij}^2} - 1\right]\exp\left[\frac{-(x_j - x_i)^2}{2h_{ij}}\right] = 0, \qquad i = 1,2,\ldots,N \tag{4.67}$$

which can be represented as a system of iteration formulas for h_i^2:

$$\begin{aligned} h_i^2 &= \frac{\sum_{j\neq i}^{N}(x_j - x_i)^2 u_{ij}^5 \exp\left[\frac{(x_j - x_i)^2}{2h_{ij}^2}\right]}{\sum_{j\neq i}^{N} u_{ij}^3 \exp\left[\frac{-(x_j - x_i)^2}{2h_{ij}^2}\right]} \\ &= G_w(x_1,\ldots,x_N,h_1,\ldots,h_i,\ldots,h_N)\,. \end{aligned} \tag{4.68}$$

Here we have introduced the definition $h_{ij} \equiv \sqrt{h_i^2 + h_j^2}$, the function $u_{ij} \equiv 1/\sqrt{1 + h_j^2/h_i^2}$ and the iteration generating function G_w. If the sample points are approximately uniformly distributed in some region of the sample space, then $h_i \approx h_j$ and we can write in the first approximation $u_{ij} = 1/\sqrt{2}$ which gives a simplified approximate iteration formula

$$h_i^2 \simeq \frac{\sum_{j\neq i}^{N}(x_j - x_i)^2 \exp\left[\frac{-(x_j-x_i)^2}{2h_{ij}^2}\right]}{2\sum_{j\neq i}^{N} \exp\left[\frac{-(x_j-x_i)^2}{2h_{ij}^2}\right]} \,. \tag{4.69}$$

The system of equations

$$\sum_{j\neq i}^{N} \int w_j \left(\frac{\partial w_i}{\partial h_i}\right) dx = 0, \qquad i = 1, 2, \ldots, N \tag{4.70}$$

permits an interesting physical interpretation. Each equation can be converted into a corresponding correlation criterion

$$C_w(h_i) \;=\; \int_{-\infty}^{+\infty} w(x - x_i, h_i)\, \phi_{N-1}(x)\, dx \;=\; \max(h_i) \tag{4.71}$$

where

$$\phi_{N-1}(x) \;\equiv\; \frac{1}{N-1} \sum_{j\neq i}^{N} w(x - x_j, h_j) \tag{4.72}$$

denotes the filtered, empirical probability density function with the i-th point truncated. The derived correlation criterion indicates that the window width corresponding to the i-th sample point can be determined by claiming that the i-th window function maximally correlates with the filtered distribution function ϕ_{N-1} which omits the i-th point. [12] A spreading of the delta function into the i-th window function is thus consistent with all the other window functions. The set of conditions

$$C_w(h_i) \;=\; \max(h_i), \qquad i = 1, 2, \ldots, N \tag{4.73}$$

therefore leads to a simultaneous or self-consistent adaptation of all widths corresponding to the complete sample vector. This generally means that a particular window function w_i essentially spans the gap between neighboring sample points. Consequently we state that the system of equations for h_i stems from the strong *Principle of Self-Consistency*.

A similar, but less strict principle is obtained by the following reasoning. The transition from a singular empirical probability density function to a smooth estimator by spreading the delta functions into the neighborhood of the points of measured data is a process that actually corresponds to the creation of information. If the probability distribution function is not known, one cannot strictly avoid the indeterminacy in the widening of the delta functions but only reasonably diminish it by assuming that a particular window function is properly spread around a particular sample point when its width is selected in accordance with the position of the other sample points. Spreading could thus be treated as a process of information diffusion whose principal purpose is to establish a self-consistent link between the measured data points. Since all the samples represent the same phenomenon, a particular window function should be widened in such a way that, in some sense, it optimally overlaps either the delta functions representing other sample points or the window functions which are spread around them. These statements represent the *Weak* and the *Strong Principle of Self-Consistency*, respectively. The overlapping of the delta functions is mathematically most simply described by a correlation. As a weak condition for consistency of the i-th window function with the other sample values, we apply the maximum of the correlation between this window function and a truncated empirical probability density function given by

$$f_{e,N-1} \;=\; \frac{1}{N-1}\sum_{j\neq i}^{N}\delta(x-x_j)\ . \tag{4.74}$$

The correlation is

$$C_\delta(h_i) \;=\; \int_{-\infty}^{+\infty} w(x-x_i,h_i)f_{e,N-1}\,dx \;=\; \max(h_i)\ . \tag{4.75}$$

This is a system of weak conditions of self-consistency. The integration yields the formulas

$$C_\delta(h_i) \;=\; \frac{1}{N-1}w(x_j-x_i,h_i) \;=\; \max(h_i)\ \ . \tag{4.76}$$

After differentiation with respect to h_i we obtain the system of equations

$$\sum_{j\neq i}^{N}\frac{1}{h_i^2}\left[\frac{(x_j-x_i)^2}{h_i^2}-1\right]\exp\left[\frac{-(x_j-x_i)^2}{2h_i^2}\right] \;=\; 0\ . \tag{4.77}$$

As in the strong case, it can again be represented in a form of iteration formulas for h_i^2:

$$h_i^2 \;=\; \frac{\sum_{j\neq i}^{N}(x_j-x_i)^2\exp\left[\frac{(x_j-x_i)^2}{2h_i^2}\right]}{\sum_{j\neq i}^{N}\exp\left[\frac{-(x_j-x_i)^2}{2h_i^2}\right]} \;=\; G(x_1,\ldots,x_N,h_i)\ ,\ i=1,\ldots,N\ . \tag{4.78}$$

Although this iteration depends on transcendental functions, the properties of the solution can be estimated from an analysis of the leading terms in the numerator and denominator in Eq. (4.78). The leading terms are determined by the sample x_{jo} which lies nearest to the value x_i. The resulting optimal kernel width is therefore approximately given by the distance to the nearest neighbor which is $d_o = \min|x_j - x_i|$. When the sample set consists of just two points it is exactly given by this distance. Since the generating function of the iteration satisfies the condition $dG(x_1,\ldots,x_N,h_i)/dh_i \approx 0$ for $h_i \approx d_o$, the iteration process converges very quickly. Numerical investigations have shown [12] that the initial rate of convergence of the iteration process is not strongly dependent on the starting values of the iteration. As a reference starting value, one can use $h_i = \sqrt{\mathrm{var}(X)/N}$. The solutions of Eq. (4.78) satisfy the approximate relation

$$h_i^2 \gtrapprox \min_j (x_j - x_i)^2 . \tag{4.79}$$

The form of the iteration generation function suggests a simple statistical interpretation of the optimal kernel width. If the exponential function appearing in the iteration formula is treated as a statistical weight assigned to a distance between points of the sample set, then the generating function $G(x_1, \ldots, x_N, h_i)$ can be interpreted as a window average of the square distance from the window center to the neighboring sample points. However, the window width must be determined self-consistently. This interpretation together with iteration formula can be easily generalized to a multi-dimensional case.

In the standard approach, the number of nearest neighbor points, K, is chosen in an *ad hoc* way. In the self-consistent approach, this number is determined by an effective number of neighbors greater than one which are inside the window of width h_i. Generally, this number, increases with an increasing total number of samples N. The properties of estimators composed of kernels with self-consistently determined width are consequently similar to the properties of estimators defined by a k-nearest-neighbor rule.

It was mentioned earlier that the adaptation of the set of widths h_i to the set of sample values $\{x_1, \ldots, x_N\}$ can be treated as the realization of a mapping of a random variable X to a new random variable H. Because this mapping is not determined by an explicit relation but by an iteration process, the properties of the probability distribution of the variable H generally cannot be described analytically. By invoking the approximate relation $h_i^2 >\approx \min_j (x_j - x_i)^2$ it can be conjectured that the width of each window diminishes with an increasing number of samples N when these become ever more densely spaced. The kernel functions then tend towards delta functions and the estimator bias decreases to zero. Since the bias depends on the width and tends to zero when $h_i \to 0$ as $N \to \infty$. The applied estimator thus automatically retains the favorable properties of Parzen's as well the nearest-neighbor estimators without special attention to the conditions governing the dependence of the widths on the sample set size N.

In the adaptation process that stems from the correlation criterion of the weak case the width of a particular window is not influenced by the widths of the other windows. The window function must therefore effectively cover the complete gap between its center and the neighboring sample points. Intuitively, this seems to be too demanding requirement, because after the spreading, the neighboring window also covers the same gap. If only the coverage of the sample points by the window function is considered, the correlation criterion of the weak case is acceptable. But when the mutual overlapping of the window functions is considered, an application of a less demanding, strong condition seems to be more reasonable. However smaller window widths obtained in the strong case generally yield less smooth estimators as those obtained in the weak case.

Although the iteration equations of the weak and strong cases are similar, there is an essential difference between the iteration generation function in them. The function for the strong case depends on all the widths and there appear additional nonlinear terms in the numerator as well as denominator of the iteration equation. As a consequence, this iteration function generally does not converge; although when there are only two samples, it still yields the correct result $h_i^2 = (x_j - x_i)^2/2$. If this result is compared with that obtained for the weak case, it is seen that the condition based on the overlapping of the window functions diminishes their widths by a factor of 2. Using a more detailed analysis of the terms in the strong case iteration equation and an appropriate modification of the iteration procedure, which is omitted here, a convergent procedure can be formulated. [12] A numerical treatment has shown that the factor of approximately 2 appears also in the case with many samples. [12] This observation is even predicted when the influence of the functions u_{ij} on the leading terms in the sums of the iteration equation is included. Numerical experiments have shown that similar results are obtained if a factor of $1/2$ is introduced on the right side of the iteration formula of the weak case. This can be guessed when comparing the approximate strong case with the exact iteration formula of the weak case. In addition to the complications that arise from the instability of the iteration formula, no essential advantage with respect to the weak case is obtained by maximizing the mutual overlapping of the window functions in the strong case. We therefore propose using in practical applications the simpler iteration formula of the weak case.

4.1.4 The Self-Consistent Method in the Multivariate Case

The weak self-consistent approach can be simply generalized to an n-dimensional distribution by using multivariate normal densities. Let us for this purpose express the window function as

$$w_i(\boldsymbol{x}) = (2\pi)^{-n/2}\sqrt{|B_i|}\,\exp\left[-\frac{1}{2}(\boldsymbol{x}-\boldsymbol{x}_i)^T\boldsymbol{B}_i(\boldsymbol{x}-\boldsymbol{x}_i)\right] \tag{4.80}$$

where $\boldsymbol{B}_i = [b_{i,kl}]$ denotes the inverse of the covariance matrix associated with the i-th point. The conditions of the weak self-consistent principle are then

$$C_\delta(\boldsymbol{B}_i) = \frac{1}{N-1}\sum_{j\neq i}^{N} w(\boldsymbol{x}_j-\boldsymbol{x}_i,\boldsymbol{B}_i) = \max(b_{i,kl}), \quad i = 1,2,\ldots,N\,. \tag{4.81}$$

Differentiation with respect to $b_{i,kl}$ and application of the formulas

$$|B| = \sum_{k=1}^{N} b_{kl} B_{kl}^c \; ; \qquad \frac{\partial |B|}{\partial b_{kl}} = B_{kl}^c \; ; \qquad \frac{[\boldsymbol{B}_{kl}^c]}{|B|} = \boldsymbol{B}^{-1} = [h_{kl}] \, ,$$

where B_{kl}^c is the cofactor of the term b_{kl} in the determinant $|B|$, leads to the system of equations for the covariances $h_{i,kl}$:

$$\sum_{j \neq i}^{N} [h_{i,kl} - (x_j - x_i)_k \, (x_j - x_i)_l] \, w(\boldsymbol{x}_j - \boldsymbol{x}_i, \boldsymbol{B}_i) = 0 \, , \qquad i = 1, 2, \ldots, N \qquad k, l = 1, 2, \ldots, n \, . \tag{4.82}$$

These equations can again be written in a form that is applicable for iteration

$$\boldsymbol{B}_i^{-1} = [\sigma_{i,kl}] = \frac{\sum_{j \neq i}^{N} (\boldsymbol{x}_j - \boldsymbol{x}_i)(\boldsymbol{x}_j - \boldsymbol{x}_i)^T \, w(\boldsymbol{x}_j - \boldsymbol{x}_i, \boldsymbol{B}_i)}{\sum_{j \neq i}^{N} w(\boldsymbol{x}_j - \boldsymbol{x}_i, \boldsymbol{B}_i)} \, . \tag{4.83}$$

This expression shows that the previously derived iteration formula of the weak case can be directly adapted to the multivariate problem by transforming h_i^2 and $(x_j - x_i)^2$ into their corresponding correlation matrices and then utilizing the multivariate normal distribution. The resulting window width is then substituted by a window covariance function. Unfortunately in this case, the result of the iteration is the covariance matrix which must be inverted prior to its input into the expression of the window function, which is generally a time-consuming task. The problem related to the singularity of the covariance matrix can generally be avoided by assuming that no covariance can be less than the experimental error of the instrumentation or $h_{kl} \geq \sigma_e$ for any pair of indices k, l.

4.1.5 Numerical Examples

The purpose of this section is to demonstrate numerically some of the characteristic features of the estimator that is based on a self-consistent determination of the window width in a one-dimensional example. Similar to the nearest neighbor estimator, this estimator is especially appropriate for the case of a small number of samples, and we are therefore demonstrating it on examples of this kind. Problems of this type are often met when the PDF is described by a set of prototype or reference samples obtained by various quantization procedures or from the self-organization of formal neurons. [10] However an essential difference between the representation of a PDF by few, independent random samples or by a set of reference values obtained by the quantization procedure should be mentioned here. In the latter case, the spacing between the prototype points is not determined by chance, as when only a few, independent samples are measured. Rather, it is determined by the limiting values of the reference samples obtained by a quantization process. The spacing of prototype points thus exhibits a systematic property of the PDF. Because this procedure was principally developed for estimating a PDF on the basis of prototype points obtained by a self-organization process [11], we usually present examples possessing sample spacings exhibiting some well-defined specific property of the PDF.

For example, one obtains an equal spacing of samples for the case of a uniform distribution and an increasing spacing with distance from the center of a normal distribution. The demonstrated cases represent a normal, a uniform, an approximately exponential, and a bimodal PDF by properly spaced reference samples. In addition, an example with a randomly changing number of samples is demonstrated for the case of a uniform distribution. For the purpose of comparison, Parzen's estimator with a global width selected according to the rule $h_o^2 = \mathrm{var}(x)/N$ is also demonstrated. In all cases, the self-consistent width was determined in ten iterations, beginning with the global width of Parzen's estimator. During the iteration, the value of the width differed from the final value by less than 5% of the limiting value after only five iterations. The results of the numerical experiments are shown in Figs. 4.1–5. In each diagram, the vertical strokes on the upper line indicate the positions of the samples or prototypes, while the adjoining table lists their values and corresponding widths. The bold line corresponds to the self-consistent estimator while the thin one corresponds to Parzen's estimator.

Figure 4.1 shows the case corresponding to a normal distribution. For purposes of comparison, the original normal distribution and the estima-

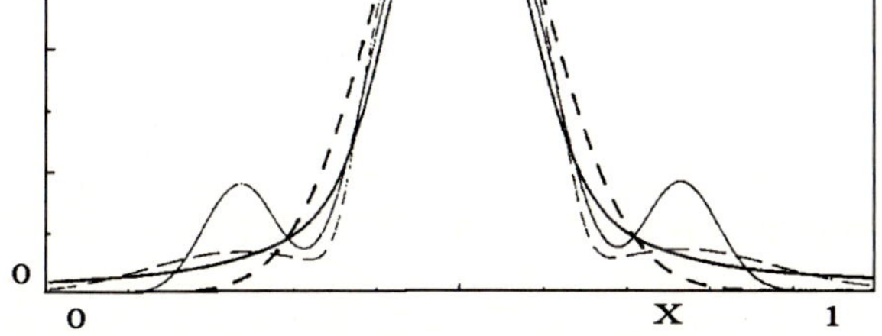

Fig. 4.1. Estimation of a normal probability density function by the self-consistent method. The positions of the prototype samples are denoted on the top line. Bold — : Self-consistent estimator, bold - - - : Original distribution, thin — : Parzen's estimator with $h_o = \mathrm{var}(x)/N$, thin - - - : Parzen's estimator with h_o = nearest-neighbor-distance

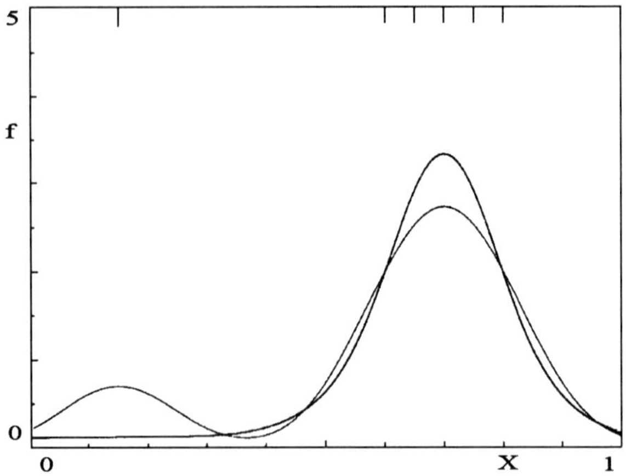

Fig. 4.2. Example of the estimation of the probability density function by the self-consistent method from a group of closely spaced samples and a single separated one. Bold —: Self-consistent estimator; thin —: Parzen's estimator with $h_o^2 = \mathrm{var}(x)/N$

tors obtained by Parzen's width or distance to the nearest-neighbor are also shown. One can estimate intuitively that the self-consistent estimator adapts better to the original distribution than the other two. The two side lobes of the Parzen's estimator indicate that this estimator assigns too great a weight to an isolated or single sample point which is separated from the rest of the data. Figs. 4.2 and 4.3 were generated to illustrate this property. In the example shown in Fig. 4.2, the width corresponding to this isolated point is much larger than the widths of the samples in the group, and therefore its kernel contributes less amplitude to the estimated PDF than the others. In Parzen's estimator, the width corresponding to each sample is the same and consequently the separated sample contributes to the estimator more than in the self-consistent case. The situation is quite different when two samples are closer, as is illustrated by the distribution depicted in Fig. 4.3. In this case, the distance to the nearest neighbor at sample points in close proximity is small and the self-consistent estimator develops an expressive peak developing into a bimodal distribution. This property corresponds to a better adaptability of the self-consistent estimator to multimodal distributions when compared with Parzen's estimator. This is also a characteristic property for nearest neighbor estimators. It becomes especially evident when a distribution with drastically changing spacing between prototypes is represented, as is for example the case of an approximately exponential spacing that is shown in Fig. 4.4. In this case, the region in which the sample spacings are small, as well as in the region where the sample spacings are large, are better described by the self-consistent estimator than by Parzen's. However,

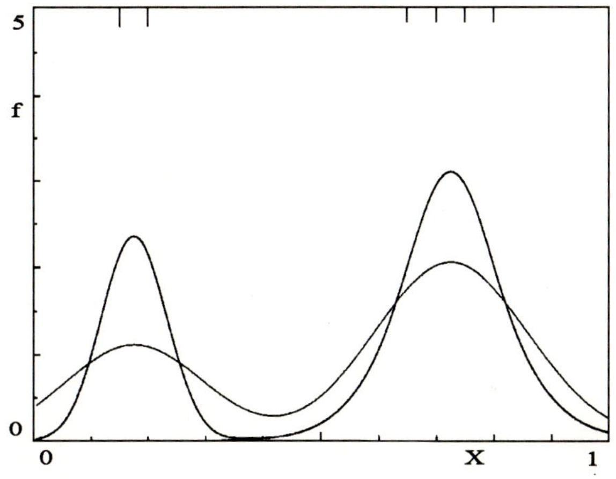

Fig. 4.3. Example of the estimation of the probability density function from two separated groups of closely spaced prototype samples. Bold —: Self-consistent estimator; thin —: Parzen's estimator with $h_o^2 = \mathrm{var}(x)/N$

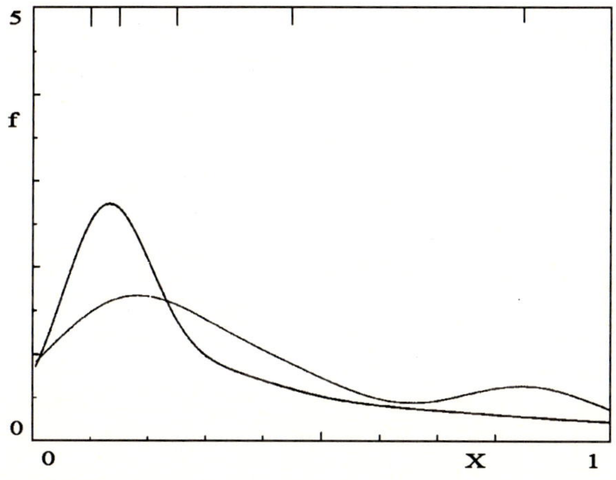

Fig. 4.4. Estimation of an approximately exponential probability density function. The samples are denoted by strokes with increasing spacing on the top line. Bold —: Self-consistent estimator; thin —: Parzen's estimator with $h_o^2 = \mathrm{var}(x)/N$

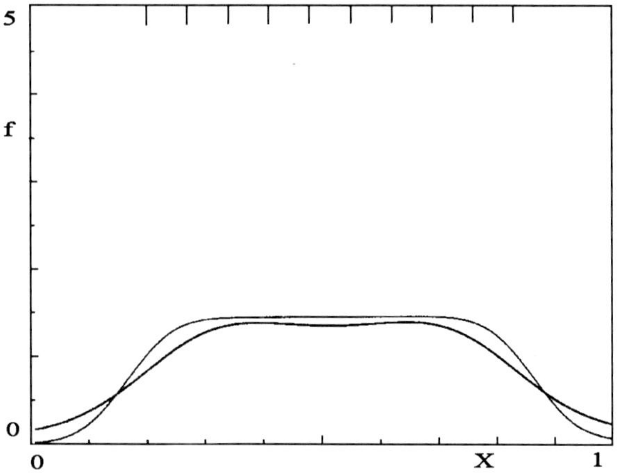

Fig. 4.5. Estimation of a uniform probability density function. The samples are denoted by equally spaced strokes on the top line. Bold —: Self-consistent estimator; thin —: Parzen's estimator with $h_o^2 = \mathrm{var}(x)/N$

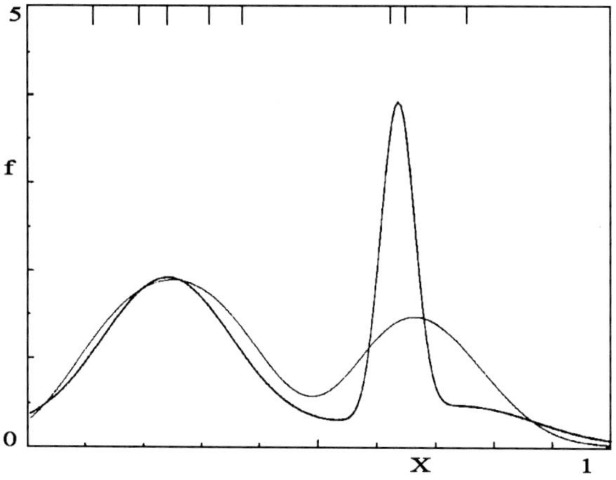

Fig. 4.6. Estimation of a uniform probability density function in the interval from 0.1 to 0.9 from the random samples. The samples are denoted by randomly spaced strokes on the top line. Bold: self-consistent estimator, thin: Parzen's estimator with $h_o^2 = \mathrm{var}(x)/N$

for the case of a uniform distribution of sample points that is shown in Fig. 4.5, Parzen's estimator performs better. This follows because those sample points which are at the edges of the distribution are, on average, more distant from the other points comprising the distribution as those lying near its center. Because the edge samples influence the width of the samples in their vicinity inside of the distribution, the smallest width corresponds to samples nearest to the edges yet inside the distribution. This causes maxima of the distribution close to the edges, which resembles the Gibbs phenomenon that is observed when representing a square wave function by a finite number of terms of a Fourier series.

The final example, shown in Fig. 4.6, demonstrates the applicability of the self-consistent estimator to the estimation of the PDF from independent, randomly distributed samples. The samples were generated randomly with a corresponding uniform PDF in the interval from 0.1 to 0.9. Because of the random selection of samples, their spacing is not uniform, but varies randomly. As the self-consistent width is approximately determined by the nearest neighbor, larger fluctuations are generally observed in the corresponding estimator than in Parzen's, which is based on the global width as determined from the variance of the samples. This difference has previously been observed when the properties of kernel-type and nearest-neighbor estimators were compared and so will not be analyzed further here. [5] We simply mention that despite the larger fluctuations, the nearest-neighbor estimators, like Parzen's estimator, are consistent and asymptotically unbiased. [1, 2, 3, 8] The same conclusion is reached for the self-consistent estimator, which can be treated as a joint version of both types.

4.1.6 Conclusions About Filtering of the Empirical PDF

The main purpose of our approach of estimation of the PDF was to propose a principle by which an optimal, global kernel function or local kernel width could be determined. The optimization of the mean square error functional yields in the global case for the optimal kernel an approximately normal distribution function with the approximate width $\sigma_w \approx \sigma_x \sqrt{2/N}$. The starting point for the optimization of the local kernel width was the modification of a kernel-type k-nearest neighbor estimator. [7] Its optimization leads to the formulation of the principle of maximal self-consistency and to a simple iteration formula, which yields the kernel widths that are approximately equal to the distance to the nearest neighbor. An *ad hoc* specification of the number of nearest neighbors, or an arbitrary selection of the kernel width in Parzen's approach is thus avoided. The examples shown, demonstrate that our approach is especially appropriate for the estimation of the PDF from prototype samples that are, for example, obtained by quantization procedures. This is of advantage for the application of the self-consistent approach in the field of neural networks, where the prototypes can be extracted from the parameters of self-organized neurons. [10, 11] In our treatment we have not mentioned the other methods of estimation of probability densities. The best known are the modes of parametric estimation, orthogonal expansion and minimum entropy methods. All of these, however, require some *a priori* specification about the phenomenon under consideration in order to be optimally applicable. Although these alternate methods can be very effective in specific cases, we have not discussed them, since our task has been to focus on those methods which can be applied when none of the properties or characteristics of the phenomenon under observation are known in advance. The interested reader can find a detailed description of these alternate methods in the current literature on probability and statistics. [1]

5. Information

5.1 Some Basic Ideas

The concept of information is intuitively related with the change of our knowledge about natural phenomena caused by observations. [1, 2] An observation can generally be treated as an experiment, performed with the help of our sensual perception or a measuring instrument. A less intuitive description of the concept of information can be obtained by relating it with the properties of measuring systems and experimentation. For this purpose we mathematically define a variable by which the change of knowledge, and with it the information acquired, can be quantitatively characterized. We therefore try to incorporate the characteristic properties that are intuitively assigned to information into the mathematical definition of a corresponding variable. However, information is usually intuitively further related to a "meaning" or a "significance" which is generally specific to a particular observer but which also depends on the entire complex of phenomena preceding the actual acquisition of the information. Obviously the "meaning" of information is conceptually much more involved than the information itself and it cannot be easily or generally described. We therefore shall not define this concept quantitatively.

The most important property of information stems from the interpretation of knowledge. Intuitively we understand the growth of our knowledge as our diminished uncertainty about the world we are observing. [5, 6, 7] Consequently, it is convenient to first give a quantitative description of uncertainty. For this, let us imagine an experiment whose outcomes are characterized by a discrete sample space S containing N sample points. A sample point can generally be treated as a representative of the communication link between the observer and that part of Nature in which the examined phenomenon under observation is occurring. From this view, the sample space S represents a communication channel with N parallel links. At the start of the experimentation we assume *a priori* that any of the available sample points could be equally well utilized to obtain a description of the phenomenon that is characterized by the experiment. The greater the number of sample points, the greater is our *a priori* uncertainty about the experimental outcome. We therefore assume that the corresponding measure of uncertainty can be mathematically described by a monotonic increasing function H of the number of possible outcomes N. Without loss of generality, we assume that this function is non-negative. The corresponding variable is called the entropy of *a priori* information and it is denoted by

$$H = H(N) \ . \tag{5.1}$$

If the sample space includes just one point, then there is no uncertainty related with the experiment and we have $H(N) = 0$ for $N = 1$.

Contrary to the situation prior to carrying out the experiment, there is no uncertainty after the experiment, provided that instrumentation is operating properly. There was just one result and we must associate with it the *a posterior* uncertainty $H' = 0$. When we perform an experiment, the uncertainty about the examined state of Nature is thus decreased. We therefore describe the information acquired through an experiment by the variable I that is equal to the diminished uncertainty

$$I = H(N) - H' = H(N) \ . \tag{5.2}$$

Another property of information which can be intuitively accepted stems from the treatment of combined experiments. If we perform the same experiment independently n-times then the uncertainty of the combined experiment is n-times the uncertainty of an individual one. The number of all possible outcomes of the combined experiment is N^n and we obtain the relation

$$H(N^n) = nH(N) \ . \tag{5.3}$$

This is a functional equation for the unknown function $H(x)$. We can obtain an idea for its solution by substituting

$$N^n = x \to n = \frac{\log x}{\log N} \to H(x) = \frac{H(N)}{\log N} \log x \ . \tag{5.4}$$

The functional dependence of H is thus determined by the function $\log x$ while the factor $k = H(N)/\log N$ can be treated as a constant which is specifically related to the given combined experiment. By its selection we specify the unit of the entropy of information. The following units and expressions are commonly used

$k = 1/\log 2$	$H = \log_2 N$	with unit *binary digit* or *bit*
$k = 1$	$H = \log N$	with unit *natural digit* or *nat*
$k = 1.3810^{-26}$J/K	$H = k \log N$	with *physical* or *Boltzmann unit.*

In our further treatment we shall use the second case with $k = 1$.

5.2 Entropy of Information

If we describe the result of an experiment by the random variable X then we must assign to each sample point before the experiment is carried out, the experiment the same *a priori* probability of occurrence $p(x_i) = 1/N = p_1$ for $i = 1, \ldots, N$. The logarithmic measure of the uncertainty can then be expressed as

$$H = \log N = -\log p_1 = -\sum_{i=1}^{N} p_i \log p_i \tag{5.5}$$

which suggests how one can proceed to the definition of the measure of uncertainty in cases when the outcomes of an experiment are not equally probable. For this purpose let us assume that the experiment is again characterized by the random variable X with the discrete sample space $S = \{x_1, \ldots, x_i, \ldots, x_N\}$ and the corresponding probability distribution $\{p_1, \ldots, p_i, \ldots, p_N\}$. With each possible outcome of the experiment, there is associated the uncertainty

$$H_i = -\log p_i \ . \tag{5.6}$$

However, in an experiment, various outcomes may appear. If the experiment is repeated many times, then we can assign to the experiment an average uncertainty that is given by the hypothetical mean value

$$H = \sum_{i=1}^{N} H_i \, p_i = -\sum_{i=1}^{N} p_i \log p_i \ . \tag{5.7}$$

This equation is the definition of the so-called Boltzmann entropy and it deserves additional elaboration. First of all, we must point out that $H_i = -\log p_i$ must be interpreted as the uncertainty related with the variable X which disappears when the experimental result is x_i. In this context H_i can be interpreted as a function of the random variable as well as the acquired information $I_i = H_i$. When the *a priori* probability p_i is small then the acquired information is high and vice versa. If the uncertainty or the associated information gain is treated as a random variable then it is possible to define the average uncertainty of the phenomenon by the hypothetical average. The corresponding average entropy H is no longer a random variable because it does not depend on a particular realization of the variable X, but it characterizes the probability distribution as a whole. From an experiment we can obtain information that is on average equivalent to the entropy H that measures the uncertainty prior to the execution of the experiment.

5.3 Properties of Information Entropy

The random entropy is positive because $p_i \leq 1$ and therefore $-\log p_i \geq 0$ for all i. This property is therefore also a characteristic for the average entropy. If associated with the i-th sample point is the probability $p_i = 0$ then the corresponding term in the expression Eq. 5.7 for the average entropy H becomes indeterminate and must be interpreted as a limit

$$\lim_{p_i \to 0} p_i \log p_i = 0 \ . \tag{5.8}$$

This result shows that the sample space $S = \{x_1, \ldots, x_i, \ldots, x_N\}$ can always be supplemented by an arbitrary set of sample points to which corresponding to zero probability which will not affect the entropy. If the examined phenomenon is deterministic, one of the probabilities p_i equals 1 while all the others are 0. In this case we obtain $H = 0$ which corresponds to minimum entropy. We can therefore conclude that the entropy associated with the phenomenon is not influenced by the observation of any deterministic variable that one associates with the phenomenon.

Let us respond to the question: "For which distribution is the average entropy maximal?" The average entropy characterizes the probability distribution which is the function defined on the sample space. Therefore this question defines a variational problem subject to the constraint

$$\psi = 1 - \sum_{i=1}^{N} p_i = 0 \tag{5.9}$$

the solution of which yields the optimal distribution. In order to proceed to the solution we first multiply the constraint by the Lagrange multiplier λ and form the composite information functional

$$\Psi = H + \lambda \psi \ . \tag{5.10}$$

Let us describe the corresponding optimal probability distribution by $\{p_{1,o}, \ldots, p_{i,o}, \ldots, p_{N,o}\}$ and let $\{\delta p_1, \ldots, \delta p_i, \ldots, \delta p_N\}$ denote its arbitrary variation. The variation of the composite information functional must equal zero for the case of an optimal distribution

$$\delta\Psi = \delta H + \lambda\, \delta\psi = -\sum_{i=1}^{N} \delta p_i (\log p_{i,o} + 1 - \lambda) = 0 \ . \tag{5.11}$$

This equation is fulfilled for an arbitrary variation δp_i if for any i

$$\log p_{i,o} + 1 - \lambda = 0 \ . \tag{5.12}$$

The solution is then

$$p_{i,o} = \exp(\lambda - 1) \ . \tag{5.13}$$

Using the constraint, we obtain the expression for the multiplier λ and for p_i

$$\exp(\lambda - 1) = \frac{1}{N} , \tag{5.14}$$

$$p_{i,o} = \frac{1}{N} \quad \text{for any } i . \tag{5.15}$$

The optimal solution thus corresponds to the uniform probability distribution. The corresponding optimal entropy of information

$$H_o = \log N \tag{5.16}$$

represents the maximal possible value of H for a random variable with a discrete sample space consisting of N sample points. The result that the optimum corresponds to an absolute maximum can be proved by considering the fact that the logarithm is a convex function. [3]

5.4 Relative Information

Let us assume that we are investigating a phenomenon which can be described by the sample space $S = \{x_1, \ldots, x_i, \ldots, x_N\}$ and let us *a priori* assign to all possible outcomes of experiment an equal uncertainty $H_{oi} = -\log p_{oi}$. The corresponding probability distribution is called *non-informative* because it is not related to the phenomenon but rather, it stems from the preparation of the experiment. Most often the uniform distribution with a maximal possible uncertainty is selected as the non-informative distribution. Let us further assume that the results of experiments reveal that the phenomenon is correctly described by the probability distribution $\{p_1, \ldots, p_i, \ldots, p_N\}$. When we ascertain that to the i-th outcome there corresponds the true uncertainty $H_i = -\log p_i$ instead of that which was assigned *a priori*, then we have obtained at the i-th point, information equal to the reduction of the corresponding uncertainty

$$\Delta I_i = -(H_i - H_{o,i}) \geq 0 . \tag{5.17}$$

On average, we obtain from the experiment the so-called *Kullback information* which is given by [2, 4]

$$\Delta I = \sum_{i=1}^{N} p_i \log p_i - \sum_{i=1}^{N} p_i \log p_{oi} = \sum_{i=1}^{N} p_i \log \frac{p_i}{p_{oi}} \geq 0 . \tag{5.18}$$

It is defined with respect to the *a priori* assumed distribution and it therefore describes the relative information of the true distribution with respect to the non-informative one. In the second term of the above expression we have utilized the true probability distribution determined by p_i because it determines the expected value of a random variable related to the phenomenon under consideration. The relative information is more fundamental for practical applications than the bare entropy of information. This is especially important when the concept of information is generalized to the case of continuous variables.

5.4.1 Information of Continuous Distributions

The measure of uncertainty is defined for discrete random variables and it cannot be converted into a corresponding measure for continuous variables without proper interpretation. [4, 5] If the interpretation of a continuous random variable is related to an increasingly dense partition of the sample space into smaller intervals, then one can expect that the number of intervals diverges and consequently the entropy of a continuous random variable should also diverge, at least as $\log N$. This is the first problem that must be solved by a proper interpretation of uncertainty. The next problem arises because the random properties of continuous variables are characterized by a probability density function, and therefore we cannot simply substitute the probabilities appropriate for the discrete case by a probability density function. This is already evident since the function $\log f$ cannot be defined, because the measurement unit of the probability density function f is $[1/x]$. However the relative information permits a generalization to the continuous case. For this, let us assume that the range of a continuous random variable X is divided into N disjoint intervals Δx_i centered at x_i and let us estimate the corresponding probabilities by $p_i \approx f(x_i)\Delta x_i$. We can then express the corresponding relative information of the true distribution with respect to the non-informative one by

$$\Delta I = \lim_{\substack{N\to\infty \\ \Delta x \to 0}} \sum_{i=1}^{N} f(x_i)\Delta x_i \log\left[\frac{f(x_i)\Delta x_i}{f_o(x_i)\Delta x_i}\right] \tag{5.19}$$

or

$$\Delta I = \int f(x) \log\left[\frac{f(x)}{f_o(x)}\right] dx = \Delta I_{ff_o} \tag{5.20}$$

with $\int f(x)\, dx = 1$ and $\int f_o(x)\, dx = 1$. In order to obtain a mathematically regular expression for ΔI the non-informative density f_o must differ from zero in those regions where $f(x) \neq 0$. If the true distribution equals the non-informative one: $f = f_o$, then we have $\Delta I = 0$ while for $f \neq f_o$ it is generally true that $\Delta I \geq 0$. The above expression shows that it is not possible to define the relative information with respect to a uniform distribution on an infinite interval, because it is ill-conditioned, and so, $f_o = 0$, because of the infinite span of the interval. However, the non-informative distribution is often approximately uniform $f_o(x) = \text{const}$ in the region in which the probability distribution $f(x)$ essentially differs from 0. In this case it is convenient to express the relative information in the form

$$\begin{aligned}\Delta I &= \int f(x)\,\log[f(x)]\,dx - \int f(x)\log[f_o(x)]dx \\ &\approx \int f(x)\log[f(x)]\,dx - \log[f_o]\;. \end{aligned} \tag{5.21}$$

The last term in this equation appears to be constant and it is therefore often omitted as an inessential quantity in the literature. This is however incorrect for the following reasons. First the last term provides for a proper compensation of the logarithm of the PDF $f(x)$ having the measuring unit $[1/x]$ in the first integral and thus making it possible to define an invariant of the information with respect to the selection of units. And second, the function f_o plays a role of reference for the description of information. If the last term is omitted, this reference is lost which has important consequences for the interpretation of information. For instance, in this case f_o must be much smaller than $f(x)$ which further implies $f(x)/f_o \gg 1$ and $\Delta I > 0$. Without the last term, this conclusion is not possible. However, if only changes of information resulting from variations of $f(x)$ are considered, then the last term is not essential.

5.4.2 Information Gain from Experiments

In relation with relative information we can present the following example which is of interest for empirical sciences. Let us suppose that a phenomenon is characterized by a random variable X and we infer that the corresponding probability density is $f_o(x)$. We can confirm our knowledge about the observed phenomenon by estimating the probability distribution empirically. Let us assume that experiments yield a different probability distribution $f(x)$. We can then describe the corresponding change or gain of our information about the phenomenon by the relative information ΔI_{ff_o}.

The concept of entropy and relative information are easily generalized to multivariate cases by introducing multivariate probability densities and integrals with respect to the multidimensional vector variables. In relation with such a generalization we mention here how one can quantitatively characterize the extent of the statistical dependency of the components of a two-dimensional random vector $\boldsymbol{Z} = (X, Y)$ with the joint probability density $f(x, y)$. For statistically independent components, the joint probability density is a product of the marginal densities $f(x, y) = f_x(x)\; f_y(y)$. Using the definition

$$\Delta I = \int\!\!\int f(x,y)\log\left[\frac{f(x,y)}{f_x(x)\; f_y(y)}\right]\; dx\; dy \tag{5.22}$$

we obtain a quantity that is zero for statistically independent components and greater than zero otherwise. It can be applied as a quantitative measure of dependence between x and y. More specifically, when this dependence proceeds towards a linear relationship, the above information diverges from 0 towards $+\infty$.

In order to present an another interesting interpretation of the relative information, let us determine it for a normal probability density

$$f(x) = \frac{1}{\sqrt{2\pi}\sigma} \exp\left[-\frac{(x-m)^2}{2\sigma^2}\right] \tag{5.23}$$

relative to the uniform distribution extending around m for several σ. In this case we can express the uniform non-informative density as $f_o = 1/L$ in the interval of width $L \gg \sigma$. By using this function in the expression of the relative information given by Eq. 5.20 and taking into account the definition of the variance we find

$$\Delta I \approx \log\left[\frac{L}{2\sigma}\right] - \frac{1}{2}\left[1 + \log(\frac{\pi}{2})\right] . \tag{5.24}$$

The last term is a small constant (≈ -0.73) that is related to the geometrical properties of the normal distribution and is not essential for the subsequent interpretation which is chiefly governed by the first term. The quotient in the first term describes how many times wider the width of the uniform distribution is than $2\,\sigma$, that is, the effective width of the normal distribution. The first term thus determines how much more uncertain the non-informative distribution is than the true distribution of the phenomenon under examination. It also indicates how much information we can obtain from a prepared experiment when studying the specific phenomenon being characterized by a normal distribution. We can conjecture that a similar interpretation can also be obtained with other distributions and therefore we conclude that above result can be utilized in the design of experiments.

The last example shows that the relative information increases with decreasing width of the normal distribution describing the phenomenon under consideration. Let us now take into account that the accuracy of the measuring instruments used in an experiment is limited. We describe the complete inaccuracy of the experiment by a probability distribution with a certain width σ_o, which describes the scattering of the data in measurements of a non-random value x_o. In this case, the expression

$$\Delta I \approx \log\left[\frac{L}{2\,\sigma_o}\right] - \frac{1}{2}\left[1 + \log\left(\frac{\pi}{2}\right)\right] \tag{5.25}$$

describes the maximal possible information that can be provided by the prepared experiment, independent of the phenomenon under examination. It is mainly determined by the dynamic span of the quotient $L/2\sigma_o$ that describes the total dynamical span of the experimental arrangement.

In a slightly different manner we can characterize the influence of the resolution of the measurement scale on the informational properties of the measurements. Let us assume that we are characterizing the random variable X by an instrument whose smallest scale division is described by the unit

u. The probability that an experiment yields a result in the i-th interval $(x_i, x_i + u]$ for a smooth PDF is given by

$$p_{i,u} = \int_{x_i}^{x_i+u} f(x)\,dx \approx f(x_i)\,u\ . \tag{5.26}$$

The measure of the uncertainty of variable X with respect to its measurement by the scale with a given scale division u is then defined by

$$\begin{aligned} H_u &= -\sum_i p_{i,u} \log p_{i,u} \\ &= -\sum_i f(x_i) \log[f(x_i)\,u]\,u \approx -\int f(x) \log\left[\frac{f(x)}{f_u}\right] dx \end{aligned} \tag{5.27}$$

where $f_u = 1/u$ indicates a constant that equals a uniform distribution inside an interval of small width u at x. This expression shows that $-H_u = \Delta I_u$ can be interpreted as the relative information about the probability distribution $f(x)$ with respect to the constant reference PDF f_u. If we compare the last expression with the corresponding one that was defined with respect to a non-informative distribution on a wide interval L, we establish the principal difference between the relations $f(x) \geq f_o$ and $f(x) \leq f_u$ that hold on average over the range of the variable X. Contrary to the previous inequality $\Delta I \geq 0$, we now have $\Delta I_u \leq 0$. With respect to these inequalities we can conclude that the non-informative distribution f_o describes the maximal possible indeterminacy, while f_u describes the minimal possible one. Therefore, when one of them is taken as a reference for the description of the indeterminacy of a certain variable X then we can observe either smaller or greater indeterminacy, respectively.

5.5 Information Measure of Distance Between Distributions

When comparing various phenomena, we often encounter the problem of quantitatively describing the difference between two probability densities f and f'. The most direct way is to use the *mean square difference* corresponding to the *Euclidean distance*

$$D(f, f') = \int_{-\infty}^{+\infty} [f(x) - f'(x)]^2 dx\ . \tag{5.28}$$

One problem that should be mentioned related to this definition is that it has a unit of measurement $[1/x]$ and consequently it is not invariant with respect to a change of units. In order to compare differences of distributions under various circumstances we need an invariant measure which can be obtained from the above definition by changing one factor of the square into a non-dimensional function as shown by the following definition [4]

$$\begin{aligned} D_i(f, f') &= \int_{-\infty}^{+\infty} [f(x) - f'(x)][\log f(x) - \log f'(x)]\, dx \\ &= \int_{-\infty}^{+\infty} [f(x) - f'(x)] \log \left[\frac{f(x)}{f'(x)}\right] dx \;. \end{aligned} \quad (5.29)$$

Because the logarithm is a monotonic function, the most characteristic property of the distance $D(f, f') \geq 0$ is preserved by this definition and $D_i(f, f') \geq 0$ as well. This quantity is not an exact measure of distance because it does not satisfy the triangle inequality that is normally stated with the definition of a distance while all the other characteristic properties of a distance or a metric defined in topology are preserved. In order to point out this difference it is better to call the distance the *divergence* of f and f'. It has many interesting properties for applications, the most outstanding, among these is its invariance with respect to a change of measuring units of x and its symmetry with respect to the position of f and f' or $D_i(f, f') = D_i(f', f)$. A comparison of the divergence with the relative information reveals that it is the sum of two relative informations: f with respect to f' and f' with respect to f:

$$D_i(f, f') = \Delta I_{ff'} + \Delta I_{f'f} \;. \quad (5.30)$$

An interesting expression is obtained from the divergence when the involved probability densities do not differ appreciably, that is for $f' = (1 + \epsilon)\, f$ with $\epsilon = (f' - f)/f \ll 1$ corresponding to a small relative difference between both densities at a given x. In this case, we obtain the approximation which is a version of *Fisher's information* measure [4]

$$D_i(f, f') \approx \int_{-\infty}^{+\infty} \frac{[f'(x) - f(x)]^2}{f(x)} dx = \int_{-\infty}^{+\infty} \epsilon^2\, f(x)\, dx \;. \quad (5.31)$$

This represents the divergence as an expected value of the square relative difference ϵ^2. If this formula is compared with that obtained from the mean square difference $D(f, f')$ we find

$$D(f, f') = \int_{-\infty}^{+\infty} \left[\frac{f'(x)}{f(x)} - 1\right]^2 f^2(x)\, dx = \int_{-\infty}^{+\infty} \epsilon^2\, f^2(x)\, dx \;. \quad (5.32)$$

We thus establish that the latter assigns a greater statistical weight to the deviations where $f(x)$ is large than to the deviations where $f(x)$ is small. This result shows that the mean square difference is in a sense contradictory with our intuition that generally permits greater differences when comparing large quantities than when comparing smaller ones.

Another, simply interpretable measure of distance between probability densities can be defined using the absolute value of the probability difference:

$$
\begin{aligned}
D_a(f,f') &= \int_{-\infty}^{+\infty} |dP'(x) - dP(x)| = \int_{-\infty}^{+\infty} |f'(x) - f(x)|\, dx \\
&= \int_{-\infty}^{+\infty} \left|\frac{f'(x) - f(x)}{f'(x)}\right| f'(x)\, dx = \int_{-\infty}^{+\infty} |\epsilon'|\, dP'(x) \\
&= \int_{-\infty}^{+\infty} \left|\frac{f(x) - f'(x)}{f(x)}\right| f(x)\, dx = \int_{-\infty}^{+\infty} |\epsilon|\, dP(x) \\
&= \int_{-\infty}^{+\infty} |\epsilon|\, f(x) dx\ . \qquad (5.33)
\end{aligned}
$$

The quantity D_a represents the mean value of the absolute relative difference between both probability densities and it satisfies the relation

$$
0 \le D_a \le 1\ . \qquad (5.34)
$$

When the probability densities are equal, then $D_a = 0$. While when the probability densities are not overlapping, the value of D_a is 1.

From an intuitive aspect it seems more appropriate to apply the divergence of the probability density functions or D_a rather than the mean square difference to characterize the distance between two probability density functions. But the mean square difference is more convenient to implement because the quadratic form leads to easily expressible and interpretable results in subsequent calculations.

6. Maximum Entropy Principles

6.1 Gibbs Maximum Entropy Principle

The concept of information can be successfully utilized for the adaptation of a probability distribution to empirical data. In order to proceed to the formulation of the corresponding principle, let us first recall the expression for the empirical probability density for the case when all the samples are distinct

$$f_e(x) = \frac{1}{N}\sum_{i=1}^{N} \delta(x - x_i) = \sum_{i=1}^{N} p_i \, \delta(x - x_i) \ . \tag{6.1}$$

The last sum in this equation shows that the same probability $p_i = 1/N$ can be ascribed to each of the experimentally measured sample values x_i. A uniform distribution corresponds to the maximum possible entropy of information. The assignment of the probability distribution to a discrete random variable when the sample space consists of experimentally measured data can be obtained by insisting that the corresponding entropy of information must be maximum. We shall now generalize this simple statement to the so-called *maximum entropy principle* in order to solve the following problem. [4, 5, 6, 8, 9, 16] Consider a discrete random variable X with the sample space $S = \{x_i, i = 1 \ldots N\}$ and let there be given some measured data in terms of the expected values of several functions g_k of the random variable X

$$G_k = \sum_{i=1}^{N} p_i \, g_k(x_i) = \sum_{i=1}^{N} p_i \, g_{ki} \ . \tag{6.2}$$

The problem is to assign the probability distribution $\{p_i\}$ to the random variable X in such a way that it yields the given average values G_k. For example, let there be given the mean value m and the variance of X. We want to find a probability distribution that yields the same values. If no other information is given about the associated phenomenon then we can only use the expressions for the first and second moments

$$m = \sum_{i=1}^{N} p_i \, x_i \qquad ; \qquad \mathrm{var}(X) = \sum_{i=1}^{N} p_i \, [x_i - m]^2 \tag{6.3}$$

to determine the probability distribution. If $N > 2$ this is not sufficient to determine the probabilities $\{p_i\}$ exactly as there can be arbitrarily many solutions of the problem. However, each solution has its own entropy of information associated with it and hence, requires for its specification, information. If there is no other information about a phenomenon than the given set of expected values, it seems reasonable to select from all the possible distributions the one which requires the least information. This is the *maximum entropy principle.* [4, 5, 8, 9, 6] The corresponding distribution is called 'least-biased' and must yield the maximum entropy of information. Of course, there is no guarantee that the maximum entropy principle will produce that distribution which, in fact, corresponds to the phenomenon under consideration. Rather, it simply includes our ignorance of the true characteristics of the phenomenon in the presentation of a possible distribution.

In a practical application of the maximum entropy principle, we look for the maximum of the information entropy

$$H = -\sum_{i=1}^{N} p_i \log p_i \tag{6.4}$$

subject to constraints

$$G_k = \sum_{i=1}^{N} p_i\, g_k(x_i) = \sum_{i=1}^{N} p_i\, g_{ki} \quad ;\; k = 1, \dots, K < N \;. \tag{6.5}$$

As a constraint with index 0 we also treat the normalization condition

$$G_0 = \sum_{i=1}^{N} p_i - 1 \equiv 0 \;. \tag{6.6}$$

We proceed towards the solution of the maximum entropy problem using the calculus of variations. For this purpose we first multiply the constraints by the Lagrange multipliers λ_k and form the functional

$$\mathcal{F} = -\sum_{i=1}^{N} p_i \log p_i + \sum_{k=0}^{K} \lambda_k \left(G_k - \sum_{i=1}^{N} p_i\, g_{ki} \right) \;. \tag{6.7}$$

In order to find its extremum, we set to zero the variation of this functional with respect to the probability p_i

$$\delta\mathcal{F} = -\sum_{i=1}^{N} \left(1 + \log p_i + \sum_{k=0}^{K} \lambda_k\, g_{ki} \right) \delta p_i = 0 \;. \tag{6.8}$$

If this condition is to be fulfilled for arbitrary variations δp_i, then the expression in parenthesis must equal zero for $i = 1, \dots, N$:

$$1 + \log p_i + \sum_{k=0}^{K} \lambda_k \, g_{ki} = 0 \; . \tag{6.9}$$

The optimal solution can be then written as

$$p_i = \exp\left[-\lambda - \sum_{k=1}^{K} \lambda_k \, g_{ki}\right] \; , \tag{6.10}$$

where we have introduced $\lambda = \lambda_o + 1$. By inserting this solution into the normalization condition we obtain the equation

$$1 = \sum_{i=1}^{N} \exp\left[-\lambda - \sum_{k=1}^{K} \lambda_k \, g_{ki}\right] \tag{6.11}$$

that can be transformed into the equation for λ

$$\lambda = \log Z \tag{6.12}$$

in which Z denotes the so-called partition function

$$Z = \sum_{i=1}^{N} \exp\left[-\sum_{k=1}^{K} \lambda_k \, g_{ki}\right] \; . \tag{6.13}$$

In order to determine λ we must first find all the multipliers, λ_k. Equations for these are obtained by inserting the expression for p_i into the constraints

$$G_k = e^{-\lambda} \sum_{i=1}^{N} g_{ki} \exp\left[\sum_{l=1}^{K} \lambda_l \, g_{li}\right] \quad ; \; k = 1, \ldots, K \; . \tag{6.14}$$

Comparing this expression with the previous one shows that they are of similar form which enables us to rewrite the last one as

$$G_k = -\frac{1}{Z} \frac{\partial}{\partial \lambda_k} \sum_{i=1}^{N} \exp\left[-\sum_{l=1}^{K} \lambda_l \, g_{li}\right] \tag{6.15}$$

or in an abbreviated form

$$G_k = -\frac{\partial \log Z}{\partial \lambda_k} \; . \tag{6.16}$$

This is a system of equations for $\{\lambda_k\}$. When these multipliers are known we can evaluate the partition function Z and finally the probabilities $\{p_i\}$. However, this procedure generally represents a tedious nonlinear problem because of the exponential function in the equations. In spite of this nonlinearity, the entropy of information can be readily expressed by using the equation for probabilities

$$H = -\sum_{i=1}^{N} p_i \log p_i = \sum_{i=1}^{N} p_i \left[\lambda + \sum_{k=1}^{K} \lambda_k \, g_{ki}\right] . \tag{6.17}$$

When using the expressions for the constraints this reduces to the simple form

$$H = \lambda + \sum_{k=1}^{K} \lambda_k G_k . \tag{6.18}$$

It is noteworthy that the Lagrange multipliers and consequently also the entropy of information are completely specified by the empirical data expressed by the quantities G_k which can be treated as the source of information.

The maximum entropy principle presented above is reminiscent of essentially the same principle introduced into statistical mechanics by Gibbs. [2, 4, 5, 8, 6] The only difference is in the interpretation of entropy which in the case of statistical mechanics is not related to information but to a statistically-defined quantity that can be interpreted as the physical entropy.

The maximum entropy principle can be generalized without difficulties to the case of continuous variables by merely substituting the entropy with a corresponding expression for the relative entropy of continuous variables. One instructive example of such a case is related with the above-mentioned problem of finding a distribution when the expected value m and variance σ^2 of a continuous variable are known. For the continuous variable we define the functional $\mathcal{F}$ by the expression

$$\mathcal{F} \equiv -\int f(x) \log\left[\frac{f(x)}{f_o(x)}\right] dx + \sum_{k=0}^{K} \lambda_k \left(G_k - \int g_k(x)\, f(x) dx\right) . \tag{6.19}$$

By utilizing a uniform reference function f_o in this equation we obtain

$$\mathcal{F} = -\int f(x) \log[f(x)] dx + \sum_{k=0}^{K} \lambda_k \left(G_k - \int g_k(x)\, f(x) dx\right) + \log[f_o] . \tag{6.20}$$

Here $\log[f_o]$ represents an essentially constant term which drops out during the process of variation. Therefore the results of the previous analysis can be directly applied provided that the sum over the sample space is replaced by the corresponding integral over x. The constraints in this case are

$$G_1 = \int x\, f(x)\, dx = m , \tag{6.21}$$

$$G_2 = \int x^2 f(x)\, dx = \sigma^2 + m^2 . \tag{6.22}$$

An expression for the probability density similar to the discrete case is now obtained from the variation of the functional given by Eq. (6.20)

$$f(x) = \exp\left[-\lambda - \sum_{k=1}^{K} \lambda_k\, g_k(x)\right] . \tag{6.23}$$

From this, we obtain the expression for the G_k of the continuous case:

$$G_k = e^{-\lambda} \int g_k(x) \exp\left[-\sum_{l=1}^{K} \lambda_l\, g_l(x)\right] dx \quad ; \; k = 1, \ldots, K \tag{6.24}$$

which yields the following system of equations

$$\int \exp[-\lambda - \lambda_1 x - \lambda_2 x^2)]\, dx = 1 , \tag{6.25}$$

$$\int x \exp[-\lambda - \lambda_1 x - \lambda_2 x^2)]\, dx = m , \tag{6.26}$$

$$\int x^2 \exp[-\lambda - \lambda_1 x - \lambda_2 x^2)]\, dx = \sigma^2 + m^2 . \tag{6.27}$$

The integrals appearing in these equations can be analytically evaluated which subsequently yields the solutions for the multipliers. Using these we find that the maximum-entropy solution is the normal distribution.

$$f(x) = \frac{1}{\sqrt{2\pi}\sigma} \exp\left[-\frac{(x-m)^2}{2\,\sigma^2}\right] . \tag{6.28}$$

Another procedure for obtaining this result is to first express the partition function analytically

$$Z = \sqrt{\frac{2\pi}{\lambda_2}} \exp\left[-\frac{\lambda_1^2}{4\,\lambda_2}\right] \tag{6.29}$$

and then following Eq. (6.16) obtain from its logarithm the multipliers by appropriate differentiation.

This example illustrates that we can obtain rather complicated integrals in the equations for the multipliers that cannot generally be expressed analytically. Unfortunately, this limits the possibility of expressing the solutions of the maximum-entropy approach in explicit analytical form.

In experimental work we often describe a measured variable by its mean value and its variance. The probability density of this variable is optimally described with respect to the maximum entropy principle, using a normal distribution.

6.2 The Absolute Maximum Entropy Principle

The assignment of the probability distribution to a random variable by utilizing some given data can be interpreted as the central problem in the empirical modeling of natural phenomena. Hence, it is not surprising that so many methods have been derived from the maximum entropy principle which, as stated above, offers an acceptable treatment of this problem. It is rather surprising that in spite of intensive research in this field, it was only recently that an essentially new version of the maximum entropy principle has emerged which is, in a sense, complementary to the above formulation. [3, 9] Because it leads to the interesting possibility for the automatic modeling of natural phenomena, it merits further description. We use it in the following chapter to obtain an explanation of a self-organized information processing scheme. In order to provide for its formulation, let us first discuss some problems related to the key concepts of the maximum entropy principle that stem from its applicability.

Let us assume that we wish to utilize a discrete sample space whose points satisfy the condition: $x_i > 0\,; i = 1, \ldots, N$ to obtain a description of a particular phenomenon. Further, let the measurements yield a negative first moment: $G_1 = \sum_{i=1}^{N} x_i\, p_i < 0$. It is evident that in this case we cannot find a set of positive values p_i which would reproduce the given experimental data when we are using the *a priori* specified sample space. In this example, the measured data and the sample space are incompatible and we can solve the problem only if we properly adapt the sample space to the given data. In other words, we cannot arbitrarily select the sample space to give a description of the phenomenon if we also wish to satisfy the given experimental conditions. When utilizing the maximum entropy principle, we usually avoid this problem by implicitly assuming that the experimental data are calculated using the sample space which has been selected *a priori* for describing the phenomenon. The question whether the selected sample space is applicable for the description of the phenomenon under consideration is thus excluded from the maximum entropy principle. However, if we wish to describe a phenomenon properly, we generally must utilize empirical information not only for determining the probabilities but also for specifying the sample space. This conclusion is not completely consistent with the formulation of the maximum entropy principle and has far-reaching consequences.

When formulating the maximum entropy principle, we have assumed that there is given a discrete sample space $S = \{x_i\,;\, i = 1, \ldots, N\}$ and some empirical data $\{G_k\,;\, k = 1, \ldots, K\}$ representing the expected values of a set of given reference functions. The problem is then to find that probability distribution which implies the empirical data without any further information about the phenomenon. It is for this reason that we have assumed that the entropy of information must be maximal, which leads us, along the standard route using the calculus of variations, to the solution of the problem. The resulting probability distribution is generally not uniform and consequently it generally does not correspond to the absolute maximum of information entropy but to a relative one. How then might we assign to the phenomenon under observation a probability distribution that generally corresponds to the absolute maximum of information entropy? This appears impossible since the entropy is obtained only in the case where the same probability $1/N$ is assigned to each of N sample points, that is, when the probability distribution is uniform over the sample space. The above formulation of the maximum entropy principle is logically closed and does not rely on the absolute but rather on the relative maximum. If we want to base our treatment on the absolute maximum, we must change the fundamental assumptions of the Principle. However, we cannot change the assumption about the given empirical data, because it represents the experimentally-determined facts. The only possibility is to remove the assumption about the given fixed sample space S and to accept only those sample points which can be adapted. At first, this sounds strange, but if the experimental data are stated only in terms of averages of some variables, then the sample space need not be specified *a priori*. We are, in fact, looking for that particular distribution of points in the sample space which will permit the assignment of a uniform probability distribution which, in turn, corresponds to the absolute maximum of information entropy. We expect that such a sample space optimally fits the description of the phenomenon under study. If we want to include in our formulation as little information as possible, we must use only the given experimental data and presume that the sample space, rather than the probability distribution itself, can be adapted to it. In the formulation we can either assume that the number of sample points is determined by the number K of empirical data or that this number is specified separately. Here we shall use the first assumption; in a later chapter when considering the optimization of this number, we shall change to the second case.

The assumptions stated above differ significantly from those utilized in our earlier formulation of the Gibbs maximum entropy principle. We can merge all these assumptions into a new formulation that is called the *second* or *absolute maximum entropy principle*. The statements of this principle are as follows:

1. *Let there be given a set of experimental data* $\{G_k \; ; \; k=1,\ldots,K\}$ *that can be treated as the expected values of a given set of reference functions* $\{g_k(x)\}$ *of a random variable* X.
2. *Let the sample space* S *of the random variable be comprised of* K *disjoint sample points and let there be assigned to each sample point the same probability* $1/K$.
3. *The distribution of sample points in the sample space* S *which yields the averages described by Statement 1 and which satisfies the condition of Statement 2 is said to be optimal with respect to the absolute maximum entropy principle.*

In order to determine such a distribution of sample points, the reference functions $g_k(x)$ must first be specified and then the system of equations

$$G_k = \frac{1}{K}\sum_{i=1}^{K} g_k(x_i) \quad ; \; k = 1,\ldots,K \tag{6.30}$$

must be solved for $x_i\,;\, i = 1...K$. This system is generally nonlinear and requires numerical procedures for its solution.

Let us show two illustrative examples which are obtained by specifying only the mean value m or the mean value and the variance σ^2. In the first case, we have only one sample point at the mean value: $x_1 = m$, while in the second example there are two points spaced by σ from the mean value: $x_1 = m - \sigma$ and $x_2 = m + \sigma$.

A similar, but more complex case arises when the K initial moments are specified

$$m_k = \frac{1}{K} \sum_{i=1}^{K} x_i^k \quad ; \; k = 1, \ldots, K \; . \tag{6.31}$$

In this case, an iterative, approximate treatment can be formulated by initially assuming an appropriate probability distribution, for example a uniform one with the mean value m and variance $m_2 - m^2$. Initial spacings are then successively changed: first, at the boundaries of the distribution in such a manner that the highest moments are adapted to the true values and then in the interior of the distribution by adapting moments of ever decreasing order as the center of the distribution is approached. One cycle of the iteration is accomplished when the lowest moment is corrected by the position of one or two central points. Although this iteration method is only approximate, it can quickly yield an approximate solution for small values of K. The following table lists the sample points that have been determined from the corresponding moments in ten initial cycles during such an adaptation. The index c in the first column denotes the central moment.

Table 6.1. Numerical example of moments and sample points

Given moments			Estimated points			Correct points		
m_1	=	-0.1	x_1	=	-1.9	x_1	=	-2.0
m_{2c}	=	2.4	x_2	=	-1.4	x_2	=	-1.5
m_{3c}	=	0.9	x_3	=	0.0	x_3	=	0.0
m_{4c}	=	10.1	x_4	=	0.8	x_4	=	0.7
m_{5c}	=	10.0	x_5	=	2.4	x_5	=	2.3

A slightly more complex procedure is obtained if we assume that the distribution corresponding to the given set of moments can be obtained from a certain initial distribution $\{x_{io}\}$ by small changes Δx_i of the initial points that have been estimated all at once. By using a linear approximation and denoting $\Delta m_k = m_k - m_{ko}$, we obtain a linear system for the changes:

$$\Delta m_k = \frac{k}{K} \sum_{i=1}^{K} x_i^{k-1} \Delta x_i \quad ; \; k = 1, \ldots, K \; . \tag{6.32}$$

In this case, the solution can be written explicitly by using standard techniques of linear algebra. We note that a solution similar to that above can be generalized without great difficulties to the multivariate case.

Application of a large number of moments is, however, somewhat problematic with respect to a numerical treatment of the problem. With increasing K, functions that are increasingly more sensitive to small changes of arguments are included into the procedures and this generally leads to numerical difficulties. To apply the absolute maximum entropy principle efficiently, attention must be paid to the selection of the reference functions. It is for this reason that we use Gaussian functions in our development of *self-organization* which is given in the following chapter. Such functions possess computationally more convenient properties than the moments. As in other applications of nonlinear equations, the problem of a proper numerical treatment is not avoided in applications of the absolute maximum entropy principle. Instructions for dealing with a particular example can be obtained from the field of numerical mathematics and will not be further discussed here.

6.3 Quantization of Continuous Probability Distributions

In the second statement of the absolute maximum entropy principle, we assumed that the random variable X was discrete. However, a phenomena may require application of a continuous variable for its description. In this case, the discrete sample space and the probability distribution defined on it can only approximately represent the properties of the random variable that is needed to describe the phenomenon under observation.

The representation of a continuous probability density function by a discrete one generally corresponds to a reduction of information. If the discrete variable corresponds to the absolute maximum of information entropy, then this reduction is minimal. Our intention is therefore to modify the absolute maximum entropy principle in order to use it for specification of a discrete probability distribution that, with respect to information loss, optimally represents a continuous variable under consideration. The corresponding procedure is called the *quantization of a continuous variable.* [10, 1, 13, 14, 15]

The presentation of a continuous probability density by a set of discrete data, as described above, can be interpreted as the inverse operation of adapting a continuous variable to a discrete probability density which was earlier described using Parzen's approach. From the information processing point of view, the first task represents a mapping of a continuous random variable onto a set of representative data which are stored as an internal picture in some information processing system that is comprised of discrete memory units. The inverse task corresponds to the recovery of a continuous distribution from the stored, discrete set of typical sample points. The optimization of these operations is of importance for the development of corresponding optimal devices intended for the transfer of data or communication and therefore the quantization of a variable deserves further elaboration.

In the initial statement of the absolute maximum entropy principle, we presumed the existence of a set of experimental data $\{G_k\ ;\ k = 1, \ldots, K\}$. In view of the above discussion, we can interpret these data as characteristics of the probability distribution of some continuous random variable X which we wish to describe using a representative discrete random variable further referred to as Q. Accordingly, the statements of the absolute maximum entropy principle can be generalized to provide a description of the quantization problem as follows:

1. *Let there be given a probability density function $f(x)$ of a continuous random variable X.*
2. *Let the sample space S_q of a representative random variable Q be comprised of K sample points $\{q_k\ ;\ k = 1, \ldots, K\}$ representing disjoint elementary events, and let there be assigned to each sample point the same probability $1/K$. The probability density of the variable Q is then expressed as*

$$\phi_q(x) = \frac{1}{K} \sum_{k=1}^{K} \delta(x - q_k)\ . \tag{6.33}$$

3. *Let there be specified a measure of discrepancy between both probability densities: $\mathcal{D}(f, \phi)$. The discrete random variable Q optimally represents the continuous variable X if this measure of discrepancy is minimized.*

In the above statements the absolute maximum of information entropy is assumed *a priori* while the positions of the representative sample points q_k are not specified. Their adaptation to the probability distribution of the variable X corresponds to the basic problem associated with representing a continuous random variable by a discrete one. This problem must be treated specifically with respect to the properties of the measure of discrepancy between both distributions.

6.3.1 Quadratic Measure of Discrepancy Between Distributions

Specification of the discrepancy measure represents a critical step of the quantization procedure and consequently it deserves special attention. At first, it appears to be easiest to follow the examples treated in the previous section and to calculate from the given probability function $f(x)$, a set of moments $\{m_{k,x}\ ;\ k = 1, \ldots, K\}$. These can be then equalized with the moments of the discrete variable: $\{m_{k,q}\ ;\ k = 1, \ldots, K\}$. By equalization the discrepancy between both kinds of moments is set to zero. This corresponds to the adaptation of the discrete variable to the continuous one. Using the given moments, the prototype points q_k are then determined as described above. But the powers and corresponding statistical moments are not those most appropriate for numerical work. Our next task, therefore, is to find more suitable reference functions to describe the discrepancy measure. With this as our goal, we first turn to the properties of the probability density of a discrete variable which is specified in Eq. (6.33). Because of the singularity of this function, we cannot directly compare it with the continuous probability density $f(x)$. To overcome this, we introduce a regularization procedure. In order to provide for our subsequent treatment of problems in which the function $f(x)$ may also include delta-type singularities, we first transform both functions by convolving them with some non-singular function $g(s-x)$ to obtain the following averages

$$g * f = \int g(s-x)\, f(x)\, dx = \mathrm{E}\,[g] = G(s) \ , \tag{6.34}$$

$$\begin{aligned} g * \phi &= \int g(s-x)\, \phi(x)\, dx = \mathrm{E}\,[g]_r = G_r(s) \\ &= \frac{1}{K}\sum_{k=1}^{K} g(s-q_k) \ . \end{aligned} \tag{6.35}$$

Both of these averages depend on the parameter s. If this parameter is fixed then the last two expressions represent two functionals which characterize the corresponding probability density functions. By minimizing the square of their difference, we can obtain a tool for comparing the corresponding distributions. However, the parameter s can be more generally interpreted as a variable. In that case, the above averages correspond to two functions of s and we therefore utilize as a commonly used measure of their discrepancy, the mean square distance

$$\mathcal{D}(f,\phi) = \int [G(s) - G_r(s)]^2 ds \ . \tag{6.36}$$

If this integral does not exist, then a weight $\eta(s)$ must be introduced in order to improve the convergence. This weight is here arbitrarily set to 1.

The quantity $\mathcal{D}(f,\phi)$, in fact, represents a measure of the discrepancy between both probability distribution functions and it generally depends on the position of the sample points in the sample space S_q. The fundamental system of equations for the set $\{q_k \ ; \ k = 1, \ldots K\}$ is obtained by minimizing the discrepancy $\mathcal{D}(f,\phi)$ as a function of q_k:

$$\frac{\partial \mathcal{D}}{\partial q_k} = -2\int [G(s) - G_r(s)]\, \frac{\partial G_r(s)}{\partial q_k}\, ds = 0 \quad ; \ k = 1, \ldots, K \tag{6.37}$$

or

$$\int [G(s) - G_r(s)]\, \frac{\partial g(s-q_k)}{\partial q_k}\, ds = 0 \quad ; \ k = 1, \ldots, K \ . \tag{6.38}$$

A similar system of equations is also obtained for the multivariate case in which s denotes a vector and ds, the corresponding differential of volume. In order to proceed to the solution of this system, the reference function $g(s-x)$ must first be specified.

To demonstrate the applicability of the proposed method, let us describe an illustrative example in which we select for the reference function the unit step function $g(s-x) = U(s-x)$. Its average yields the cumulative probability distribution

$$F(s) = \int U(s-x)\, f(x)\, dx = \int_{-\infty}^{s} f(x)\, dx\ , \tag{6.39}$$

$$\Phi_q(s) = \frac{1}{K} \sum_{k=1}^{K} U(s - q_k)\ . \tag{6.40}$$

The discrepancy between both distributions is then defined by the mean square difference between F and Φ_q, or

$$\mathcal{D} = \int [F(s) - \Phi_q(s)]^2\, ds \tag{6.41}$$

which after differentiation with respect to the sample point values yields the system of equations

$$\int [F(s) - \Phi_q(s)]\, \delta(s - q_k)\, ds = 0\ . \tag{6.42}$$

We further define the value at the point of discontinuity of the unit step function by $U(0) = 1/2$ and sort the sample points in increasing order: $q_1 < \ldots < q_k < \ldots < q_K$. Equation (6.42) is then transformed into

$$F(q_k) = \Phi_q(q_k) = \frac{k - \frac{1}{2}}{K}\quad ;\ k = 1, \ldots, K\ . \tag{6.43}$$

This equation is well-known in applied probability where it is used for the determination of fractiles, such as for example, the median. [11] It is interesting that uncoupled equations for sample point values emerge in this case. This advantageous property is a consequence of a fortuitously selected reference function and not a general characteristic of the fundamental system of equations.

Equation (6.43) is generally non-linear and it can be most easily solved when the graph of $F(x)$ is known. The solution q_k is determined by the intersection of a horizontal line at height $(k-1/2)/K$, as is shown in Fig. 6.1. For a number of characteristic distributions, such as a normal one, the solutions can also be obtained by using tables or standard numerical techniques. If the function $F(x)$ is constant on certain intervals of the range of the variable X and it coincides there with Φ, then the solution of Eq. (6.43) can be

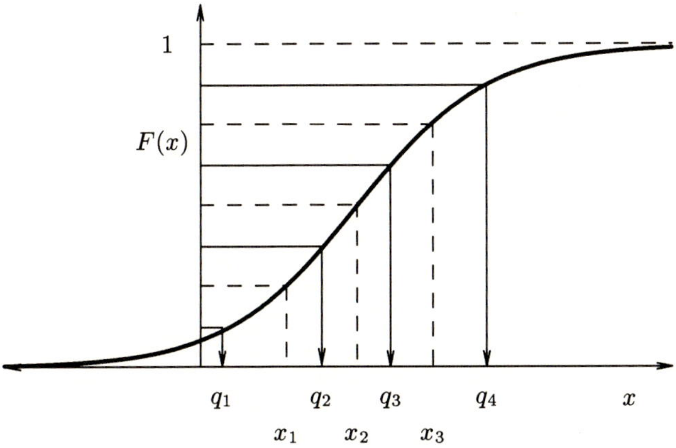

Fig. 6.1. Graphical determination of prototype sample points. Dashed lines denote partition of probability range to equal subintervals

represented also by intervals and the problem need not be solved in terms of discrete set of representative points q_k.

In relation to Eq. (6.43) we can consider the equation

$$F(x_k) = \frac{k}{K} \quad ; \; k = 1, \ldots, K \tag{6.44}$$

by which a point x_k is determined between two successive sample points q_k and q_{k+1}. We can use these points to partition the range of the continuous random variable X into K disjointed, continuous intervals, given by $A_1 = (x_o = -\infty, x_1], \ldots, A_k = (x_{k-1}, x_k], \ldots, A_K = (x_{K-1}, x_K = +\infty]$. To each interval there is corresponding a probability $1/K$ which is expressed as

$$P[A_k] = \int_{x_{k-1}}^{x_k} f(x)\,dx = \frac{1}{K} = F(x_k) - F(x_{k-1}) \tag{6.45}$$

where the sample point q_k is situated inside the interval. The widths of the intervals $\Delta x_k = (x_k - x_{k-1})$ are generally finite, except at the boundaries of the distribution where an effective width is usually limited. If the function $F(x)$ is smooth, the sample points lie approximately at the mean value or at the center of the interval A_k and therefore the following equation is approximately satisfied

$$q_k \approx K \int_{x_{k-1}}^{x_k} x\, f(x)\,dx \; . \tag{6.46}$$

By assigning to each sample point a probability of the corresponding interval $1/K$ we define a random variable Q to which there corresponds a discontinuous cumulative distribution function $\Phi_k(x)$ with a singular density ϕ_k. For a continuous random variable X, the distribution function $F(x)$, and consequently also its density $f(x)$, are generally smooth. Thus, rather than using the singular probability density ϕ_q, it is more appropriate to use a histogram or an another approximation to represent the probability density $f(x)$. In the case presented here, the histogram is most easily defined. For a smooth distribution, the probability density inside each interval, except those at the edges of the distribution, can be treated as approximately constant f_k and we obtain

$$f_k \int_{x_{k-1}}^{x_k} dx = f_k(x_k - x_{k-1}) = f_k \Delta x_k \approx \frac{1}{K} , \tag{6.47}$$

$$q_k \approx K\, f_k \int_{x_{k-1}}^{x_k} x\, dx = K\, f_k(x_k^2 - x_{k-1}^2)/2 \approx (x_k + x_{k-1})/2 . \tag{6.48}$$

Except at the outermost intervals, the last equation implies that

$$\Delta q_k = (q_k - q_{k-1}) \approx (x_k - x_{k-2})/2 \approx \Delta x_k \tag{6.49}$$

and

$$f_k \approx \frac{1}{K\, \Delta q_k} . \tag{6.50}$$

A smooth probability density function can then be represented by

$$f(x) \approx \sum_{k=1}^{K} f_k I(A_k) = \frac{1}{K} \sum_{k=1}^{K} w(x - q_k, \Delta x_k) . \tag{6.51}$$

Here $I(A_k)$ is the indicator function of the interval A_k that covers the sample point q_k and the window function $w(x - q_k, \Delta x_k)$ is a histogram approximation of the delta function

$$w(x - q_k, \Delta x_k) = \frac{I(A_k)}{\Delta x_k} . \tag{6.52}$$

This result is essentially identical to the Parzen filtering of empirical probability densities that was described in Chap. 4.

This treatment could be generalized to multivariate variables, but unfortunately, instead of obtaining an equation analogous to Eq. (6.43), divergent integrals appear if a unit step function is used as a reference in the definition of the discrepancy. If the divergence is improved by a proper weight function, such as $\eta(x) = f(x)$, then there will still result a set of complicated coupled equations for the sample point values which cannot be simply interpreted. This indicates that the unit step function is not an appropriate reference function for comparing multivariate distributions.

Based on the result of the last example, that is expressed by Eq. (6.51), we conclude that from the outset, one should use a function representing a smooth approximation of the delta function, such as the Gaussian

$$g(s-x;\sigma)=\frac{1}{\sqrt{2\pi}\sigma}\exp\left[-\frac{(s-x)^2}{2\,\sigma^2}\right] . \tag{6.53}$$

As explained in Chap. 4 on filtering of empirical distributions, we can properly represent the continuous probability density by a filtered, discrete one, provided we select the width as $\sigma \approx \sigma_x/\sqrt{K}$. The filtered densities are then given by

$$G(s)=\int g(s-x;\sigma)\,f(x)\,dx\;, \tag{6.54}$$

$$G_r(s)=\frac{1}{K}\sum_{k=1}^{K} g(s-q_k;\sigma)\;. \tag{6.55}$$

These are expected to be smooth functions with converging integrals. The discrepancy between both distributions can then be defined by the mean square difference

$$\mathcal{D}=\int [G(s)-G_r(s)]^2\,ds \tag{6.56}$$

which yields the fundamental system of equations

$$\int [G(s)-G_r(s)]\left[\frac{\partial g(s-q_k;\sigma)}{\partial q_k}\right]\,ds=0 \tag{6.57}$$

or

$$\int\int g(s-x;\sigma)\left[\frac{\partial g(s-q_k;\sigma)}{\partial q_k}\right]\,ds\,f(x)\,dx$$
$$-\frac{1}{K}\sum_{i=1}^{K}\int g(s-q_i;\sigma)\left[\frac{\partial g(s-q_k;\sigma)}{\partial q_k}\right]\,ds\;=0\;. \tag{6.58}$$

An advantage of this system of equations is that the integrals with respect to s can be carried out analytically and that convergence is assured because of the convenient properties of Gaussian function. After utilizing the expression for the convolution of a Gaussian function given by

$$\int g(s-x;\sigma)\,g(s-q_k;\sigma)\,ds=g(x-q_k;\sqrt{2}\sigma) \tag{6.59}$$

we obtain for $k\in\{1,\ldots,K\}$

$$\int (x-q_k)g(x-q_k;\sqrt{2}\sigma)f(x)dx-\frac{1}{K}\sum_{i=1}^{K}(q_i-q_k)g(q_i-q_k;\sqrt{2}\sigma)=0\,, \tag{6.60}$$

and

$$q_k\,f_q(q_k)=\frac{1}{K}\sum_{i=1}^{K}q_i\,g(q_i-q_k;\sqrt{2}\sigma)-\int (x-q_k)g(x-q_k;\sqrt{2}\sigma)f(x)dx\;, \tag{6.61}$$

where

$$f_q(q_k) = \frac{1}{K}\sum_{i=1}^{K} g(q_i - q_k; \sqrt{2}\sigma) \tag{6.62}$$

is the estimated density function at point q_k. Although the equations of the basic system Eq. (6.58) are non-linear and coupled, the Gaussian function in them permits an efficient iterative approach to find an approximate solution. However, for this the probability density function $f(x)$ must first be specified. The fact, that the equations are coupled, indicates that a position of an individual sample point is not determined by the probability density function $f(x)$ alone, but it also depends on the position of all the other sample points. This characteristic property will be further utilized in developing a formulation of a self-organized information processing system that is given in the next chapter.

The basic system of equations, given by Eq. (6.58) can easily be generalized to the multivariate case by replacing the sample values by the corresponding vectors and replacing the Gaussian functions with the multivariate normal distributions.

6.3.2 Information Divergence as a Measure of Discrepancy

Relative to the previous section, one may ask two questions. The first is related to the selection of the width of the reference function which appears as an artificially introduced parameter. One might argue that a more general treatment would result if the width is determined by minimizing the discrepancy. The second question is related to the description of the discrepancy between the two filtered probability densities using the mean square distance. Although the mean square distance leads to a very simple treatment, a slightly more complicated information measure might be a more appropriate basis for the criterion used in the information processing. We therefore describe in the following section a modified procedure which avoids these questions.

Let us consider again Eq. (6.51) which shows that the solution of the problem originating from the generalized absolute maximum entropy principle can be related to Parzen's window representation of a singular probability density of a discrete variable. The task is more complex here because neither the widow functions, nor their centers or widths are specified. According to the above derivations as well as the earlier treatment of the nearest-neighbor approach, we conjecture that the distances between neighboring centers determine the widths of the window functions. We therefore expect that the problem can be essentially reduced to a determination of the window centers. This generally corresponds to an inverse procedure for estimating a smooth distribution function from a set of discrete data. An additional problem is a proper selection of window functions which emerge as a by-product in the treatment of cumulative probability distribution functions. Quite intuitively, the corresponding discontinuous histogram approximation of the window function seems to be generally less convenient for representing continuous distributions than a smooth one such as, for example, a Gaussian one. We shall use this in the subsequent treatment. With all this in mind, we now propose a slightly modified quantization procedure which is based on the absolute maximum entropy principle.

Let us first select a continuous window function $g(x-q,h)$ in which q and h denote the window center and its width, respectively. Then, associated with the discrete variable Q is a continuous, representative probability density

$$f_q(x) = \frac{1}{K}\sum_{k=1}^{K} g(x-q_k, h_k) \ . \tag{6.63}$$

With this representation, the initial step of the regularization of the singular probability density is performed. In fact, this corresponds to a parametrization of the distribution function. In order to permit the adaptation of the discrete variable Q to the continuous distribution of the variable X, we now describe the discrepancy between both variables by the divergence between $f_x(x)$ and $f_q(x)$:

$$\mathcal{D}_i(f_x, f_q) = \int_{-\infty}^{+\infty} [f_x(x) - f_q(x)] \left[\log \frac{f_x(x)}{f_q(x)}\right] dx \ . \tag{6.64}$$

Although this is not simplest measure of the discrepancy [7, 12, 17, 18] it should be noted that the application of the simpler, relative information does not lead to an essentially different derivation in the subsequent treatment. The variable Q is then said to represent the variable X optimally if the centers q_k and the widths h_k are taken such that the measure of the discrepancy $\mathcal{D}_i(f_x, f_q)$ is minimized. This statement leads to the following system of equations for q_k and h_k

$$\frac{\partial \mathcal{D}_i}{\partial q_k} = -\frac{1}{K}\int \left\{\log\left[\frac{f_x(x)}{f_q(x)}\right] + \frac{f_x(x)-f_q(x)}{f_q(x)}\right\}\left[\frac{\partial g(x-q_k,h_k)}{\partial q_k}\right] dx = 0 \ , \tag{6.65}$$

$$\frac{\partial \mathcal{D}_i}{\partial h_k} = -\frac{1}{K}\int \left\{\log\left[\frac{f_x(x)}{f_q(x)}\right] + \frac{f_x(x)-f_q(x)}{f_q(x)}\right\}\left[\frac{\partial g(x-q_k,h_k)}{\partial h_k}\right] dx = 0 \ . \tag{6.66}$$

This system is generally nonlinear and we can obtain its solution only if the functions appearing in these equations are first specified. However, a simplification results if some assumptions are made. We can generally assume that the process of adaptation of the representative variable Q to the observed one will reduce the difference between both of the distributions. As a consequence, we write $f_x(x) = f_q(x) + \Delta f(x)$ and assume that $|\Delta f(x)/f_q(x)| \ll 1$.

This further implies $\log[f_x(x)/f_q(x)] \approx \Delta f(x)/f_q(x)$. We then obtain from Eqs. (6.65) and (6.66) the following system of equations

$$\int \left[\frac{f_x(x) - f_q(x)}{f_q(x)}\right] \left[\frac{\partial g(x - q_k, h_k)}{\partial q_k}\right] dx = 0;\; k = 1, \ldots, K \;, \tag{6.67}$$

$$\int \left[\frac{f_x(x) - f_q(x)}{f_q(x)}\right] \left[\frac{\partial g(x - q_k, h_k)}{\partial h_k}\right] dx = 0;\; k = 1, \ldots, K \;. \tag{6.68}$$

It is interesting to note that if we describe the discrepancy by the mean square difference, then the function $f_q(x)$ in the denominator is absent, and thus only the difference rather than the relative difference appears in the equations. This system of equations can be further simplified if the normalization condition and its derivatives are taken into account. That is

$$\begin{aligned} &\int g(x - q, h)\, dx = 1 \;, \\ &\int \frac{\partial g(x - q, h)}{\partial q}\, dx = 0 \;, \\ &\int \frac{\partial g(x - q, h)}{\partial h}\, dx = 0 \;, \end{aligned} \tag{6.69}$$

$$\int \left[\frac{f_x(x)}{f_q(x)}\right] \left[\frac{\partial g(x - q_k, h_k)}{\partial q_k}\right] dx = 0;\; k = 1, \ldots, K \;, \tag{6.70}$$

$$\int \left[\frac{f_x(x)}{f_q(x)}\right] \left[\frac{\partial g(x - q_k, h_k)}{\partial h_k}\right] dx = 0;\; k = 1, \ldots, K \;. \tag{6.71}$$

The same system is also obtained without any approximation if the relative information of $f_x(x)$ with respect to $f_q(x)$ is used to describe the discrepancy between the distributions f_x and f_q. The two equations obtained by differentiating the discrepancy with respect to q and h for each k depend on f_q, which is expressed by the sum of all window functions. This means that the equations obtained for various k are coupled and consequently it is not possible to analyze the general behavior of the solutions.

An interesting comparison of the method presented in this section with the previously mentioned histogram approach can be made if we apply the Gaussian window function. By taking into account that the probability in this case is effectively concentrated in the interval $q \pm h$ and that $\partial g/\partial q \propto (x - q)\, g(x)$ we obtain from the first equation

$$\int \frac{f_x(x)}{f_q(x)} (x - q_k)\, g(x - q_k, h_k)\, dx = 0 \;. \tag{6.72}$$

We presume that in the neighborhood of the k-th center, the representative function is mainly determined by the k-th window function. Therefore $g_k \propto f_q$ and we obtain approximately

$$\int_{q_k \pm h_k} f_x(x)\,(x - q_k)dx \approx 0 \tag{6.73}$$

which corresponds to Eq. (6.46). Unfortunately this equation is insufficient to determine q_k because we still do not know h_k, but it nevertheless shows that similar results can be obtained if in the expression of the discrepancy between distributions, we utilize the information divergence rather than the mean square distance.

It is, however, instructive to point out the difference between the two cases in which the equations are obtained from a mean square difference or the information measure of discrepancy. In the first case, the filtered probability density $G(s) = g(s - x) * f(x)$ is used while in the second case $f(x)$ is used directly. In the first case, there is only one type of equation that determines the positions of the sample points.

$$\int [G(s) - G_r(s)] \left[\frac{\partial g(s - q_k; \sigma)}{\partial q_k}\right] ds = 0\;. \tag{6.74}$$

The corresponding equations of the second case are given by

$$\int \left[\frac{f_x(x) - f_q(x)}{f_q(x)}\right] \left[\frac{\partial g(x - q_k, h_k)}{\partial q_k}\right] dx = 0\;;\; k = 1, \ldots, K\;. \tag{6.75}$$

The equations corresponding to Eq. (6.75) differ from the previous equation principally in the relative expression for the difference between the probability densities. These differences in the properties generally do not significantly influence the properties of the solutions, provided that the window width is kept fixed. However, an essential difference between both these cases is the result of the adaptable window widths in the second case. Because of this property, the representative PDF of the second case is more adaptable to the true density and, consequently, in this case the positions of the sample points can be quite different from those in the first case. For example, one question that cannot be answered is whether the solutions of Eqs. (6.74) and (6.75) differ. This is equivalent to the question whether the window functions for various k approximately represent disjoint events. For instance, if the function $f_x(x)$ is normal with its center at m and with standard deviation σ_x then we could expect that the solution of the above equations is trivial $q_1 = \ldots = q_k = \ldots = q_K\;;\; h_1 = \ldots = h_k = \ldots = h_K = \sigma_x$. This, however, does not correspond to K different sample points and the interpretation of q_k in terms of the absolute maximum entropy principle is doubtful. At present, the only way of avoiding this dilemma is to find numerically the solution for a given $f_x(x)$ and then to interpret its properties. Related to this problem are some additional difficulties with the interpretation of the filtering process. In the first case, we used the average $(g * f)$ to define the given data which should be described as accurately as possible by the corresponding average $(g * \phi)$ of the discrete variable. In contrast, in the second case, the representative PDF $f_q(x)$ corresponds to a parametric representation of the true density $f(x)$, which itself is treated as a given datum related to the phenomenon. The statements of the absolute maximum entropy principle which were initially very clear become less applicable for interpreting the second case. It should therefore be treated more than a parametric method of the probability density description. In view of the above-mentioned properties, the application of the mean square difference of densities appears to be simpler for the interpretation of results and in applications and it will therefore principally be used henceforth in this monograph.

6.3.3 Vector Quantization and Reconstruction Measure of Discrepancy

In the third statement of the quantization problem we have left open the question of how one specifies the measure of discrepancy between a continuous and a discrete random variable. This is a weak point in the formulation and is reminiscent of the arbitrariness of the selection of empirical data in terms of averages of arbitrarily selected reference functions in the formulation of the maximum entropy principle of Gibbs. Because of this, the method based on the maximum entropy principle depends on subjective arguments for selection of the reference functions. [16] We try to avoid subjective arguments by utilizing a concept of reproduction which is defined in the method of vector quantization that has been developed in the field of optimal coding. The method is similar to the quantization that originated from the absolute maximum entropy principle with one exception. In the vector quantization method, the maximum absolute entropy of a representative variable is not assumed *a priori*. Its fundamental concepts are briefly explained in the following paragraphs. [13, 1, 14]

Suppose that there is given a vector random variable $\boldsymbol{X}$ which is mapped to another variable $\boldsymbol{Q}$ possessing a discrete sample space $S_y = \{\boldsymbol{q}_1, \ldots, \boldsymbol{q}_K\}$ which is usually called the *reconstruction codebook.* Any vector quantization procedure is associated with a partition of the sample space S_X into K disjoint regions or cells $\{C_k,\ 1 \le k \le K\}$. This partition can be described by the cell indicator function

$$I_k(\boldsymbol{x}) = \begin{cases} 1, & ;\ \boldsymbol{x} \in C_k \\ 0, & ;\ \boldsymbol{x} \notin C_k \end{cases} . \tag{6.76}$$

The mapping from the variable $\boldsymbol{x}$ to the reference or code vectors $\boldsymbol{q}_k$ is generally described by a quantization function $\boldsymbol{q}(\boldsymbol{x}) = \boldsymbol{q}_k$ which yields the k-th code vector when $\boldsymbol{x} \in C_k$:

$$\boldsymbol{q}(\boldsymbol{x}) = \boldsymbol{q}_k . \tag{6.77}$$

Related to each vector quantization procedure is a reconstruction function

$$\boldsymbol{y}(\boldsymbol{x}) = \sum_{k=1}^{K} \boldsymbol{q}_k I_k(\boldsymbol{x}) . \tag{6.78}$$

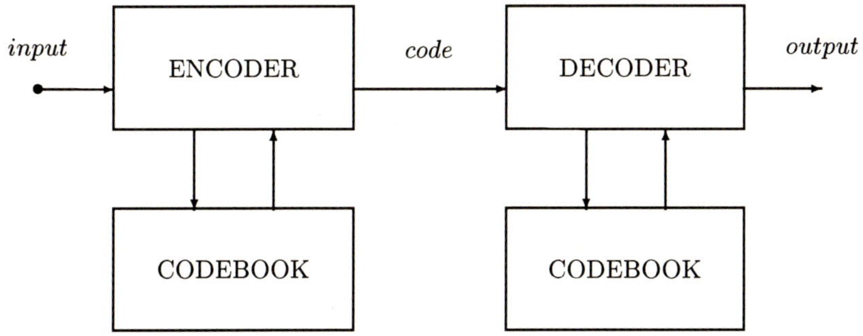

Fig. 6.2. Scheme of a quantizer

The last two functions indicate that a quantization procedure can be interpreted as the operation of an information processing system which is comprised of two main building blocks called the encoder and the decoder which are connected by K lines. The complete system is called a *quantizer*. The input variable X is transformed in the encoder into K responses, with only one having value 1 while all the others equal 0. These responses are then transferred over K-lines of the information transmission channel to the decoder where the variable $\boldsymbol{y}$ is reconstructed from the given responses and the reference or code vectors and presented as the output of the quantizer. In view of this interpretation, the applicability of the quantization procedure in communication systems is self-explanatory. The main purpose of the quantizer is therefore to compress the information presented by a continuous random variable into K representative code vectors and, after transmission, to reproduce it again for the needs of the user. It is, therefore, quite natural to look for such a procedure that ensures a good reproduction or one that minimizes the discrepancy between the input variable and the reproduced one. Although various measures of discrepancy have been utilized in practice, the most natural seems to be the expected square of the Euclidean distance

$$\mathcal{D} = \int ||\boldsymbol{y}(\boldsymbol{x}) - \boldsymbol{x}||^2 f(\boldsymbol{x})\, d^M x \; . \tag{6.79}$$

We define that system which minimizes this discrepancy between the input and the reproduction as an *optimal quantizer*. The last formula shows that this optimality is task-dependent because the criterion depends on the properties of the input variable which is described by the probability density function and the corresponding partition of the sample space S_x into the cells of quantizer.

Let us now assume that the reference vectors are not given and we want to find that set which minimizes the above discrepancy measure. By taking into account that the cells of the partition represent disjoint events and that they all together represent the complete sample space we can rewrite the previous equation in the form

$$\mathcal{D} = \int || \sum_{k=1}^{K} (\boldsymbol{q}_k - \boldsymbol{x})\, I_k(\boldsymbol{x}) ||^2 f(\boldsymbol{x})\, d^M x \ . \tag{6.80}$$

For the optimal set of reference vectors, the gradient of $\mathcal{D}$ with respect to each reference vector must be zero which yields the result

$$\int \left[\sum_{i=1}^{K} (\boldsymbol{q}_i - \boldsymbol{x})\, I_i(\boldsymbol{x}) \right] I_k(\boldsymbol{x})\, f(\boldsymbol{x})\, d^M x = 0 \ , \tag{6.81}$$

$$\boldsymbol{q}_k = \frac{1}{p_k} \int \boldsymbol{x}\, I_k(\boldsymbol{x})\, f(\boldsymbol{x})\, d^M x = \mathrm{E}\left[\boldsymbol{x} | \boldsymbol{x} \in C_k\right] \ , \tag{6.82}$$

$$p_k = \int I_k(\boldsymbol{x})\, f(\boldsymbol{x})\, d^M x \ . \tag{6.83}$$

The optimal reference vectors are thus determined by the conditional averages of the input vector variable over the cells of the partition. However, in order to determine them, the partition must first be specified. In the literature on quantization procedures, the form of the cells is usually not fixed but is chosen by the designer of the quantizer. However, one can find a great deal of advice and instructions how to form the cell shapes in order to obtain an efficient transmission of information. [13, 14]

If we compare the description of the vector quantization procedure given above and the quantization based on the absolute maximum entropy principle described earlier, in particular the one-dimensional case with comparison of cumulative functions, we can conclude that both procedures are similar in some characteristic features. However, in the absolute maximum entropy approach, one assumes a uniform probability distribution over the representative points, which in the quantization procedure can be only obtained by a proper partition. This assumption seems advantageous, so we wish to retain it in our subsequent analysis. By contrast, in the quantization procedure, the *a priori* selection of the reconstruction function and the corresponding measure of the discrepancy is very advantageous. In order to preserve both properties we now propose a unified problem of the absolute maximum entropy principle and the quantization procedure by joining the following statements.

1. *Let there be given a probability density function $f_x(x)$ of a random variable X.*
2. *Let $\{z_k(\boldsymbol{x});\ k = 1, \ldots, K\}$ denote a set of positive reference functions satisfying the conditions*

$$p_k = \int z_k(\boldsymbol{x})\, f(\boldsymbol{x})\, d^M x = \frac{1}{K} \ ; \ k = 1, \ldots, K \tag{6.84}$$

and let it be possible to define a centroid of each function by the equation

$$\boldsymbol{q}_k = \frac{1}{p_k} \int \boldsymbol{x}\, z_k(\boldsymbol{x})\, f(\boldsymbol{x})\, d^M x < \infty \;. \tag{6.85}$$

It is then possible to define the reconstruction function by the expression

$$\boldsymbol{y}(\boldsymbol{x}) = \sum_{k=1}^{K} \boldsymbol{q}_k\, z_k(\boldsymbol{x}) \;. \tag{6.86}$$

3. *The problem is to find the set of reference functions which satisfy the conditions expressed by Eq. (6.84) and which minimize the mean Euclidean distance between the variable* $\boldsymbol{x}$ *and its reconstruction* $\boldsymbol{y}$

$$\mathcal{D} = \int ||\boldsymbol{y}(\boldsymbol{x}) - \boldsymbol{x}||^2 f(\boldsymbol{x})\, d^M x \;. \tag{6.87}$$

We have thus represented the quantization based upon the absolute maximum entropy principle as an optimization problem with constraints. To proceed towards its solution, we first multiply the conditions specified by Eqs. (6.81) by Lagrange multipliers λ_k and form the Lagrange functional

$$\mathcal{F} = \int ||\boldsymbol{y}(\boldsymbol{x}) - \boldsymbol{x}||^2 f(\boldsymbol{x})\, d^M x - \sum_{k=1}^{K} \lambda_k \left\{ \int z_k(\boldsymbol{x})\, f(\boldsymbol{x}) d^M x - \frac{1}{K} \right\} \;. \tag{6.88}$$

The problem can be then treated by the standard calculus of variations. Unfortunately the resulting equations are complicated and it is not known how a general solution might be analytically expressed. Some insight related to the general properties of the solutions may be obtained by a qualitative analysis of the terms appearing in the conditions and the discrepancy measure. We present the following arguments for the formulation of the above statements:

First, we have not assumed *a priori* that the reference functions correspond to indicator functions. Further, work with the conditional average estimator, which will be described later, has convinced us that the application of a smooth approximation of the indicator function, such as a Gaussian function, can result in a more accurate reproduction than the box-like indicator function itself. Therefore, we conjecture that, just as in the solution of the previous problem, a set of smooth reference functions should emerge. We have also not assumed a continuous random variable $\boldsymbol{X}$ to facilitate a more general description. As a true condition, only Eq. (6.84) is used, while the expression given in Eq. (6.85) serves mainly as a definition that is subsequently utilized in the description of the reconstruction function. The condition specified by Eq. (6.84) is indeed the generalization of the corresponding overlapping integral defined by the indicator function given in Eq. (6.83). We conjecture that this condition leads to a shrinking of the optimal reference functions, while the minimization of the mean-square Euclidean distance between the reproduction and the original random variable leads to a proper distribution of centroids.

Additional basic properties of the reference functions can be estimated from the following treatment. By forming the sum of the conditions described by Eq. (6.84) we obtain the following expression

$$\int \sum_{k=1}^{K} z_k(\boldsymbol{x})\, f(\boldsymbol{x})\, d^M x = \sum_{k=1}^{K} p_k = 1\,. \tag{6.89}$$

If we compare this equation with the normalization condition

$$\int f(\boldsymbol{x})\, d^M x = 1 \tag{6.90}$$

we expect that in the region where $f(\boldsymbol{x}) \neq 0$ there must be

$$\sum_{k=1}^{K} z_k(\boldsymbol{x}) = 1\,. \tag{6.91}$$

It is characteristic that in the region where $f(\boldsymbol{x}) = 0$ the reference functions can be arbitrarily changed without affecting either the conditions specified by Eq. (6.84) or the discrepancy. From summation of terms given in Eq. (6.85) we obtain

$$\begin{aligned} \int \sum_{k=1}^{K} z_k(\boldsymbol{x})\, \boldsymbol{x}\, f(\boldsymbol{x})\, d^M x &= \int \boldsymbol{x}\, f(\boldsymbol{x})\, d^M x \;=\; \boldsymbol{M}_x \\ &= \sum_{k=1}^{K} p_k \boldsymbol{q}_k \;=\; \frac{1}{K} \sum_{k=1}^{K} \boldsymbol{q}_k\,. \end{aligned} \tag{6.92}$$

This expression shows that the centroids can be treated as the prototype points which are suitable for representing the random variable $\boldsymbol{X}$. Each of the points is associated with an equal probability $P_k = 1/K$ and the average of the centroids corresponds to the mean of the variable $\boldsymbol{X}$. In addition, the conditions of Eq. (6.84) can only be fulfilled if the reference functions and their centroids lie in the region where $f(\boldsymbol{x}) \neq 0$, or inside of the distribution of the variable $\boldsymbol{X}$.

Let us consider the results obtained from an analysis of the effective widths of the reference function which provide a quantitative explanation of the properties of these functions. For the sake of simplicity, we restrict ourselves to the one-dimensional case described by the scalar variable X. The expression for the discrepancy is written as

$$\mathcal{D} = \int \{y(x) - x\}^2 f(x)\, dx = \int \left\{ \sum_{k=1}^{K} (q_k - x)\, z_k(x) \right\}^2 f(x)\, dx$$

$$= \sum_{k=1}^{K} \int \{(q_k - x)\, z_k(x)\}^2 f(x)\, dx$$

$$+ \sum_{k=1}^{K} \sum_{n \neq k}^{K} \int \{(q_k - x)\, z_k(x)\}\{(q_n - x)\, z_n(x)\} f(x)\, dx \ . \qquad (6.93)$$

To obtain the second integral in the first line we have taken into account $\sum z_k = 1$. In the second line, the integral appearing in the first term represents the square of an effective width of the reference function. Because of the presence of the quadratic terms, the complete sum is positive. If the discrepancy $\mathcal{D}$ is to approach 0 with an optimization of the reference functions z_k, then the double sum must yield a negative value. The integrals in it represent the overlapping of the reference functions. If these functions are effectively different from zero only in the interval spanning an effective width around the centroid, then these must be separated for approximately one such width so as to obtain sufficient overlapping as well as a decrease in the discrepancy. However, as stated previously, all the centroids cannot be placed at the same point, but must be spread over the span of the variable X. We thus conclude that the condition $\int z_k(x)\, f(x)\, dx = 1/k$ assures the proper overlapping of the reference functions with the source probability density. The minimization of the discrepancy provides for a proper mutual overlapping of the reference functions and the corresponding proper positioning of the centroids. It is instructive to analyze what occurs when the reference functions are assumed *a priori* to be the indicator functions with no mutual overlapping. In that case, the discrepancy is determined principally by the sum of the effective widths and consequently, it is generally larger than when overlapping is permitted. A smooth transition from the region of one reference function to another can therefore provide a continuous change of the reproduction function. This is generally not the case when using non-smooth indicator functions. This is also the principal reason for not assuming that the reference functions must be described by the indicator function, as is normally done in the treatment of quantization problems. To support the above discussion, let us represent an approximately optimal solution that can be obtained by inspection. It is the simple case of a uniform distribution in one-dimension and a large number of data, K. A simplified example is shown in Fig. 6.3. Let L denote the span of the variable X with a uniform distribution $f(x) = 1/L$ for $0 < x < L$. The centroids must be uniformly distributed over this interval with the mutual spacing $\Delta q = L/K$. Let the i-th centroid be at q_i and the closest neighbors are at $q_{i-1} = q_i - \Delta q$ and at $q_{i+1} = q_i + \Delta q$. If the reference function is of triangular form that is specified by

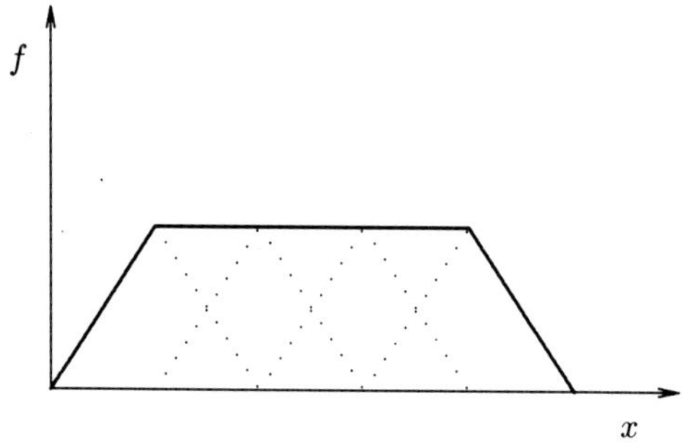

Fig. 6.3. Approximate description of a uniform PDF by triangular reference functions

$$z_i(x) = \begin{cases} 1 - \frac{|x-q_i|}{\Delta q} & ; x \in [q_i - \Delta q, q_i + \Delta q] \\ 0 & ; x \notin [q_i - \Delta q, q_i + \Delta q] \end{cases} \tag{6.94}$$

then the conditions specified by Eq. (6.84) are fulfilled and we obtain a good reproduction $y(x) = x$ everywhere, except on the sub-intervals of the outermost reference functions. With increasing K, the width of these sub-intervals tends to zero as does the contribution of both sub-intervals to the total discrepancy $\mathcal{D}$. For large K the triangular reference functions therefore approximately represent an optimal solution with $\mathcal{D} \approx 0$, $y(x) = x$ and $z_i = 1$ almost everywhere over the span of the random variable X. The half-width h of the reference functions corresponds to the spacing Δq between the centroids, which is an expected result. Similar functions are known from the theory of splines, and as was described in Chap. 2, they are used for interpolation purposes. It is characteristic that the triangular functions efficiently reduce the discrepancy that might be obtained when the indicator function is selected *a priori* for representing the reference functions. A similar reduction of the discrepancy is also obtained when using properly adapted Gaussian functions which are more appropriate for analytical work than the triangular function described here.

The general formulation of the quantization problem in terms of reference functions that must be determined by the variational procedure can be relaxed to a certain extent by utilizing the parametric form of these functions. We avoid the non-smooth indicator function and utilize the intuitively more convenient Gaussian. Then the problem is reduced to a determination of the centers and widths. Let us represent this in the one-dimensional case. First we define

$$z(q - x; h) \equiv \exp\left[-\frac{(q-x)^2}{2\,h^2}\right] . \tag{6.95}$$

We then utilize this function in the expression of the Lagrange functional whose minimization leads to the equations

$$\frac{\partial \mathcal{F}}{\partial q_i} = 0 \quad ; \quad \frac{\partial \mathcal{F}}{\partial h_i} = 0 \quad ; \quad \frac{\partial \mathcal{F}}{\partial \lambda_i} = 0 \ , \tag{6.96}$$

$$2q_i \left[\sum_{k=1}^{K} q_k \int z_i \frac{\partial z_i}{\partial q_i} f(x)\, dx - \int x \frac{\partial z_i}{\partial q_i} f(x)\, dx \right] - \lambda_i \int \frac{\partial z_i}{\partial q_i} f(x)\, dx = 0 \ , \tag{6.97}$$

$$2q_i \left[\sum_{k=1}^{K} q_k \int z_i \frac{\partial z_i}{\partial h_i} f(x)\, dx - \int x \frac{\partial z_i}{\partial h_i} f(x)\, dx \right] - \lambda_i \int \frac{\partial z_i}{\partial h_i} f(x)\, dx = 0 \ , \tag{6.98}$$

$$\int z_i\, f(x)\, dx - \frac{1}{K} = 0 \ . \tag{6.99}$$

This system is non-linear; we cannot proceed towards a general solution because the integrals cannot be expressed analytically. However, an instructive example can be analyzed when the distribution is uniform, i.e. $f(x) = f = 1/L$, and the number K is large. For this simple case, the solution can be guessed, just as in the case of the triangular reference functions. It corresponds to an approximately uniform distribution of centroids with a mutual spacing $\Delta q = L/K$. From the overlapping integral of the reference functions and the probability density we obtain the equation for the width

$$f \int z_i(x)\, dx = \frac{1}{K} \ , \tag{6.100}$$

$$h_i \approx \frac{1}{\sqrt{2\pi} f\, K} = \frac{L}{\sqrt{2\pi} K} = h \tag{6.101}$$

which shows that $h \ll L$. The relationship between the spacing of the reference functions and the width is then

$$\Delta q = \sqrt{2\pi} h \approx 2.5\, h \ . \tag{6.102}$$

Let us now analyze what follows from the above system of equations Eqs. (6.97) - (6.99). In the first equation, the integrals for almost all values of i, become

$$\int \frac{\partial z_i}{\partial q_i} f(x)\, dx = \frac{\partial}{\partial q_i} f \int z_i(x)\, dx \approx 0 \ , \tag{6.103}$$

$$\int x \frac{\partial z_i}{\partial q_i} dx = \frac{\partial (q_i \sqrt{2\pi} h_i)}{\partial q_i} \approx \sqrt{2\pi} h \ , \tag{6.104}$$

$$\begin{aligned} \int z_i(x)\, z_k(x)\, dx &= \frac{\sqrt{2\pi} h_i h_k}{h_{ik}} \exp\left[-\frac{(q_i - q_k)^2}{2\, h_{ik}^2} \right] \\ &\approx \sqrt{\pi} h \exp\left[-\frac{(q_i - q_k)^2}{4\, h^2} \right] \ . \end{aligned} \tag{6.105}$$

In the last expression we have used the approximation $h_{ik}^2 = h_i^2 + h_k^2 \approx 2h^2$ that holds for nearly all i. By inserting these integrals into Eq. (6.97) we obtain for almost all values of i

$$\sum_{k \neq i} q_k (q_i - q_k) \exp\left[-\frac{(q_i - q_k)^2}{4\,h^2}\right] \approx \sqrt{\pi} h\ . \tag{6.106}$$

Let us consider the i-th centroid q_i and denote the left and the right neighbor to it by q_- and q_+ respectively. Because of the uniform distribution, the distance to the left neighbor is equal to that on the right and we can therefore write

$$q_+ - q_i = -(q_i - q_-) = \Delta q\ . \tag{6.107}$$

By taking into account only the four nearest neighbors of the i-th centroid in the sum appearing in Eq. (6.106), it can be estimated as follows

$$\exp\left[-\frac{\Delta q^2}{4\,h^2}\right] + 4\exp\left[-\frac{4\Delta q^2}{4\,h^2}\right] \approx \frac{1}{\sqrt{8}}\frac{4\,h^2}{\Delta q^2}\ . \tag{6.108}$$

The first term corresponds to the nearest neighbors situated $|\Delta q|$ from the centroid and the second term is the result of the two next-nearest neighbors, separated by $|2\Delta q|$. There are two solutions of this equation: $\Delta q \approx 2.5\,h$ and $\Delta q \approx 1.5\,h$. For the first solution, the sum of the two neighbor reference functions is near 1.0, yielding in the middle of the centers the value 0.92, and 1.08 at the centroid, while for the second solution the sum yields the value 1.5 in the middle. The first solution thus closely corresponds to the previously-mentioned condition $\sum z_k = 1$ and this results in very good reproduction $y(x) \approx x$ over the entire region over which $f(x) \neq 0$. It also corresponds to the intuitively guessed solution.

A similar treatment can be carried out for the case when the probability density is a smooth function and the number K is large. Then the widths of the reference functions can be locally adapted to the PDF using the equation

$$h_i \approx \frac{1}{\sqrt{2\pi} f(q_i)\,K} \tag{6.109}$$

while the centroids can be positioned by using the spacing

$$\Delta q_i = \sqrt{2\pi} h_i \approx 2.5\,h\ . \tag{6.110}$$

However, this procedure cannot be easily extended to multivariate distributions because for these, the width must be properly substituted by a covariance matrix.

The examples in this chapter have shown that finding an optimal set of reference functions for a generalized quantization problem is generally a formidable task-dependent problem and for this reason, the indicator functions are most often assumed *a priori*. Many algorithms and techniques have been developed for the approximate determination of an optimal partition with respect to a distortion of the reproduction. [14, 15] In one dimension, the indicator function is determined by the interval, while in two dimensions, a square yields greater reproduction distortion than the hexagon. At large K and with an increasing dimension of the continuous vector variable $\boldsymbol{X}$, one can expect that the optimal form of an individual cell becomes ever more hyperspherical. [10] If the centroids are known, then the width of the indicator function can be approximately determined by halving the distance from the centroid to the nearest neighbors. This procedure is also applicable to the case when the Gaussian function is used rather than the indicator function.

At the end of this chapter let us emphasize the characteristic difference between the procedures of vector quantization and the method based on the absolute maximum entropy principle. We have seen that the fundamental problem of quantization when it is based on the absolute maximum entropy principle is the determination of the set of representative reference sample points $\{q_k \; ; \; k = 1, \ldots, K\}$ that represent the distribution of the observed random variable. We found in this case, that we must use in the description of the discrepancy measure, some reference functions that can be arbitrarily selected. This arbitrariness can be avoided to a certain extent by introducing the distortion measure of the reproduction. On the other hand, the fundamental problem in the vector quantization method is related to specifying the set of representative functions $\{z_k(x) \; ; \; k = 1, \ldots, K\}$ that properly span the probability density of the random variable. However, the complexity of both approaches is approximately equivalent.

7. Adaptive Modeling of Natural Laws

7.1 Probabilistic Modeler of Natural Laws

In the second chapter we mentioned that a natural law is usually expressed in terms of a relation between the variables used to characterize the state of Nature. We also pointed out that a natural law can be represented by a model formed by a mapping from the observed part of Nature to the properties of some device in such a way that the properties of Nature can later be retrieved from the model. In order to carry out a modeling automatically, the corresponding system, or the modeler, must include an array of sensors, a processor with a memory, such as an electronic computer, and an array of actuators. Our purpose here is to try to answer the previously-posed question: What are the general characteristics of an operating program that makes such a system into an automatic modeler of natural phenomena? To answer this question, we must first take into account the random character of physical variables and generalize the concept of a natural law. We then describe the fundamental operations of the system that are needed for the formation and application of natural laws.

Let us assume that in a particular description of natural phenomena a state of the Nature is characterized by the vector $\boldsymbol{X} = (x_1, x_2, \ldots, x_M)$. Let us consider first a two-dimensional case with $\boldsymbol{X} = (x, y)$. A natural law is usually described by the function $y = y(x)$ which associates with a given value of x a unique value of y. If we take into account the random character of physical variables, then we can no longer insist that a unique value of y is associated with a given value of x but rather that one may observe various values of y, provided that the value of x is simultaneously determined in the measurements. This means that we must replace the relation $y = y(x)$ by a function describing the probability that the value of y is observed in a certain interval provided that a fixed value of x is given. The corresponding function is thus a conditional probability density $f(y|x)$. The natural law $y = y(x)$ is then to be extracted from $f(y|x)$ by some arithmetic procedure. This procedure can be arbitrarily selected in principle. For example, it might be equal to the argument of the maximal value of $f(y|x)$, a mean value of y at given x, etc. To avoid this arbitrariness, we shall rather postpone the demand of uniqueness of y and state that the function $f(y|x)$ represents a generalization of the description of the natural law $y = y(x)$. By using the relationship $y = y(x)$ we assume *a priori* that the variable y is dependent on x which is generally an arbitrarily assumed distinction as well. To avoid this, we write the function $y = y(x)$ in an implicit form $\psi(x, y) = 0$ which avoids specifying which variable is dependent and which is independent. The corresponding generalization of such a description is then performed by specifying the joint probability density function $f(x, y)$. The generalization of these ideas to higher dimensional cases is straightforward.

We formulate as a basis of our further treatment the following hypothesis:

> *Any natural law, describing the properties of a phenomenon that is characterized by a vector variable $\boldsymbol{X}$, can be most generally described by a probability density function $f(\boldsymbol{X})$.*

The validity of this statement is not restricted to macroscopic phenomena but it also applies to microscopic ones and including those pertaining to a quantum mechanical description of Nature.

We then establish that a general automatic modeler of natural phenomena must perform the following fundamental tasks:

1. *The autonomous, simultaneous measurement of various variables.*
2. *The estimation of a joint probability density from the measured variables.*
3. *The storage of information about the probability density.*
4. *The transformation of the stored information and its transmission to the user.*

Each of these tasks has specific requirements which are dependent on the properties of the units composing the modeler as well as the characteristics of the phenomenon being modeled. Some of them have been mentioned in the preceding chapters with theoretical suggestions for their solutions. The aim in this chapter is to focus on the automatic modeling itself which is not task-dependent. Consequently, we shall omit discussion of the sensory unit performing the first task which is most typically task-dependent. The properties of the units performing the other three tasks can to certain extent be treated quite generally.

The first problem which we have not yet discussed is how one can generally describe the performance of the units composing the modeler and how one can optimize this performance. We presume that a modeler must be adaptable to a rather general class of natural phenomena, and consequently the units composing it must be treated as adaptive systems. We are thus faced with the problem of optimizing the performance of an adaptive system. The next problem that must be mentioned stems from the discrete nature of the elements composing modern electronic information processing units. In the previous chapter we developed the theoretical foundation for the treatment of this problem by formulating the absolute maximum entropy principle, which will be utilized in this and subsequent chapters. But, we have not yet discussed how to utilize this principle in a recursive and adaptive formation of the representative sample points and their subsequent storage. The last problem is related to the transformation and the preparation of the stored information for the user. All these problems will now be treated step-by-step on the basis of probability, statistics, and information theory.

7.2 Optimization of Adaptive Modeler Performance

In the following we treat the units performing the fundamental tasks of the modeler as adaptive physical systems. Generally, such a system is described by specifying its structure, that is, its building blocks and their mutual connections. We assume that a building block, as well as its connections, can be described by a finite set of K physical parameters.

A physical system is always subject to influences from its surroundings. We assume that these influences can be described by a finite set of physical variables, commonly expressed by a vector $\boldsymbol{X} = (x_1, \ldots, x_i, \ldots, x_M)$. This vector is then treated as the input to the system under consideration. The input generally depends on time t which is usually described by a discrete parameter. A phenomenon influencing the system is treated as a process generating this variable. A probabilistic description of processes and their modeling will be treated in later chapters, therefore we will consider here only vector random variables as inputs to the system. That is, at time t the input to our system is described as a sample of a vector random variable expressed as $\boldsymbol{X}$.

The properties of the system at a certain moment are described by the parameter vector $\boldsymbol{Q} = (\boldsymbol{q}_1, \ldots, \boldsymbol{q}_i, \ldots, \boldsymbol{q}_K)$ which has been concatenated from the sub-vectors $\boldsymbol{q}_i$ having the dimension D. This dimension is most frequently taken to be equal to the dimension M of the state vector $\boldsymbol{X}$ but for the present, we treat the general problem with $D \neq M$. For an adaptive system we assume that the parameter vector can be influenced by the inputs, and consequently, it must be generally treated as a function of time t. That is $\boldsymbol{Q} = \boldsymbol{Q}(t)$. The behavior of the system under influences from its environment is described by a set of physical variables that form the state vector $\boldsymbol{Y} = (y_1, \ldots, y_i, \ldots, y_S)$. This vector is often treated as a response of the system to the influences from its surroundings. This is formally expressed as

$$\boldsymbol{Y} = \boldsymbol{F}(\boldsymbol{X}; \boldsymbol{Q}) \ . \tag{7.1}$$

The corresponding response function may be quite complex. For instance, it need not be linear nor instantaneous, meaning that the output may also depend on the past history of the input. Similarly, the output $\boldsymbol{Y}$ can be treated as a random process as well, but for now, we treat it only as an ordinary random vector variable.

Each unit composing an automatic modeler can be treated as a system that performs a specific task and we are generally interested in a description of its performance. Let the input to the unit be given by the vector $\boldsymbol{X}$ subject to the probability distribution described by the density $f(\boldsymbol{X})$. The task of the unit can be described by specifying some hypothetical or desired response $\boldsymbol{Y}_d(\boldsymbol{X})$ for each possible sample of $\boldsymbol{X}$. However, the true response $\boldsymbol{Y}(\boldsymbol{X})$ may differ from the desired one, and we generally observe a discrepancy or error which is described by the square norm of the corresponding difference $\|\boldsymbol{Y}(\boldsymbol{X}) - \boldsymbol{Y}_d(\boldsymbol{X})\|^2$. The complete performance of the unit can then be characterized by the expected value of this discrepancy called a *cost function* [2, 12, 21]

$$\begin{aligned} \mathcal{D} &= \int \|\boldsymbol{Y}(\boldsymbol{X}) - \boldsymbol{Y}_d(\boldsymbol{X})\|^2 dP(\boldsymbol{X}) \\ &= \int \|\boldsymbol{Y}(\boldsymbol{X}) - \boldsymbol{Y}_d(\boldsymbol{X})\|^2 f(\boldsymbol{X})\, d^M X \, . \end{aligned} \tag{7.2}$$

This functional generally depends on the properties of the system being observed as well as on the probability density. We therefore write

$$\mathcal{D} = \mathcal{D}[\boldsymbol{Q};\, f(\boldsymbol{X})] \, . \tag{7.3}$$

Although the above measure of performance was introduced quite naturally, it is, in fact, not always the most appropriate. In the literature on optimal and adaptive systems one can find several other examples, a wide class of which, can be written using the generalized form of the above expression [3, 15, 5]

$$\mathcal{D} = \mathcal{D}[\boldsymbol{Q};\, f(\boldsymbol{X})] = \int \Psi[\boldsymbol{Q};\, f(\boldsymbol{X})]\, d^M X \, . \tag{7.4}$$

Here $\Psi[\boldsymbol{Q};\, f(\boldsymbol{X})]$ is assumed to represent some positive definite function of the variables indicated. In a stationary environment, the probability density is assumed to be fixed and we can specify the optimal system as that for which the discrepancy measure is minimal with respect to the parameters of the system. For the case in which the above discrepancy measure is defined by a differentiable function, we can require that the gradient of the discrepancy measure in the parameter space must equal zero for the optimal system. That is

$$\nabla_q \mathcal{D}[\boldsymbol{Q};\, f(\boldsymbol{X})] = \int \nabla_q \Psi[\boldsymbol{Q};\, f(\boldsymbol{X})]\, d^M X = 0 \, . \tag{7.5}$$

This generally represents a system of $D \times K$ nonlinear equations for the components of the parameter vector. In most cases, this system cannot be solved analytically and we must apply recursive approximate methods to obtain its solution.

Among various methods for the treatment of nonlinear systems of equations, the gradient or steepest descent method is most often applied. [19, 5] It is an iterative method based on the assumption that one can proceed towards the relative local extremum of the cost function from an arbitrarily-selected point by stepping in the direction of the gradient of cost function. For the case in which the minimum is sought, the steps must proceed in the direction of the negative gradient

$$\Delta \boldsymbol{Q} = -\alpha \nabla_q \mathcal{D}(\boldsymbol{Q};\, f) \;. \tag{7.6}$$

The iteration then yields the sequence

$$\boldsymbol{Q}_{i+1} = \boldsymbol{Q}_i - \alpha_i \nabla_q \mathcal{D}(\boldsymbol{Q}; f) \tag{7.7}$$

which represents the most generally applied adaptation law. The iteration is usually stopped when the norm of the parameter vector change $||\Delta \boldsymbol{Q}||$ reaches in one step some prescribed measure of accuracy. This measure can be related to the experimental error of the measurements or the specification of the parameter vector.

It is generally advised [19] that the step size determined by the constant α is reduced as the minimum of the cost function is approached, or $\alpha \Rightarrow \alpha_i$ and $\alpha_{i+1} < \alpha_i$. Suggestions related to the proper selection of the step size can be obtained from the analysis of the iteration error and the rate of approach to the minimum. [19, 13, 5] If α_i is too small, the convergence is slow, while if α_i is too large, the iteration may overshoot and the recursive process may then diverge. In the vicinity of a local extremum, the cost function can usually be approximated by a series expansion

$$\begin{aligned} \mathcal{D}(\boldsymbol{Q}; f) &\approx \mathcal{D}(\boldsymbol{Q}_i; f) + (\boldsymbol{Q} - \boldsymbol{Q}_i) \cdot \nabla_q \mathcal{D}(\boldsymbol{Q}_i; f) \\ &+ \frac{1}{2}[(\boldsymbol{Q} - \boldsymbol{Q}_i)^T \boldsymbol{G} (\boldsymbol{Q} - \boldsymbol{Q}_i)] \end{aligned} \tag{7.8}$$

where the matrix $\boldsymbol{G}$ is defined by the outer product $\nabla_q \otimes \nabla_q \mathcal{D}(\boldsymbol{Q}; f)$ at $\boldsymbol{Q} = \boldsymbol{Q}_i$. If the iteration formula is used in this expansion, we obtain

$$\begin{aligned} \mathcal{D}(\boldsymbol{Q}_{i+1}; f) &\approx \mathcal{D}(\boldsymbol{Q}_i; f) - \alpha_i ||\nabla_q \mathcal{D}(\boldsymbol{Q}_i; f)||^2 \\ &+ \frac{1}{2}\alpha_i^2 \nabla_q \mathcal{D}(\boldsymbol{Q}_i; f)^T \boldsymbol{G}\, \nabla_q \mathcal{D}(\boldsymbol{Q}_i; f) \;. \end{aligned} \tag{7.9}$$

The minimum of $\mathcal{D}(\boldsymbol{Q}_{i+1}; f)$ as function of α_i is then obtained at

$$\alpha_i = \frac{||\nabla_q \mathcal{D}(\boldsymbol{Q}_i; f)||^2}{\nabla_q \mathcal{D}^T(\boldsymbol{Q}_i; f) \boldsymbol{G} \nabla_q \mathcal{D}(\boldsymbol{Q}_i; f)} \;. \tag{7.10}$$

This formula shows that an optimal determination of the step size requires considerable calculation and consequently it is often more appropriate to work with a properly decreasing, simple function of the iteration index i such as

$$\alpha_i = \frac{\alpha_1}{i} \;. \tag{7.11}$$

By using the second-order expansion of the cost function, one can even avoid specification of α, by directly searching for the parameter vector $\boldsymbol{Q}$ that minimizes the expression (7.8). This yields the so-called Newton's algorithm

$$\boldsymbol{Q}_{i+1} = \boldsymbol{Q}_i - \boldsymbol{G}^{-1}\nabla_q \mathcal{D}(\boldsymbol{Q}_i; f) \tag{7.12}$$

which could be even more efficient than the gradient descent method if it did not involve the calculation of the inverse of the matrix $\boldsymbol{G}$. Therefore, this algorithm is likely not the most useful for automatic modeling applications.

The sequence of vectors $\boldsymbol{Q}_i$ can either approach a local or a global minimum. This means that the starting point can influence the final solution as well as the rate of the convergence to it. [25] Unfortunately, the method does not tell us how to select the starting point or which type of minimum is reached and we must estimate the character of the limit point by additional analysis. In spite of this deficiency, the gradient descent method is widely applied because of its simplicity. [5, 13, 18, 20]

7.3 Stochastic Approach to Adaptation Laws

In formulating the performance measure in the preceding section we have utilized the probability density function of the input random variable $\boldsymbol{X}$. Although not explicitly stated, we have assumed that it is given as a characteristic of the phenomenon under observation. However, this assumption is in reality only a hypothesis which is helpful for analytical work, while the appertaining distribution function can only be estimated from the measured samples. The optimization of an adaptive system can then proceed in accordance with the previously described methods in which the empirical distribution is utilized instead of the hypothetical one. We are thus faced with two optimization steps: the first is related to an optimal estimation of the probability density and the second one corresponds to the optimization of the system parameters. The question therefore arises whether it is possible to merge both steps into a single, computationally more effective procedure.

It is often the case that the probability density is not needed. Consequently, one might first think of avoiding specification of the probability density and instead apply the samples of the random variable directly in optimizing the system. This is even the case when the system is used for a representation of the probability density, as is for example in an automatic modeler of natural phenomena. In this case, during adaptation the system parameters must approach values which can be taken as prototype sample points. If the probability density is not known, the fundamental problem is how to describe the performance measure and how to determine its gradient that is required to optimize the system parameters. In the following we shall present a method for the solution of this problem that resembles the gradient descent approach described in the previous section. The method is called the *stochastic approximation* which was developed for the solution of problems related to the determination of the root and maximum of regression functions, identification and control, etc. This technique has been described in detail [22, 14, 4, 1] and it has also been successfully applied. [5, 6, 24, 13, 18, 20] The stochastic approximation is based on the application of an arbitrarily selected step size. We therefore shall later describe a relatively new perturbation method by which this arbitrariness can be avoided. We shall see that it leads to an application of the absolute maximum entropy principle in the description of a self-organizing modeler of natural phenomena. [7, 8, 9]

Here we briefly review the stochastic approximation method. For this purpose we consider a class of optimal systems that can be characterized by the performance measure defined by the expected value of some function $\Theta(\boldsymbol{Q};\boldsymbol{X})$:

$$\mathcal{D} = \mathcal{D}[\boldsymbol{Q}; f(\boldsymbol{X})] = \int \Theta(\boldsymbol{Q};\boldsymbol{X})\, f(\boldsymbol{X})\, d^M X = \mathrm{E}\,[\Theta(\boldsymbol{Q};\boldsymbol{X})]\ . \tag{7.13}$$

The optimization criterion of such systems is given by

$$\begin{aligned}\nabla_q \mathcal{D}[\boldsymbol{Q}; f(\boldsymbol{X})] &= \nabla_q \int \Theta(\boldsymbol{Q};\boldsymbol{X})\, f(\boldsymbol{X})\, d^M X \\ &= \nabla_q \mathrm{E}\,[\Theta(\boldsymbol{Q};\boldsymbol{X})] \ = \ 0\ .\end{aligned} \tag{7.14}$$

This generally corresponds to a system of nonlinear equations written in terms of the parameter vector $\boldsymbol{Q}$. We solve these by an iteration procedure. Following the gradient descent method, we formulate the iterative procedure using the expression

$$\boldsymbol{Q}_{i+1} = \boldsymbol{Q}_i - \alpha_i \nabla_q \mathrm{E}\,[\Theta(\boldsymbol{Q}_i;\boldsymbol{X})] = \boldsymbol{Q}_i - \alpha_i \mathrm{E}\,[\nabla_q \Theta(\boldsymbol{Q}_i;\boldsymbol{X})] \tag{7.15}$$

in which $\{\alpha_i, i = 1, 2 \ldots\}$ is a scalar sequence that characterizes the adaptation rate. A crucial point in the stochastic approximation is that the expected value of the gradient is substituted by its particular value determined by a sample $\boldsymbol{X}_i$ measured in the i-th experiment. The argument for this step follows from the following consideration. Because the gradient $\nabla_q \Theta(\boldsymbol{Q};\boldsymbol{X})$ depends on the random variable $\boldsymbol{X}$, it is also a random variable. We can express this gradient as a sum of two terms representing the mean value and a random fluctuation γ about the mean. That is

$$\nabla_q \Theta(\boldsymbol{Q};\boldsymbol{X}_i) = \mathrm{E}\,[\nabla_q \Theta(\boldsymbol{Q};\boldsymbol{X})] + \gamma\ . \tag{7.16}$$

This equation is then utilized to express the gradient of the mean value as

$$\mathrm{E}\,[\nabla_q \Theta(\boldsymbol{Q};\boldsymbol{X})] = \nabla_q \mathrm{E}\,[\Theta(\boldsymbol{Q};\boldsymbol{X})] = \nabla_q \Theta(\boldsymbol{Q};\boldsymbol{X}_i) - \gamma\ . \tag{7.17}$$

In the iteration formula, the mean gradient is then replaced by the sum of the random value and the fluctuation term

$$\boldsymbol{Q}_{i+1} = \boldsymbol{Q}_i - \alpha_i [\nabla_q \Theta(\boldsymbol{Q}_i;\boldsymbol{X}_i) - \gamma]\ . \tag{7.18}$$

We seek such an iteration procedure in which the effect of the random fluctuations γ vanishes. In this case, the limiting solution corresponds to the equation with the mean value of the gradient even though we are using at each iteration step the function $\nabla_q \Theta(\boldsymbol{Q}_i; \boldsymbol{X}_i)$ which is determined by an individual sample $\boldsymbol{X}_i$. The limiting value is then the result of many contributions of individual steps in which the influence of the zero-mean fluctuation is cancelled. We thus require that $\boldsymbol{Q}_{i+1}$ approaches $\boldsymbol{Q}_i$ at large i. That is

$$\lim_{i\to\infty} \boldsymbol{Q}_{i+1} = \lim_{i\to\infty} \boldsymbol{Q}_i \ . \tag{7.19}$$

We therefore require

$$\lim_{i\to\infty} \alpha_i = 0 \tag{7.20}$$

for otherwise, the iterative sequence need not terminate on a constant value. This condition guarantees a smoothing effect. However, the α_i sequence must not approach zero too rapidly in order not to diminish the effect of driving the iteration process by the average value. The adaptation that follows the sequence of measured samples is generally called *stochastic learning.* In the corresponding formula the fluctuating term is not taken into account because it is not known. Hence, the learning follows the rule

$$\boldsymbol{Q}_{i+1} = \boldsymbol{Q}_i - \alpha_i \nabla_q \Theta(\boldsymbol{Q}_i; \boldsymbol{X}_i) \ . \tag{7.21}$$

In the literature on the stochastic approximation,[22, 4, 1, 24] it is reported that the iteration procedure proposed above, leads to a correct parameter estimation with probability 1, provided that the following conditions are met: The condition expressed by Eq.(7.20) and a requirement that the variance of the fluctuating term is limited and further, the following conditions:

$$\sum_{i=1}^{\infty} \alpha_i = \infty \ , \tag{7.22}$$

$$\sum_{i=1}^{\infty} \alpha_i^2 < \infty \ . \tag{7.23}$$

The first of these conditions stems from the requirement that the expected value of the gradient must drive the iteration process to the correct point, while the second removes the cumulative effect of the fluctuating term. Since the fluctuating term has a zero mean, the first condition implies that

$$\sum_{i=1}^{\infty} \alpha_i \gamma_i = 0 \tag{7.24}$$

while the limited variance of the fluctuation yields for each component index j the inequality

$$\sum_{i=1}^{\infty} \alpha_i^2 \gamma_{j,i}^2 < \infty \ . \tag{7.25}$$

There are many sequences that satisfy the conditions described by Eqs. (7.20), (7.22), and (7.23), the simplest of them appears to be the harmonic sequence

$$\alpha_i = \alpha_1 / i \ . \tag{7.26}$$

The method described above does not indicate how to select the initial value, except that it must be positive. In contrast to this arbitrariness, we obtain a suggestion for this value in the treatment of the perturbation method. This will be presented later in this chapter.

7.4 Stochastic Adaptation of a Vector Quantizer

We now demonstrate the application of stochastic approximation method for the optimization of a vector quantizer which was described in the previous chapter. The method for the adaptation of reference vectors to the samples of a random variable $\boldsymbol{X}$ has a long and interesting history [10, 13] and it is now often named after Kohonen who has made significant contributions to it, in relation to studies of topological mapping and self-organization in biological and simulated neural networks. [16, 17, 18]

Let us consider a vector random variable $\boldsymbol{X}$ that is mapped onto the discrete variable $\boldsymbol{Q}$ by a vector quantizer as described in the previous chapter. Let the quantization procedure be associated with a partition to cells $\{C_k;\ 1 \leq k \leq K\}$ described by the cell indicator function

$$I_k(\boldsymbol{x}) = \begin{cases} 1 \ , & \boldsymbol{x} \in C_k \\ 0 \ , & \boldsymbol{x} \notin C_k \end{cases} \ . \tag{7.27}$$

The mapping is described by the quantization function

$$\boldsymbol{q}(\boldsymbol{x}) = \boldsymbol{q}_k \quad \text{if } \boldsymbol{x} \in C_k \tag{7.28}$$

which permits the reconstruction

$$\boldsymbol{y}(\boldsymbol{x}) = \sum_{k=1}^{K} \boldsymbol{q}_k I_k(\boldsymbol{x}) \ . \tag{7.29}$$

Let us suppose that we can sequentially measure the samples of the variable $\boldsymbol{X}$ and we want to optimize the performance of the quantizer by minimizing the reconstruction error defined by

$$\mathcal{E} = \int ||\boldsymbol{y}(\boldsymbol{x}) - \boldsymbol{x}||^2 f(\boldsymbol{x})\, d^M x \tag{7.30}$$

as a function of the reference vectors. Using the stochastic approximation method we construct the iteration formula as

$$\boldsymbol{Q}_{i+1} - \boldsymbol{Q}_i = \Delta \boldsymbol{Q}_{i+1} = -\alpha_i \nabla_q ||\boldsymbol{y}(\boldsymbol{x}_i) - \boldsymbol{x}_i||^2 \tag{7.31}$$

which yields for the k-th sub vector $\boldsymbol{q}_k$ of the concatenated parameter vector $\boldsymbol{Q}$ the iteration equation

$$\Delta \boldsymbol{q}_{k,i+1} = \alpha_i(\boldsymbol{x}_i - \boldsymbol{q}_k)\, I_k(\boldsymbol{x}_i)\ . \tag{7.32}$$

This equation shows that a particular sample x_i affects only the centroid of the indicator function covering this sample, while all other centroids are left intact. In other words, the adaptation proceeds by selecting only that centroid which is most similar to the sample vector and shifting it towards the sample value. In the preceding chapter it was shown that a reproduction can generally be improved by substituting the box-like indicator function by a hill-like function representing a smooth transition between the regions pertaining to different centroids. In this case, the indicator function in the above iteration formula is also transformed into a function that extends to the neighboring centroids. Consequently not only the closest centroid of the sample, but also its neighbors are changed at each step of the iteration. This generally results in a more refined adaptation of the complete set of reference vectors than can be obtained from the indicator function.

The above derivation of the adaptation rule is not very rigorous, because it does not take into account that changing the centroids also influences the changes of the indicator functions. This influence has been completely ignored here; one can hope that the resulting problem can be avoided by an interpretation based on a description of the similarity between samples and centroids which need not rely on the indicator function but only on a proper specification of the neighborhood. A significant step in this direction has been made by Kohonen [16, 17, 18] and this is explained in the next chapter dealing with the self-organization of neurons.

7.5 Perturbation Method of Adaptation

Let us presume that the parameter vector that minimizes the cost function $\mathcal{D}(\boldsymbol{Q}; f)$ for a given probability density function $f(\boldsymbol{X})$ is known. We now seek the optimal parameter vector $\boldsymbol{Q}_1$ which corresponds to the perturbed probability density function $f_1(\boldsymbol{X}) = f(\boldsymbol{X}) + \Delta f(\boldsymbol{X})$. If the perturbation $\Delta f(\boldsymbol{X})$ is small compared to $f(\boldsymbol{X})$ we can assume that the new optimal vector $\boldsymbol{Q}_1$ can be obtained by a small perturbation $\Delta \boldsymbol{Q}$ of the previous one. An equation for it can be obtained by expanding the performance measure into a power series with respect to the perturbations of the vector $\boldsymbol{Q}$ and the function $f(\boldsymbol{X})$

$$\begin{aligned}\mathcal{D}_1 &= \mathcal{D}[\boldsymbol{Q} + \Delta \boldsymbol{Q}\, ;\, f(\boldsymbol{X}) + \Delta f(\boldsymbol{X})] \\ &\cong \int \Big[\Theta(\boldsymbol{Q}; f) + (\Delta \boldsymbol{Q} \cdot \nabla_q + \Delta f \partial_f)\, \Theta(\boldsymbol{Q}; f)\Big]\, d^M X \\ &+ \frac{1}{2} \int \Big[(\Delta \boldsymbol{Q} \cdot \nabla_q + \Delta f \partial_f)^2 \Theta(\boldsymbol{Q}; f)\Big]\, d^M X\ . \end{aligned} \tag{7.33}$$

By taking into account that $\mathcal{D}(\boldsymbol{Q}; f)$ has a minimum for $\boldsymbol{Q}$ at a given probability density function $f(\boldsymbol{X})$ we can obtain the minimum of $\mathcal{D}_1$ by requiring

$$\nabla_{\Delta q} \mathcal{D}_1 = 0 \ . \tag{7.34}$$

In the previous expansion, the linear term vanishes and we obtain from the remaining portion the following equation

$$\int (\Delta \boldsymbol{Q} \cdot \nabla_q + \Delta f \partial_f) \nabla_q \Theta(\boldsymbol{Q}; f) \, d^M X = 0 \tag{7.35}$$

which can be written as a linear system of equations for the perturbation $\Delta \boldsymbol{Q}$

$$\boldsymbol{C} \cdot \Delta \boldsymbol{Q} = \boldsymbol{B} \ . \tag{7.36}$$

Here the matrix $\boldsymbol{C}$ and the vector $\boldsymbol{B}$ are defined by the expressions

$$\boldsymbol{C} \equiv \nabla_q \nabla_q \int \Theta(\boldsymbol{Q}; f) \, d^M X \quad , \tag{7.37}$$

$$\boldsymbol{B} \equiv -\nabla_q \int \Delta f \partial_f \Theta(\boldsymbol{Q}; f) \, d^M X \tag{7.38}$$

where $\nabla_q \nabla_q \equiv \nabla_q \oplus \nabla_q$ denotes the matrix of the outer product of two nabla vectors. The last equation can be generalized to the case in which $\Theta(\boldsymbol{Q}; f)$ represents a functional with respect to f. In that case the term $\Delta f \partial_f \Theta(\boldsymbol{Q}; f)$ must be interpreted in terms of a functional derivative as is done in the calculus of variations.

The largest terms in the matrix $\boldsymbol{C}$ are often the diagonal ones. In this case, the matrix is first split into a diagonal and an off-diagonal part

$$\boldsymbol{C} = \boldsymbol{C}_d + \boldsymbol{C}_o \ . \tag{7.39}$$

The inverse of the diagonal part $\boldsymbol{C}_d^{-1}$ can be simply determined and the linear system can then be represented in a form that is convenient for an iterative treatment

$$\Delta \boldsymbol{Q} = \boldsymbol{C}_d^{-1} \boldsymbol{B} - \boldsymbol{C}_d^{-1} \boldsymbol{C}_o \Delta \boldsymbol{Q} \ . \tag{7.40}$$

The iteration begins with

$$\Delta \boldsymbol{Q}_1 = \boldsymbol{C}_d^{-1} \boldsymbol{B} \tag{7.41}$$

and proceeds according to the equation

$$\Delta \boldsymbol{Q}_{i+1} = \boldsymbol{C}_d^{-1} \boldsymbol{B} - \boldsymbol{C}_d^{-1} \boldsymbol{C}_o \Delta \boldsymbol{Q}_i \ . \tag{7.42}$$

Quite often the first term already yields an acceptable result.

Equations (7.41) and (7.42) are more important than might appear at first glance. In practical applications we often encounter phenomena which are not thoroughly stationary and consequently the probability density function is not exactly constant but can vary slightly with time. That is, $\Delta f = \Delta f(\boldsymbol{x}, t)$. In this case, the treatment presented above is applicable and we can use it efficiently to describe adaptation of the modeler to variations of the properties of the phenomenon under observation. However, for this purpose we must analytically specify how the probability density changes dynamically by providing $\Delta f(\boldsymbol{x}, t)$. A similar treatment is also applicable when the probability density is estimated empirically. This treatment is of greatest importance for the automatic modeling of natural phenomena and it will therefore be elaborated in more detail in the following paragraph.

When estimating the empirical probability density we observe that each newly presented sample of the random variable has ever lesser statistical weight. The corresponding changes of the probability density therefore decrease with an increasing number of samples. If the empirical probability density is estimated as

$$f_N(\boldsymbol{X}, \sigma) = \frac{1}{N} \sum_{n=1}^{N} w(\boldsymbol{X} - \boldsymbol{X}_n, \sigma) \tag{7.43}$$

where σ denotes an acceptable experimental measurement error and w is an appropriate window function, then the change of f caused by the $(N+1)$-st sample is determined by

$$\begin{aligned} \Delta f(\boldsymbol{X}, \sigma) &= f_{N+1}(\boldsymbol{X}, \sigma) - f_N(\boldsymbol{X}, \sigma) \\ &= \frac{1}{N+1}[w(\boldsymbol{X} - \boldsymbol{X}_{N+1}, \sigma) - f_N(\boldsymbol{X}, \sigma)] \ . \end{aligned} \tag{7.44}$$

With an increasing number of samples, the empirical probability density becomes smoother and its fluctuations, described by the contribution of the window function pertaining to an individual sample, decrease in amplitude approximately monotonically. This indicates that the corresponding adaptation of an optimal system can be described by the above-derived perturbation procedure and we obtain for the driving term of the corresponding equation, the expression

$$\boldsymbol{B} = -\frac{1}{N+1} \int [w(\boldsymbol{X} - \boldsymbol{X}_{N+1}, \sigma) - f_N(\boldsymbol{X}, \sigma)] \partial_f \nabla_q \Theta(\boldsymbol{Q}; f)\, d^M X \ . \tag{7.45}$$

When the window function is a narrow approximation of a delta function, then the integral of the first term can be simply approximated and we obtain

$$\begin{aligned} \boldsymbol{B} = &- \frac{1}{N+1}\Big[\ \partial_f \nabla_q \Theta(\boldsymbol{Q}; f(\boldsymbol{X}_{N+1}, \sigma)) \\ &- \int f_N(\boldsymbol{X}, \sigma) \partial_f \nabla_q \Theta(\boldsymbol{Q}; f)\, d^M X \ \Big] \ . \end{aligned} \tag{7.46}$$

The first term shows that the driving of the iteration is principally described by the gradient of the function $\Theta(.)$ determined at the particular sample. This indicates a similarity with the stochastic approximation method. The principal difference is that the harmonic adaptation rate stems directly from the perturbation expansion. The second term differs from that in the stochastic approximation of the gradient descent and, as the numerical simulations show, generally leads to a refinement of the complete adaptation. A similar treatment and interpretation as presented above is possible when $\Theta(\boldsymbol{Q}; f)$ is expressed by a functional. Such a case will be treated in the chapter on self-organization.

A convenient characteristic of this method is that the true probability density is not needed. The empirically estimated one is used instead. We shall use the perturbation approach for the derivation of the self-organization algorithm in the next chapter on self-organization of the memory of the adaptive modeler. This is the focus of the theory of automatic modeling of natural phenomena.

7.6 Evolution of an Optimal Modeler and Perturbation Method

Up to now we have assumed that the properties of the system utilized for the modeling of natural phenomena can be described by a parameter vector $\boldsymbol{Q} = (\boldsymbol{q}_1, \ldots, \boldsymbol{q}_i, \ldots, \boldsymbol{q}_K)$ that has been concatenated out of a fixed number of sub-vectors $\boldsymbol{q}_i$ whose dimension is D. This generally means that we apply a fixed structure of the system while its structural elements are permitted to adapt. That is, the parameters can be influenced by the inputs from the surroundings and consequently, they are treated as functions of time t: $\boldsymbol{Q} = \boldsymbol{Q}(t)$. Observation of biological systems which are capable of modeling natural phenomena, such as the brains of human beings, reveals that these systems are generally not fixed structures but evolve in time. The evolution proceeds in each individual as well as on the species as a whole. The first type might be termed micro- and the second, macro-evolution. Micro-evolution proceeds simultaneously with adaptation, but generally it is much slower than adaptation. Slower still is the macro-evolution of a species that proceeds with small mutations of many individuals. In our description, we try to include micro-evolution by assuming that while the structure of an individual modeler is specified, the number K of available parameters is slightly changeable. In contrast, in order to permit a description of macro-evolution we must allow for a changeable structure as well as a variation of the number of parameters. In our systems, the structure is determined by the number of sensors used to observe the natural phenomena, that is, by the dimension of the vector variable $\boldsymbol{X}$, and by the structural connections between the building blocks of the modeler. We assume that by changing the structure we can vary the applicability of the modeler to the description of a specific phenomenon, while by changing the number of parameters, we can correct the adaptability of the modeler, that is the performance of an adapted system. We are thus, in fact, introducing a kind of hierarchy into the description of natural phenomena.

The division into micro- and macro-evolution described above suggests that we should assume that the set of natural phenomena can be divided into subsets of mutually similar phenomena that represent clusters in some abstract space of many dimensions. For a given cluster, various common properties are specific and to describe them we must utilize sensors of corresponding physical variables. For modeling of a certain cluster, a common modeler can be used, while for the description of various subsets, various modelers should be employed.

With an increasing number of parameters K we would expect that the system is capable to ever better model a natural phenomenon under observation. But when considering the physical realization of system, its cost is also generally increasing. A criterion for the description of the optimal system applicability thus may not only include the performance measure, but also a properly estimated cost related to the complexity of the system. The simplest measure of the complexity $\mathcal{K}$ takes into account only the number of units, i.e. the number of the applied parameters K. A simple criterion for the estimation of the system optimality can then be described by the sum of the performance and the complexity measure

$$\mathcal{F} = \mathcal{D} + \mathcal{K} . \tag{7.47}$$

When forming this functional, we must utilize the same physical units of both terms on the right side, as well as provide for a common interpretation. This is most easily done if the measure of information discrepancy is applied. Let the system be capable representing the probability density $f(x)$ by its parameters

$$\phi(\boldsymbol{x}, \boldsymbol{Q}, K) = \frac{1}{K} \sum_{k=1}^{K} w(\boldsymbol{x} - \boldsymbol{q}_k, \sigma) \tag{7.48}$$

with w being an appropriate window function and σ a constant window width. As a measure of discrepancy we can utilize the divergence of information which is given by

$$\mathcal{D} = \int [f(\boldsymbol{x}) - \phi(\boldsymbol{x})] \, [\log f(\boldsymbol{x}) - \log \phi(\boldsymbol{x})] \, d^D x . \tag{7.49}$$

We can then introduce the complexity measure in a way which is similar to that used in the definition of information entropy. Let us for this purpose consider a system consisting of K equivalent elements that are described by K parameters. Consider also that two such systems are joined together. We then try to determine the number of different configurations of wiring between the elements of both sub-systems. We presume that the elements can be mutually connected in different ways. By intuition we usually relate the complexity of the compound system to the total number of different wirings that can be established between the constitutive elements. Suppose that each element of the first system can be connected to each element of the second one so that the number of different possible connections is $N_2 = K^2$. Similarly, if we are joining in the compound system, n sub-systems of K elements then the number of different possible wirings is increased to $N_n = K^n$. The greater this number, the more information we must provide in order to describe a particular wiring among all possible connecting points. If all connections are equally probable, then we can associate with each of them a probability $p_n = 1/N_n$ and the entropy of information will be given by $H_n = -\log(p_n)$. If we connect n equal sub-systems, we presume that the complexity is increased n times with respect to the initial sub-system. This further suggests that we should express the complexity measure by the number of all possible interconnections using the equation

$$\mathcal{K} = c \log(N_n) = c\, n \log(K) \; . \tag{7.50}$$

The constant c determines the *units* of the complexity which is selected such that it coincides with the unit of the information entropy or $c = 1$. To a sub-system with K elements we thus assign the complexity measure

$$\mathcal{K} = \log(K) \tag{7.51}$$

which coincides with the entropy of information needed to specify the wiring. Such a specification can be further generalized to cases possessing non-equally probable connections.

In relation to the description of system complexity, Turing introduced an algebraic complexity measure. [11] He assumed that the systems can be described by a sequence of data and for their creation, he introduced the concept of a universal computer. He then defined a measure of the algebraic degree of complexity by the minimum length of a program by which such a sequence can be created. However, there arises a question of how to find a minimum-length general program which, as has been shown by Goedel in his famous theorem, represents a drawback of this definition. [11] In our definition, we have avoided utilization of the program length and instead specified the complexity measure by specification of the information entropy that follows from consideration of the connections between ensembles of similar systems.

With the above definition of the complexity measure, we obtain the following expression for the criterion functional

$$\begin{aligned} \mathcal{F} &= \int [f(\boldsymbol{x}) - \phi(\boldsymbol{x})]\,[\log f(\boldsymbol{x}) - \log \phi(\boldsymbol{x})]\, d^D x + \log K \\ &= \int \Theta[f(\boldsymbol{x}), \phi(\boldsymbol{x})]\, d^D x + \log K \; . \end{aligned} \tag{7.52}$$

The optimal system is then defined by a minimization of this functional with respect to the values of the system parameters as well as to their number. K is a positive integer, therefore the minimization of the criterion functional cannot be performed by differentiation. The problem is thus more complicated than in cases when it is possible to set a derivative to zero.

In any case, the fundamental equation for the optimization is obtained by requiring

$$\nabla_q \mathcal{F} = 0 \, . \tag{7.53}$$

This equation generally gives different solutions $\boldsymbol{Q}(K)$ for different numbers of parameters. It also yields different values of the criterion functional $\mathcal{F}(\boldsymbol{Q}(K))$ and we must find among them the smallest one. This can generally be done by selecting an ensemble of systems, estimating the functional value for each member of the ensemble and finding in the ensemble that system with the minimal value of $\mathcal{F}$. However, a question arises as to how we can proceed toward an optimal system by modifying just one particular system. A perturbation treatment can be of advantage in this case.

Let us assume that we know the optimal solution which is described by the joint set of parameters $\{\boldsymbol{Q}, K\}$ for a certain probability distribution f and we are looking for the solution corresponding to a slightly perturbed probability density $f_1 = f + \Delta f$. Let the corresponding optimal solution be expressed as $\{\boldsymbol{Q}_1, K_1\} = \{\boldsymbol{Q}, K\} + \{\Delta \boldsymbol{Q}, \Delta K\}$. Using the perturbation method we expand the functional $\mathcal{F}_1$ in powers of perturbations $\Delta f, \Delta \boldsymbol{Q}, \Delta K$ and obtain

$$\begin{aligned}
\mathcal{F}_1 &= \int \Theta[f + \Delta f, \phi + \Delta\phi]\, d^D x + \log(K + \Delta K) \approx \\
&\approx \mathcal{F} + \int \left(\Delta f \frac{\partial}{\partial_f} + \Delta\phi \frac{\partial}{\partial \phi} \right) \Theta \, d^D x + \frac{1}{2} \int \left(\Delta f \frac{\partial}{\partial_f} + \Delta\phi \frac{\partial}{\partial \phi} \right)^2 \Theta \, d^D x \\
&\quad + \frac{\Delta K}{K} - \frac{1}{2} \left(\frac{\Delta K}{K} \right)^2 .
\end{aligned} \tag{7.54}$$

We then assume that $\mathcal{F}$ is already optimal for $\boldsymbol{Q}, K$ and search for the minimum of $\mathcal{F}_1$ by retaining in the above equation the lowest-order terms that are different from zero and use these to adjust $\Delta \boldsymbol{Q}$ and ΔK. Here one must be careful because the vector $\boldsymbol{Q}$ may be modified for two reasons: First, the positions of the representative points are perturbed and second, their number can be perturbed as well. A perturbation of K generally corresponds to the creation or annihilation of representative points. In the following, we are principally interested in the creation because one usually develops an optimal system by creating it with the addition of new elements. In this case, we obtain

$$\boldsymbol{Q} = (\boldsymbol{q}_1, \ldots, \boldsymbol{q}_K), \tag{7.55}$$

$$\boldsymbol{Q}_1 = (\boldsymbol{q}_1 + \Delta \boldsymbol{q}_1, \ldots, \boldsymbol{q}_K + \Delta \boldsymbol{q}_K, \boldsymbol{q}_{K+1}, \ldots, \boldsymbol{q}_{K_1}) \tag{7.56}$$

and the corresponding perturbation can be expressed as a concatenation of two terms

$$\Delta \boldsymbol{Q} = \Delta \boldsymbol{Q}_K \oplus \Delta \boldsymbol{Q}_{\Delta K} \, , \tag{7.57}$$

with $$\Delta \boldsymbol{Q}_K = (\Delta \boldsymbol{q}_1, \ldots, \Delta \boldsymbol{q}_K) \, , \tag{7.58}$$

and $$\Delta \boldsymbol{Q}_{\Delta K} = (\boldsymbol{q}_{K+1}, \ldots, \boldsymbol{q}_{K1}) \, . \tag{7.59}$$

The perturbation of ϕ is then expressed as

$$\Delta\phi = \boldsymbol{c}_1 \cdot \Delta\boldsymbol{Q}_K + \boldsymbol{c}_2 \cdot \Delta\boldsymbol{Q}_{\Delta K} + c_3\, \Delta K/K \tag{7.60}$$

where $\cdot$ denotes scalar product and the coefficients $\boldsymbol{c}_i$ must be determined from the expression for ϕ. The expressions for the perturbations $\Delta\boldsymbol{Q}_K$, $\Delta\boldsymbol{Q}_{\Delta K}$, and $\Delta K/K$ are then obtained from Eq. (7.54) by letting the gradient of $\mathcal{F}_1$ with respect to this perturbation equal zero. However differentiation with respect to $\Delta K/K$ requires some explanation. If the number K is large, then $\Delta K/K$ can be assumed to be a small quantity and differentiation with respect to it can be introduced. This yields a linear system of equations for $\Delta\boldsymbol{Q}_K, \Delta\boldsymbol{Q}_{\Delta K}$, and $\Delta K/K$.

$$\int \left[\boldsymbol{c}_1 \frac{\partial\Theta}{\partial\phi} + \left(\Delta f \frac{\partial}{\partial_f} + \Delta\phi \frac{\partial}{\partial\phi}\right) \boldsymbol{c}_1 \frac{\partial\Theta}{\partial\phi}\right] d^D x = 0 \tag{7.61}$$

$$\int \left[\boldsymbol{c}_2 \frac{\partial\Theta}{\partial\phi} + \left(\Delta f \frac{\partial}{\partial_f} + \Delta\phi \frac{\partial}{\partial\phi}\right) \boldsymbol{c}_2 \frac{\partial\Theta}{\partial\phi}\right] d^D x = 0 \tag{7.62}$$

$$\int \left[c_3 \frac{\partial\Theta}{\partial\phi} + \left(\Delta f \frac{\partial}{\partial_f} + \Delta\phi \frac{\partial}{\partial\phi}\right) c_3 \frac{\partial\Theta}{\partial\phi}\right] d^D x = \frac{\Delta K}{K} - 1\,. \tag{7.63}$$

This system is rather complicated because of the non-fixed K and the number of components of $\boldsymbol{Q}$. At present, no general method for its solution is available. Therefore we still must determine the solutions $\Delta\boldsymbol{Q}_K, \Delta\boldsymbol{Q}_{\Delta K}$, and the value of $\mathcal{F}_1$ for various fixed ΔK and select that solution which corresponds to the minimum of $\mathcal{F}_1$. For a weak perturbation $\Delta f/f \ll 1$. Also, we generally expect a small value of ΔK of about 0 or 1 so that this task does not appear overwhelming. It appears that a solution of this problem would be to permit a smooth inclusion of new elements into the system. For this purpose, we must assign to each vector parameter $\boldsymbol{q}_k$ a weight or a probability p_k that can change continuously from 0 up to 1/K. Unfortunately, in such a process the concept of absolute maximum entropy cannot be directly applied. Rather, it could only be expected to result from the procedure of optimization.

7.7 Parametric Versus Non-Parametric Modeling

At the beginning of this chapter we accepted the probability density function as a general basis for the description of any natural law. Following the maximum entropy concept, we have quietly assumed that the probability density can be most directly adapted to empirical observations by utilizing representative sample points. Such a description is generally called non-parametric, to differentiate it from the parametric one which is based on the utilization of a certain functional form for the presentation of the probability density. The latter is well-suited when we know *a priori* that a phenomenon under observation pertains to a certain class of phenomena and it can therefore be described by one of the known functional forms describing the probability densities, for example, the normal one.

A similar situation occurs when we know *a priori* that the description of a phenomenon must be invariant with respect to some transformation of variables. In this case, the probability density is first expressed by a set of parameters which are then adapted to the phenomenon using empirical data. For this purpose, the above-mentioned methods are again applicable, provided that the performance measure is replaced by some measure of fitness. In fact, there is no clear computational distinction, except for the specific formulas, between the parametric and non-parametric treatments. In each case, a set of parameters is adapted to the empirical data. The distinction appears only in the interpretation. The parameters $\boldsymbol{q}_k$ of the non-parametric approach, which were introduced by the concept of the absolute maximum entropy, are interpreted as the representative sample points while in the parametric case the parameters can be related with the invariances. We conjecture that the non-parametric approach is more general than the parametric one, because it does not introduce *a priori* a certain property expressed by a specific form of the probability density. Because we are interested in a description of a general modeler of natural phenomena, we thus admit the non-parametric approach, although in certain specific cases it may be inferior to a parametric one. A similar situation occurs in the application of the empirical modeler which we shall consider in the following chapter.

8. Self-Organization and Formal Neurons

8.1 Optimal Storage of Empirical Information in Discrete Systems

The ultimate goal of our study is to obtain suggestions for the development of devices capable of the automatic modeling of natural phenomena. It is therefore of fundamental importance to develop a theoretical basis for the description of their optimal performance. In the previous chapters it was stated that an empirical modeling of natural phenomena includes three main tasks: Estimation, storage, and application of a probability distribution. Each of these tasks can be optimized by using the methods described in the previous chapter. The aim of this section is to present the problems and the solutions related with an optimal storage of empirical information about a continuous probability distribution in a system comprised of a finite number of discrete memory units. Such a system can be considered as a basic building block of an automatic modeler of natural phenomena. For example, we can imagine the brain of a biological organism or the digital memory of a computer that continually obtains signals from its surroundings and it optimally stores the corresponding empirical information. The first problem is the estimation of the probability density function of a continuous variable from the empirical data. We have already seen that this can be solved using Parzen's window. [2] The second problem is the storage of the continuous probability density in a discrete system. This task is generally related with a loss of information and thus the question arises, how one can minimize this loss. Therefore, an important task in the experimental sciences and in biological evolution is the development of an optimal, discrete representation of continuous variables. [1, 8, 14, 15, 16, 22] Research into neural networks has shown that neurons are capable of self-organizing by lateral interaction, so that they become selectively sensitized to sensory excitations. [20, 14, 15] Further, the study of competitive learning in simulated neural networks has revealed that the resulting self-organization leads to an approximately uniform distribution of the excitation probabilities of neurons, [16, 22] which characteristically corresponds to the maximum of information entropy. Similarly, Linsker [19] discovered that generalized Hebb-type learning [11] in some typical models of neural networks is equivalent to the maximum preservation of information. This led him to propose a principle of maximum information preservation and to explain some characteristic properties of neural networks. In the statistical sciences, the maximum-entropy principle, as formulated by Gibbs [3], has been successfully applied. [12, 9, 10, 13, 21] Our intention is to simultaneously solve both of the above-mentioned problems using the methods developed in the previous chapters related to the second maximum-entropy principle and the optimal adaptation, especially as part of an empirical approach for processing information. Before we proceed to this task let us briefly review the fundamental concepts of Kohonen's work on vector quantization and topological mappings in neural networks because it is essential for our interpretation of the self-organization process in an optimal memory which is comprised of a finite number of memory units. It has initiated our development of the Second Maximum-Entropy Principle for the description of the self-organization of neurons

8.2 Adaptive Vector Quantization and Topological Mappings

In this section we present an interpretation of the stochastic learning rule of a vector quantizer presented in the previous chapter. This was proposed and developed by Kohonen [14, 15, 16] in relation to his study of self-organized feature mappings of biological and simulated neural networks.

We assume a set of K information processing units called formal neurons, each one provided with a vector parameter $\boldsymbol{q}_k$. The input sample of the random variable $\boldsymbol{X}$ detected at time t is broadcast to each neuron where the similarity of this input and the parameter of the neuron is estimated. Any measure of similarity can be applied; however, for the sake of simplicity the norm of the distance $\|\boldsymbol{x}_t - \boldsymbol{q}_k\|$ is most often utilized. Let the excitation of the neuron by the input be described by some monotonically increasing function of similarity measure and let it be possible to compare the excitations of the neurons. Based on biological observations,[14, 20] Kohonen conjectured that this comparison can be performed in a neural network by some collective competitive interaction between the neurons, and this can be simply programmed into a network that is simulated on a computer. The most excited neuron, called the "winner" and further denoted by the index "w", is then taken as the origin with respect to which the adaptation is performed in the network. For this purpose it is advantageous to introduce some topological relation that describes the "position" of a particular neuron with respect to the others in the network. It can either be the distance in the space used to describe the variable $\boldsymbol{X}$, or, what is surprising and often advantageous for applications, also the distance in some other index-space, for example a real space where the network is situated. [14, 16] Using such a relation, a topological neighborhood $\mathcal{N}_w$ of the most excited neuron can be described. The adaptability of the neurons is then described by a variable η. Its value is 1 if a particular neuron lies in the neighborhood of the winner, and 0 if it is outside:

$$\eta_k = \begin{cases} 1 & \text{for } k \in \mathcal{N}_w \\ 0 & \text{for } k \notin \mathcal{N}_w \ . \end{cases} \tag{8.1}$$

The simplest adaptation law that leads to an acceptable formation of reference vectors can then be described by the equation

$$\Delta \boldsymbol{q}_k = \alpha(t)\,(\boldsymbol{x}_t - \boldsymbol{q}_k)\,\eta_k \ , \tag{8.2}$$

in which t denotes the index of the sample vector and $\alpha(t)$ is the variable that describes the adaptation rate. The similarity of the above relations with the stochastic learning rule of a vector quantizer is evident, although the interpretation as well as the motivation for its formulation are not exactly the same.

When the variable $\boldsymbol{X}$ describes a time-dependent process then t represents the discrete time. The neighborhood, as well as the adaptation rate, must be determined by trial and error. Generally it is convenient to describe the adaptation rate by some decreasing function of time. The adaptation can begin with the reference vectors initialized by the first K sample vectors, or even from a randomly initialized set. During the adaptation every neuron becomes maximally sensitized to samples of $\boldsymbol{X}$ that lie in the neighborhood of its reference vector $\boldsymbol{q}$. The established distribution of reference vectors exhibits properties similar to the probability distribution of the input random variable. [15, 16] In addition, the topological ordering of the neurons is preserved during adaptation. This observation can be explained in the following way. Let us imagine the k-th neuron which is characterized by $\boldsymbol{q}_k$ and consider then some other neurons that are, according to the selected topological relation, its nearest neighbors. It is then characteristic that the adapted neighbors also have the most similar reference vectors of all the neurons in the network. The reference vectors are thus ordered in the index space according to their similarity in the space utilized for the presentation of the variable $\boldsymbol{X}$. Therefore the applied learning rule induces a mapping in which the topological properties of the distribution of reference vectors in the index space are similar to the probability distribution in the space of variable $\boldsymbol{X}$ and hence the name *topological feature mapping.* This property has been introduced with respect to the experimental observations of the formation of localized activity patterns in biological neural networks which are the result of the lateral interaction of neurons. Numerical study of the adaptation based on the above formula has shown [14, 15, 16] that introduction of topological ordering is generally of advantage for the implication of the corresponding algorithm on a computer because it reduces the number of operations needed in calculations of distances in the space of the vector variable $\boldsymbol{X}$. By selecting a topological relation between the neurons, the problem of the selection of indicator function width that was mentioned in the previous chapter is avoided.

It also introduces some generalized interaction between the neurons that, because of collective competitive effects, leads to a self-organized adaptation. However, the arbitrariness of the selection of the neighborhood and the adaptation rate as well, shows that while many applications using this approach have been published, work on this research topic is not yet completed [16]

8.3 Self-Organization Based on the Absolute Maximum-Entropy Principle

Numerical studies of the Kohonen learning rule have shown that the excitation frequency is, on average, equal for all neurons of the network and that the neuron parameters can be treated as prototype sample points representing the probability distribution. [16] We therefore conjecture that this property is in fact a fundamental one and should therefore be derived from some fundamental principle. For this purpose, the second–maximum entropy principle appears to be applicable. Using such a derivation, we try to avoid the arbitrariness introduced by modeling the learning process on the basis of biologically-motivated models. We also do not wish to accept *a priori* the scheme of a quantizer as the basis of our description of the information processing system because it is specifically arranged for the purpose of data transmission. Our aim is to prepare a firm theoretical basis for the description of an adaptive memory that is optimal for the memorization of a wide class of random variables. With all this in mind, we formulate the basic optimization problem related to the application of a discrete memory system in a modeler of natural phenomena as follows:

> *Let there be given an experimental procedure by which an infinite sequence of samples of a continuous random vector variable* $\boldsymbol{X}$ *can be obtained. The problem is how the appertaining probability density function* $f(\boldsymbol{x})$ *can be optimally estimated and stored in a storage device comprised of* K *discrete memory units that are capable of accepting only one presented datum vector* $\boldsymbol{q}_k$.

For the solution of the first part of this problem we can apply the methods mentioned in relation with the estimation of probability densities from empirical data, especially Parzen's window approach. [2] The last part of this problem is theoretically equivalent to a representation of a continuous random variable $\boldsymbol{X}$ by a discrete one, here denoted by $\boldsymbol{Q}$, and this problem can thus be solved by generalizing the second maximum-entropy principle to the multivariate case. In fact, both methods are based on similar procedures and we join them in a unified approach. For this purpose we must form from the given empirical samples a set of representative data $\{G_k,\ k = 1, \ldots, K\}$, specified as the averages of properly selected reference functions and then find from them the corresponding sample points $\{\boldsymbol{q}_k\}$ by minimizing the discrepancy between the empirical and the representative data. These points and their related probabilities $p_k = 1/K$, can be interpreted as an empirical distribution representing the associated probability density function $f(x)$ in an optimal way. For this we must develop a procedure by which the variable $\boldsymbol{Q}$ can be optimally adapted to an ever increasing number of samples. Following the stochastic perturbation method of optimal system adaptation, we derive in the next section an adaptive algorithm which is later interpreted as a self-organized process of interaction between memory units or formal neurons. Our intention is to prepare a general basis for the practical application of the absolute-maximum entropy principle. We therefore assume a multivariate case and apply a similar performance measure as that used in the one-dimensional case presented in Chap. 6 that was based on maximum entropy principles. [4, 6] The complete derivation is represented as a self-consistent unit in order to show how the second maximum-entropy principle and the stochastic perturbation method can be joined in a development of an optimal information processing system.

8.4 Derivation of a Generalized Self-Organization Rule

Let $\boldsymbol{X}$ represent a continuous M-dimensional vector random variable with a corresponding set of N samples $\{\boldsymbol{x}_1, \ldots, \boldsymbol{x}_N\}$ The associated appertaining empirical probability distribution density is:

$$\phi_e(\boldsymbol{x}) = \frac{1}{N} \sum_{n=1}^{N} \delta(\boldsymbol{x} - \boldsymbol{x}_n) \ . \tag{8.3}$$

Instead of employing this empirical distribution function to describe the properties of the random variable $\boldsymbol{X}$, we prefer to introduce a finite set of K M-dimensional representative prototype or reference vectors $\{\boldsymbol{q}_1, \ldots, \boldsymbol{q}_K\}$ which should optimally represent the typical sample points of this variable.

A representative discrete random variable $\boldsymbol{Q}$ is defined by associating a probability to each reference vector: $\{P(\boldsymbol{q}_i) : i = 1 \ldots K\}$. With an optimal mapping from the continuous to the discrete random variable $\boldsymbol{X} \to \boldsymbol{Q}$, the entropy of information should be minimally reduced. This occurs when the probability distribution of the map corresponds to the maximal entropy of information or when the distribution is uniform: $\{P(\boldsymbol{q}_i) = 1/K : i = 1 \ldots K\}$. *By assigning a uniform distribution to the variable* $\boldsymbol{Q}$ *we in fact utilize the absolute maximum-entropy principle in the determination of the optimal mapping.* The corresponding density function:

$$\phi_r(\boldsymbol{x}) = \frac{1}{K} \sum_{k=1}^{K} \delta(\boldsymbol{x} - \boldsymbol{q}_k) \ , \tag{8.4}$$

is then used to represent the density of the probability distribution of the continuous variable $\boldsymbol{X}$. However, first the reference vectors must be determined. For this we compare the representative probability density ϕ_r with the empirical one ϕ_e and diminish their discrepancy by adaptation of the reference vectors. Because of the singularity of both probability density functions, functionals are employed for this comparison. Following Parzen's approach, we first introduce an appropriate window function $g(\boldsymbol{x}, \boldsymbol{s})$, in which $\boldsymbol{s}$ is an M-dimensional vector parameter, and then define two functionals by the empirical and the representative average:

$$G_e(\boldsymbol{s}) = \langle g \rangle_e = \int_{-\infty}^{+\infty} g(\boldsymbol{x}, \boldsymbol{s})\, \phi_e(\boldsymbol{x})\, d^M x = \frac{1}{N} \sum_{n=1}^{N} g(\boldsymbol{x}_n, \boldsymbol{s}) \, , \tag{8.5}$$

$$G_r(\boldsymbol{s}) = \langle g \rangle_r = \int_{-\infty}^{+\infty} g(\boldsymbol{x}, \boldsymbol{s})\, \phi_r(\boldsymbol{x})\, d^M x = \frac{1}{K} \sum_{k=1}^{K} g(\boldsymbol{q}_k, \boldsymbol{s}) \, . \tag{8.6}$$

The continuous vector parameter $\boldsymbol{s}$ is used here in order to generalize the discrete parameter reference functions introduced in the initial formulation of the maximum-entropy principle. The principal difference between the procedure followed here and the general treatment is that here the hypothetical expected value is replaced by the empirical average. This step thus introduces the stochastic approach. If the filtering of the probability density presented in Chap. 6 on maximum-entropy principles may appear to have been artificially introduced, here it places the empirical and the representative distributions on an equal basis so that in this case it is convenient and in fact unavoidable. Following the steps of the optimal system adaptation, we use the difference:

$$\epsilon = \epsilon(\boldsymbol{q}_1, \ldots, \boldsymbol{q}_K \; ; \; N \; ; \; \boldsymbol{s}) = \langle g \rangle_r - \langle g \rangle_e \tag{8.7}$$

to define the measure of discrepancy between both distributions by its mean square:

$$\mathcal{D}(\phi_r, \phi_e) = \overline{\epsilon^2} = \int_{-\infty}^{+\infty} \epsilon^2 d^M s \, . \tag{8.8}$$

This discrepancy can be treated as the performance measure of the system consisting of memory units characterized by the reference vectors and applied for representation of the probability density function. *Using the set $\{\boldsymbol{q}_1, \ldots, \boldsymbol{q}_K\}$ which minimizes the discrepancy $\mathcal{D}$, we define the optimal mapping $\boldsymbol{X} \rightarrow \boldsymbol{Q}$ of the continuous onto the discrete random variable* . The corresponding optimal set of reference vectors satisfies the following set of equations:

$$\frac{\partial \mathcal{D}}{\partial q_{lm}} = 2 \int_{-\infty}^{+\infty} \epsilon \frac{\partial \epsilon}{\partial q_{lm}} \, d^M s = 0 \quad ; \quad l = 1 \ldots K \quad ; \quad m = 1 \ldots M \, , \tag{8.9}$$

which are the fundamental equations describing an optimal system.

In two trivial cases, when $N = 1$ or when $N = K$ the optimal mapping corresponding to the absolute minimum in which the discrepancy is equal to zero: $\mathcal{D} = 0$ is obtained immediately from consideration of the form of both averages. These trivial results are:

$$\text{for} \quad N = 1 \ : \ \boldsymbol{q}_1 = \boldsymbol{q}_2 = \ldots = \boldsymbol{q}_k = \boldsymbol{x} \ , \tag{8.10}$$

$$\text{for} \quad N = K : \ \boldsymbol{q}_1 = \boldsymbol{x}_1, \boldsymbol{q}_2 = \boldsymbol{x}_2 \ldots \boldsymbol{q}_k = \boldsymbol{x}_k \ . \tag{8.11}$$

In other cases, the window function $g(\boldsymbol{x}, \boldsymbol{s})$ must first be specified and then the system of equations must be solved. A linear function $g(\boldsymbol{x}, \boldsymbol{s})$ is not adequate for obtaining a system of independent equations, therefore non-linear window functions must be employed. As a consequence, a closed form solution cannot be found and we are forced to apply an iterative treatment. For this purpose we follow the perturbation method described in Chap. 7.

Of special interest is the determination of the reference vectors when $N \gg K$. Since with increasing N the empirical average $\langle g \rangle_e$ converges to a fixed value, the addition of successive samples at large N causes only minor changes of the reference vectors and therefore a linear approximation is used to express the changes of the difference ϵ:

$$\begin{aligned} \epsilon_1 &= \epsilon(\boldsymbol{q}_1 + \Delta\boldsymbol{q}_1, \ldots, \boldsymbol{q}_K + \Delta\boldsymbol{q}_K \ ; \ N+1) \\ &= \epsilon(\boldsymbol{q}_1, \ldots, \boldsymbol{q}_K \ ; \ N) + \Delta\epsilon \ . \end{aligned} \tag{8.12}$$

Where

$$\Delta\epsilon = \sum_{k=1}^{K} \sum_{i=1}^{M} \frac{\partial \langle g \rangle_r}{\partial q_{ki}} \Delta q_{ki} - \frac{1}{N+1} [g(\boldsymbol{x}_{N+1}, \boldsymbol{s}) - \langle g \rangle_e] \ . \tag{8.13}$$

If the discrepancy $\mathcal{D}(\boldsymbol{q}_1, \ldots, \boldsymbol{q}_K \, ; N)$ is already minimal for $\{\boldsymbol{q}_1, \ldots, \boldsymbol{q}_K\}$, then the minimum of $\mathcal{D}_1 = \overline{\epsilon_1^2}$ is approached by changing the reference vectors for $\{\Delta\boldsymbol{q}_1, \ldots, \Delta\boldsymbol{q}_K\}$ in accordance with the conditions:

$$\frac{\partial \mathcal{D}_1}{\partial q_{lm}} = 0 \ ; \ l = 1 \ldots K \ ; \ m = 1 \ldots M \ . \tag{8.14}$$

These conditions yield the following system of linear equations:

$$\sum_{k=1}^{K} \sum_{i=1}^{M} C_{lmki} \, \Delta q_{ki} = B_{lm} \quad ; \quad l = 1 \ldots K \ ; \ m = 1 \ldots M \ , \tag{8.15}$$

in which we recognize the fundamental elements of the perturbation method. The coefficients are determined by the expressions:

$$C_{lmki} = \int_{-\infty}^{+\infty} \frac{\partial g(\boldsymbol{q}_k, \boldsymbol{s})}{\partial q_{ki}} \, \frac{\partial g(\boldsymbol{q}_l, \boldsymbol{s})}{\partial q_{lm}} \, d^M s \ , \tag{8.16}$$

$$B_{lm} = \frac{K}{N+1} \int_{-\infty}^{+\infty} [g(\boldsymbol{x}_{N+1}, \boldsymbol{s}) - \langle g \rangle_e] \frac{\partial g(\boldsymbol{q}_l, \boldsymbol{s})}{\partial q_{lm}} \, d^M s \ . \tag{8.17}$$

Here l or k denote prototypes and m or i denote components of the prototypes. By introducing the vector $\boldsymbol{B}_l = (B_{l1}, \ldots, B_{lM})$ and the matrix $\boldsymbol{C}_{lk} = ||C_{lmki}||$ we can write Eq.(8.15) in an abbreviated form

$$\boldsymbol{C}_{lk}\,\Delta\boldsymbol{q}_k = \boldsymbol{B}_l \quad ; \quad l \text{ or } k \;=\; 1\ldots K\;. \tag{8.18}$$

Its solution is

$$\Delta\boldsymbol{q}_l = \boldsymbol{C}_{lk}^{-1}\boldsymbol{B}_k \quad ; \quad l \text{ or } k \;=\; 1\ldots K\;. \tag{8.19}$$

This equation shows that the vector $\boldsymbol{B}_l$ can be interpreted as a driving variable that describes the influence of a new sample $\boldsymbol{x}_{N+1}$ on the changes of the prototypes. Accordingly, the matrix $\boldsymbol{C}_{lk}$ depends on the position of pairs of prototypes and thus describes their joint influence on the changes of these prototypes. This interaction can be further interpreted as a kind of self-organization.

The system of equations represented by Eq.(8.15) is the foundation for a practical adaptive application of the absolute maximum-entropy principle. In order to avoid utilization of the inverse matrix $\boldsymbol{C}_{kl}^{-1}$ in expressing its solution it is convenient to turn to an iterative treatment. If the conditions

$$C_{lmki} = 1 \;\text{ for } k = l\;,\; m = i \;\text{ and }\; |C_{lmki}| \ll 1 \;\text{ for } k \neq l\;,\; m \neq i \tag{8.20}$$

can be satisfied by a proper selection of the function $g(\boldsymbol{x}, \boldsymbol{s})$, then it is reasonable to represent the system of equations in the form:

$$\Delta\boldsymbol{q}_l^{(i+1)} \doteq \boldsymbol{B}_l - \sum_{k\neq l}^{K}\sum_{i\neq m}^{M} C_{lk}\,\Delta\boldsymbol{q}_k^{(i)} \quad ; \quad l \text{ or } k = 1\ldots K\;, \tag{8.21}$$

which is convenient for iteration. It starts with $\Delta\boldsymbol{q}_l^{(0)} = \boldsymbol{B}_l$. This system forms a basis for the formation of an algorithm by which the adaptation of representative variable $\boldsymbol{Q}$ to empirical samples $\{\mathbf{x}_i\}$ is described. For this purpose specification of the window function is unavoidable.

For a comparison of the probability densities, the most convenient window function is the Gaussian one:

$$g(\boldsymbol{x}, \boldsymbol{s}) = \exp\left[\frac{-\|\boldsymbol{x} - \boldsymbol{s}\|^2}{2\sigma^2}\right]. \tag{8.22}$$

We have omitted the normalizing factor which is not important here. This function fulfills the conditions given by Eq.(8.20) and yields analytically expressible integrals in Eqs.(8.16) and (8.17):

$$C_{lmki} \;=\; \left[\delta_{mi} - \frac{(q_{lm} - q_{km})\,(q_{li} - q_{ki})}{2\sigma^2}\right]\exp\left[\frac{-\|\boldsymbol{q}_l - \boldsymbol{q}_k\|^2}{4\sigma^2}\right] \tag{8.23}$$

$$\begin{aligned} B_{lm} = \frac{K}{N+1}\Big\{&(x_{N+1,m} - q_{lm})\exp\left[\frac{-\|\boldsymbol{x}_{N+1} - \boldsymbol{q}_l\|^2}{4\sigma^2}\right] \\ &- \frac{1}{K}\sum_{k=1}^{K}(q_{km} - q_{lm})\exp\left[\frac{-\|\boldsymbol{q}_k - \boldsymbol{q}_l\|^2}{4\sigma^2}\right]\Big\}\,, \end{aligned} \tag{8.24}$$

with $\delta_{mi} = 1$ for $m = i$ and $\delta_{mi} = 0$ for $m \neq i$.

In order to avoid storing all the empirical data we have replaced in the latter expression the empirical average $\langle g \rangle_e$ by the representative one $\langle g \rangle_r$. The width σ of the Gaussian function should approximately correspond to the estimated distance between representative reference vectors. If L^M denotes the range of variable $\boldsymbol{X}$, then the Parzen window approach suggests selecting $\sigma = L/\sqrt{K}$.

8.5 Numerical Examples of Self-Organized Adaptation

The general linear system of equations expressed by Eq.(8.21) together with the expressions (8.23) and (8.24), which have been specifically adapted to the Gaussian window function, represents a practically applicable algorithm for the adaptation of a representative variable to empirical data. [4, 6]

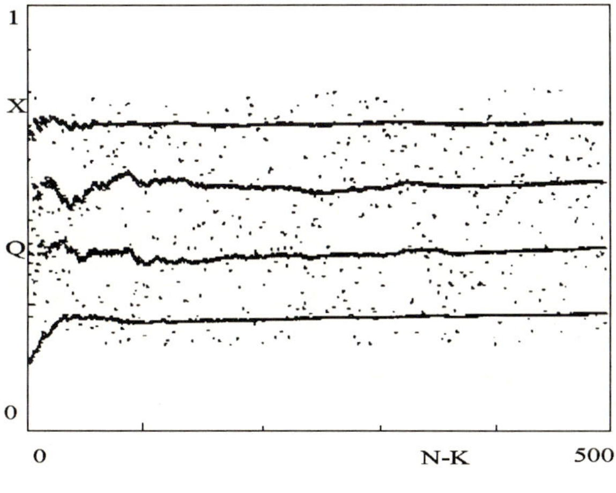

Fig. 8.1. Position of four reference vectors (*bold*) and sample points of a uniform probability distribution (*thin*) versus number of adaptations $N - K$. $\sigma = 0.1$

In Fig. 8.1 we illustrate an adaptation of four reference values q_k to a one-dimensional normal random variable X. In this case the vector $\boldsymbol{B}$ turns to a scalar variable. The generated random samples x_N are shown as thin points while the reference values q_k are shown as thick ones. Starting from the trivial solution, the reference values accommodate to the decreasing fluctuation of the empirical distribution of X. The initial fluctuations of the reference values are not a consequence of an incorrect adaptation, but are the result of describing the random phenomenon by the estimated empirical probability distribution at low N. The empirical and the representative cumulative probability distribution functions of variables X and Q are shown in Fig. 8.2.

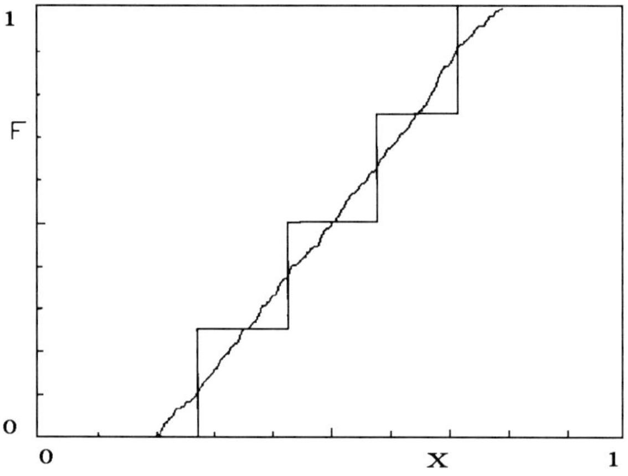

Fig. 8.2. Empirical (*line*) and representative (*staircase*) cumulative probability distribution functions, recorded at $N = 500$

The adaptation of the reference values to the random variable X is evident. In order to demonstrate the adaptation quantitatively, the averages $\langle g \rangle_e$ and $\langle g \rangle_r$ are shown in Fig. 8.3. A slight discrepancy of both averages is principally the consequence of the low chosen number K of representative reference values. The results were calculated by three iterations, but in this case a similar agreement is obtained by just a single iteration.

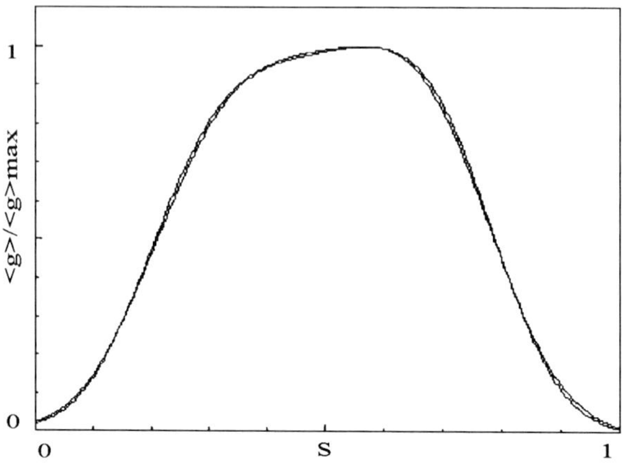

Fig. 8.3. Agreement between empirical and representative averages $\langle g \rangle_e$ and $\langle g \rangle_r$. The unit on the ordinate corresponds to the maximal value of each average

The second example illustrates the maximum-entropy mapping of a two-dimensional random variable, of uniform probability distribution on a circle, to four reference vectors. Fig. 8.4 shows generated random points on a circle and the four streams of reference vector points. The approximately equal distances between the limit points indicate the correct adaptation of the two-dimensional reference vectors $\boldsymbol{q}_k$ to a uniform circular distribution. Closed geometrical forms of stochastically distributed points in multi-dimensional space are often observed when chaotic dynamical phenomena are analyzed. This example demonstrates that the method developed is suitable for an optimal representation of chaotic attractors by a finite set of representative points.

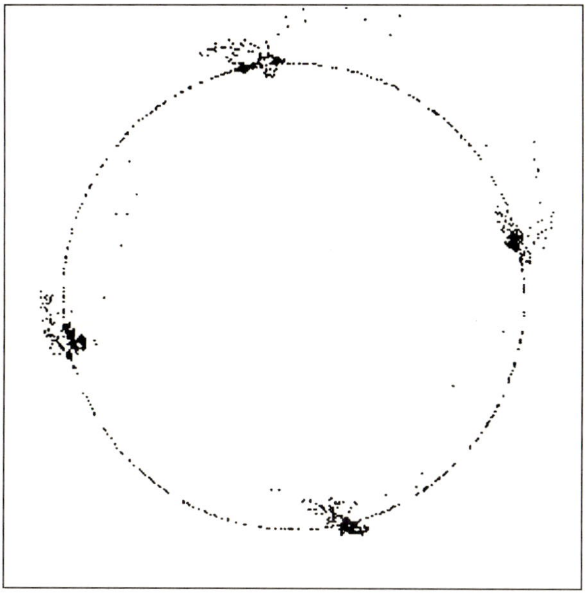

Fig. 8.4. Adaptation of four reference vectors to a circular distribution. $\sigma = 0.2$

8.6 Formal Neurons and the Self-Organization Process

The adaptation of prototypes described by Eq.(8.21) is the consequence of the specific form of coefficients $\boldsymbol{B}_l$ and $\boldsymbol{C}_{kl}$ and it renders possible the following interpretation. Let us consider that a variable $\boldsymbol{X}$ is detected by an array of sensors which send the signals to the discrete information processing system. This system first forms the samples of the input variable and then generates the set of reference vectors $\boldsymbol{q}_1, \ldots, \boldsymbol{q}_K$. We interpret these reference vectors as the contents of the memory units. The reference vectors are stored by adapting parameters of response characteristics of the memory units. We further call these memory units *formal neurons.* The sample vector transmitted from the sensors to each neuron, results in adaptive changes of the parameters according to Eq.(8.21). In this context, the parameter N can be treated as a discrete description of time and the adaptation of the reference vectors can be considered as a dynamical process. Eqs. (8.23) and (8.24) show that the coefficients $\boldsymbol{B}_l$ and $\boldsymbol{C}_{kl}$ are non-linearly dependent on sample values and reference vectors. Non-linearity is related to the Gaussian function which describes the neuron excitation by the input sample vector. The first term in Eq. (8.24), which determines the vector $\boldsymbol{B}_l$, contains the function

$$v_l(\boldsymbol{x}) = \exp[-\|\boldsymbol{x} - \boldsymbol{q}_l\|^2/4\sigma^2] \ . \tag{8.25}$$

This function describes the excitation of the l-th neuron by the input sample $\boldsymbol{x}$. That neuron whose parameter $\boldsymbol{q}_l$ is most similar to the sample value is that most excited by the input; the neurons are thus selectively sensitized. The excitation state of all neurons is described by the vector

$$\mathbf{V}(\boldsymbol{x}) = (v_1, \ldots, v_K) \tag{8.26}$$

which can be treated as an encoded parallel system output.

Each sample of the sensory signal results in the excitation of the output as well as the adaptation of the reference vectors. Their changes are not only dependent on the excitation caused by the input signal but also on the mutual excitations between neurons described by the function

$$v_{kl} = \exp[-\|\boldsymbol{q}_k - \boldsymbol{q}_l\|^2/4\sigma^2] \ . \tag{8.27}$$

This function provides a basis for the description of the interactions between formal neurons.

In order to qualitatively explain how the adaptation proceeds, let us analyze the properties of the terms appearing in Eqs. (8.23) and (8.24). Consider the start of the iteration, when the change of reference vectors $\Delta\boldsymbol{q}_l$ is given by coefficient $\boldsymbol{B}_l$ only, and take into account the index l for which the reference vector is closest to the input $\boldsymbol{x}$. The corresponding excitation function v_l is then the largest component of the excitation vector $\boldsymbol{V}$. It is also much greater than any of the mutual excitation functions v_{kl} , so that the sum in Eq. (8.24) can be replaced with only the first term. For the most excited neuron we expect $v_l \approx 1$ and its adaptation can be approximately described by:

$$\boldsymbol{B}_l \approx \frac{K}{N+1}(\boldsymbol{x}_{N+1} - \boldsymbol{q}_l) \ . \tag{8.28}$$

This indicates that the input signal influences the most excited neurons by attracting their reference vectors towards the input sample vector. This is also the most conspicuous effect of the adaptation.

Let us consider now the second term of the coefficient $\boldsymbol{B}_l$ which is represented by the sum. It includes the mutual excitation functions v_{kl} and thus describes the *synergy or self-organization* of neurons. With this term the adaptation of neurons which are strongly excited by the input is included. Because of the negative sign before the sum in Eq.(8.24) the interaction between the reference vectors is repulsive. For a reference vector lying in the middle of the range of the variable $\boldsymbol{X}$, the influences from all reference vectors approximately cancel. In contrast to this, the reference vectors at the edges of the range are repelled away from its center by this interaction. The mutual interaction thus counter-balances the attractive influence of the input signal and maintains the correct distribution of reference vectors at the edges of the range of the variable $\boldsymbol{X}$. With an increasing number of reference vectors, the importance of this term decreases.

Additional computer experiments have shown that a rough adaptation is already obtained by the initial iteration by which only the coefficient $\boldsymbol{B}_l$ is taken into account in Eq.(8.21). However, a more accurate adaptation is obtained by including also the second term in the right side of Eq.(8.21). This term describes the mutual adjustment of changes of the reference vectors. It depends on the coefficient $\boldsymbol{C}_{kl}$ which resembles in one dimension the so called "Mexican-hat function" qualitatively described by Kohonen in the study of the competitive interaction between the neurons. [14, 15, 16] Let us assume that during an iteration, the value of a certain reference vector is increased by $\Delta \boldsymbol{q}_k$. In the next iteration this change results in a reduction of the adjacent reference vectors, with similar values, while increasing the more distant reference vectors. The successive iterations thus describe the competition and cooperation between neurons leading to a refined adaptation. Similar properties have also been found in the lateral interaction research of neurons in some biological networks and their artificial models. [14, 16, 20, 22] We can therefore conjecture that the absolute maximum-entropy principle could be utilized when explaining the basis for the evolution of biological neural networks.

Our results are independent of the labeling order of the neurons. Therefore the reference vectors can be labeled in any topological order. For example, in a one-dimensional case, we may have $q_1 < q_2 < \ldots < q_K$, by which the neighborhood of each neuron is defined. Since the coefficients $\boldsymbol{B}_l$ and $\boldsymbol{C}_{lk}$ depend only on the differences of the input signal and the reference vectors or the differences of the reference vectors, the topological order is not influenced by the self-organization and it is thus preserved during adaptation of the neurons. This property is valid also in a multi-dimensional case and Kohonen [14, 16] used it in his formulation of topological mappings. His rule is based on the assumption that only the most excited neuron and its closest neighbors are adapted according to the rule

$$\Delta \boldsymbol{q}_l \approx \alpha(N)\,(\boldsymbol{x}_{N+1} - \boldsymbol{q}_l)\,\eta_l \,, \tag{8.29}$$

where the neighborhood as well as the coefficient $\alpha(N)$ must be determined from experience. This rule corresponds to the first approximation according to the absolute maximum entropy principle which is obtained when $\alpha(N) = K/(N+1)$ is selected in the above equation. Calculation of the excitation functions and the iterative treatment of Eq.(8.21) is then not needed which is an advantage for simulations of self-organization on a sequential digital computer. However, Kohonen's rule asymptotically leads only to an approximately optimal solution. Since the mutual interaction term of the coefficient $\boldsymbol{B}_l$ is missing in this rule, the resulting self-organization is not accurate at the edges of the distribution, as has already been observed by Kohonen himself. [14] On the other hand, the influence of competition and cooperation can be approximately taken into account by a proper selection of neighbors. This is confirmed by Figs. 8.5 and 8.6 which illustrate the results obtained using Kohonen's self-organization procedure described by Eq.(8.28) when the same random variable was used, as that used to generate Figs. 8.1 and 8.2. Comparison of both examples also shows that when Kohonen's rule is applied, the reference vectors converge to their final values more slowly . Similar conclusions follow from a comparison of Figs. 8.4 and 8.7 which show numerical results obtained in a two-dimensional case.

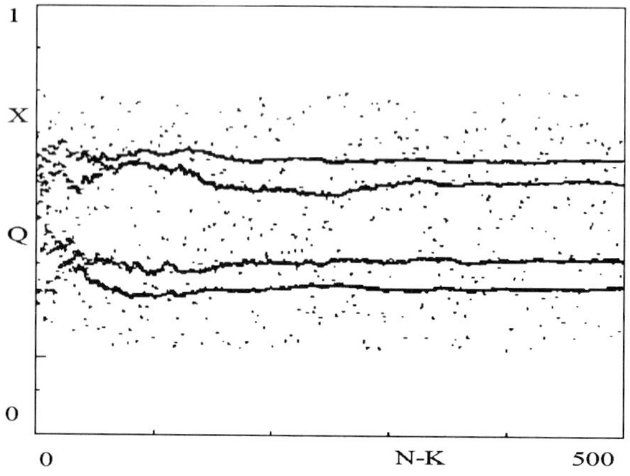

Fig. 8.5. The problem corresponding to Fig. 8.1 solved by Kohonen's rule

The advantage of Kohonen's algorithm, which stems from the topological connection between neurons, can be preserved by using the perturbation treatment of the maximum-entropy problem. For this purpose, one can determine the distances between neural parameters in the initial period of adaptation. The adaptation of an individual neuron is influenced principally by the other neurons having similar parameters. Therefore we can define its

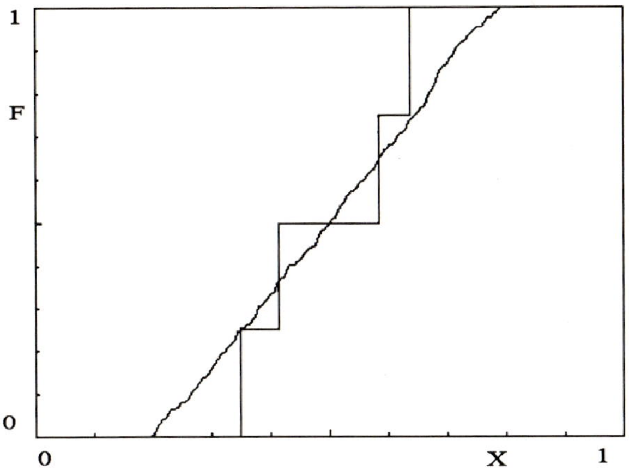

Fig. 8.6. The problem corresponding to Fig. 8.2 solved by Kohonen's rule

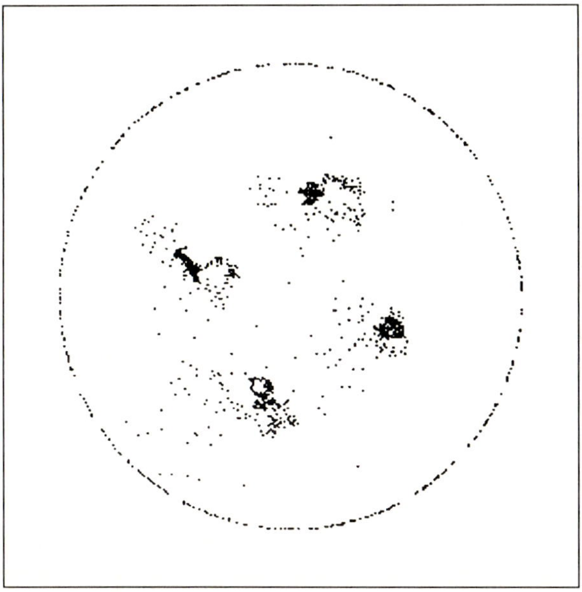

Fig. 8.7. The problem corresponding to Fig. 8.4 solved by Kohonen's rule

"neighbors" with respect to the distance of the representative parameters and in subsequent calculations take into account only a few nearest neighbors. Usually this efficiently reduces the computing time that is needed to carry out the adaptation on a sequential computer. However, on a parallel computer this step is not needed. By such a selection of the neighborhood we are, in fact, defining an adaptation of the neural network in accordance to the task it is performing and not *a priori* as is most often done. According to the self-organization rule stemming from the absolute maximum entropy principle, the excitation and interaction of formal neurons is governed by three functions represented by the Gaussian function and the terms $\boldsymbol{B}_l$ and $\boldsymbol{C}_{kl}$. For a one-dimensional case, the dependence of these functions on the mutual distance between the neural parameters is shown in Fig. 8.8.

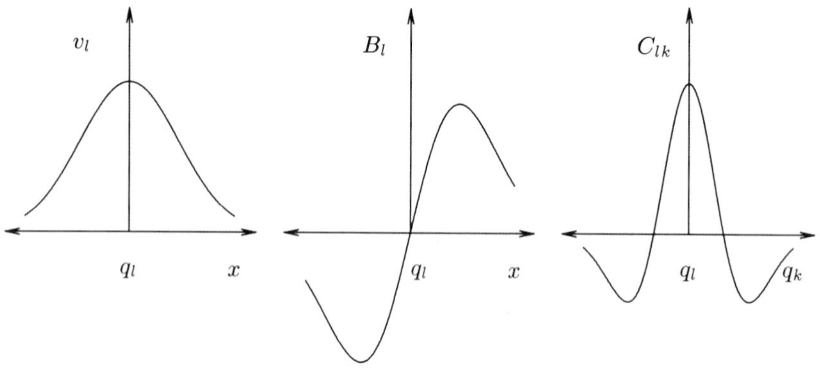

Fig. 8.8. Functions describing the excitation and interaction terms in self-organized adaptation rule of formal neurons in one dimension

The derivation of the self-organization rule based on the perturbation treatment of the maximum-entropy problem yields the factor $K/(N+1)$ for the adaptation rate in the expression for $\boldsymbol{B}_l$. In a real system, the number of samples N corresponds to the time of adaptation which is a measure of the "age" of the system. Our derivation shows that with increasing age, the system becomes ever less adaptable; which is a well-known property of biological systems as well. In our description, we consider the number of prototypes to be determined by the construction of the adaptive system. Adaptation of the system to clustered distributions has revealed that the prototypes tend to concentrate in these clusters, but the important question about the optimization of the number of prototypes is still open. If an analogy with biological neural networks is considered, then we might conjecture that the solution of this problem can be obtained by the evolution of a population of similar networks intended for a specific task.

An algorithm for adapting a one-dimensional set of reference vectors, similar to that presented here but employing sigmoidal reference functions, such as, for example the error-function or the tanh-function, was derived. No essential difference was observed in the numerical results obtained when using the sigmoidal or Gaussian functions. A detailed analysis shows that the sigmoidal function is favourable when explaining this algorithm for the one-dimensional case in terms of the cumulative probability function, while application of Gaussian function enables a more lucid explanation of the algorithm in a multi-dimensional case. The Gaussian filter function is more appropriate for the further application of reference vectors in the modeling of natural phenomena and the prediction of chaotic processes. [5, 7] Until now, no criterion has been found for choosing the optimal filter function although based on the derivations presented in the subsequent chapters, it appears that the Gaussian function is most appropriate for most practical applications.

In the literature on neural networks the excitation of a neuron by the input signal is usually described by a sigmoidal response function of a scalar product between the input vector and the reference vector of the neuron. [8, 15, 22] In contrast to this, we describe the excitation by the Gaussian function of the difference between input and the reference vector. In order to emphasize this difference, we shall call the memory units described in this chapter as *Gaussian formal neurons.* In practical applications, the signals obtained from the sensors are often normalized. This property is then transferred also to the reference vectors. Without essential loss of generality we can assume that

$$||\boldsymbol{x}||^2 = ||\boldsymbol{q}||^2 = 1 \tag{8.30}$$

and obtain

$$||\boldsymbol{x} - \boldsymbol{q}||^2 = 2\,(1 - \boldsymbol{x} \cdot \boldsymbol{q}) \;, \tag{8.31}$$

where the dot indicates the scalar product. Because of the normalized vectors, the value in parenthesis is limited:

$$0 \leq (1 - \boldsymbol{x} \cdot \boldsymbol{q}) \leq 2 \;. \tag{8.32}$$

Therefore the excitation function

$$v_l(\boldsymbol{x}) = \exp[-(1 - \boldsymbol{x} \cdot \boldsymbol{q})/2\,\sigma^2] \tag{8.33}$$

increases monotonically from $\exp[-1/\sigma^2]$ to 1 when the vector $\boldsymbol{x}$ changes from $-\boldsymbol{q}$ to $\boldsymbol{q}$, that is from the least to the greatest possible similarity. This property is qualitatively the same as at the sigmoidal function of the scalar product of $\boldsymbol{x} \cdot \boldsymbol{q}$.

Various applications of self-organization have been reported in the literature. [14, 15, 16, 8] Another application was recently proposed, relative to the automatic modeling of physical phenomena, on the basis of empirical data. [5] In an optimal automatic modeler, a self-organizing memory is one of its most essential elements. It is needed to store the empirical information obtained during learning and to make it available for the optimal estimation or prediction. We shall formulate some basic principles for its construction in the forthcoming chapters.

One problem that must be mentioned relative to the above, is related to the selection of the window function. In order to facilitate a comparison of singular distributions, we use a window function whose width can be arbitrarily selected. This arbitrariness is the principal weakness of Parzen's approach and it can be avoided by utilizing kernel functions which depend on the data. In the case presented here, we partially avoided this problem by selecting the parameter σ in accordance with the standard deviation of the variable $\boldsymbol{X}$ estimated from the empirical data. As has been shown in Chap. 4 on the estimation of probability densities from empirical data, a more appropriate selection of window width stems from the principle of maximal self-consistency. This, however, results in a more complicated treatment and equations and does not permit a simple interpretation in terms of the maximum-entropy principle. The application of spherical Gaussian functions does not appear to be the most suitable for describing normal, multivariate probability distributions whose covariances differ from 0. It has recently been demonstrated that the self-organization formalism presented in this chapter can be generalized to the case with elliptical Gaussian functions spread around prototypes. [17, 18]

9. Modeling by Non-Parametric Regression

9.1 The Problem of an Optimal Prediction

In the previous chapter on the adaptive modeling of natural laws it was stated that tasks associated with such modeling included the estimation and storage of the probability distribution, as well as the development of a method for its effective application. The fundamental problem related to the application of a natural law is the formulation of a method by which some unknown property of Nature can be predicted on the basis of information obtained by partial observation and a known natural law. If a natural law which is expressed as a functional relationship between various physical variables is to be applied, the task is to find the values of some variables from the given values of others by arithmetic procedures. If a natural law is represented by a model contained in a certain physical system then this task corresponds to the projection of a set of inputs into a set of outputs. However, this concept must be generalized when a natural law is represented by a probability density. To do this, we use some of the fundamental concepts of prediction theory which will be briefly reviewed in the next part of this section. The main goal of this chapter is to demonstrate the solution of several problems related with an optimal application of empirical information, stored as a set of representative points in a discrete memory of a modeler.

Let us consider a phenomenon that can be characterized by a two-dimensional vector random variable $\boldsymbol{Z} = (X, y)$ and let us assume that the corresponding joint probability density $f(\boldsymbol{Z})$ has large values along some line that can be described by a function $\widehat{y} = \widehat{y}(x)$. If we associate with each measured value x of the variable X the value $\widehat{y}(x)$ then we are, in fact, mapping the random variable X to a new random variable $\widehat{Y}$ called the *predictor* of the variable Y. However, because of the random nature of the variable Y there can exist various predictors that better or worse correspond to the possible realizations of this variable. The problem thus arises how one selects among all possible predictors the optimal one which provides the best possible prediction. The standard problem about the meaning of "the best possible prediction" is avoided by introducing as a reference an acceptable measure of discrepancy between the predictor and the variable being predicted. The meaning of the words "best" or "optimum" is then always to be with respect to the accepted measure of discrepancy. Among various measures of the discrepancy, the mean-square difference between the random variable Y and its predictor $\widehat{Y}$ is most often utilized. It is defined as the value given by statistical average

$$\mathcal{D} = \mathrm{E}\left[(\widehat{Y} - Y)^2\right] = \int [\widehat{y}(x) - y]^2 \, f(x, y) \; dx \; dy \; . \tag{9.1}$$

In accordance with the deterministic treatment of natural laws, it is assumed that there exists a deterministically describable functional relationship between the variables X and Y and that measurements yield disturbed values of these variables. The probability distribution then describes how likely a certain disturbance or measurement error occurs. Consequently, the above measure is called the mean-square error. However, in our treatment we have adopted the probability distribution itself as the basis for the description of the natural phenomena and consequently we can interpret the above measure only as a measure of goodness or the applicability of the predictor for the description of the observed properties of Nature.

An optimal predictor is the function $\widehat{y}(x)$ that minimizes the functional $\mathcal{D}$. We can determine it using the calculus of variations. For the optimal predictor, the variation $\delta\mathcal{D}$, resulting from the variation $\delta\widehat{y}(x) = \epsilon\eta(x)$, with the amplitude $\epsilon > 0$ and the form of variation $\eta(x)$ described by an arbitrary selected function, must satisfy the expression

$$\delta\mathcal{D} > 0 \quad \text{for} \quad \epsilon \to 0 \tag{9.2}$$

regardless of the function $\eta(x)$. This means that at $\epsilon = 0$, $\delta\mathcal{D}$ must have a minimum as a function of the amplitude ϵ, which yields the equation

$$\frac{\partial \delta\mathcal{D}}{\partial \epsilon}\Big|_{\epsilon=0} = 2 \int [\widehat{y}(x) - y] \; \eta(x) \; f(x, y) \; dx \; dy = 0 \; . \tag{9.3}$$

If we express the joint probability density in terms of the density of the conditional probability distribution as $f(x, y) = f(y|x) \; f(x)$, this equation can be rewritten in the form

$$\int \left\{ \int [\widehat{y}(x) - y] \; f(y|x) \; dy \right\} \; \eta(x) \; f(x) \; dx = 0 \; . \tag{9.4}$$

This equation must be satisfied regardless of the function $\eta(x)$. This is generally the case when the first integral equals zero. That is

$$\int [\widehat{y}(x) - y] \; f(y|x) \; dy = 0 \; . \tag{9.5}$$

Since the integral extends only over y, the first term can be brought out and we obtain for the optimal estimator, the following conditional average of y, at a given x

$$\widehat{y}(x) = \int y \; f(y|x) \; dy \; . \tag{9.6}$$

Using the definition of the conditional probability density we obtain

$$\widehat{y}(x) = \frac{\int y\ f(x,y)\ dy}{\int f(x,y)\ dy}\ . \tag{9.7}$$

This result can readily be generalized to the multi-variate case as follows. Let us consider a phenomenon that can be characterized by a multi-dimensional vector random variable $\boldsymbol{Z} = (X_1,\ \ldots,\ X_D)$ and let the corresponding joint probability density be written as $f(\boldsymbol{Z})$. Let us assume that partial information about the phenomenon is given by a truncated sub-vector which is comprised of m measured components $\boldsymbol{x} = (x_1,\ \ldots,\ x_m,\ \emptyset)$ and we wish to predict the remaining components comprising the next truncated vector, that is $\boldsymbol{y} = (\emptyset,\ x_{m+1},\ \ldots,\ x_D)$ where the sign $\emptyset$ denotes the missing components. The complete datum about a state of the phenomenon is then represented by the concatenation of both truncated sub-vectors as $\boldsymbol{z} = \boldsymbol{x} \oplus \boldsymbol{y} = (x_1,\ \ldots,\ x_m,\ x_{m+1},\ \ldots,\ x_D)$. In accordance with the previous treatment, we presume that there exists a relationship between two vector random variables represented by both sub-vectors. It is described by the vector function

$$\widehat{\boldsymbol{Y}} = \widehat{\boldsymbol{Y}}(\boldsymbol{X})\ . \tag{9.8}$$

This function can be optimally described by the conditional average of the variable $\boldsymbol{Y}$ at a given realization of the variable $\boldsymbol{X}$

$$\widehat{\boldsymbol{y}}(\boldsymbol{x}) = \int \boldsymbol{y}\ f(\boldsymbol{y}|\boldsymbol{x})\ dx_{m+1}\ \ldots\ dx_D \tag{9.9}$$

or

$$\widehat{\boldsymbol{y}}(\boldsymbol{x}) = \frac{\int \boldsymbol{y}\ f(\boldsymbol{x},\boldsymbol{y})\ dx_{m+1}\ \ldots\ dx_D}{\int f(\boldsymbol{x},\boldsymbol{y})\ dx_{m+1}\ \ldots\ dx_D}\ . \tag{9.10}$$

In these expressions the particular value of variable $\widehat{\boldsymbol{Y}}$ is denoted as $\widehat{\boldsymbol{y}}$. The corresponding optimal predictor

$$\widehat{\boldsymbol{Y}} = \mathrm{E}[\boldsymbol{Y}|\boldsymbol{X}] \tag{9.11}$$

is called the general non-parametric regression of the dependent variable $\boldsymbol{Y}$ on the independent variable $\boldsymbol{X}$.

In an application, the main task of any modeler of natural phenomena is the estimation of the conditional probability density from the stored joint probability density and the estimation of the conditional average corresponding to the given partial data $\boldsymbol{x}$ about the phenomenon. These partial data can generally be treated as the input to the modeler, while the conditional average $\boldsymbol{y}$, or equivalently, the completed vector $\boldsymbol{z} = \boldsymbol{x} \oplus \boldsymbol{y}$ can be treated as the particular output from the modeler. The corresponding completion of data is generally called *associative recall* from the memory. In this approach the truncation of the vector $\boldsymbol{Z}$ into the subsets of given or estimated data is decided by the user of the modeler. This is not the same as the biological modeling that is performed in the brain in which the properties of the data itself dictate the division into the given and estimated portions. How this can be automatically performed in a modeler that is simulated on an information processing system is still not known.

9.2 Parzen's Window Approach to General Regression

In practical applications of data regression analysis for predictive purposes, the non-parametric approach which is based on a conditional average is often not utilized. The reason for this is that empirical data are usually described by a finite number of sample points. In this case, the conditional average can also be determined only for a finite number of given data and thus it cannot be utilized to describe the relation between an arbitrary given datum $\boldsymbol{x}$ and the corresponding predicted value $\boldsymbol{y}$. If, for example, the datum $\boldsymbol{x}$ does not coincide with at least one of the data specified in the set of samples, then the conditional average cannot be determined without additional assumptions about the properties of the phenomenon. In order to avoid this problem, a parametric regression is most often used, but this diminishes the generality of the description of relations between data. The problem is the same as the estimation of the probability density of a continuous random variable from a finite set of sample values that can be efficiently avoided by following Parzen's window approach. Therefore, the conditional average predictor can be treated as the basis of an automatic modeler of natural phenomena and related neural network-like information processing systems. [5, 6, 7, 27, 28, 26] Our task is now to present the corresponding derivation, to show how it can be generalized to self-organized systems and to demonstrate some important applications of practical importance.

Let us assume that $\boldsymbol{Z}$ represents a continuous D-dimensional random vector to which corresponds the data base comprised of N samples $\{\boldsymbol{z}_1, \ldots, \boldsymbol{z}_N\}$. Further let $w(\boldsymbol{z}, \sigma)$ represent a spherical Gaussian window function of width σ

$$w(\boldsymbol{z}, \sigma) = \frac{1}{(2\pi)^{D/2}\sigma^D} \exp\left[\frac{-||\boldsymbol{z}||^2}{2\sigma^2}\right] . \tag{9.12}$$

The appertaining singular empirical probability distribution density

$$\phi_e(\boldsymbol{z}) = \frac{1}{N} \sum_{n=1}^{N} \delta(\boldsymbol{z} - \boldsymbol{z}_n) \tag{9.13}$$

can be represented by a smooth Parzen's estimator

$$f_e(\boldsymbol{z}) = \frac{1}{N} \sum_{n=1}^{N} w(\boldsymbol{z} - \boldsymbol{z}_n, \sigma) . \tag{9.14}$$

According to Eq. A.36 from Appendix A, the conditional probability density is expressed by this estimator as

$$f(\boldsymbol{y}|\boldsymbol{x}) = \frac{f(\boldsymbol{x},\, \boldsymbol{y})}{\int f(\boldsymbol{x},\, \boldsymbol{y})\, dx_{m+1}\, \ldots\, dx_D}$$

$$= \frac{\sum_{n=1}^{N} w(\boldsymbol{z} - \boldsymbol{z}_n,\, \sigma)}{\sum_{n=1}^{N} \int w(\boldsymbol{z} - \boldsymbol{z}_n,\, \sigma)\, dx_{m+1}\, \ldots\, dx_D} . \tag{9.15}$$

The integral in the denominator can readily be evaluated by expressing a spherical Gaussian window function as the product of Gaussian functions representing normal distributions in subspaces

$$w(\boldsymbol{z},\, \sigma) = \frac{1}{(2\pi)^{m/2}\sigma^m} \exp\left[\frac{-||\boldsymbol{x}||^2}{2\sigma^2}\right] \frac{1}{(2\pi)^{(D-m)/2}\sigma^{(D-m)}} \exp\left[\frac{-||\boldsymbol{y}||^2}{2\sigma^2}\right]$$

$$= w(\boldsymbol{x},\, \sigma)\, w(\boldsymbol{y},\, \sigma) . \tag{9.16}$$

Integration of the denominator of Eq. 9.15 with respect to the components representing the variable $\boldsymbol{y}$ yields factor 1 so that we obtain

$$f(\boldsymbol{y}|\boldsymbol{x}) = \frac{\sum_{n=1}^{N} w(\boldsymbol{y} - \boldsymbol{y}_n,\, \sigma)\, w(\boldsymbol{x} - \boldsymbol{x}_n,\, \sigma)}{\sum_{n=1}^{N} w(\boldsymbol{x} - \boldsymbol{x}_n,\, \sigma)} . \tag{9.17}$$

By using this function we can express the conditional average predictor as

$$\widehat{\boldsymbol{y}}(\boldsymbol{x}) = \frac{\sum_{n=1}^{N} \boldsymbol{y}_n w(\boldsymbol{x} - \boldsymbol{x}_n,\, \sigma)}{\sum_{k=1}^{N} w(\boldsymbol{x} - \boldsymbol{x}_k,\, \sigma)} = \sum_{n=1}^{N} \boldsymbol{y}_n C_n(\boldsymbol{x}) . \tag{9.18}$$

Here the coefficients $C_n(\boldsymbol{x})$ are defined by the expression

$$C_n(x) = \frac{w(\boldsymbol{x} - \boldsymbol{x}_n,\, \sigma)}{\sum_{k=1}^{N} w(\boldsymbol{x} - \boldsymbol{x}_k,\, \sigma)} . \tag{9.19}$$

These coefficients represent the *measure of similarity* between a given vector $\boldsymbol{x}$ and the truncated sample vector $\boldsymbol{x}_n$. They satisfy the relations

$$\sum_{n=1}^{N} C_n(\boldsymbol{x}) = 1 , \tag{9.20}$$

$$0 \leq C_n(\boldsymbol{x}) \leq 1 \tag{9.21}$$

which indicates that the $C_n(\boldsymbol{x})$ can be treated like a probability that is defined on the subspace used for presentation of the truncated vectors $\boldsymbol{x}$. When the given sample data is separated for much more than σ from the sample values the above expression for the coefficients $C_n(\boldsymbol{x})$ can lead to numerical difficulties in practical applications. In this case it is convenient to express the coefficients by the formula

$$C_n(\boldsymbol{x}) = \left\{ \sum_{k=1}^{N} \frac{w(\boldsymbol{x} - \boldsymbol{x}_k,\ \sigma)}{w(\boldsymbol{x} - \boldsymbol{x}_n,\ \sigma)} \right\}^{-1}$$

$$= \left\{ \sum_{k=1}^{N} \exp\left[\frac{||\boldsymbol{x} - \boldsymbol{x}_n||^2 - ||\boldsymbol{x} - \boldsymbol{x}_k||^2}{2\sigma^2} \right] \right\}^{-1} \tag{9.22}$$

which is less sensitive to numerical treatment. The alternative is to increase the value of σ.

The conditional average estimator expressed by the window functions exhibits several convenient properties that are especially advantageous for an automatic modeling of natural phenomena. It is most significant that we do not require any prior information about the phenomenon; we only need the sample values of the variable $\boldsymbol{Z}$. This demonstrates that the conditional average acts as a general predictor rather than a specific one which is usually the case when a parametric approach is utilized. Another advantageous property is that we can arbitrarily select those components of the vector $\boldsymbol{Z}$ which represent the given datum $\boldsymbol{X}$. This means that the separation into dependent and independent variables occurs only during the estimation and is thus task-dependent and it is therefore not incorporated into the modeling. This is not the case when a parametric approach is used. Although the estimator is expressed as a linear combination of truncated vectors $\boldsymbol{y}_n$, the coefficients $C_n(\boldsymbol{x})$ are highly non-linear functions of the given data. Consequently essentially non-linear phenomena can be modeled by this approach. These characteristics make the application of the conditional average useful for recursive estimation and simulation of stochastic processes which is one of the most challenging problems in the modeling of natural phenomena. [5, 7, 8]

Using the derived optimal predictor $\widehat{\boldsymbol{Y}}$, the complete data vector can be estimated by the concatenation

$$\boldsymbol{Z}_c = \boldsymbol{X} \oplus \widehat{\boldsymbol{Y}}(\boldsymbol{X}) = \mathcal{Z}_y(\boldsymbol{X}) \ . \tag{9.23}$$

Here the index c denotes a concatenation and y indicates that portion of the vector $\boldsymbol{Z}$ that is estimated. Determination of the function $\mathcal{Z}_y(\boldsymbol{X})$ represents the main task which must be performed by the modeler. To do this, the modeler must be able to store the samples of the variable $\boldsymbol{Z}$, to accept partial data and to estimate from the partial data the missing components of the complete vector. In doing this, the modeler can either store all of the data or only a reduced set of representative prototypes which have been obtained by the self-organization procedure. For the reduced set of data vectors the predictor is described by the equations

$$\widehat{\boldsymbol{y}}(\boldsymbol{x}) = \frac{\sum_{n=1}^{N} \boldsymbol{q}_{y,n}\ w(\boldsymbol{x} - \boldsymbol{q}_{x,n},\ \sigma)}{\sum_{k=1}^{N}\ w(\boldsymbol{x} - \boldsymbol{q}_{x,k},\ \sigma)} = \sum_{n=1}^{N} \boldsymbol{q}_{y,n} C_n(\boldsymbol{x}) \ , \tag{9.24}$$

where $\boldsymbol{q}_{x,n}$ and $\boldsymbol{q}_{y,n}$, denote the truncated prototype vectors corresponding to the set of given and estimated components.

9.3 General Regression Modeler, Feedback and Recognition

In order to implement the procedure described in the previous section to obtain a general modeler of natural phenomena, we must include into the system an array of sensors, a memory, and a central processing unit that is capable carrying out the following tasks:

1. *Obtain signals from the array of sensors, preprocess them and transform them into data vectors.*
2. *Control the self-organized storing of empirical data.*
3. *Truncate the data.*
4. *Estimate the missing data by the general regression.*
5. *Complete the partial data.*
6. *Transmit the completed data to the user.*

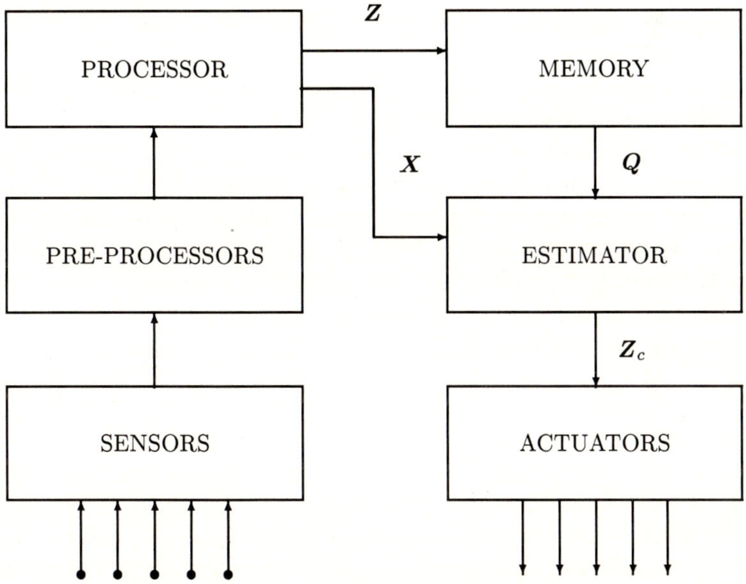

Fig. 9.1. Scheme of the general regression modeler

The structure of such a modeler is shown in Fig. 9.1. According to this scheme a modeler should be capable of operating in two essentially different operating modes:

1. Adaptation or learning and,
2. Prediction or estimation.

During learning, the internal representation of the sensory signals is formed by adaptation of the prototype sample points stored in the memory by a self-organization process. While during prediction, the partial input data and the stored prototypes are utilized for the prediction of the missing data. The completed data vector $\boldsymbol{Z}_c$ can then be interpreted as the output from the central processing unit or as the main result of the application of the modeler. However, only the given partial data $\boldsymbol{X}$ which is input to the system influences the system output. The components of $\boldsymbol{X}$ must be directly transferred to the corresponding components of the output $\boldsymbol{Z}_c$, while the complementary components are determined by the optimal predictor $\widehat{\boldsymbol{Y}}(\boldsymbol{X})$ and are concatenated to the input data to obtain the complete data vector.

Based on the interpretation of a self-organized process in terms of formal neurons we can interpret the operation of the general modeler as follows: During the learning mode of operation the sensors provide complete data about the observed phenomenon. Those neurons in the memory which are most excited by the input signals will most efficiently change their characteristic parameters that correspond to stored prototype sample points of the random variable $\boldsymbol{Z}$ being detected. During the prediction, the sensors provide only some signals. These represent the truncated input data vector $\boldsymbol{X}$ which excites the neurons of the memory as described by the partial excitation functions

$$v_n(\boldsymbol{x}) = w(\boldsymbol{x} - \boldsymbol{q}_{x,n},\ \sigma) \ . \tag{9.25}$$

The complete set of these functions $\{v_n(\boldsymbol{x}),\quad n = 1,\ \ldots,\ N\}$ represents the encoded information about the input signal. When normalized by their sum these functions are changed into the coefficients $C_n(\boldsymbol{x})$ that represent the measure of the similarity between the input signal and the corresponding parameters stored in the neurons of the memory. The excitations then determine the composition of the predictor. The general modeler operating in this way corresponds to an artificial neural network comprised of formal Gaussian neurons. [7, 27, 28]

The expression of the completed output of the system, as described by Eqs. (9.23) and (9.24), is very similar to the reconstruction of the output of a vector quantizer system. A general regression modeler thus appears as a kind of a generalized vector quantizer. However, there is an essential difference. In the vector quantizer the output is completely reconstructed from the encoded data that represents a complete input signal, although it may be noise corrupted, while here the input information is incomplete and only the missing components of the output are estimated. If we compare the application of the modeler with one based on biological neural networks, we find that healthy biological neural networks, just as vector quantizers, always receive a complete set of signals from all the sensors and that the situation of a truncated input vector is never encountered although it is intuitively acceptable. For instance, we can imagine observing the world around us with only one open eye. In this case our brain by an internal operation excludes the dummy signals provided by the closed eye so that only the truncated set of signals provided by the open eye is used in the subsequent observations. In our approach we have left it to the user of the modeler to decide which sensory channels are active, or in other words, how to truncate the input data. Contrary to this, it seems evident that the selection into *active* and *non-active* information channels is regulated in the brain by the signals transmitted over it, or by the modeler itself. In this case, the brain, or more abstract, an *ego*, can be treated as a hidden user or a kind of *supervisor* of the information. The question therefore arises as how one avoids the *a priori* specification of the number of components in a given truncated vector. This, in a sense restricts the applicability of the modeler when compared to biological neural networks. A related problem is then how one develops a method of an automatic truncation of data. It seems that this problem leads to task-dependent solutions and consequently we will therefore not further discuss it in relation to the properties of a general modeler. We note that when human beings observe the world, they usually associate perception with abstract objects such as a tree, a house, etc., which are subsequently used to describe situations in the Nature. In the process of associating, a set of input signals must somehow be mapped to the notions that can be treated as stored prototypes. It is still not known how the brain forms an appropriate set of prototypes or how the recognition proceeds. The self-organization process in an artificial neural-network could be treated as an aid for the explanation of the formation of prototypes while steps towards the modeling and understanding of recognition are taken in the next sections.

The similarity between the modeler and that part of a brain which forms an internal representation of the world can be further emphasized by adding feedback to the modeler. For this purpose let us define first a complementary system to the system specified by Eqs. (9.23) and (9.24). In the complementary system the components of $\boldsymbol{y}$ are treated as given and are used to estimate the components of $\boldsymbol{x}$. Its operation is then described by the equation

$$\boldsymbol{Z}_{cc} = \widehat{\boldsymbol{Y}} \oplus \widehat{\boldsymbol{X}}(\widehat{\boldsymbol{Y}}) = \mathcal{Z}_x(\widehat{\boldsymbol{Y}}) \,, \tag{9.26}$$

with

$$\widehat{\boldsymbol{x}}(\boldsymbol{y}) = \frac{\sum_{n=1}^{N} \boldsymbol{q}_{x,n}\, w(\boldsymbol{y} - \boldsymbol{q}_{y,n},\, \sigma)}{\sum_{k=1}^{N} w(\boldsymbol{y} - \boldsymbol{q}_{y,k},\, \sigma)} = \sum_{n=1}^{N} \boldsymbol{q}_{x,n} C_n(\boldsymbol{y}) \,. \tag{9.27}$$

By successively linking two systems specified by the functions $\mathcal{Z}_y(\boldsymbol{X})$ and $\mathcal{Z}_x(\widehat{\boldsymbol{Y}})$, a compound system is obtained in which the input vector $\boldsymbol{X}$ is transformed into the output $\boldsymbol{Z}_{cc}$. The compound system generally includes a nonlinear mapping from the space of truncated input data to the same space, or in other words, a reconstruction of the input. A similar mapping is also characteristic for a vector quantizer with the exception that in this case a complete data vector is reconstructed. By such a mapping, an iteration can be defined if the vector $\widehat{\boldsymbol{X}}$ is treated as an input vector as shown in Fig. 9.2.

This corresponds to establishing a feedback in the compound system. The iteration can then be represented by the sequence

$$\boldsymbol{X} \rightarrow \widehat{\boldsymbol{Y}} \rightarrow \widehat{\boldsymbol{X}}^{(1)} \rightarrow \widehat{\boldsymbol{Y}}^{(1)} \rightarrow \ldots \rightarrow \widehat{\boldsymbol{X}}^{(i)} \rightarrow \widehat{\boldsymbol{Y}}^{(i)} \rightarrow \ldots \tag{9.28}$$

which can equivalently be represented by the sequence

$$\boldsymbol{Z}_{cc} \rightarrow \boldsymbol{Z}_{cc}^{(1)} \rightarrow \boldsymbol{Z}_{cc}^{(2)} \rightarrow \ldots \rightarrow \boldsymbol{Z}_{cc}^{(i)}\,, \tag{9.29}$$

where $\boldsymbol{Z}_{cc}^{(i)}$ indicates a composite step of iteration, in which $\boldsymbol{Y}$ is estimated from $\boldsymbol{X}$ and vice versa. This iteration corresponds to a discrete, nonlinear, dynamical process evolving in a multi-dimensional sample space of variable $\boldsymbol{Z}$. The question then arises: Where, if at all, does this process converge?

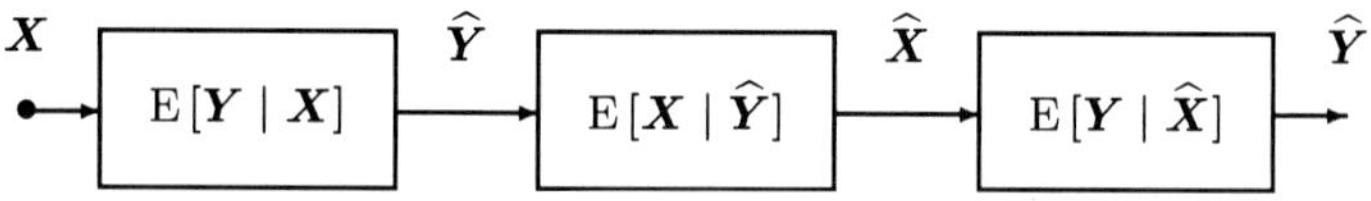

Fig. 9.2. Scheme of iterative mapping

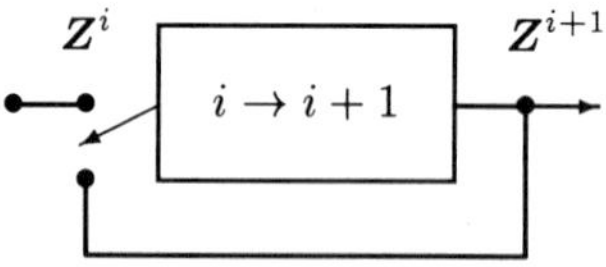

Fig. 9.3. Scheme of a compound system performing iterative mapping through feedback

Numerical simulations using signals obtained from real systems have shown [5, 6] that the vectors $\boldsymbol{Z}_{cc}^{(i)}$ obtained after successive iteration correspond increasingly close to the properties represented by the prototype vectors comprising the data base. If the given data vector $\boldsymbol{X}$ is represented as a noise corrupted version of a corresponding part in one vector of the data base, then by iteration, the noise can be reduced. This means that the prototype vectors stored in the memory can be treated in those examples as attractors in the sample space of variable $\boldsymbol{Z}$ to which the vectors $\boldsymbol{Z}_{cc}^{(i)}$ converge by this iteration. In such a case, the iteration resembles the process of *recognition* of a prototype on the basis of incomplete information provided by a truncated, noisy data set. However, we conjecture that the iteration converges properly only if there is a monotonic relationship between the components of the vector variable used for the description of the phenomenon under consideration.

A problem similar to that described above is met in pattern recognition if the information is specified in terms of the complete vector $\boldsymbol{Z}$ and we wish to associate it with some particular prototype stored in the data base. In this case, the k-th prototype for which the norm of the distance $||\boldsymbol{Z} - \boldsymbol{Z}_k||$ is the minimum is most often taken as the result of recognition. However, various other measures of similarity could be utilized for this purpose as well. [2, 18, 25]. In this case, the pattern recognition can be treated as a dynamical process which begins at some given point $\boldsymbol{Z}$ and proceeds in time towards the attractor point $\boldsymbol{Z}_k$. Such an interpretation was used by Hopfield [16, 17] to explain the dynamics of neural networks that incorporated feedback. He assumed in his formulation that the set of attractor points $\{\boldsymbol{Z}_k\}$ of the dynamical process in a neural network can be simply prescribed by the properties of the neurons in the network. This assumption is not needed in our treatment because it directly follows from the interpretation of the prototypes of the data base stored in the memory of the modeler.

However, a distinction between the case with complete and incomplete given information should be mentioned here. The case of incomplete information appears to be equivalent to the case with complete information only after the first completion of the data vector by prediction of the missing components from the given ones. Such prediction of missing information can generally be called *cross-associative recall.* This is in contrast to an *auto-associative recall* in which a complete vector, like that specifying an attractor point, is obtained from another vector that might eventually be noise corrupted. If we add to the vector utilized in the cross-associative recall an arbitrary disturbance or noise and apply it in an auto-associative recall then the distinction between both cases nearly disappears, although the case with a truncated input vector is logically not equivalent to the case of a complete input vector. Both cases are equivalent only if the noise added to the truncated vector is weak enough so that it does not influence the convergence of the iteration process to the final point. Consequently, a system capable of cross-associative recall appears to be, in a sense, more general than the corresponding auto-associative one although there is the need of a supervisor that dictates how the vectors should be truncated. With respect to this reasoning, we can conclude that a general modeler of natural phenomena can be changed into a general recognizing device by including in it a supervising system, that controls the truncation of vectors, and feedback that enables the iteration procedure.

In accordance with the above discussion it appears that truncation is needed only at the input state of the recognition. After the first prediction, the vector $\boldsymbol{Z}$ is completed and we can carry out the recognition process by using the complete vectors only if we properly define one step of the iteration process. For this purpose we slightly change the predictor formula Eq. (9.24) into the expression

$$\boldsymbol{Z}^{(i+1)} = \sum_{k=1}^{N} \boldsymbol{Q}_k C_k(\boldsymbol{Z}^{(i)}) \, , \tag{9.30}$$

with C defined as a measure of similarity of the input vector $\boldsymbol{Z}$ with the complete prototype vectors $\boldsymbol{Q}$

$$C_n(\boldsymbol{Z}) = \frac{w(\boldsymbol{Z}-\boldsymbol{Q}_n,\ \sigma)}{\sum_{k=1}^{N} w(\boldsymbol{Z}-\boldsymbol{Q}_k,\ \sigma)} \ . \tag{9.31}$$

Here again w denotes a properly selected window function of width σ. Equation (9.30) coincides with the reproduction formula of a vector quantizer and describes an auto-associator. It replaces the two-step mutual estimation from truncated vectors introduced previously by Eqs. (9.24) and (9.27) and yields the one-step iteration process

$$\boldsymbol{Z} \to \boldsymbol{Z}^{(1)} \to \boldsymbol{Z}^{(2)} \to \ldots \to \boldsymbol{Z}^{(i)} \to \ldots \ . \tag{9.32}$$

which is shown schematically in Fig. 9.3.

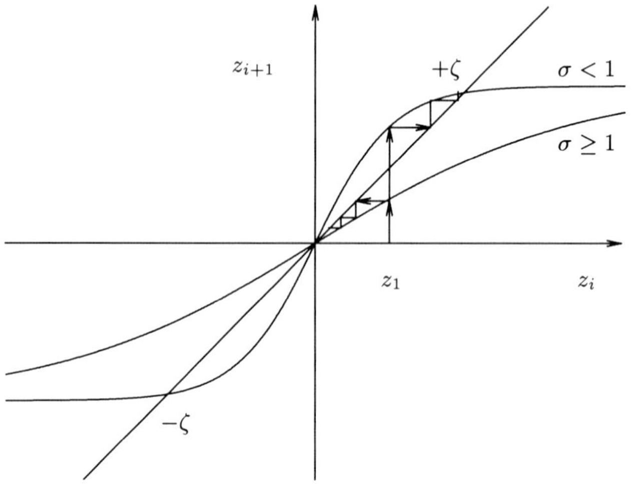

Fig. 9.4. Scheme of the iteration process $z^{(i+1)} = \tanh(z^{(i)}/\sigma^2)$ for $\sigma \geq 1$ and $\sigma < 1$

Example. In order to obtain some idea about the convergence of this iteration process let us treat a one-dimensional example consisting of two prototypes at $q_1 = -1$ and $q_2 = +1$. The iteration process is then described by the function

$$z^{(i+1)} = \sum_{k=1}^{2} q_k C_k(z^{(i)}) = \frac{w(z^{(i)}-1,\ \sigma) - w(z^{(i)}+1,\ \sigma)}{w(z^{(i)}-1,\ \sigma) + w(z^{(i)}+1,\ \sigma)} \ . \tag{9.33}$$

For a Gaussian window function this becomes

$$z^{(i+1)} = \tanh(z^{(i)}/\sigma^2) \ . \tag{9.34}$$

The limit points of the iteration process are determined by the solutions of the equation

$$z = \tanh(z/\sigma^2) \ . \tag{9.35}$$

The iteration process is shown schematically in Fig. 9.4 for $\sigma \geq 1$ and $\sigma < 1$. For $\sigma \geq 1$ there is only one solution $z_0 = 0$. For $\sigma < 1$ there are three solutions: one at $z_0 = 0$ and two at $z_\pm = \pm\zeta(\sigma)$ where $\zeta(\sigma) > 0$ and $\lim_{\sigma\to 0} \zeta(\sigma) = 1$. The solution at $z = 0$ is unstable for $\sigma < 1$ and cannot be reached by the iteration process. The limit point at $+\zeta$ is reached from the starting point $z^{(1)} > 0$ while $-\zeta$ is reached from $z^{(1)} < 0$. As σ changes from $\sigma > 1$ to $\sigma < 1$ a symmetrical single solution becomes unstable and two stable asymmetrical solutions appear. The iteration process thus exhibits so called symmetry breaking.

The broken symmetry in recognition has often been mentioned in relation to oscillations in the perception of ambiguous patterns. [14, 15] What does this symmetry breaking in the above example really mean? Let us assume that there are two prototypes q_1 and q_2 which have been memorized in the modeler and subsequently we present to the modeler an input value. The probability density is estimated by the prototypes as

$$f(z) = \frac{1}{2}\left[w(z - q_1, \sigma) + w(z - q_2, \sigma)\right] \ . \tag{9.36}$$

In order to find its extremal points, we set its derivative equal to zero and obtain the equation $z = \tanh(z/\sigma^2)$ which is identical to Eq. (9.35) that determines the limit points $\pm\zeta$ of the iteration process. If the smoothing parameter σ is greater than half of the distance between prototypes, then the filtered, smooth probability density exhibits only one maximum that lies in the middle between the two prototypes. An input value cannot then be associated with either of the prototypes, but one is led by the iteration process to the most probable value which is lying symmetrically between both prototypes. Let us now assume that the smoothing parameter is reduced to values less than 1. In this case, the corresponding filtered probability density has two maxima in the vicinity of the prototypes. The input value is then drawn by the iteration process to one of the regions of higher probability. One thus associates the final value with that prototype which lies closer to the input value. The iteration process then resembles the nearest neighbor classifier that has been developed for pattern recognition. [2] In this case, the middle value corresponds to a local minimum of the probability density and it therefore represents an unstable point in the iteration process which draws the input value towards the regions of maximal probability. Thus, the recognition ability is usually dependent on the connection between the prototypes represented by the smoothing parameter σ. The limiting point of the iteration process lies approximately in the center of gravity of the prototypes which is in the region extending by approximately σ around the center. We have thus obtained a completely new basis for the interpretation of the window width. This parameter should be further treated as a variable resolution parameter of the modeler by which its recognition ability can be adjusted. A treatment similar to the one-dimensional case with just two stored prototypes can be extended to the multi-point, multi-dimensional case. We conjecture that for $\sigma \to 0$ this process generally converges to the prototype point which is closest to the starting point $\boldsymbol{Z}$ and thus facilitates the recognition of input data in terms of stored prototypes. For $\sigma \neq 0$ we can expect that the iteration process converges to the point that can be taken as the center of a cluster of prototypes that is encompassed by a hyper-sphere of radius σ.

The iteration process we have described above can also be interpreted in terms of the excitations of formal neurons. We have described the iteration by the sequence of vectors $\boldsymbol{Z}^{(i)}$. But each vector $\boldsymbol{Z}$ can be expressed by the excitations of neurons defined as $v_n(\boldsymbol{z}) = w(\boldsymbol{z} - \boldsymbol{q}_n, \sigma)$, which altogether results in the excitation vector

$$\boldsymbol{V}(\boldsymbol{z}) = (v_1(\boldsymbol{z}), \ldots, v_N(\boldsymbol{z})) . \tag{9.37}$$

By recalling the definition of constants C_n we obtain the expression

$$C_n(\boldsymbol{z}) = \frac{v_n(\boldsymbol{z})}{\sum_{k=1}^{N} v_k(\boldsymbol{z})} \tag{9.38}$$

which shows that these coefficients can be interpreted as the relative excitations of neurons. The vector of relative excitations is then defined as

$$\boldsymbol{C}(\boldsymbol{z}) = [C_1(\boldsymbol{z}), \ldots, C_K(\boldsymbol{z})] . \tag{9.39}$$

The iteration process can be then described as

$$\boldsymbol{V}(\boldsymbol{z}^{(i)}) \to \boldsymbol{V}(\boldsymbol{z}^{(i+1)}) , \tag{9.40}$$

or in terms of relative excitations

$$\boldsymbol{C}(\boldsymbol{z}^{(i)}) \to \boldsymbol{C}(\boldsymbol{z}^{(i+1)}) . \tag{9.41}$$

The reconstruction of the vector $\boldsymbol{z}$ is described by the iteration formula

$$\boldsymbol{z}^{(i+1)} = \sum_{k=1}^{N} \boldsymbol{Q}_k C_k(\boldsymbol{z}^{(i)}) \tag{9.42}$$

which can be treated as an auxiliary equation indicating that the iteration can be treated as an entirely internal excitation process developing on the neural network but excited by an external initial input. For the empirical modeling of natural phenomena, Eq. (9.42) appears to be directly interpretable. But when considering an application of empirical modeling, as, for example, related to communication or control, then the encoded presentation of information in terms of excitations may be more advantageous, although both interpretations are non-separable. It also appears that an interpretation in terms of excitations directly invokes the analogy with the operation of biological intelligent systems. Quite intuitively we may also conjecture that the expression of the internal picture about Nature in terms of excitations that represent encoded information is more appropriate for the further description of various cognitive processes and logical reasoning.

9.4 Application of the General Regression Modeler

In this section we show how the general empirical modeler of natural phenomena and the non-parametric regression can be applied to solve a variety of problems in technology or which may be encountered in fields of human activity. We begin with the characterization of sources of sound based on emitted or scattered ultrasonic waves. These examples are drawn from engineering where they form the basis of the non-destructive testing of materials, the identification of manufacturing processes and quality assurance. An example is then presented from the field of civil engineering where the seismic capacity is estimated on the basis of data about structural components. We close this section by describing the empirical modeling of the healing process of a periodontal disease following a surgical procedure. For each of the aforementioned problems an exact analytical description is difficult, if not impossible, but yet, as we shall show, each of these cases is relatively easily amenable to empirical modeling.

9.4.1 Empirical Modeling of Acoustic Phenomena

It is well known that animals and humans are able to recognize, practically instantly, various events in their surroundings from the sound they perceive. For this purpose their brains must associate the sound signals, provided by the sensors in the ears, with the notions which are in their imagination representing various properties of Nature. Similar is the problem of recognizing objects on the basis of touch and the recognition of patterns on the basis of visual perception. It is well known that the ability of recognition is a consequence of learning during the initial phases of the development of an individual. Despite the well-developed theory of sound and a great variety of sensors, signal processors, and computers, at present we are still not able to develop a technical system possessing an equivalent ability. The problem is thus not in the detection and amplification of sound signals but in the proper internal modeling of natural phenomena in an information processing system. Our technical inability in these fields is remarkable and it, in fact, limits our progress in the development of advanced technical systems. For instance, there are many apparently simple manufacturing, transportation and communication operations which are performed by workers on the basis of perceived sound, touch or a visible image. Because of the lack of corresponding *intelligent* devices that could transform relatively easily detectable signals in a particular situation into machine actions, we cannot readily replace human operators which may have to work under difficult or dangerous environmental conditions. Further, we often encounter in technical practice, problems that can be analytically well-described but we still cannot solve them properly using existing theoretical methods, while humans simply find solutions on the basis of their knowledge. As a typical example of this, we cite the non-destructive testing of structures which is based on the analysis of acoustic emissions that are generated in stressed materials. A similar problem is the recognition of objects or discontinuities in materials from scattered ultrasound. This phenomenon is a basis of a well-established ultrasonic testing method for checking the quality of materials and structures and provides the basis for non-invasive imaging techniques that have important medical applications.

Recognition of Acoustic Emission Sources. Practically all loaded materials generate sounds when failure occurs. The sudden, transient redistribution of internal stresses propagate through a structure as acoustic signals. This phenomenon is generally called acoustic emission, although the signals of interest most often possess frequencies in the ultrasonic region. [20] With only a limited experience, an operator can predict the rupture of a loaded specimen from the perceived cracking. It is also possible to estimate from detected acoustic emission bursts, the development of micro-cracks in a loaded specimen and to estimate its strength prior to final catastrophic failure. It is well-known that animals living in trees are capable of avoiding the breaking of a branch by perceiving the cracking of the stressed wood. Our aim is, therefore, to develop a non-destructive testing instrument possessing the same abilities. [20] Here we limit our treatment to cases of so-called discrete acoustic emissions which are represented by a series of distinct ultrasonic impulses that are usually associated with discrete micro-failures of materials. A treatment of continuous acoustic emission requires the inclusion of invariances into the empirical modeling of time series which will be described in later chapters.

The emission and propagation of sound in a solid specimen can be analytically described by the partial differential equations of elastodynamics. They relate the force acting at a source to the emanating displacement or velocity field that represents a sound wave. The basis for the physical description is thus a system of coupled field equations that include time- and space-dependent vectors and tensors. The estimation of the source properties from the detected sound generally represents an inverse problem. [29] When considering the measurement situation that exists in testing a stressed material, we recognize that the inverse problem of determining the characteristics of the emission source is ill-posed because of the incomplete detection of the emitted displacement field as well as the instrument noise that is introduced during the test. Even if we neglect the difficulties associated with finding the solution of a set of coupled partial differential equations governing tensor variables, we must introduce some method for the regularization of such ill-posed problems which introduces some ambiguity into the solution. An additional ambiguity is introduced if we take into account the statistical and nonlinear character of the sound generation in the source region that is the result of the degradation and failure process occurring in the material. A theoretical approach to the exact description of the acoustic emission source properties from the detected signals is generally beyond of the scope of practical non-destructive testing application and consequently qualitative or semi-quantitative procedures are principally utilized today. [5, 6, 20] In view of this, it is rather surprising that an intelligent being can so quickly and apparently so easily recognize the sounds produced by the degradation and rupture of materials. We expect that in the future a proper modification of the general regression modeler could lead to the development of an automatic device possessing the recognition ability of intelligent beings.

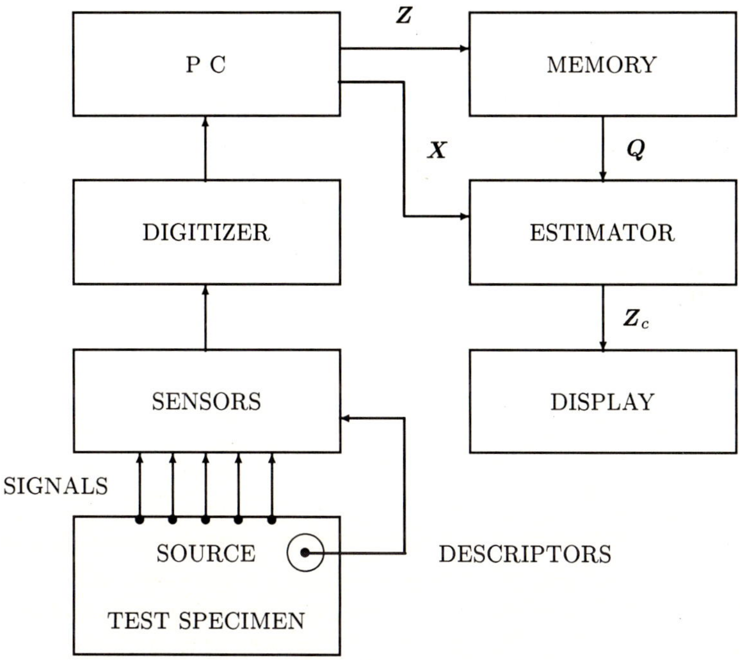

Fig. 9.5. Scheme of the system applied for the modeling of acoustic emission and ultrasonic scattering phenomena

The schematic diagram of the modeler applied to the description of acoustic emission and ultrasonic scattering phenomenon is shown in Fig. 9.5. The acoustic emission (AE) events were simulated by dropping steel balls onto the surface of a plate-like specimen. The experiment included two steps: Training and testing. In the first step, the data base of prototype samples $\{\boldsymbol{Z}_1, \boldsymbol{Z}_2, \ldots, \boldsymbol{Z}_N\}$ was developed and in the second step, the set of test samples was acquired. The detected signals were digitized using a 10-bit waveform digitizing system and transferred to a personal computer where the data base was formed from the prototype sample signals. The data base was subsequently used to calculate the optimal estimator during the testing portion of the experiment, by which the performance of the modeler could be studied. The standard deviation of the signals in the data base was used as the smoothing parameter σ.

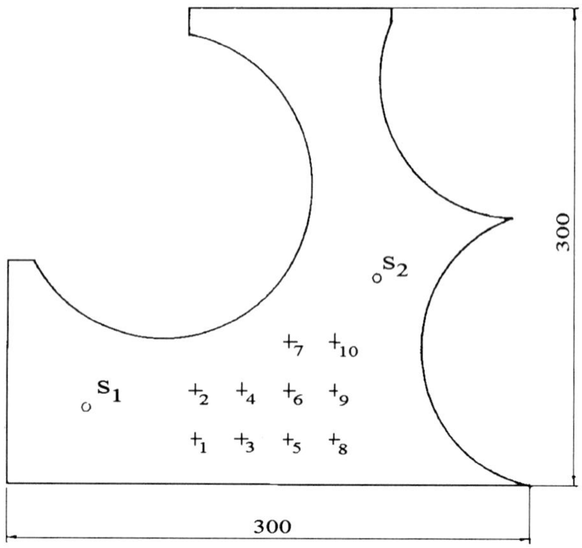

Fig. 9.6. Scheme of the specimen utilized in the AE experiment. S_1 and S_2 denote the locations of the sensors.

The specimen was an aluminum plate of irregular shape and gross dimensions $(300 \times 300 \times 46)$ mm^3 as shown in Fig. 9.6. Two miniature, broadband piezoelectric acoustic emission sensors of 1.3 mm diameter active area were mounted on the plate at arbitrarily selected positions. The AE events were excited by impacts of steel balls of diameter 1.9 or 3.2 mm dropped from a height of 25 mm at various positions on the surface of the plate. To form the data base, ten points in a net of squares of dimension $33 \times 33\text{mm}^2$ were

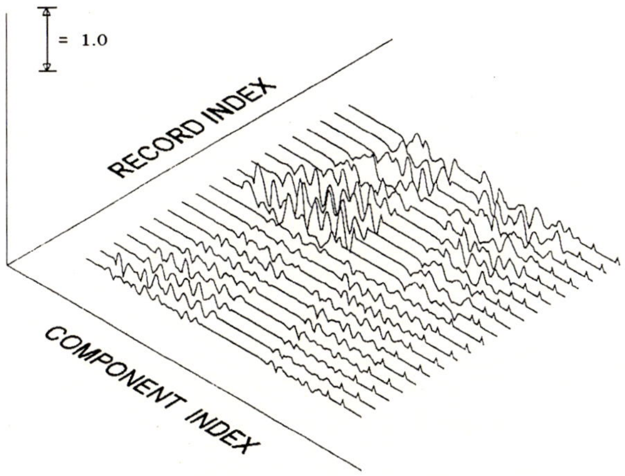

Fig. 9.7. The data base of AE experiment

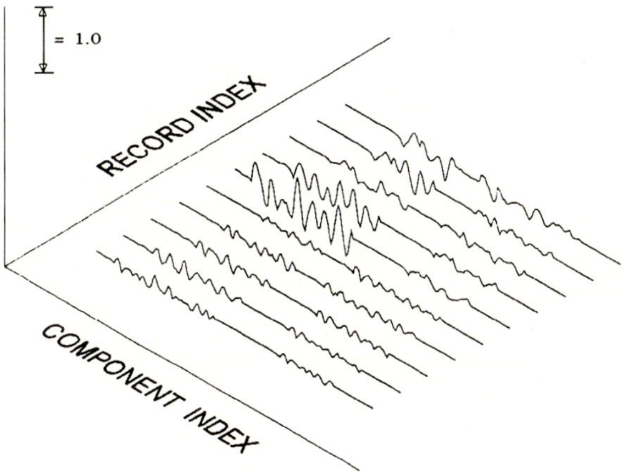

Fig. 9.8. The test signals comprised of AE records without source descriptors

selected. The signals from the sensors were amplified by 40 dB and recorded using a two-channel waveform digitization system. The digitization of both signals commenced with a synchronization pulse from a third sensor placed 15 mm from the source. A 5 MHz sampling rate was used. From 1024 data points in each record only each 16th point was retained to generate 64 signal components. To the compound 128 data points generated by the signals from both sensors another 22 components were appended. They were used to encode the information about the source position and the radius of each ball. In one half of this set the components with the index equal to the po-

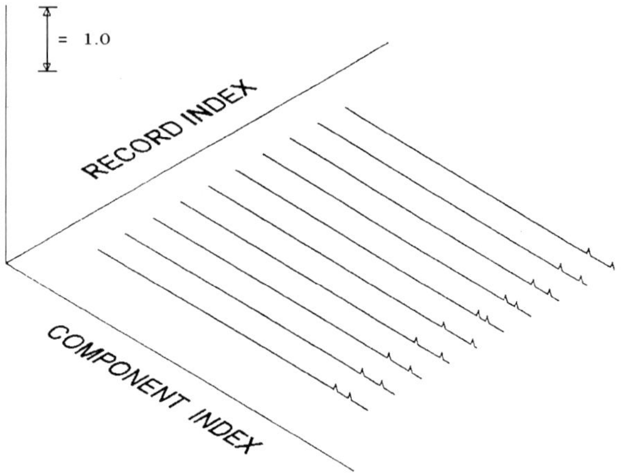

Fig. 9.9. The AE source descriptors of test signals presented in Fig. 9.8

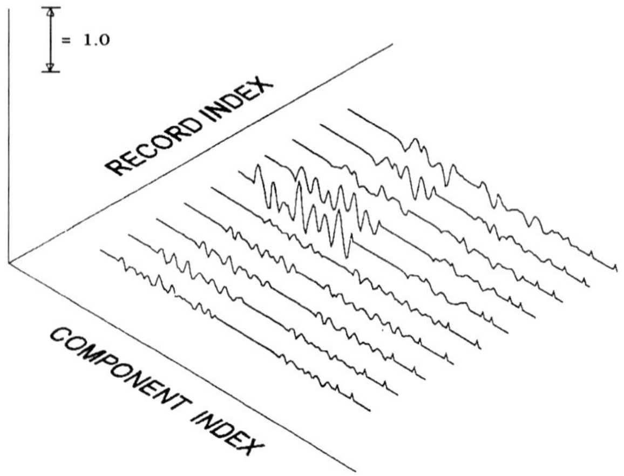

Fig. 9.10. The set of completed test signals. The descriptors on the right side of the records are estimated from the test AE signals presented in Fig. 9.8

sition descriptor was set to constant values, different from zero. In the same way, the radius of the ball and hence the amplitude of the source force, was encoded by the other 11 components. The experimental samples were thus described by compound vectors of 150 components. The data base included 20 samples, describing the impacts of the ball at each preselected position. They are represented by the records shown in Fig. 9.7. In this example each record, representing the value of the data components as a function of index, consists of four segments. The first two, are the AE sensor data while the two peaks toward the right of each trace correspond to the encoded source descriptors. The rightmost peak corresponds to the encoded impact position. The standard deviation calculated from all the values of the AE signals on the data base is approximately 0.07 while the maximal fluctuation from the average has a value of 0.42. For the parameter σ we further applied the standard deviation S_x.

After formation of the data base, it can be used to recover the missing components in experimental data. The acoustic emission test signals are shown in Fig. 9.8 while the corresponding source descriptors, shown in Fig. 9.9, were stored separately, to be used later for comparison. Each signal from the set in Fig. 9.8 was then taken as a given data vector from which the descriptors were estimated. The given data appended with the estimated descriptors are shown in Fig. 9.10. The agreement between the original and estimated descriptors is almost perfect. The coefficient C_n in the expression of the conditional average estimator of the most similar signal was for all the samples greater than 0.9999. This result demonstrates that the modeler has correctly recognized the characteristics of the source from the acoustic emission signals. Thus, the modeler has been used to obtain the solution of an AE inverse problem. If the encoded source descriptors are taken as the given data, then the coefficients C_n are all equal zero, except one which is equal to 1. In this case the data base serves as a look-up table from which the signals are used to match the encoded descriptors. The corresponding records with the estimated signals are shown in Fig. 9.11 and they represent the complete test data. With this example we have demonstrated that the modeler can also be used to obtain the solution of the so-called AE forward problem. That is,

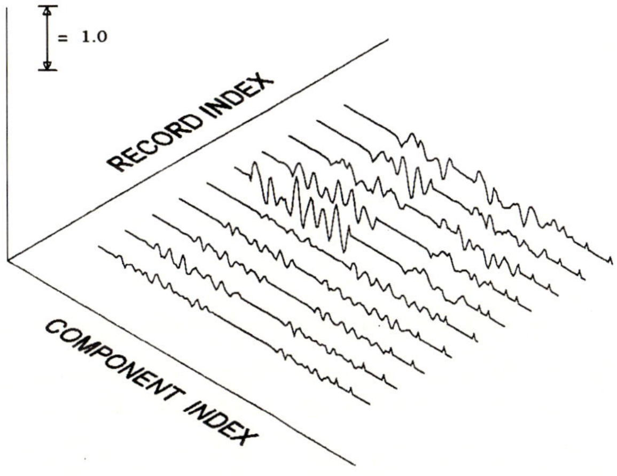

Fig. 9.11. The set of completed test signals. The AE signals on the left side of the records are estimated from the source descriptors presented in Fig. 9.9

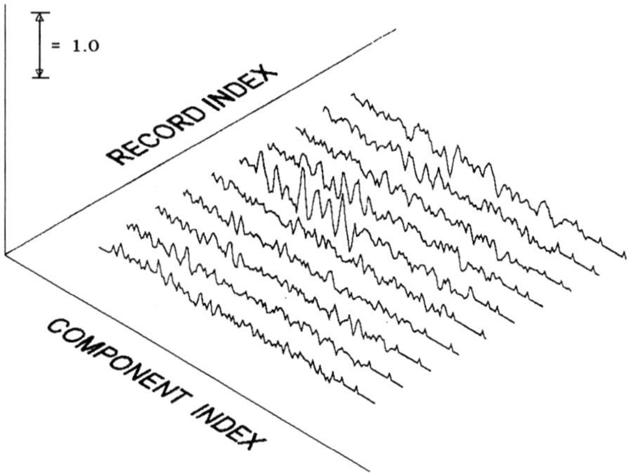

Fig. 9.12. Noisy test signals complemented by source descriptors. The signal–to–noise ratio is approximately SNR ≈ 1.0

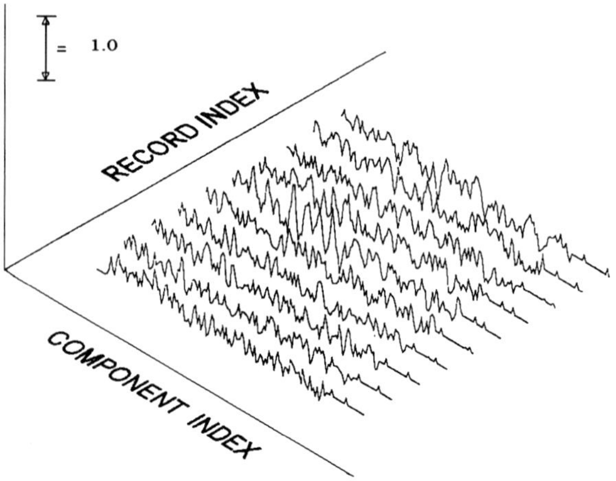

Fig. 9.13. Noisy test signals complemented by source descriptors. The signal–to–noise ratio is approximately SNR ≈ 0.5 ; $\sigma = 3S_x$

it can be used to predict the acoustic signals from the characteristics of source.

The estimation of missing data is quite noise insensitive. In order to demonstrate this property, we have added to the test data, random noise of uniform amplitude distribution. Two examples with maximal noise amplitude of 0.1 and 0.2 and standard deviations of 0.06 and 0.12, respectively, are presented here. The completed data with the estimated descriptors are shown in Figs. 9.12 and 9.13. In order to describe the correctness of the estimation process, we present the mean value of the descriptor index that is obtained by averaging the index values separately over each characteristic portion of 11 components. If the descriptors are correctly estimated then both average values must agree with the corresponding correct values calculated separately from the known corresponding data. The values of the correct and estimated descriptors are shown in Table 9.1. In the first case, the standard deviation of the noise is of the same value as the root-mean-square (rms) amplitude of the AE signal for which the signal-to-noise ratio (SNR) is ~1 and the signal is therefore buried in noise. Nevertheless, the estimated descriptor values are still in good agreement with the correct values. In this case the width σ of the window function in the estimator was taken to be equal to the standard deviation of the signals in the data base. If the same width is also selected for estimating the descriptors in the second case in which the standard deviation of noise is already two times greater than the rms-value of the signals (corresponding to a signal to-noise-ratio ~ 0.5), then the value of the window function becomes too small and the results are uncertain. It is interesting that the estimation can be improved by selecting a larger σ or a smoother window function. In the example shown here, the presented estimations of the descriptors were determined by choosing $\sigma = 3\,S_x$. These examples demonstrate that a satisfactory estimation can generally only be obtained by this procedure when the signal-to-noise ratio is not less than approximately 0.3.

Table 9.1. Comparison of actual and estimated AE source descriptors. First column describes the ball size and the second the impact position. Underlined values indicate an incorrect recall caused by the noise disturbance

Row	Actual		Estimated		Estimated SNR:1.0		Estimated SNR:0.5		Estimated SNR:1.0 (Using iteration)		Estimated SNR:0.5 (Using iteration)	
1	4.00	1.00	4.00	1.00	4.00	1.00	4.00	1.03	4.00	1.00	4.00	1.00
2	4.00	4.00	4.00	4.00	4.00	4.00	4.00	3.95	4.00	4.00	4.00	4.00
3	4.00	5.00	4.00	5.00	4.00	5.00	4.00	5.06	4.00	5.00	4.00	5.00
4	4.00	7.00	4.00	7.00	4.00	6.97	4.00	4.20	4.00	7.00	4.00	5.00
5	4.00	9.00	4.00	9.00	4.00	9.00	4.00	8.45	4.00	9.00	4.00	9.00
6	8.00	2.00	8.00	2.00	8.00	2.00	8.00	2.00	8.00	2.00	8.00	2.00
7	8.00	3.00	8.00	3.00	8.00	3.00	8.00	3.00	8.00	3.00	8.00	3.00
8	8.00	6.00	8.00	6.00	6.98	7.01	8.00	7.62	8.00	6.00	4.00	10.0
9	8.00	8.00	8.00	8.00	8.00	8.00	8.00	8.00	8.00	8.00	8.00	8.00
10	8.00	10.0	10.0	10.0	8.00	10.0	8.00	10.0	8.00	10.0	8.00	10.0

In order to demonstrate the noise correction property of the estimation that can be obtained by an iterative application of the conditional average, the estimated source descriptors were taken as given data by which the signal

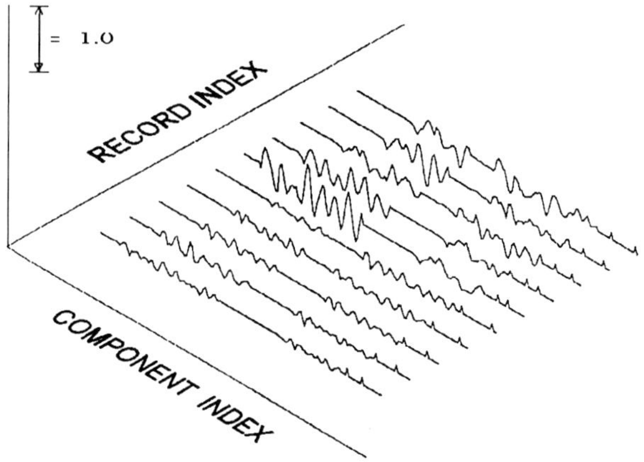

Fig. 9.14. Signals obtained by two complete steps of iteration from records of AE signals presented in Fig. 9.12 at SNR $\approx$ 1.0. The test acoustic signals are correctly recovered from the noisy input signal

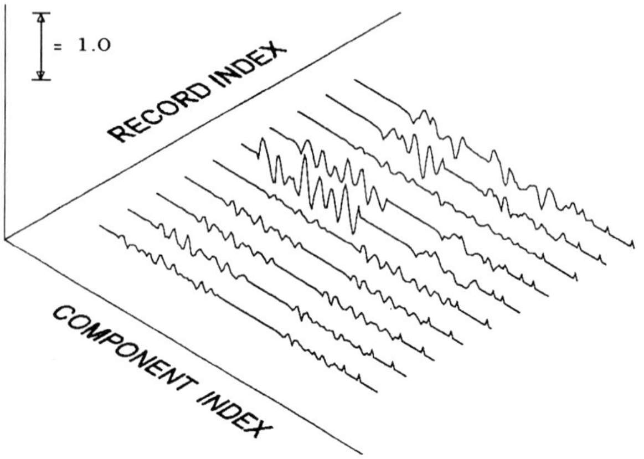

Fig. 9.15. Signals obtained by two complete steps of iteration from records of acoustic emission signals presented in Fig. 9.13 at SNR $\approx$ 0.5. The fourth and eighth record of the test acoustic signal is not correctly recovered from the noisy input signal because of the high noise amplitude

portion was recovered. Using this new signal, a new set of descriptors was estimated. The vectors estimated by these three iteration steps are shown in Figs. 9.14 and 9.15. Each of the recovered signals is practically identical with one of the vectors of the original data base. The corresponding coefficient C_n is within 10^{-4} of 1.0. The iteration procedure obtained by the conditional average thus serves as an efficient matched filter, the attractor points of which are determined in the corresponding D-dimensional sample space of the given data base. However the initial partial input vector $\boldsymbol{X}$ can cause convergence to a "correct" attractor only if the signal-to-noise ratio is not too small. For a signal-to-noise ratio of 0.5, an incorrect recall was obtained on lines No. 8 and 4. If the added noise is similar to the vector $\boldsymbol{Z}_n$ in the data base and if its amplitude is larger than the amplitude of the signal to which it is added, then one can quite naturally expect that as a result of the iteration, $\boldsymbol{Z}_n$ is recognized as the most appropriate, stored data vector.

If two signals of the test series are summed with equal weights and the descriptors are then estimated from the sum, the recalled vector contains the weighted sum of descriptor vectors of the applied signals with the weight 0.5.

One problem that should be mentioned relative to the recognition of acoustic signals is as follows: Human beings recognize events in nature over a broad range of intensity levels independent of the intensity of perceived sound. This means that the information is mainly represented by the "structure" of the sound signal and less by its amplitude. If we wish to achieve such a performance by a general regression modeler then we must train it with all possible sound amplitudes. This appears to be an overwhelming task and at once raises the question how one might avoid it. A solution of this problem can be immediately suggested. By incorporating a normalization constant of the detected signals, the amplitude can be extracted. That is, if we train the modeler with normalized signals, it can later correctly recognize the source descriptors regardless of the sound intensity. However, the intensity often represents an important piece of information and one may ask how one can develop a modeler that could automatically control the extent to which the information "hidden" in the sound intensity is utilized in the recognition and identification of acoustic sources. We recall that human beings identify sources of sound regardless by which ear an acoustic event is perceived. Similarly the modeler should act independent of the way in which it obtains its input information. All these topics pertain to the field of invariant modeling of natural phenomena that will be treated in the next chapter.

Recognition of Defects from Ultrasonic Scattering. Quite different from the acoustic emission phenomenon is the scattering of ultrasound by discontinuities or inhomogenities in materials. Here, the ultrasound is generated by an external transducer coupled either directly or via a fluid to a test specimen and it propagates into the material where it is scattered by the defect. The scattered sound is then detected and used to characterize the defect which acts as a secondary source. The method is widely applied in non-destructive quality control for detecting and characterizing defects such as cracks or voids in various products and in medicine for characterizing the properties of tissue. The difference between acoustic emission and ultrasonic defectoscopy is the recognition of the primary or secondary sources of sound respectively. However, there exist specific problems in the application of each method. This will become evident from the following example.

An experiment was performed using an experimental arrangement similar to that used to recognize acoustic emission sources. The specimen was an aluminum block of dimensions $127 \times 38 \times 51\ \mathrm{mm}^3$ in the middle of which were four horizontal, side-drilled holes of diameters 3.8, 3.2, 2.2, and 1.6 mm as shown in Fig. 9.16. The holes could be filled with various fluids to vary their scattering characteristics. The fluids used were air, iso-alcohol, water, carbon

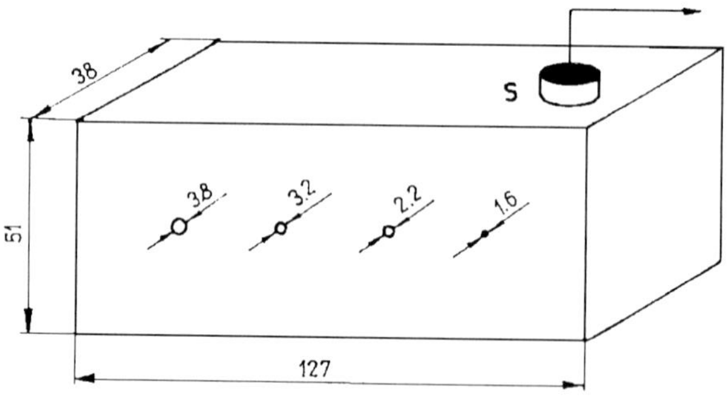

Fig. 9.16. Scheme of the specimen used in empirical modeling of scattering of ultrasonic signals from a cylindrical inclusion in aluminum

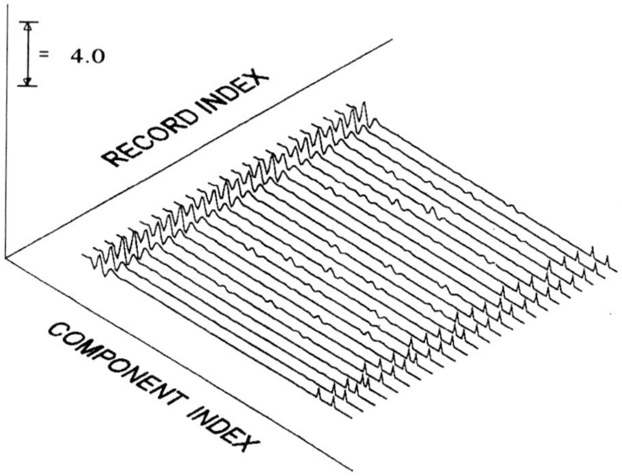

Fig. 9.17. Data base of scattering experiment

tetrachloride, glycerin, and mineral oil. The ultrasonic pulses were generated by a 2.25 MHz transducer of 12.7 mm diameter active area. The transducer was placed directly above each hole being interrogated. The transducer excitation pulse was also used to trigger the waveform acquisition system, which in these experiments was operated at a sampling frequency of 60 MHz. The waveforms were recorded on one channel of the digitizing system.

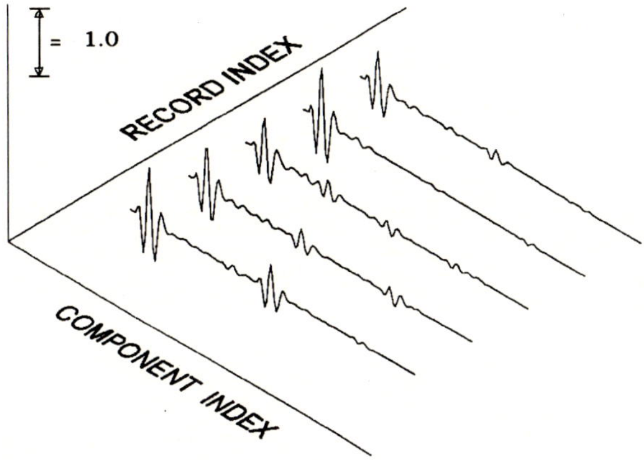

Fig. 9.18. Ultrasonic test signals of the scattering experiment

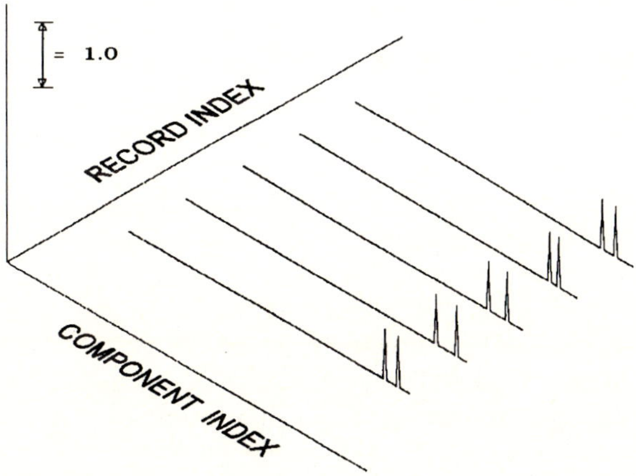

Fig. 9.19. Test descriptors of the scattering experiment

In generating the input data, the 4096-point digitized waveforms were reduced to 128 points. The reflector radius and the inclusion material were encoded by two descriptors in the field of the last 22 components of the data vector $\boldsymbol{Z}_n$. The acquired data base is shown in Fig. 9.17, while the test signals and the corresponding descriptors are shown in Figs. 9.18 and 9.19.

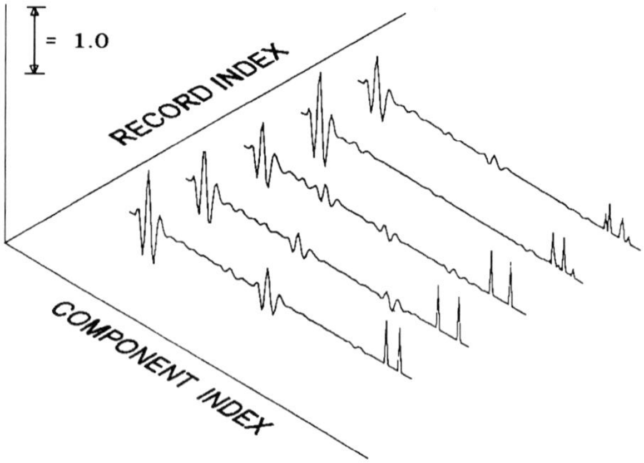

Fig. 9.20. Test signals complemented by estimated descriptors of the scattering experiment

The first descriptor on the right portion of each data vector designates the hole diameter. The amplitude-time records of the ultrasonic signals scattered by the side-drilled hole exhibit characteristic features which have previously been extensively investigated. [22, 24] Each pulse consists of several wave packets possessing different amplitudes and time-delays with respect to the trigger signal. The first received pulse is the large, speculary reflected signal from the front surface of a hole. Its amplitude increases with the radius of the cylindrical inclusion, though it is to a lesser degree also affected by the properties of the inclusion material. After this first signal there are several observable pulses that represent the successive arrivals of wavefronts which have been diffracted around the circumference of the inclusion or diffracted and then refracted around and through the medium in the hole or simply transmitted and reflected through the inclusion medium. [22] The arrival times as well as the amplitudes of these pulses depend on the size of the cylindrical hole as well as the properties of the specimen and the inclusion material. The fundamental problem here is to extract from the scattered signals an estimate of the size of the scatterer and the wave-speed (and hence a mechanical property) of the inclusion material. This problem, just as in the AE source characterization, pertains to the general class of inverse problems and cannot generally be solved. For its solution we again apply the general modeler in which complete information about the scattered pulses is utilized for the analysis. It is not surprising that an analysis of the additional signals leads to a more efficient characterization procedure for the scatterers as in the case when only the front surface echo is analyzed.

If the test signals are applied to estimate the source descriptors, the completed signals shown in Fig. 9.20 are obtained. The highest peak in the descriptor portion of the data vector always corresponds to the correctly estimated descriptor, but in the last two records additional peaks appear. This is a consequence of the similarity between the ultrasonic test signals with various vectors from the data base. For estimation of the similarity, the norm of the difference of vectors from the test and the data base is applied. The principal contribution to this norm comes from the first peak of the scattered signals, which is not significantly influenced by the flaw material properties. It is therefore reasonable to take as given data for the characterization of the material property only that portion of the signal that includes the secondary pulses. The set of truncated, partial signals is shown in Fig. 9.21, while the composed partial and estimated scatterer data are shown in Fig. 9.22. It is obvious that in this case, a much better performance of the modeler is observed. If the iteration procedure is applied, the descriptors are recovered perfectly which indicates that the proposed model can also be used to recover the wavespeed of the inclusion material. This example demonstrates that physical arguments are important for the selection of the given data from which unknown properties of the phenomenon under consideration are estimated. How such property characterization can be automatically included in the modeler is still not known. In the last example, the initial echo was also recalled from the other scattered signals. This indicates that the modeler can also be applied to estimate the properties of one portion of the signal from another portion. Because of this property, the modeler is capable of predicting even chaotic time series from the partial input and stored signals in the data base. This property is of great importance for the modeling of chaotic phenomena and it will be further elaborated in later chapters.

Related to applications of the general modeler of natural phenomena to acoustic emission and ultrasonic scattering phenomena, we must mention the problem of the formation of an appropriate data base which cannot always be avoided. The signals included in the data base must be obtained under the same experimental conditions as those utilized in the subsequent analysis. It is, for instance, practically impossible to obtain prototype signals from the nucleation and growth of a crack if the test specimen is so large that it cannot be easily stressed or it is not permitted to be damaged. This problem could be solved by utilizing data from theoretical simulations to form the data base and thus incorporate analytical information about the phenomenon into the processing system. The modeler is then only used to obtain the solution of the

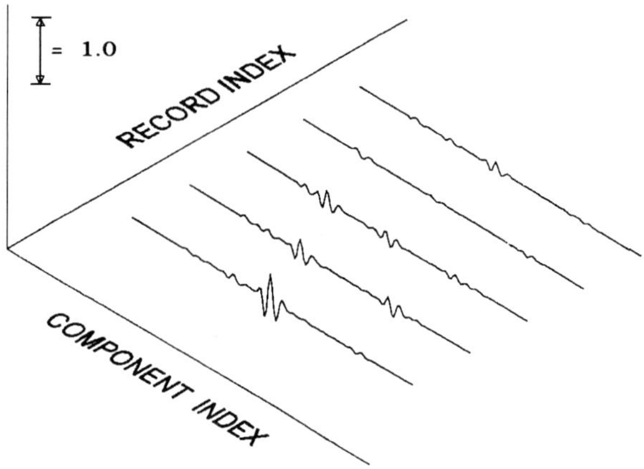

Fig. 9.21. Portion of scattered ultrasonic signals which most significantly depend on the material property of the inclusion

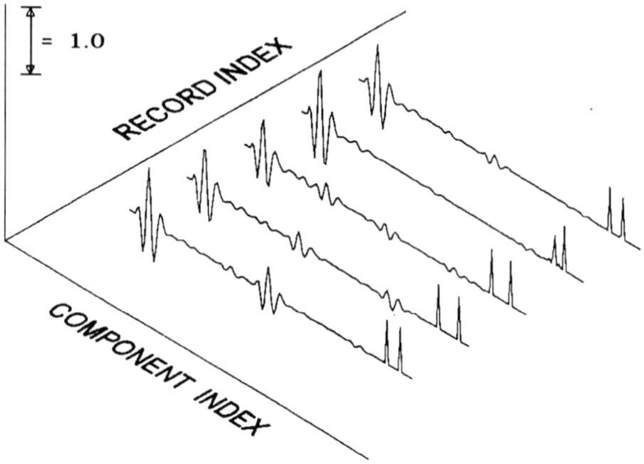

Fig. 9.22. Partial ultrasonic signals completed with the recovered scatterer descriptors and front surface echo pulses

corresponding inverse problem. Further, since the exact location of a crack in a stressed specimen is not known in advance, we should utilize a continuous set of prototypes. The problem is then how to most economically represent a continuous manifold by a data base comprised of a finite number of prototype vectors. This problem can be solved by application of the self-organization described in the previous chapter which, as we shall see, can be included into the modeler during the formation of prototypes. The regression then proceeds using the self-organized prototypes. For simplicity, the self-organization has not been utilized in the demonstration of the above examples because the corresponding effect has been obtained by a proper selection of prototype samples during the preparation of the experimental data base.

An additional problem that arises during the recognition of acoustic and ultrasonic signals is related to various invariances. For instance, when utilizing ultrasonic defectoscopy in materials testing, the absolute amplitude level of the received ultrasound signal is not important. Much of the information about the phenomenon is hidden in the relative amplitudes and the temporal structure of the scattered signals. When forming the data base, we can thus safely use arbitrarily normalized signals. However, when we use such a trained data base to analyze test signals, we must perform the same normalization. The above operation pertains to the preprocessing of signals. The modeler itself is sensitive to the amplitude of a complete signal and generally yields different results for input signals of different amplitudes. One might then ask how one can develop a modeler that could recognize signals only on the basis of their structure, regardless of their amplitudes. A similar problem can be mentioned relative to the recognition of equivalent defects located at different positions in a sample or acoustic emission events occurring at different times. In all of these cases, we are faced with the problem of the inclusion of a certain invariance into the recognition which appears to be very fundamental for the application of the modeler and will therefore be treated in a separate chapter. Before addressing these problems, let us consider several additional applications of the modeler.

9.4.2 Prediction of the Seismic Capacity of Walls

One of the most challenging problems in civil engineering is related to the construction of buildings of proper seismic resistance. The behavior of structural elements, mostly made of reinforced concrete (RC), is at high loads essentially nonlinear and because of the inhomogeneous structure, inaccessible to deterministic modeling. In addition, the elements are usually of complex geometrical form which further complicates any deterministic description. The estimation of the seismic resistance is thus principally based on experimental testing. In the existing literature on this topic are published extensive empirical data bases which describe the so-called seismic capacity in terms of shear strength, drift, ductility and failure mode of various structural elements when they are subject to static or dynamic loadings. [19, 30, 31] Both, linear and non-linear regression have been used to estimate the seismic capacity from the available data. A recent investigation [23] has shown that the non-parametric regression presented in this chapter is in many cases and in various aspects superior to existing methods. First of all, it can be carried out practically automatically without a deep theoretical analysis of the complex mechanisms that occur during the degradation of structural elements. In addition, the data base can be permanently updated and various parameters can be extracted from the same model. It is also capable of solving the inverse problem related to the specific structural parameters of an element which are needed in order to obtain a desired seismic capacity. In the following paragraph the application of the non-parametric regression to the estimation of seismic capacity of RC structural walls is described.

Based on the survey of the literature, one may include in a description of empirical information about reinforced concrete structural walls, the parameters characterizing the geometry and loading of specimens and the observed seismic capacity. [23] The list of parameters includes:

1. *Geometry and Loading Type*:
   - Type of cross section (rectangular, barbell, flanged)
   - Vertical reinforcement index in web
   - Horizontal reinforcement index in web
   - Vertical reinforcement index in boundary columns
   - Confinement index in boundary columns
   - Monotonic or cyclic loading
   - Axial load index
   - Shear span index
2. *Seismic Capacity*:
   - Shear strength
   - Drift
   - Ductility
   - Failure mode

These parameters were used as components x_i of the data vector. The first group was treated during the testing as given data while the second group was used to estimate the seismic capacity. In utilizing the data published by several authors, we have represented the structural parameters in a normalized form. Various data bases consisting of more than 100 prototypes each were utilized in this study. The evaluation of the modeler was carried out on different data included in the data base. During testing, various parameters of seismic capacity were separately estimated and compared to the experimentally observed values.

Two relative error measures were introduced to describe the performance of the modeler:

$$E_1 = \frac{\sqrt{\frac{1}{N}\sum_{i=1}^{N}(x_{pi} - x_{mi})^2}}{\overline{x}_m}, \tag{9.43}$$

$$E_2 = \sqrt{\frac{\frac{1}{N}\sum_{i=1}^{N}(x_{pi} - x_{mi})^2}{x_{mi}^2}}. \tag{9.44}$$

Here x_{pi} and x_{mi} represent the predicted and the measured parameters respectively and $\overline{x}_m$ is the mean value of the measured parameter. The index i denotes a particular member from the set of N tested samples. The predicted and measured values were presented separately for each tested parameter of seismic capacity in a common diagram.

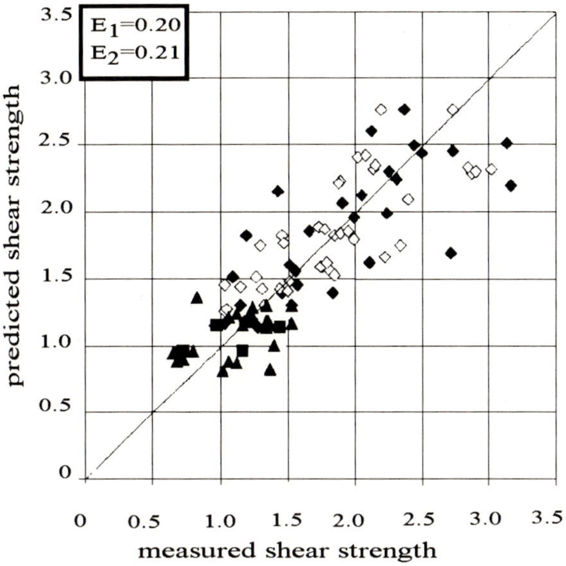

Fig. 9.23. Relation between measured and predicted shear strength

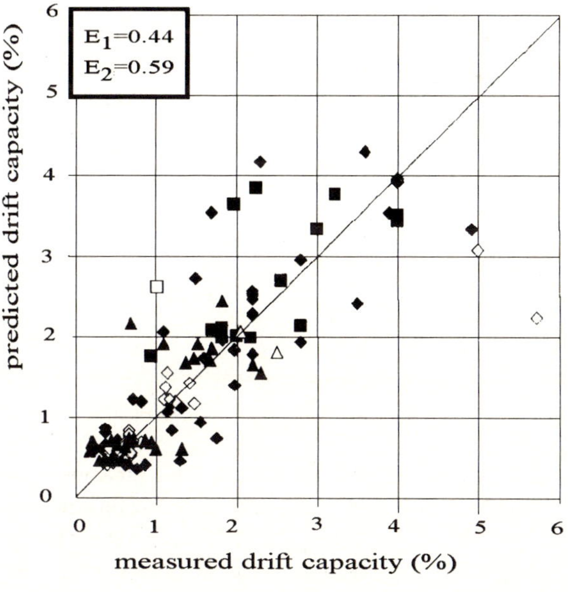

Fig. 9.24. Relation between measured and predicted drift capacity

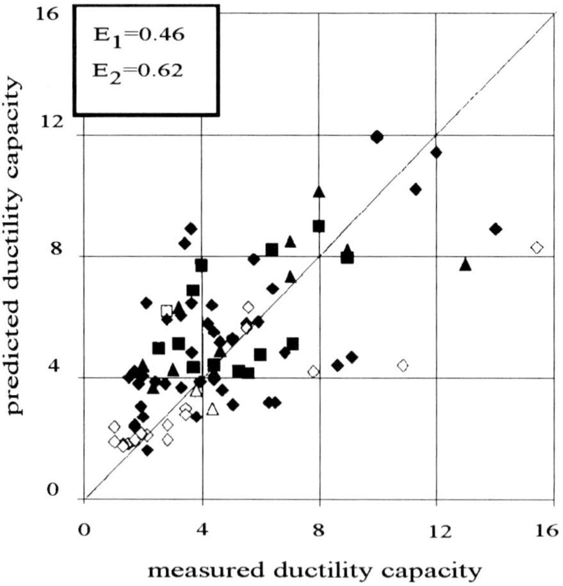

Fig. 9.25. Relation between measured and predicted ductility capacity

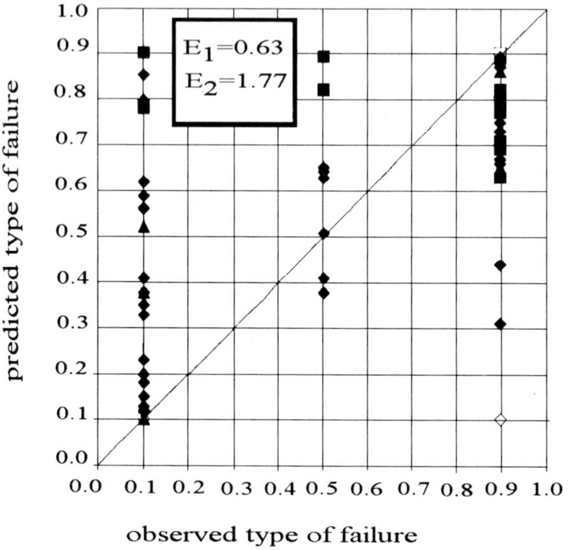

Fig. 9.26. Relation between measured and predicted type of failure

The better the agreement between the predicted and experimental value, the closer the corresponding point lies on the line $x_p = x_m$ in the diagram. Figures 9.23 to 9.26 are the corresponding diagrams, showing the agreement between the predicted and the measured values of the parameters which describe the seismic capacity. The small scatter of the points around the indicated line $x_p = x_m$ indicates that the non-parametric regression predicts with acceptable reliability the seismic capacity in terms of the shear strength, drift and ductility, while the prediction of the observed failure type is less successful. The explanation of the latter observation is that the characterization of the type of failure is subjectively determined from visual observations of the specimen surfaces. An estimation of the shear strength based on more theoretically-based but still empirical models does not achieve the performance of the non-parametric regression presented here. [23] For the shear strength, the theoretically-based model yields an error measure $E_2 = 0.37$ while the non-parametric regression gives $E_2 = 0.21$ which indicates a significant advantage of the latter method in estimating the seismic capacity from empirical parameters.

9.4.3 Modeling of a Periodontal Disease Healing Process

The diagnostic methods of modern medicine are often based on quantitative biological, chemical, and physical measurements. In contrast, a quantitative modeling of the phenomena taking place in the human body is relatively less frequently done. The reason for this difference is the extreme complexity of biological phenomena. The first problem is that a human body is, in fact, an open system living in a non-stationary environment that cannot be physically well-described. It includes processes of various kinds that evolve at various levels, in space and time. The second problem is that these processes are often essentially nonlinear, intrinsically unstable and consequently often chaotic. [21] For these reasons, an empirical modeling would appear to be especially appropriate for their description.

One of the problems that is often met in clinical practice is related to the forecasting of the development of disease or healing processes and to the description of the influence of a surgical treatment or the estimation of the effects of drugs on healing. The purpose of this sub-section is to demonstrate how the non-parametric regression can be applied to solve these problems. As an example, we describe the effect of periodontal treatment. [11, 12, 13] The tooth mobility was used as a characteristic measurable variable which is applicable to the quantitative description of the periodontal condition. It was characterized by the displacements of teeth using relatively small loads below 1 Newton. The resulting displacement ranges approximately between 0 and 100 μm. It was measured by an especially designed experimental setup including electronic sensors of tooth displacement and loading force, which are simultaneously recorded during the experiment. [12] The focus of this investigation was the healing after the surgical procedure of periodontitis called modified Widman flap. The influence of the additional drug administration on the healing process was studied. For the characterization of the healing process, the tooth mobility at a fixed loading force of 0.5 N was measured as function of time over a period of 12 months. In the investigation the incisors in the upper and lower jaw were examined.

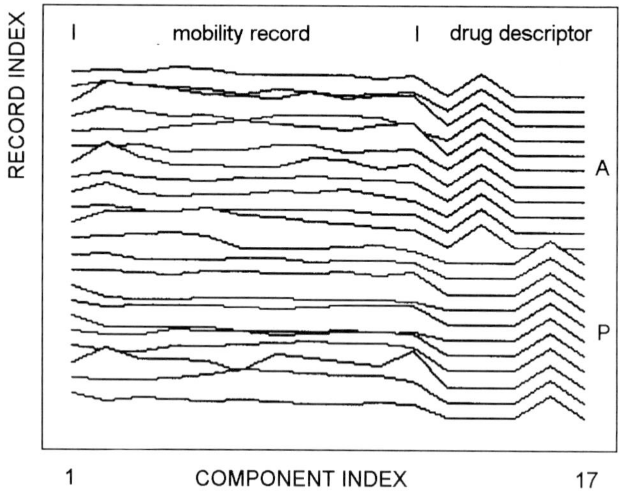

Fig. 9.27. Data base of the empirical modeling of the healing of a periodontal disease. The records on the left side describe the measured tooth mobility in dependence of time index on the abscissa. The two peaks on the right side of the records indicate the presence - (P) or absence - (A) of drug treatment during healing

Fig. 9.27 shows quantitative data for a group of 22 patients. The 11 records at the bottom correspond to the patients treated by the systemic metronidazole drug after the surgical procedure while the top 11 records represent the healing processes in patients not receiving this drug. Each record consists of two parts: On the right, a peak describes the presence or absence of the drug, and on the left a joint curve over 12 points represents the tooth mobility as function of time. The records are shifted vertically by a constant amount for visualization. The starting point on the left side of the record represents the mobility just before the surgical treatment. Many patients react with an increased tooth mobility immediately after the treatment which gradually decreases as healing process proceeds. A characteristic recovery process can be described by averaging the data at given times over the portion of the population treated similarly. An equivalent procedure stems

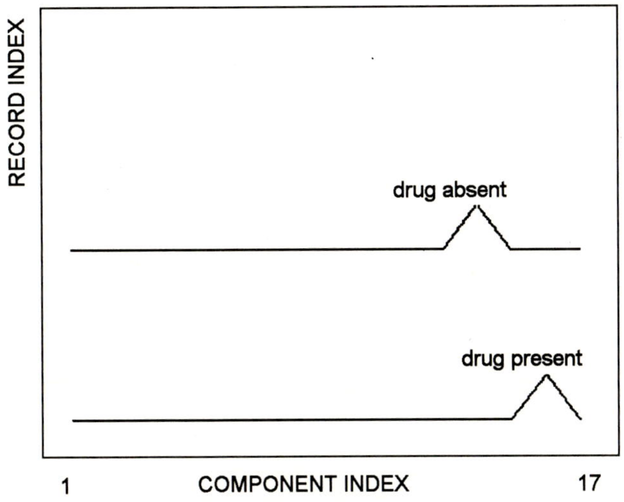

Fig. 9.28. Drug presence descriptors

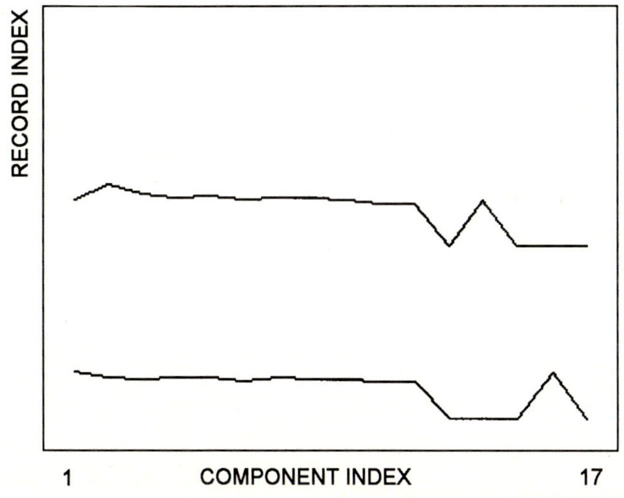

Fig. 9.29. Drug presence descriptors (*right*) complemented by the estimated tooth mobility (*left*). The estimated portions on the left represent the average tooth mobility of the sub-group specified by the descriptor

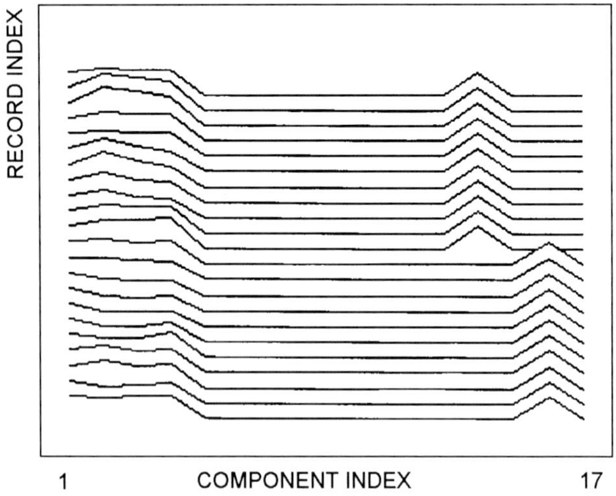

Fig. 9.30. Partial data of the test group comprised of the initial tooth mobility record on the left and the descriptor of drug presence on the right

from the non-parametric regression. For this purpose, the truncated vectors were generated that represent the drug application descriptors as shown in Fig. 9.28. The corresponding records with the estimated missing portions are shown in Fig. 9.29. The estimated portions represent the average predicted mobility as a function of time. The bottom record shows that the mobility decreases on average after the surgical procedure if the drug is administered, while it increases first and decreases only later when the drug is not used. The effect of the drug on healing can thus be simply estimated by the conditional average. The corresponding record can be further interpreted as representative of a normal process. However, all the patients do not react similarly, neither in the complete group nor in the subgroup treated equivalently. It is therefore of interest to develop a method by which the healing process could be predicted from the first few initial measurements of tooth mobility. For this purpose, we appended to the drug presence descriptors some initial data about the measured mobility. Fig. 9.30 shows examples of such partial data for a test group consisting of an additional 22 patients. The corresponding records completed with the predicted data are shown in Fig. 9.31 while the records depicted in Fig. 9.32 show the actual healing process that has been observed on the test group. Comparison of the predicted and observed recovery shown in the last two figures reveals that the observed processes generally fluctuate around these which were predicted. Because of the averaging, the predicted records are generally smoother than those actually measured. On average, the agreement between the predicted and observed

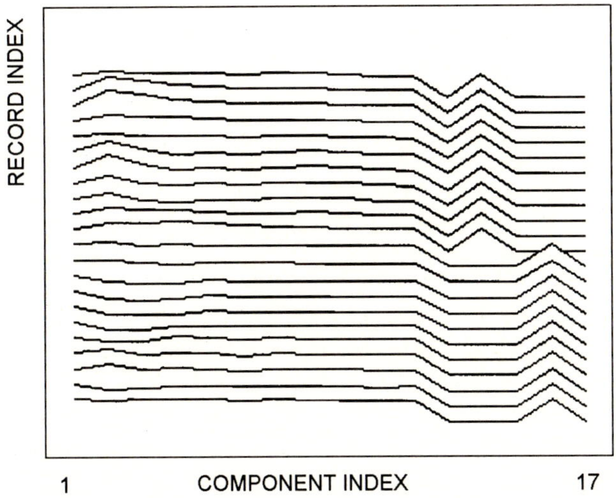

Fig. 9.31. Partial data of the test group completed by the estimated data

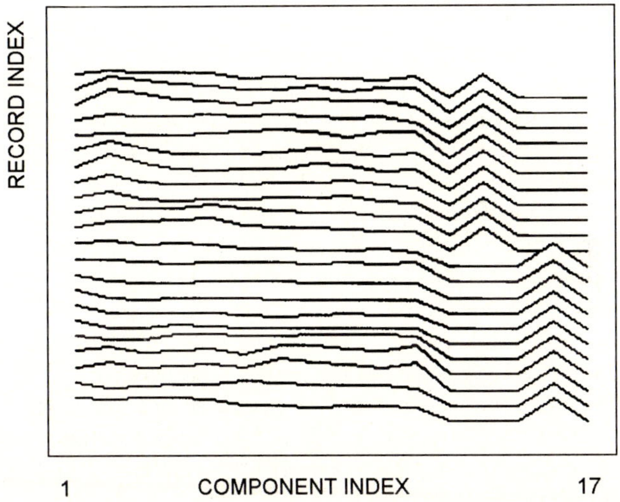

Fig. 9.32. Observed data of the test group. Because of the averaging that is included in the estimation, the estimated portions of the tooth mobility records in Fig. 9.31 exhibit smaller fluctuations as the observed data

process is good and increases with the number of samples in the data base. Such results can be of assistance to a dentist who can apply more aggressive therapy if the prediction indicates slower healing than is normal. The forecasting of the recovery process using a data base and a conditional average estimator, in fact, resembles the forecasting based on intuition by a physician, who, just as the regression modeler, extracts information from the data of previously treated patients to forecast the healing of a particular patient.

In the treatment of various prediction problems, the system is often assumed to be autonomous and the corresponding processes are assumed to be stationary and linear. In our case, the inclusion of the drug presence descriptors into the forecasting generally corresponds to the assumption of a non-autonomous system. Further, we have assumed neither linearity nor stationarity of the healing process. The prediction of the healing process thus corresponds to a general type of time-series forecasting. The applicability of non-parametric regression to such problems is of fundamental importance and it will therefore be further considered in subsequent chapters in which a quantitative measure of performance prediction will be introduced.

10. Linear Modeling and Invariances

10.1 Relation Between Parametric Modeling and Invariances

Let us consider a natural phenomenon that can be characterized by a two-dimensional variable $\boldsymbol{Z} = (X, Y)$ and let us assume that the empirical sample points ($\boldsymbol{Z}_1$, $\boldsymbol{Z}_2$, ..., $\boldsymbol{Z}_N$) lie close to the line described by a linear relation $\hat{Y} = AX$. In order to estimate the variable Y from a given value of X we can apply the conditional average estimator defined in the previous chapter or we can directly start with the linear relation and by minimizing the mean square error determine the coefficient A. The resulting expression $A = \text{cov}(XY)/\text{var}(X)$, is well-known from the literature on linear statistical estimation. It can then be utilized to estimate Y from given X by the linear regression equation $\hat{Y} = X \,\text{cov}(XY)/\text{var}(X)$.

An advantage of such an approach is immediately evident: Instead of calculating several constants C_n of the conditional average estimator for each new X, we determine just one product AX. Still another advantage stems from the following consideration. When we apply the conditional average estimator then the estimated values are placed between the sample points in such a way that the closest sample value that has the projection to the x-axis most similar to the given value X, contributes the greatest amount to the estimate. This leads to an acceptable interpolation when modeling a general functional relationship between the variables X and Y from a given cluster of points. When attempting to extrapolate using the same method to estimate the variable Y from a value X which lies outside the cluster (X_1, X_2, ..., X_N), then the conditional average estimator yields a value which lies within the cluster (Y_1, Y_2, ..., Y_N). However, in the case of a linear relationship between these variables, this result is not a good estimation and we can significantly improve the error obtained in the extrapolation if we apply a linear estimator. *A priori* given information is thus utilized in the selection of a parametric form of a model, which then looses its generality because it has been tailored to model a specific phenomenon. By assuming a linear relationship between measured variables we, in fact, constrain our modeling to the selection of a set of parameters describing the corresponding function. A similar reasoning is valid also for other functional relationships between variables.

Presuming a functional relationship, we *a priori* assume that the description of the natural phenomenon is invariant with respect to some transformation of the variables used in the description. For instance, in the case of the linear relationship, $Y = AX$, we can multiply both variables by an arbitrary constant $C : X' = CX$; $Y' = CY$ and still obtain the same relation between the transformed variables $Y' = AX'$. In other words, if it is probable that there exists a sample point $\boldsymbol{z} = (x, y)$ then it is equally probable that there also exists the sample point $\boldsymbol{z}' = (Cx, Cy)$. The probability distribution is thus invariant with respect to multiplication of the sample vector by a constant $f(\boldsymbol{z}) = f(C\boldsymbol{z})$. This property can be called an amplitude invariance. The model described by a non-parametric regression does not represent such an invariance although it can be approximately fulfilled in a limited region in which the sample points of the data base lie. Various invariances generally correspond to various relations between variables. When we assume a certain invariance we indirectly assume a corresponding relationship between the variables and vice versa. The corresponding natural law is then generally described by proper functions and parameters that make possible the adaptation of the corresponding expression to the phenomenon under consideration. The task of modeling is thus reduced to a determination of the parameters in a certain function from the empirical data and consequently the resulting empirical law is called a *parametric regression.* At present it is still not known how to make a modeler of natural phenomena that could automatically find the invariances in the data corresponding to some empirical information about a natural phenomenon and directly incorporate these invariances into a model. If this were possible, then the basis of theoretical physics could be automatically built. Among various parametric regression methods, a linear one is most frequently applied. Before we proceed to the presentation of the methods for its adaptive determination in the next section, we briefly put forth some arguments for its wide applicability.

10.2 Generalized Linear Regression Model

There are many natural phenomena that can be physically described by linear models. For example, the generation of sound by the forces acting in the source, the emission of electromagnetic waves caused by an electrical current, the diffusion of substances from a source, and others, can all be described by linear differential equations. We know *a priori* when dealing with these phenomena that a linear model is applicable. The question then arises why we need empirical modeling in these cases at all if a deterministic approach is possible and if the physical basis for the description is already given by the corresponding differential equations. The answer is found from the following consideration.

For the sake of simplicity, let us assume that the phenomenon can be described by a scalar field $Y(\boldsymbol{r}, t)$ that is emanating from a scalar source field $X(\boldsymbol{r},\, t)$, where the four-dimensional parameter $s = (\boldsymbol{r}, t)$ denotes spatial coordinates and time respectively. A corresponding differential equation is then written as

$$L(\partial_{\boldsymbol{r}},\, \partial_t)\, Y(\boldsymbol{r},\, t) = X(\boldsymbol{r},\, t)\ , \tag{10.1}$$

where L denotes a linear functional. The problem is completed by specifying a set of consistent boundary and initial conditions. The solution of the corresponding mathematical problem can then be formally represented in the form

$$Y(\boldsymbol{r},\, t) = \Psi(\boldsymbol{r},\, t) + \int G(\boldsymbol{r},\, \boldsymbol{r}';\, t,\, t')\, X(\boldsymbol{r}',\, t')\, dt'\, dV'\ , \tag{10.2}$$

where $\Psi(\boldsymbol{r},\, t)$ denotes the solution of the corresponding homogeneous equation and $G(\boldsymbol{r},\, \boldsymbol{r}';\, t,\, t')$ denotes the Green's function that satisfies the given boundary and initial conditions. If only the field emanating from the source X is considered, then the solution of the homogeneous equation is taken equal to zero. That is, $\Psi(\boldsymbol{r},\, t) = 0$, and we obtain the input-output relation

$$Y(\boldsymbol{r},\, t) = \int G(\boldsymbol{r},\, \boldsymbol{r}';\, t,\, t')\, X(\boldsymbol{r}',\, t')\, dt'\, dV' = G * X\ , \tag{10.3}$$

where the asterisk $*$ denotes the convolution in space and time. In order to determine the field $Y(\boldsymbol{r},\, t)$ form the source field $X(\boldsymbol{r},\, t)$ we must determine the Green's function that satisfies the equation

$$L(\partial_{\boldsymbol{r}},\, \partial_t)\, G(\boldsymbol{r},\, \boldsymbol{r}';\, t,\, t') = \delta(\boldsymbol{r} - \boldsymbol{r}')\delta(t - t')\ . \tag{10.4}$$

The solution of this equation is often very difficult to find analytically. For instance, we can imagine a geometrically complicated structure with an operating source that cannot be described analytically. In such a case it is only possible to describe the problem in empirically, that is, in terms of data obtained from measurements in an actual situation. The problem is then transferred to the numerical determination of the Green's function. Even this approach is often not possible. For instance, we may not know the boundary or the initial conditions but only several examples of empirically determined joint pairs of data about the source and the emanating field, say $(x_1(\boldsymbol{r},\, t),\, y_1(\boldsymbol{r},\, t);\, x_2(\boldsymbol{r},\, t),\, y_2(\boldsymbol{r},\, t);\, \ldots;\, x_N(\boldsymbol{r},\, t),\, y_N(\boldsymbol{r},\, t))$. In this case, we must determine the Green's function from these empirical pairs by the deconvolution of equation (10.3). [6, 22] A direct deconvolution generally represents an ill-posed inverse problem which requires additional conditions or assumptions in order to be regularized. [30, 5] Just as for the case of general non-parametric regression, we can use for this purpose the minimization of the mean square error. In this approach, we first assume that the set of data pairs need not exactly satisfy a linear relationship but could be corrupted by noise which is of experimental or some other character. Equation (10.3) must then be treated as the definition of a linear estimator, that is

$$\hat{Y}(s) = \int G(s, s')\, X(s')\, ds' = G * X \ , \tag{10.5}$$

where s denotes a point in space and time $s = (\boldsymbol{r},\, t)$. If we consider an empirical joint pair $\{x(s),\, y(s)\}$ then we can generally establish a discrepancy between the estimated field $\hat{Y}(s)$ and the experimentally observed one $y(s)$. The Green's function describing the estimator must be determined in such a way that the mean square error or discrepancy

$$\mathcal{D} = \mathrm{E}\left[(\hat{Y} - Y)^2\right] = \mathrm{E}\left[(G * X - Y)^2\right] \tag{10.6}$$

is minimized. In contrast to the non-parametric case, we are here not seeking an unknown function $\hat{Y}(X)$ but rather a particular linear relationship between X and Y. It is determined by a Green's function G that in this case plays a role of a generalized parameter. Our task is to select an optimal function G_o that minimizes the functional $\mathcal{D}$ with respect to the measured data. For this purpose we again apply the calculus of variation. We first represent the varied Green's function as $G = G_o + \epsilon\eta(s,\, s')$ with the amplitude $\epsilon > 0$ and the form of the variation $\eta(s,\, s')$ described by an arbitrary function. Next we require that the functional $\mathcal{D}$ must have a minimum as a function of amplitude ϵ, regardless of the arbitrary variation of $\eta(s,\, s')$.

A comment about the arbitrariness of the variation $\eta(s,\, s')$ should be mentioned here. When the field Y is the result of the field X then the Green's function must satisfy the causality condition which is a statement of *cause must precede effect.* Hence

$$G(\boldsymbol{r};\, \boldsymbol{r}';\, t, t') = 0 \quad \text{for } t < t' \ . \tag{10.7}$$

This also means that the variation $\eta(s,\, s')$ can, in fact, not be arbitrarily chosen but rather, it must also satisfy the causality condition, provided that Green's function represents the response of a real system. This requirement significantly influences and complicates the analytical treatment of this problem. [20] We therefore avoid such a strict treatment by making a weak simplification which is based on the following interpretation. There are many linear phenomena for which we do not know *a priori* which variable, if any, can be treated as the cause and which as the effect. In these cases, we do not assume *a priori* that the Green's function is causal. However, in the empirical data the causality may be hidden, and when we determine the Green's function from such data, its value at $t < t'$ will approach zero with increasing causality of the data. We shall then interpret the corresponding non-zero values of G as an error that is the result of the empirical approach. Without the strict requirement of causality $\eta(s,\, s')$, can be treated as an arbitrary variation which essentially simplifies the further treatment.

For the optimal Green's function, the variation $\delta\mathcal{D}$ caused by the variation $\delta G(s,\, s') = \epsilon\eta(s,\, s')$, must satisfy the expression

$$\delta\mathcal{D} \to 0 \quad \text{for } \epsilon \to 0 \tag{10.8}$$

regardless of the function $\eta(s, s')$. Therefore, $\delta\mathcal{D}$ must have a minimum at $\epsilon = 0$ as a function of the amplitude ϵ, or

$$\left.\frac{\partial(\delta\mathcal{D})}{\partial\epsilon}\right|_{\epsilon=0} = 2\mathrm{E}\left[(\hat{Y} - Y)(\eta * X)\right] = 0 \,. \tag{10.9}$$

Because of linearity we can exchange the order the convolution and the expected value and obtain

$$\int \eta(s, s') \int G_o(s, s'')\, \mathrm{E}\left[X(s'')\, X(s')\right] ds''\, ds' \\ - \int \eta(s, s')\mathrm{E}\left[Y(s)\, X(s')\right] ds' = 0 \,. \tag{10.10}$$

By taking into account that this equation must be satisfied for an arbitrary $\eta(s, s')$ and by introducing the auto- and cross-correlation functions by the expressions

$$R_{XX}(s, s') = \mathrm{E}\left[X(s)\, X(s')\right] \,, \tag{10.11}$$

$$R_{YX}(s, s') = \mathrm{E}\left[Y(s)\, X(s')\right] \,, \tag{10.12}$$

we obtain the following equation

$$\int G_o(s, s'')\, R_{XX}(s'', s')\, ds'' = R_{YX}(s, s') \,, \tag{10.13}$$

which can be represented in a short form as

$$G_o * R_{XX} = R_{YX} \,. \tag{10.14}$$

A similar equation with one-dimensional "s" corresponding to the time variable only, which also incorporates the causality condition is known in the theory of linear adaptive filtering as the normal or the Wiener-Hopf (WH) equation. [20, 23, 16] In our case, where s denotes the spatial and time coordinates simultaneously and where the causality condition has been omitted, the above equation represents a generalization of the Wiener-Hopf equation. Unfortunately it is an integral equation for the Green's function. The integral represents a convolution and in order to extract the Green's function from it, we need a specific treatment called an *optimal deconvolution.* The exact approach to the solution of the original Wiener-Hopf one-dimensional problem begins with application of a Fourier-transform and factorization of the complex frequency spectrum, which is a cumbersome task. [20] Approximate methods have therefore been developed which are especially useful for discrete signals in which the integral is changed into a sum and the Green's function is represented by a matrix. [21, 2, 5, 23, 16]

Before proceeding to the approximate treatment of the generalized WH equation let us briefly analyze how the properties of the correlation functions are mirrored on the Green's function. As mentioned in the previous chapter on random processes we usually identify processes as being stationary or non-stationary. The joint probability distributions for stationary processes are invariant with respect to a shifting of the parameter t which usually represents time. Similarly, we differentiate fields that depend on multi-dimensional parameter $\boldsymbol{r}$, which usually corresponds to the spatial coordinates, into homogeneous and non-homogeneous, where the homogeneous fields exhibit an invariance of joint probability densities with respect to translation of the parameter $\boldsymbol{r}$. Stationarity as well as homogeneity are further reflected in the properties of the correlation functions of all orders.

Most frequently only the lowest-order correlation functions are used in the description of natural phenomena. For a stationary process the correlation function must depend only on the difference of parameters specifying the moments of observation or $R(t, t') = R(t' - t)$. A similar observation can be made for the homogeneous field. Its correlation function must depend only on the difference of the parameters $\boldsymbol{r}$ and $\boldsymbol{r}'$ that represent the positions of field observation. In other words, $R(\boldsymbol{r}, \boldsymbol{r}') = R(\boldsymbol{r}' - \boldsymbol{r})$. Similarly, we can conclude that for a homogeneous and isotropic field, the correlation function must be independent on an arbitrary simultaneous rotation of both parameters described by the transformation $\boldsymbol{r}_1 = \boldsymbol{T}\,\boldsymbol{r}$ in which $\boldsymbol{T}$ represents the rotation matrix. In this case the correlation function can depend only on the absolute value of the difference $(\boldsymbol{r} - \boldsymbol{r}')$ or $\|\boldsymbol{r} - \boldsymbol{r}'\|$.

In developing an empirical description of natural phenomena we often encounter processes or fields whose correlation functions are invariant with respect to one of the above-mentioned transformations of parameters. As yet we do not know if the same property is also a characteristic for all joint probability distributions. We shall, therefore, consider stationarity, or homogeneity or isotropy in the wide sense. When we are dealing with time- and space-dependent fields, then the simultaneous stationarity and homogeneity results in an invariance of the probability distribution functions with respect to an arbitrary translation of the parameter $s = (\boldsymbol{r}, t)$. This means that the correlation function must be dependent only on the difference of the appertaining parameters s and s', that is $R(s, s') = R(s' - s)$. This property must hold for the auto- as well as cross-correlation functions when we are observing a linear phenomenon. However, because both functions are related by the WH equation, the invariance with respect to a shifting of the parameter s is further transmitted to the Green's function. This becomes evident if we insert the corresponding expressions for the correlation functions into the WH equation:

$$G_o(s, s'') * R_{XX}(s' - s'') = R_{YX}(s' - s) \; . \tag{10.15}$$

By introducing $s' - s'' \equiv s_1$ and $s' - s \equiv s_2$ we get

$$G_o(s' - s_2, s' - s_1) * R_{XX}(s_1) = R_{YX}(s_2) \; . \tag{10.16}$$

The cross-correlation shows that the result of the convolution does not depend on the parameter s' which can be realized only if this parameter disappears from the resulting Green's function on the left side of the last equation. This occurs when the optimal Green's function only depends on the difference of the associated parameters or

$$G_o(s' - s_2, s' - s_1) = G_o(s_2 - s_1) \; . \tag{10.17}$$

This fact drastically reduces the computational work needed in the determination of the Green's function because we must determine a function of only one parameter $\Delta s = s_2 - s_1$ instead of a function of the corresponding two parameters. The WH equation can then be re-written as

$$G_o(s_2 - s_1) * R_{XX}(s_1) = R_{YX}(s_2) \; . \tag{10.18}$$

If the fields are either stationary or homogeneous, then a similar property is reflected in the corresponding portion of the multi-dimensional parameter s.

The foregoing analysis demonstrates that a linear model can be used for the description of natural phenomena that includes amplitude, time and space invariances. The first invariance is included into description by the selection of the linear model while the next two invariances are accounted for by the functional form of the Green's function. In an empirical description of natural phenomena, the time and space invariances need not be assumed *a priori* because they are automatically included in a general treatment by the properties of the correlation functions. However, the treatment of the general Wiener-Hopf problem is more complicated than the treatment of the stationary and homogeneous problems described above. Therefore it will be of advantage to know if a phenomenon can be described by a Green's function that depends on differences of parameters alone, for then, according to Eq. (10.18) the solution of WH equation can be found more easily.

The inclusion of invariances into a physical description is of fundamental importance for the development of automatic pattern recognition devices. In addition to amplitude, temporal and spatial (translational and rotational) invariances there is also the very crucial scale invariance. If we wish to include scale invariance into a model, we must require that the probability distribution functions are invariant with respect to a simultaneous stretching or contraction of spatial coordinates. [15] This invariance is of great importance for the description of visual pattern recognition but since this topic is not the central theme of this book, we shall not discuss it further.

The properties of a random phenomenon that are described by shift-invariant correlation functions can be equivalently represented by the Fourier-transform (FT) of the correlation function that yields the spectral distribution. For analytical purposes such an approach is quite advantageous, because the Fourier-transform of the convolution corresponds to the product of the corresponding spectral distributions. This property leads to an apparently simple expression of the solution of Wiener-Hopf equation in terms of spectral distributions which are introduced by:

$$R_{XX}(s) = \frac{1}{(2\pi)^4} \int_{-\infty}^{\infty} S_{XX}(w) \exp(-i\, w \cdot s)\, d^4 w \ , \tag{10.19}$$

$$R_{YX}(s) = \frac{1}{(2\pi)^4} \int_{-\infty}^{\infty} S_{YX}(w) \exp(-i\, w \cdot s)\, d^4 w \ , \tag{10.20}$$

$$G(s) = \frac{1}{(2\pi)^4} \int_{-\infty}^{\infty} S_G(w) \exp(-i\, w \cdot s)\, d^4 w \ . \tag{10.21}$$

Here we assume that the parameters s and w are four-dimensional. That is, $s = (\boldsymbol{r},\, t)$ and the wave number $w = (\boldsymbol{k}, \omega)$. By inserting the above into the Wiener-Hopf equation and taking into account the expression for the Fourier-transform of the delta function

$$\delta(s) = \frac{1}{(2\pi)^4} \int_{-\infty}^{\infty} \exp(-i\, w \cdot s)\, d^4 w \ , \tag{10.22}$$

we obtain the expression

$$\int_{-\infty}^{\infty} S_G(w)\, S_{XX}(w)\, \exp(-i\, w \cdot s) d^4 w$$
$$= \int_{-\infty}^{\infty} S_{YX}(w)\, \exp(-i\, w \cdot s) d^4 w \tag{10.23}$$

which yields the Wiener-Hopf equation expressed in terms of the spectral distributions

$$S_G(w)\, S_{XX}(w) = S_{YX}(w) \ . \tag{10.24}$$

The Green's function can be then expressed by using its spectral distribution

$$S_G(w) = \frac{S_{YX}(w)}{S_{XX}(w)} \tag{10.25}$$

as

$$G(s) = \frac{1}{(2\pi)^4} \int_{-\infty}^{\infty} \frac{S_{YX}(w)}{S_{XX}(w)} \exp(-i\, w \cdot s)\, d^4 w \ . \tag{10.26}$$

The foregoing analytical approach is, in principle, feasible but in order to determine the Green's function we must twice Fourier-transform the correlation functions and then inverse Fourier-transform the spectral density, which may become computationally demanding when considering certain spatio-temporal problems. In the empirical treatment the correlation functions are statistically determined from sample data and thus they are not given analytically nor are they exact. We are therefore interested in an approximate treatment, that could be computationally less demanding.

Let us illustrate this approach with the trivial case of stationary, uncorrelated source field for which the solution of the Wiener-Hopf equation can be immediately found. In this case, the auto-correlation function can be described by the delta function

$$R_{XX}(s'',\, s') = C\delta(s'' - s') \ , \tag{10.27}$$

with C describing a constant proportional to the mean square of the variable X. That is: $C = \tau \mathrm{E}[X^2]$. The constant of proportionality τ represents the characteristic correlation width in the space of the parameter s. The optimal Green's function is then determined by the cross-correlation function between the emanating and the source field:

$$G_o(s,\, s') = C^{-1}\, R_{YX}(s,\, s')\ . \tag{10.28}$$

Using this approach, an iterative treatment can be introduced which is instructive for the adaptive treatment to be presented later and appropriate for a number of applications and therefore we briefly describe it here. [7, 8, 33, 34]

Often the auto-correlation function resembles an approximation of a delta function. We can then expect that the optimal Green's function resembles the cross-correlation function and therefore we begin the iteration process using the expression

$$G_o(s,\, s') = A_o R_{YX}(s,\, s') + G_1(s,\, s')\ , \tag{10.29}$$

in which the constant A_o will be later specified. By inserting this expression into the generalized Wiener-Hopf equation and introducing a new cross-correlation function by

$$\begin{aligned} R_{YX,1}(s,\, s') &= R_{YX}(s,\, s') - A_o \int R_{YX}(s,\, s'') R_{XX}(s'',\, s')\, ds'' \\ &= R_{YX}(s,\, s') - A_o Q_o(s,\, s')\ , \end{aligned} \tag{10.30}$$

we obtain an equation for the correction of the first approximation

$$\int G_1(s,\, s'') R_{XX}(s'',\, s')\, ds'' = R_{YX,1}(s,\, s')\ . \tag{10.31}$$

We have obtained the Wiener-Hopf equation again, but now with a changed forcing term $R_{YX,1}$. The new cross-correlation $R_{YX,1}$ determines the goodness of the approximation (10.29). The better the approximation, the smaller will be the amplitude of the new cross-correlation function on average. We therefore determine the constant A_o such that the correction measure, defined by the integral of $R^2_{YX,1}(s,\, s')$ with respect to s and s', is a minimum.

$$\begin{aligned} C &= \int R^2_{YX,1}(s,\, s')\, ds\, ds' \\ &= \int \left[R_{YX}(s,\, s') - A_o Q_o(s,\, s')\right]^2\, ds\, ds' \;=\; \min(A_o)\ . \end{aligned} \tag{10.32}$$

Setting $dC/dA_o = 0$ we obtain

$$A_o = \frac{\int R_{YX}(s,\, s')\, Q_o(s,\, s')\, ds\, ds'}{\int Q_o^2(s,\, s')\, ds\, ds'}\ . \tag{10.33}$$

If the cross-correlation function depends only on the difference $\triangle s = s - s'$ then the integrations appearing in this expression need to be performed just once with respect to $\triangle s$.

Utilizing expression (10.29) again to determine the correction of the first approximation, we obtain an expression for the second approximation etc. The optimal Green's function is then approximated by a series

$$G_j(s,\, s') = \sum_{i=0}^{j} A_i R_{YX,i}(s,\, s') \ , \tag{10.34}$$

where

$$R_{YX,i+1}(s,\, s') = R_{YX,i}(s,\, s') - A_i Q_i(s,\, s') \ , \tag{10.35}$$

$$Q_i(s,\, s') = \int R_{YX,i}(s,\, s'') R_{XX}(s'',\, s')\, ds'' \ , \tag{10.36}$$

$$A_i = \frac{\int R_{YX,i}(s,\, s')\, Q_i(s,\, s')\, ds\, ds'}{\int Q_i^2(s,\, s')\, ds\, ds'} \ . \tag{10.37}$$

10.2.1 An Example of Iterative Determination of a Linear Regression Model

The iterative method described above is suitable for implementation on a digital computer. For this, the fields X and Y must first be represented by sequences of digitized data and then the correlation functions must be determined as statistical averages over the given set of joint samples $\{x_n(s),\, y_n(s);\, n = 1,\, \ldots,\, N\}$. It is more advantageous to measure the auto- and cross-correlation functions directly by a correlator. The speed of calculation needed in the iteration is principally determined by the procedures required in the determination of new cross-correlation functions. The iteration is usually stopped when the amplitude of the new correlation function reaches the value that is of the same order as the statistical error found from the determination or measurement of cross-correlation function. The method is also applicable for cases in which one models only a time-dependent phenomenon that corresponds to the description of a linear discrete system. In this case the parameter s indicates time alone and the data can be simply represented. Therefore we present here an example of calculation of Green's function for the case of a discrete system. The convergence of the proposed iterative method has not yet been analyzed, but we have verified its operation on simulated and real, discrete [7, 8] and distributed noisy systems. [33]

In engineering practice one often encounters stationary random processes, the auto-correlation of which resembles the function

$$R_{XX}(t) = R_o \exp(-\alpha t) \cos(t) \ , \tag{10.38}$$

where $t = t'' - t'$ denotes the difference of two times of observation of the signal $X(t)$. In addition we often deal with systems whose response function

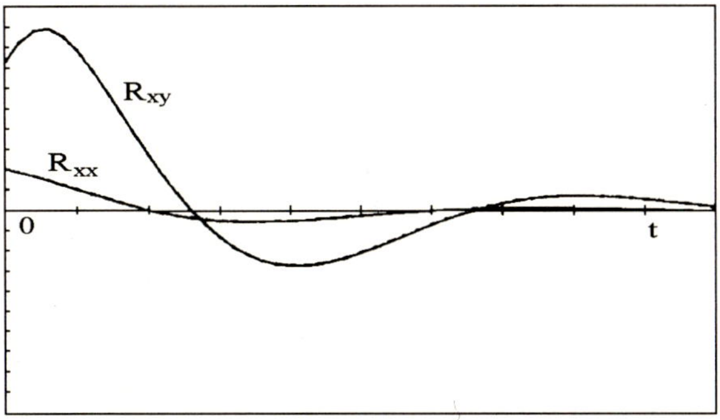

Fig. 10.1. Auto-correlation function R_{XX} of the input and the cross-correlation function R_{YX} of input and output stochastic signals of a linear system used in an optimal estimation of a response function

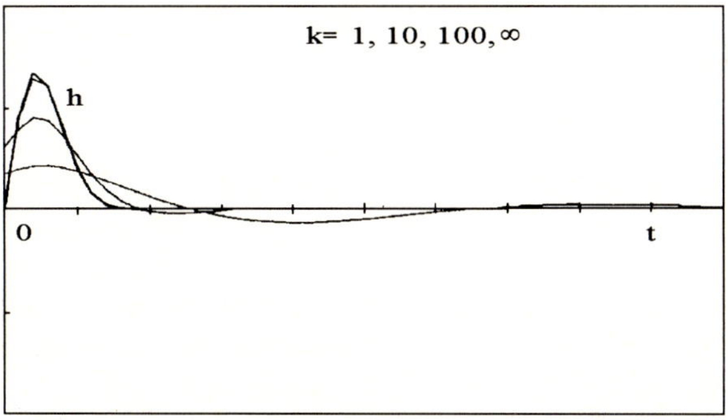

Fig. 10.2. The true response function $h(t)$ of the system and the approximations of the optimal one estimated by $k = 1, 10, 100$ steps of iteration. The maximal value of estimated response function $h_k(t)$ increases with the number of iteration steps, and the approximations converge to the true response function

is determined by some pulse-like function $h(t)$ of certain amplitude, width, and delay. For the demonstration of the iterative estimation of an optimal response function, we have simulated such an example because it permits the comparison of the approximated and the exact response functions. The auto-correlation function $R_{XX}(t)$ is shown in Fig. 10.1 and the selected response function is shown by the bold line in Fig. 10.2. Using the Wiener-Hopf equation, the cross-correlation function $R_{YX}(t)$ of the system output $Y(t)$ and input $X(t)$ has been determined and it is also shown in Fig. 10.1. Using these two correlation functions the iterative estimation of the optimal response function has been carried out. The results of the first, tenth, and the hundredth step of the iteration are presented together with the exact

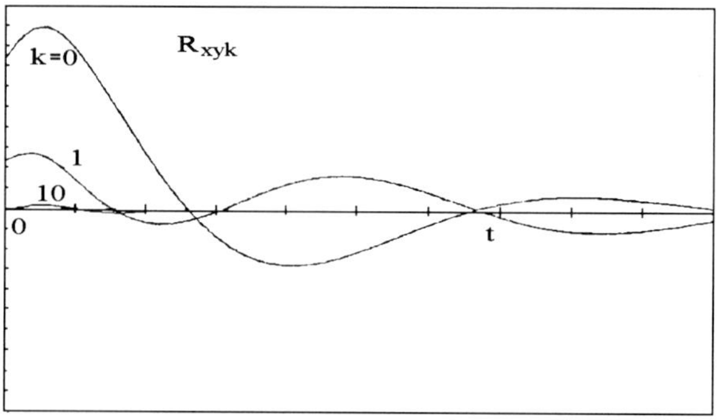

Fig. 10.3. Sequence of records of the new cross-correlation function $R_{xy,k}(t)$ in dependence of the iteration number k. The amplitude decreases with the increasing number of iteration steps

response function in Fig. 10.2. The results show that after the hundredth iteration, the estimated optimal response function apparently differs from the correct one only in one point. Close inspection shows, however, that there is a small difference at the other points, but the corresponding thin and thick lines connecting the points cannot be delineated in the figure. The error of the estimation depends on the new cross-correlation function $R_{YX}(t)$. The convergence of the iteration is illustrated in Fig. 10.3. It is seen that the amplitude of the new cross-correlation functions calculated at characteristic steps decreases with the number of iterations. It appears that already after the tenth step the amplitude of this function is within an order of magnitude of the error that is usually found in measured correlation functions ($\approx 1\%$). From numerical simulations as these, it has been concluded that in actual applications, about ten iterations usually suffice. The numerical examples have also revealed that the constant A_i does not change appreciably during the iteration procedure so that in practice it needs to be determined only in the first step of iteration.

The iteration method demonstrated above is also applicable to the modeling of linear systems on the basis of transient signals. However, in this case the correlation functions do not depend on the time difference $t = t'' - t'$ but rather on both times t' and t'' and one therefore must deal with functions depend on both these parameters. Because of this increased complexity, a procedure relying on calculations of the correlation functions to find the solution of the Wiener-Hopf equation and to determine the Green's function is avoided. For the case of discrete systems a modification of the demonstrated optimal filtering method applied to process acoustic emission signals has already been published in the literature. [8] The iteration method described in this section can be also generalized to the case of multi-component or vector fields. [33]

In applying the iteration method described here, one must specify the correlation functions and calculate the Green's function in a batch type procedure. However, in a multi-dimensional case this requires considerable computer memory and processing time. The above-mentioned fundamental problem of solving the Wiener-Hopf equation by using the joint empirical samples $\{x_1(s), y_1(s); \ldots; x_N(s), y_N(s)\}$ directly without calculating and storing the correlation functions must thus be solved in order to provide for an efficient application of a generalized linear modeling. With this goal in mind, we again follow the perturbation approach to the stochastic approximation and introduce a sequential adaptation procedure in place of the batch-type.

10.3 Sequential Adaptation of Linear Regression Model

Let us assume that the solution of the optimal Green's function, G, from the Wiener-Hopf equation is known for certain starting correlation functions R_{XX} and R_{YX} and we wish to determine the solution for the weakly perturbed couple $R'_{XX} = R_{XX} + \Delta R_{XX}$ and $R'_{YX} = R_{YX} + \Delta R_{YX}$. We assume that the corresponding Green's function is also weakly perturbed $G' = G + \Delta G$. By inserting these expressions into the Wiener-Hopf equation and taking into account that it is fulfilled for functions with index zero, and retaining only the terms linear in the perturbations, we obtain the equation for the perturbation of the Green's function. That is,

$$\Delta G * R_{XX} = \Delta R_{YX} - G * \Delta R_{XX} \; . \tag{10.39}$$

Let us assume that the starting correlation functions are specified by an empirical set of $N - 1$ samples $\{x_n(s), y_n(s); n = 1, 2, \ldots, N-1\}$ and we acquire a new sample pair $[x_N(s), y_N(s)]$. The starting correlation functions are then determined by the empirical average

$$R_{XX}(s, s') = \frac{1}{N-1} \sum_{n=1}^{N-1} x_n(s)\, x_n(s') \; , \tag{10.40}$$

$$R_{YX}(s, s') = \frac{1}{N-1} \sum_{n=1}^{N-1} y_n(s)\, x_n(s') \; , \tag{10.41}$$

which yield for the perturbations caused by the N-th sample pair the following expressions

$$\Delta R_{XX}(s, s') = \frac{1}{N} \left[x_N(s)\, x_N(s') - R_{XX}(s, s')\right] \; , \tag{10.42}$$

$$\Delta R_{YX}(s, s') = \frac{1}{N} \left[y_N(s)\, x_N(s') - R_{YX}(s, s')\right] \; . \tag{10.43}$$

If we insert these perturbations into the equation for ΔG and take into account that the Wiener-Hopf equation is fulfilled for the starting correlation functions we obtain

$$\Delta G(s,\, s'') * R_{XX}(s'',\, s') \; = \\ = \frac{1}{N}\,[y_N(s) - G(s,\, s'') * x_N(s'')]\; x_N(s') \; . \tag{10.44}$$

The asterisk $*$ denotes the convolution integral with respect to the parameter s''.

This equation describes the sequential adaptation of the Green's function to the new acquired pair of sample data. To express the perturbation ΔG explicitly, we must carry out the deconvolution. This cannot be done without specifying the properties of the auto-correlation function. For this, we further restrict our treatment to stationary and homogeneous processes and assume that the auto-correlation function can be approximated by a delta function $R_{XX}(s'',\, s') \cong \mathrm{E}[X^2]\,\tau\,\delta(s'' - s')$. The convolution then yields a term $\mathrm{E}[X^2]\,\tau\,\Delta G(s,\, s')$ and we obtain

$$\Delta G(s,\, s') \cong \alpha(N)[y_N(s) - G(s,\, s'') * x_N(s'')]\, x_N(s') \; , \tag{10.45}$$

where the adaptation constant is given by

$$\alpha(N) = \frac{1}{\tau N \mathrm{E}[X^2]} \; . \tag{10.46}$$

The second term in the above equation for ΔG can be expressed as

$$\hat{y}_N(s) = G(s,\, s'') * x_N(s'') \; . \tag{10.47}$$

It represents the estimation of the emanated field $\hat{y}_N$ because of the source field x_N. The complete equation can be then written as

$$\Delta G(s,\, s') = \alpha(N)\,[y_N(s) - \hat{y}_N(s)]\; x_N(s') \; . \tag{10.48}$$

This equation shows that the Green's function can be simply adapted to the sequence of stochastic samples by calculating the difference between the true response field and the estimated or predicted one. That is

$$\Delta y_N(s) = y_N(s) - \hat{y}_N(s) \; . \tag{10.49}$$

We then form the correction term using the product

$$\Delta G(s,\, s') = \alpha(N)\, \Delta y_N(s)\, x_N(s') \; . \tag{10.50}$$

This equation represents a version of the stochastic approximation. [1, 25] As in the derivation of the self-organization rule, we have found that during adaptation the constant $\alpha(N)$ must decrease monotonically with the number of samples N. This constant also depends on the mean square value of the source field and the correlation width τ which are usually not known in advance but they can be estimated from a few initial samples.

When we are using a one-dimensional discrete parameter s, then the Green's function can be treated as a response function described by the matrix $\boldsymbol{M} = [m_{ij}]$. Alternatively it may be called the weight or memory matrix.

The complete model can be then simply interpreted. It describes a linear system with the vector input $\boldsymbol{X} = (x_1, \ldots, x_j, \ldots, x_D)$ and a vector output $\boldsymbol{Y} = (y_1, \ldots, y_i, \ldots, y_{D'})$. Here the dimensions of both vectors D and D' need not be the same. The input-output relation and the adaptation equation are then represented in matrix notation as

$$\hat{\boldsymbol{Y}} = \boldsymbol{M}\boldsymbol{X} \,, \tag{10.51}$$

$$\triangle \boldsymbol{M} = \alpha(N) \, \triangle \boldsymbol{Y}_N \boldsymbol{X}_N^T \,. \tag{10.52}$$

The last equation is well known in the field of adaptive signal processing and in research of artificial neural networks where it is called the Widrow-Hoff learning rule or alternatively, LMS or *delta learning rule*. [31, 2, 17]. It is interesting that a similar concept of adaptation has been postulated by Hebb on the basis of a neurobiological study of learning and this has provided the basis for the development of many other learning paradigms in the recent intensive research of neural networks. [18, 19, 26] The corresponding system is usually called a *linear associator*.

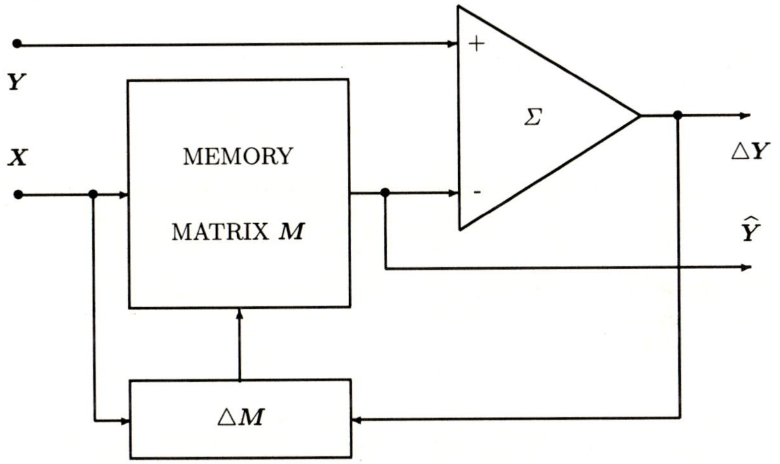

Fig. 10.4. Scheme of an information processing system capable of linear adaptive modeling

An information system capable of linear adaptive modeling or a linear associator can be represented by the scheme shown in Fig. 10.4. The difference between the estimated and the desired outputs $\triangle \boldsymbol{Y}$ is fed back to the memory where the outer product with the input $\boldsymbol{X}$ and the learning constant α is calculated according to Eq. (10.52) and this is added to the memorized response function $\boldsymbol{M}$. The system is thus learning from the error it finds during its operation. It is well-known that humans and other living beings learn in a similar way. Much research has been carried out to try to understand how biological neural networks achieve a similar capability. [3, 4, 19] It is now believed that the site of memory in neural networks is a synaptic joint between dendrites of neural cells that can be described as a matrix element m_{ij}. Hebb's explanation of learning as well as the delta rule suggest a conjecture that the synaptic joint must be changed by the pre-synaptic as well as by the post-synaptic signals. The former represents the excitation of the i-th neuron by the other neurons, while the latter represents the excitation of the other neurons by the i-th one. A memorization is thus an essentially collective phenomenon, stemming from the synergy of neurons. Although an associator can be treated as a linear system, the adaptive changes of its parameters are not described in terms of a linear operation. We have thus arrived again at a conclusion similar to that utilized to introduce the self-organization process and the general modeling that is based on a non-linear regression. From this aspect it appears that the linear associator can be treated only as an approximate, or the simplest system resembling the associative properties of biological neural networks. Nevertheless, it is quite useful in many cases of practical interest as will be later demonstrated on examples of analyzing acoustic emission signals.

10.4 Transition from the Cross- to Auto-Associator

The linear system presented above is capable of associating with an input field $\boldsymbol{X}$ the corresponding output field $\hat{\boldsymbol{Y}}$. It can therefore generally be called a cross-associator. When the Green's function is formed on the basis of given samples, then the field $\boldsymbol{Y}$ is treated as a desired output that must be achieved by the response of the system to an excitation by the corresponding input $\boldsymbol{X}$. When dealing only with time-dependent signals, then it is often possible to consider one signal to be a consequence of the other. But there are many instances in which such a distinction is not possible, although we know that the system is linear. In this case either the experimenter must decide which signal is the result of the other or we must develop a more general modeling in which this distinction is not needed.

Let us for this purpose interpret the equation

$$\boldsymbol{Y} = \boldsymbol{G} * \boldsymbol{X} \tag{10.53}$$

as the solution of the problem of mapping the given field $\boldsymbol{X}$ to the corresponding field $\boldsymbol{Y}$. Let us arbitrarily call this the "direct" problem. We can then pose the "inverse" problem by providing the field $\boldsymbol{Y}$ and seeking the corresponding $\boldsymbol{X}$. Let us formally represent the corresponding solution of this inverse problem as

$$\boldsymbol{X} = \boldsymbol{H} * \boldsymbol{Y} \ . \tag{10.54}$$

By this expression we have not violated the fact that $\boldsymbol{X}$ and $\boldsymbol{Y}$ are linearly related, but if we apply the adaptation procedure described in the previous section, then the resulting optimally estimated functions $\boldsymbol{G}$ and $\boldsymbol{H}$ will generally not satisfy the identity relation $\boldsymbol{G} * \boldsymbol{H} = \boldsymbol{I}$ which follows from the last two equations. This inconsistency indicates that we cannot generally assume both equations simultaneously since the result is a dilemma to select which equation corresponds better to the phenomenon under consideration. A suggestion for avoiding this dilemma can be found in the solution to the following problem: let there be given partial information about the field $\boldsymbol{X}$ and partial information about the field $\boldsymbol{Y}$. How can one determine the missing information? In this case neither the first nor the second of the above view points and the corresponding equations is applicable. In order to solve the problem we will have to change our view completely and to treat both fields as parts of a more general joint field. The corresponding variable, called a *pattern field*, can be represented by the concatenation of both fields. That is,

$$\boldsymbol{Z} = \boldsymbol{X} \oplus \boldsymbol{Y} = (\boldsymbol{X}, \boldsymbol{Y}) = \boldsymbol{Z}(\boldsymbol{r}, t, i) \ . \tag{10.55}$$

Here $\boldsymbol{r}, t$ denotes that the pattern field can depend on spatial coordinate and time, while i indicates that it may also be comprised of various components.

We now develop a linear model to process the pattern field. For the sake of simplicity let us first treat a two-dimensional discrete example in which the field $\boldsymbol{Z}$ is substituted by a vector $\boldsymbol{z}$ which is represented by only two components x, y. Let us further assume that the set of joint empirical samples $(\boldsymbol{z}_1, \boldsymbol{z}_2, \ldots, \boldsymbol{z}_N)$ is scattered in the corresponding sample space along some "line" passing through the origin and having its direction specified by the unit vector $\boldsymbol{e}$. An arbitrary vector $\boldsymbol{z}$ is then represented by the components parallel and orthogonal to this line:

$$\boldsymbol{z} = \boldsymbol{z}_{\|} + \boldsymbol{z}_{\perp} = \boldsymbol{u} + \boldsymbol{v} \ . \tag{10.56}$$

The parallel component is determined by the expression

$$\boldsymbol{u} = \boldsymbol{e}(\boldsymbol{e} \cdot \boldsymbol{z}) = \boldsymbol{W} \boldsymbol{z} \ . \tag{10.57}$$

The dot in this equation denotes the inner or scalar product while the matrix $\boldsymbol{W}$ is determined by the outer product

$$\boldsymbol{W} = \boldsymbol{e}\,\boldsymbol{e}^T \ . \tag{10.58}$$

The orthogonal component is expressed as

$$\boldsymbol{v} = \boldsymbol{z} - \boldsymbol{e}(\boldsymbol{e} \cdot \boldsymbol{z}) = \boldsymbol{z} - \boldsymbol{W} \boldsymbol{z} \ . \tag{10.59}$$

We can interpret this component as the discrepancy between the selected vector $\boldsymbol{z}$ and the corresponding vector $\boldsymbol{u}$ lying on the regression line. For the case in which the vector $\boldsymbol{z}$ is obtained by measurements, the discrepancy is usually treated as an error between an observed value and the value predicted by a model. Our task is to determine the unit vector $\boldsymbol{e}$, and with it the matrix $\boldsymbol{W}$ in such a way that the mean square error is minimized. Hence,

$$\mathcal{D} \equiv \mathrm{E}[\boldsymbol{v}^2] \equiv \mathrm{E}[||\boldsymbol{z} - \boldsymbol{e}(\boldsymbol{e} \cdot \boldsymbol{z})||^2] \to \min(\boldsymbol{e}) \qquad (10.60)$$

subject to the condition

$$\mathcal{C} \equiv ||\boldsymbol{e}||^2 - 1 \equiv 0 \,. \qquad (10.61)$$

By multiplying this condition with the Lagrange multiplier λ and adding the product to the mean-square-discrepancy, we obtain the Lagrange functional

$$\mathcal{F} = \mathrm{E}[||\boldsymbol{z} - \boldsymbol{e}(\boldsymbol{e} \cdot \boldsymbol{z})||^2] + \lambda(||\boldsymbol{e}||^2 - 1) \,. \qquad (10.62)$$

The extreme value is obtained for $\boldsymbol{e}$ satisfying the conditions

$$\frac{\partial \mathcal{F}}{\partial e_k} = 0 \qquad (10.63)$$

that yield the following system of equations

$$\sum_{i=1}^{2} \mathrm{E}[z_i\, z_k]\, e_i - \lambda e_k = 0, \;\; k = 1, 2 \,. \qquad (10.64)$$

This can be represented in matrix notation as

$$(\boldsymbol{R} - \lambda \boldsymbol{I})\, \boldsymbol{e} = 0 \,, \qquad (10.65)$$

where $\boldsymbol{R}$ and $\boldsymbol{I}$ denote the correlation matrix of the vector $\boldsymbol{z}$ and the identity matrix respectively. The determination of the regression line is thus converted into the solution of an eigenvalue problem whose eigenvalues are λ_1 and λ_2. In the non-degenerate, two-dimensional case, the resulting eigenvectors $\boldsymbol{e}_1$, $\boldsymbol{e}_2$ correspond to the maximum and the minimum eigenvalues λ_1, λ_2 of the correlation matrix. For each eigenvalue, the mean square discrepancy can be expressed by the equation

$$\mathcal{D} = \mathrm{E}[||\boldsymbol{z}||^2] - \lambda \,. \qquad (10.66)$$

This shows that the discrepancy is smallest for the maximal eigenvalue of the correlation matrix. It is therefore reasonable to select as the direction of the regression line the principal axis of the correlation matrix as determined by $\boldsymbol{e}_1$ which corresponds to the largest eigenvalue λ_1. Along this line, the probability distribution is most expanded. However, if the problem is degenerate, both eigenvalues are equal and there is thus no means for identifying any specific direction.

Similar conclusions also apply when modeling phenomena in which multi-dimensional random vectors are used as the characteristic variables. Further, the conclusion can be generalized to the case in which random vector variables are substituted by random processes or fields. When the distribution of empirical data is concentrated along a line that does not pass through the origin of the coordinate system $\boldsymbol{z} = 0$, the shifting transformation $x' = x - \mathrm{E}[x]$, $y' = y - \mathrm{E}[y]$ permits the same description.

Let us now consider an application of the formalism we have derived. By using the eigenvector $\boldsymbol{e}_1$, or the corresponding memory matrix $\boldsymbol{W} = \boldsymbol{e}_1\,\boldsymbol{e}_1^T$, we can associate with each given vector $\boldsymbol{z}$ a projection to the regression line $\boldsymbol{u} = \boldsymbol{e}_1(\boldsymbol{e}_1 \cdot \boldsymbol{z}) = \boldsymbol{W}\boldsymbol{z}$. If the measured vector $\boldsymbol{z}$ can be considered as a sum of two contributions, with the first representing the phenomenon that can be described by a linear regression line through the origin and the second, representing a random disturbance, then the projection $\boldsymbol{u}$ can be thought of an "improvement" of the data used to characterize the phenomenon. Such an improvement is generally called filtering because it makes the observed data more consistent with the empirical linear model.

Following this explanation, we can solve the following problem related to the completion of missing information. Let there be given the component x_o and we wish to estimate the corresponding value of y. We can put $\boldsymbol{z}_x = (x_o, 0)$ and find the corresponding $\boldsymbol{u}$. However, because of the orthogonal projection to $\boldsymbol{e}_1$, the vector $\boldsymbol{u}$ now has a smaller component than $\boldsymbol{z}_x$ in the direction of the x-axis. That is $u_x < x_o$. This apparent deficiency can be improved by multiplying the vector $\boldsymbol{u}$ by the factor $c = x_o/u_x$. Using this correction we obtain the vector $\boldsymbol{u}' = \boldsymbol{u}x_o/u_x = \boldsymbol{e}_1 x_o/(\boldsymbol{e}_1 \cdot \boldsymbol{e}_x)$ which has a projection to the x-axis that is equal to x_o and it is on the regression line. The missing information is then represented by the correction vector $\boldsymbol{v}' = \boldsymbol{u}' - \boldsymbol{z}_x$. If we associate with the vector $\boldsymbol{z}_x$ the orthogonal projection $\boldsymbol{u}$, we then make the smallest change $\boldsymbol{v} = \boldsymbol{u} - \boldsymbol{z}_x$, which also corresponds to the smallest error, though the projection to the x-axis is then not strictly preserved in the representative vector $\boldsymbol{u}$. Contrary to this, we can strictly preserve the value of given projection x_o by associating with it the vector $\boldsymbol{u}'$. But in this case, the correction $\boldsymbol{v}' = \boldsymbol{u} - \boldsymbol{z}_x$ is not minimal. The projection we use usually depends on the interpretation of the accuracy with which the given data are determined. Most frequently it is reasonable to select the smallest correction because it is most easily found.

If we compare the derivation presented above with that given in the previous section, we find that the principal difference between them is in the expression of the discrepancy or error. In the first derivation we have used $\boldsymbol{v} = \boldsymbol{y} - \boldsymbol{W}\boldsymbol{x}$ while above, wee used $\boldsymbol{v} = \boldsymbol{z} - \boldsymbol{W}\boldsymbol{z}$. In the first derivation the source vector $\boldsymbol{x}$ and the estimated or desired vector $\boldsymbol{y}$ do not pertain to the same class. They generally represent different quantities and can have different dimensions and units. In contrast, in the derivation in this section the estimated as well as the source vector pertains to the same class. The operator $\boldsymbol{W}$ is correspondingly called the cross- or auto-associator, respectively.

An advantage of the latter approach should be mentioned. When the auto-associator is formed, we can arbitrarily select the given data and estimate from it the missing components. This is not the case with the cross-associator, in which the distinction between given and estimated data is fixed prior to adaptation. If in a later application the interpretation of the data is changed, we a new training procedure must be performed. The auto-associator is thus capable of solving either forward of inverse problems as well as mixed problems by using the same memory matrix. In this aspect the auto-associator appears to be more general and exhibits the same characteristics as the general non-parametric regression discussed in the previous chapter.

In our treatment of the sequential adaptation of the cross-associator, we derived a simple learning rule $\triangle \boldsymbol{M} = \alpha(N)\,\triangle \boldsymbol{Y}_N\, \boldsymbol{X}_N^T$. This can be readily transformed into the sequential learning rule of the auto-associator, provided that the above-mentioned similarity between both modelers is taken into account:

$$\triangle \boldsymbol{W} = \alpha(N)\,\triangle \boldsymbol{z}_N\, \boldsymbol{z}_N^T\ . \tag{10.67}$$

By inserting $\triangle \boldsymbol{z}_N = \boldsymbol{z}_N - \widehat{\boldsymbol{z}}_N = \boldsymbol{z}_N - \boldsymbol{W}\,\boldsymbol{z}_N$ we obtain the sequential learning rule of the auto-associator

$$\triangle \boldsymbol{W} = \alpha(N)\,(\boldsymbol{z}_N - \boldsymbol{W}\,\boldsymbol{z}_N)\,\boldsymbol{z}_N^T\ , \tag{10.68}$$

with

$$\alpha(N) = \frac{1}{\tau N \mathrm{E}[||z||^2]}\ . \tag{10.69}$$

This represents the stochastic approximation of the solution of the eigenvalue problem. The auto-associator can also be used to determine the novelties in the presented samples by merely calculating the difference $\triangle \boldsymbol{z}_N$.[19] The presented learning rule of the auto-associator leads to the formation of the memory matrix $\boldsymbol{W}$ that acts as a projector to the principal axis of the correlation matrix corresponding to the largest eigenvalue λ_1. [19] Therefore the operator $\boldsymbol{W}$ may also be called the *correlation memory matrix.* Because the properties of the auto-associator are well-understood and possess a firm analytical foundation it is used as a fundamental paradigm for the study of the properties of artificial neural networks. [19, 17] Numerical treatments lead to similar final results if the learning rule expressed by Eq. (10.67) is replaced by the following

$$\triangle \boldsymbol{W} = \alpha(N)\triangle \boldsymbol{z}_N \triangle \boldsymbol{z}_N^T\ , \tag{10.70}$$

but in this case, learning proceeds more slowly, especially at low values of $\triangle \boldsymbol{z}$.

Let us mention the characteristic properties of the auto-associator. From the definition $\boldsymbol{v} = \boldsymbol{z} - \boldsymbol{e}(\boldsymbol{e}\cdot\boldsymbol{z}) = \boldsymbol{z} - \boldsymbol{W}\boldsymbol{z}$ it follows that $\boldsymbol{W} = \boldsymbol{e}\,\boldsymbol{e}^T$. According to the above explanation the eigenvector of the correlation matrix corresponding to the largest eigenvalue λ_1 must be selected as the unit vector $\boldsymbol{e}$. Because the memory matrix $\boldsymbol{W}$ is expressible as the outer product of two unit vectors, it represents a non-dimensional physical quantity that acts as a projector to the selected principal axis of the correlation. In order to obtain the non-dimensional variable via sequential learning, the important term $1/\mathrm{E}[||z||^2]$ must be included in the adaptation constant α. This point is often ignored in the literature pertaining to the properties of the auto-associator. It is, therefore, not very critical what the average amplitude of the signals is which are used to train the auto-associator. However, from an experimental viewpoint, one usually seeks a signal-to-noise ratio which is adequate for reliable measurements. One approach for increasing this ratio is to use high amplitude excitation signals. Because of the linear operation on the vector $\boldsymbol{z}$, it is, in principle, irrelevant when applying the auto-associator matrix how high the signal amplitude is in order to obtain the proper projection and the corresponding filtering effect of the associator. However, some attention must also be paid to the signal-to-noise ratio. When the auto-associator is implemented using analog techniques, the signal must be well above the instrument noise. When using digital techniques the signals must be appreciably above the least significant bits. The fact that the operation of the auto-associator is, in principle, insensitive to the amplitude of the signals being presented to it, is convenient because one does not need to carry out the measurements in the analysis phase under the same conditions as those which were used during the training of the auto-associator. One may ask how the same invariance can be obtained in the general modeler that is based on the non-parametric techniques.

One problem that appears in applying the auto-associator is related with the concatenation of the variables used to describe the phenomenon under observation. By concatenating the variables X and Y in our derivation we have implicitly assumed that both variables possess the same physical meaning, or at least that both have the same physical units. This is, however, not always the case and we must provide for a common interpretation by a proper processing of signals from sensors. For this purpose we often utilize the mean value and the standard deviation of the signals and express the field components as standardized fluctuations around the mean value

$$Z_x = \frac{X - \mathrm{E}[X]}{\sqrt{\mathrm{E}[(X - \mathrm{E}[X])^2]}} \, . \tag{10.71}$$

Such a normalized component has a mean value of zero and a standard deviation 1.0. Normalization can also be simultaneously performed over a set of components representing the same vector quantity, as for example

$$\boldsymbol{Z}_x = \frac{\boldsymbol{X} - \mathrm{E}[\boldsymbol{X}]}{\sqrt{\mathrm{E}[||\boldsymbol{X} - \mathrm{E}[\boldsymbol{X}]||^2]}} \, . \tag{10.72}$$

When forming the field by concatenating several portions, it is of advantage if the contribution of each portion to the norm of $\boldsymbol{Z}$ is proportional to the number of components composing each portion. This can be realized by a proper modification of the normalization constant for each portion.

10.4.1 Application of the Auto-Associator to Analysis of Ultrasonic Signals

As has been already mentioned in the previous chapter ultrasonic methods are widely applied for the non-destructive testing of materials and structures. The principal goal of such measurements is to extract from the signals detected at the surface of the structure, information about the source, the medium, or the sensors. [12, 13] In active ultrasonic measurements, the source of the ultrasound is typically a piezoelectric transducer, although there are a large number of other possible excitations based on mechanical, electromagnetic, thermal, or other excitations. [27, 28] The elastic waves propagate from the source through the medium to the surface of the specimen where they are detected by one or more sensors. A major emphasis of recent research dealing with quantitative ultrasonic techniques has been towards the development of signal processing procedures by which the detected ultrasonic signals can be processed to recover the characteristics of the source, the properties of propagating medium, or the characteristics of sensors. [29, 24] The foundation for the analytical treatment of the related problems is elastodynamic theory. It begins with the assumption that the properties of a solid material do not change during an observation and it treats an ultrasonic source as a consequence of the force applied from outside the specimen. For its description the distribution of the force density $\gamma(\boldsymbol{r}, t)$, that may also be time-dependent, is utilized. The elastic waves emanating from the given source are then described by the displacement field vector $\boldsymbol{u}(\boldsymbol{r}, t)$. From elastodynamic theory it follows that the excited displacement field components u_i at location $\boldsymbol{r}$ and time t can be related to the force density by a linear equation

$$u_i(\boldsymbol{r}, t) = \int G_{ij}(\boldsymbol{r}, \boldsymbol{r}'; t - t')\, \gamma_j(\boldsymbol{r}', t')\, dt'\, d^3\boldsymbol{r}' \ , \tag{10.73}$$

in which G_{ij} represents the dynamic Green's tensor. By introducing the parameter $s = (\boldsymbol{r}, t, i)$ into the above equation, it can be re-written in the general form

$$Y(s) = \int G(s, s')\, X(s')\, ds' = G * X \ , \tag{10.74}$$

to which we can apply all the formalism developed for the cross- or auto-associator.

However, such an approach excludes proper description of nonlinear mechanisms resulting in internal changes of the materials which may also lead to the generation of ultrasound, such as, the development and growth of cracks in stressed materials or the martensitic transformation in metals. [9, 10, 11] But even in these cases the effect of a mechanism taking place in a limited region can be often described by an equivalent distribution of forces or moments called strain nuclei so that the propagation of the elastic waves from the source region can be described by elastodynamic theory. Quite often the operation of a source can be adequately described by strain nuclei so that only the wave phenomenon needs to be considered when one wishes to calculate the signals which are detected at a particular receiver point. Under these conditions it is appropriate to apply linear modeling of the phenomenon.

The principal problem with the empirical approach is how one should represent the relationship between the force and the field in terms of experimentally obtained records of waveform parameter and force signals. For this, one must first decide whether to apply the cross- or auto-associator. At first glance, it may seem that Eq. (10.74) dictates the use of the cross-associator and we can prove that it indeed works. But in this case, we loose the possibility of applying the same model for the solution of forward and inverse problems that is offered by application of the auto-associator. For this reason we prefer to use it. The use of the auto-associator also enables the estimation of ultrasonic wave field at one point from observation of the signals detected elsewhere. This property is of interest and importance for the prediction of signals from their past records, which is treated more generally in Chap. 11.

In an actual experimental situation, the input field can never be completely detected; moreover no computers possess a continuous distributed memory by which the tensor $G_{ij}(\boldsymbol{r}, \boldsymbol{r}', t, t')$ could be calculated without an analytical representation of the data. We therefore assume that the pattern field is represented by a vector concatenated of a finite number of digitized experimental data points of the displacement field and the corresponding source force:

$$\boldsymbol{Z}(\boldsymbol{r},t,i) \mapsto \boldsymbol{Z} = (u_1, \ldots, u_k; \gamma_1, \ldots, \gamma_l) . \tag{10.75}$$

According to this assumption, the memory tensor is represented by a matrix:

$$G_{ij} \mapsto \boldsymbol{W} \tag{10.76}$$

and the convolution becomes a matrix multiplication while the fundamental scheme of the auto-associator and its learning rule is not changed. At the start of learning we choose $\boldsymbol{W} = 0$.

After learning, the system can be used as a trained analyzer, provided the feedback to the memory is turned off. In this case an input vector $\boldsymbol{Z}$, partially filled with experimental data, excites the estimator $\hat{\boldsymbol{Z}} = \boldsymbol{W}\,\boldsymbol{Z}$ which also includes estimated missing data. Such an operation corresponds to the auto-associative recall of missing information. [12, 13] Experiments were performed on a setup consisting of a specimen, sensors, amplifiers, digitizers and a computer shown in Fig. 10.5. The sample was a steel block of dimensions $50 \times 80 \times 175\,\mathrm{mm}^3$. AE phenomena were simulated on it by breaking a pencil lead of 0.5 mm diameter. Acoustic signals were detected by two broadband sensors of 1.3 mm diameter. The record of the AE signal from each sensor was represented by 50 components while 18 components were used to encode the source data.

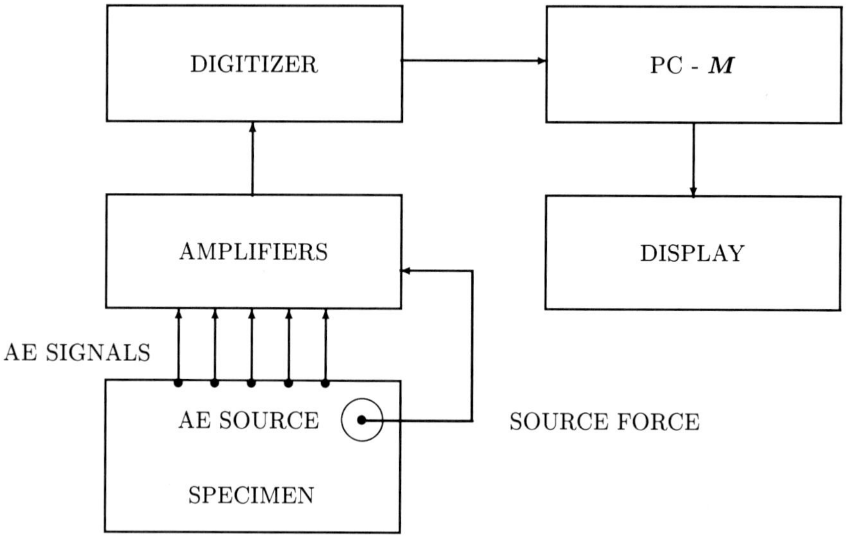

Fig. 10.5. Scheme of experimental setup applied in linear adaptive modeling of AE phenomena

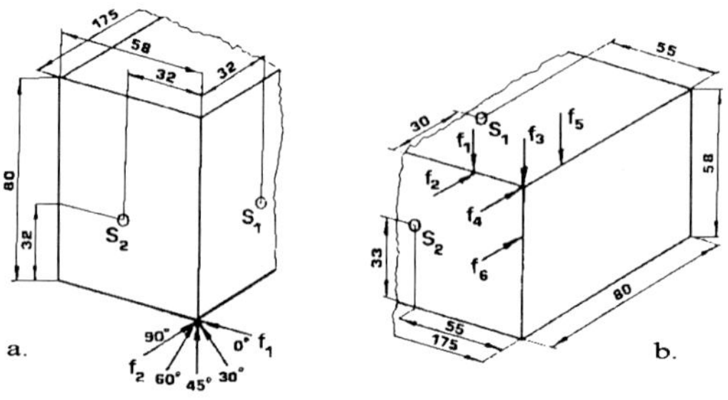

Fig. 10.6. Scheme of sensor and source distributions on the specimen for both types of experiment

Two types of experiments were performed. In the first, the ability of proposed system to learn and to analyze signals excited at a fixed source point was investigated. In the second one, the AE signals were excited at various places by forces of various directions and the auto-associator was used to solve forward and inverse AE problems. The sample vectors were concatenated using digitized data from both sensors and data inscribed via the keyboard provided encoded information about the orientation of the released force and the position of the source. The position of the sensors and sources in both cases are shown in Fig. 10.6a and 10.6b.

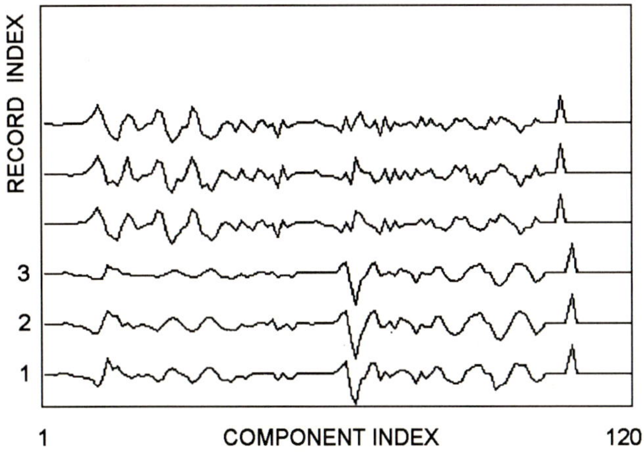

Fig. 10.7. The set of concatenated experimental records used in training of the linear modeler

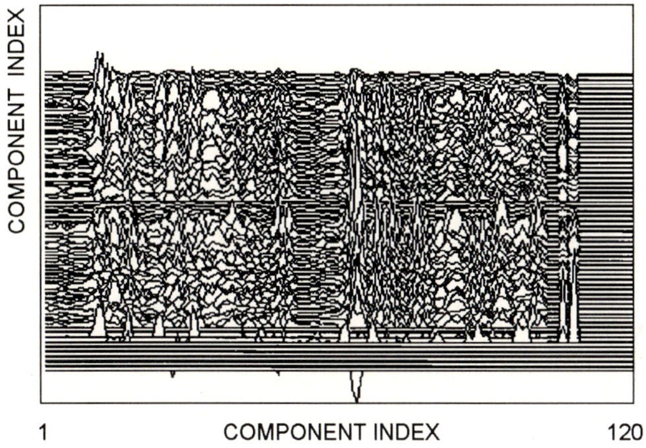

Fig. 10.8. The memory matrix

In the first experiment two mutually perpendicular forces were applied three times at the same point. The training set is shown in Fig. 10.7. The repetition of excitations was applied to take into account relatively great variability of sources simulated by a breaking pencil lead by hand and to demonstrate the noise insensitivity of the applied system. The sample vectors were first normalized and then presented to the system in ten iterations. During the adaptation the norm of discrepancy between the input vector and the estimated vector, that is $\|\widehat{\boldsymbol{Z}} - \boldsymbol{Z}\|$ fell below 10^{-2}. Each row of the memory matrix can be represented as a record of a signal. The complete memory matrix is then graphically represented by a set of signals as shown in Fig. 10.8.

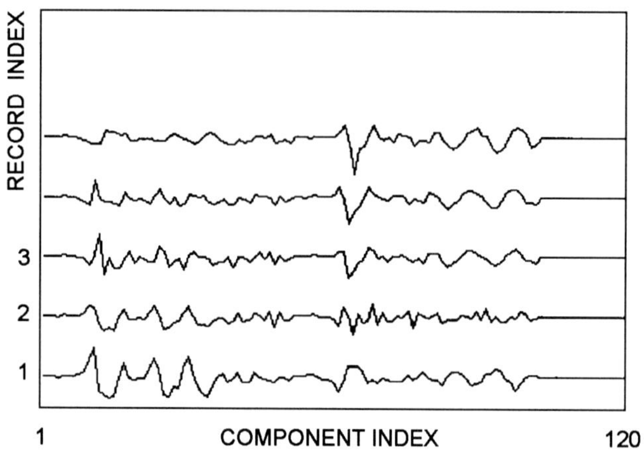

Fig. 10.9. The set of test ultrasonic signals excited by oblique forces used as the system input without source force descriptors

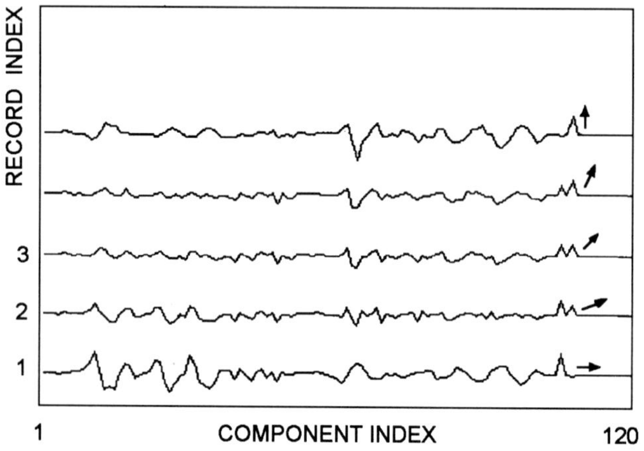

Fig. 10.10. The set of output signals from a trained system for the case of oblique forces. The source force descriptors on the right side of records are complemented by linear estimation

It consists of nine characteristic regions corresponding to three auto- and six cross-correlation fields of the sample vector sectors. After the completed learning, a transient ultrasonic wave was excited by forces of various directions but in the same plane as the training sources. The angles of inclination of the test force were 30, 45, 60 and 90 degrees with respect to the first applied force as shown in Fig. 10.6. The input samples of the pattern vector $\boldsymbol{Z}$ were comprised in this case of only the detected AE signals as shown in Fig. 10.9. The corresponding output vector $\hat{\boldsymbol{Z}}$ also contained the missing source descriptors as indicated by the recorded descriptor peaks appearing on the right of records in Fig. 10.10. The recorded descriptor peaks appear in that portion of the sample vector where the descriptors of the source force in the training set were inscribed.

If the source is treated as a composition of all training sources, then each peak represents the amplitude of the corresponding training force. Even though this experiment is very crude, the recalled descriptors correctly represent the angle of the applied force. This example shows that a trained system is capable of solving a simple inverse problem related to the determination of source forces from detected acoustic signals. For this purpose no elastodynamic theory is needed since the method is completely empirical.

It is important to note that the forces generated during the analysis mode of operation have not been used in the training. Because the auto-associator possesses a linear structure, it correctly estimates the applied force vector by two corresponding components. The determination of the orientation of the force from the wave signal corresponds to a cross-associative recall. In a similar manner, one portion of the ultrasonic signal can be recalled from the other. This corresponds to an auto-associative recall. In order to demonstrate this feasibility, a portion of the test signals excited by oblique forces was simply set to zero as shown in Fig. 10.11. The corresponding recalled signals are shown in Fig. 10.12. Comparison of these signals with these ap-

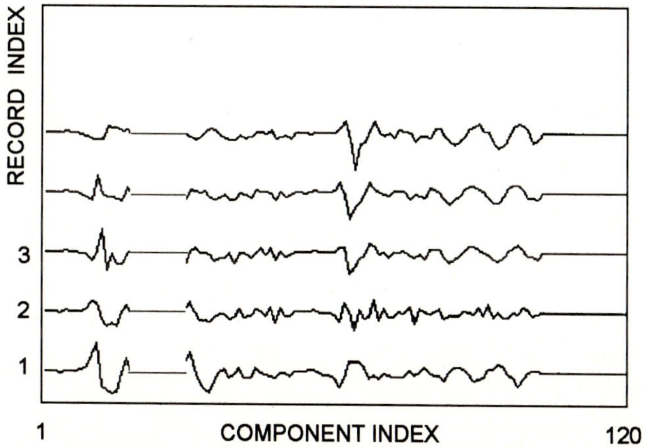

Fig. 10.11. The set of truncated test records

pearing in Fig. 10.9 reveals that the structure of the recalled signals resembles the structure of the original signals although the amplitude is smaller. A reduction of the signal amplitude is a consequence of the projection performed by the auto-associator during recall. However, the accuracy of the recall decreases as the deformed portion increases. In connection with this example we mention that the recall represents a signal that has been forecast from the presented segment. Quite analogously, a portion of the signal that occurred in the past can be estimated. This corresponds to a retrodiction. A similar performance of the auto-associator is observed when the signals are excited at different places and directions or are noise-corrupted which the second experiment demonstrates. In it the ultrasonic sources were simulated by breaking a pencil lead in various directions at various positions as shown in Fig. 10.6. The training set and the corresponding memory matrix are shown in Figs. 10.13 and 10.14 respectively.

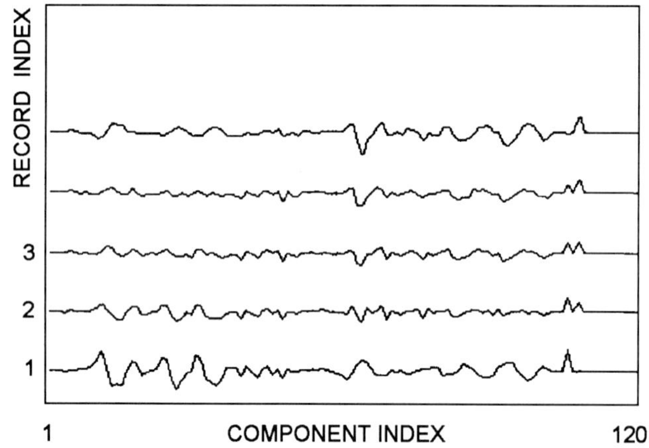

Fig. 10.12. The set of signals estimated from truncated test records. The truncated signal portion and source descriptors are complemented by associative recall from the trained correlation memory

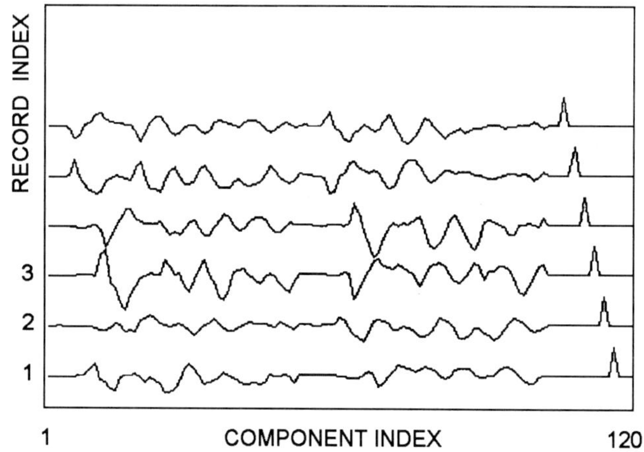

Fig. 10.13. The set of training signals excited at various places by forces in various directions

The source position and the force orientation were again described by a peak at different positions in the descriptor portions of the records. After training the system, the test signals were obtained by adding white noise to the signals of the training set. The corresponding records of the input signals are shown in Fig. 10.15. In this case, the output, shown in Fig. 10.16, is again completed by the force descriptor part, which describes the characteristics of the applied sources well, albeit with some superimposed noise. At the same time, the noise which was in the original signals is now efficiently filtered out. In the example

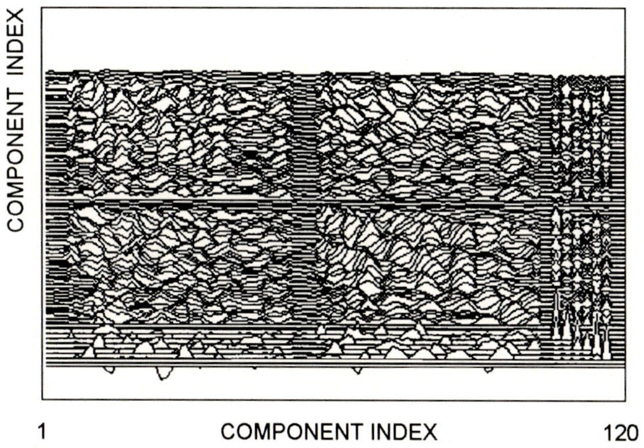

Fig. 10.14. The memory matrix corresponding to the training set in Fig. 10.13

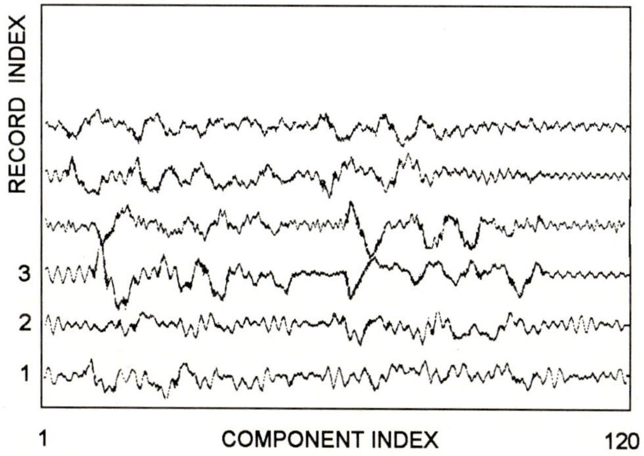

Fig. 10.15. The set of noisy test signals of the second experiment

presented here, the signal-to-noise ratio is of the order 0.1. With decreasing signal-to-noise ratio, the performance of the recall decreases but the characteristics of the source property can still be correctly estimated from the recall until the signal-to-noise ratio reaches order 1.

If only the source descriptors of the training set are used as the input signals, then in the output of the associator we find the ultrasonic signals with a structure similar to that of the training set, as shown in Fig. 10.17. What is surprising, is the appearance of additional peaks in the descriptor portion. These are a consequence of the non-zero components

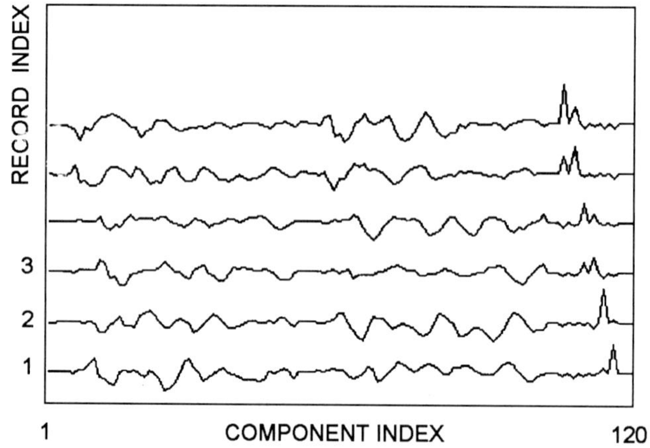

Fig. 10.16. The set of signals recovered from noisy test signals of the second expeiiment. The noise is filtered out and the source descriptors are recalled

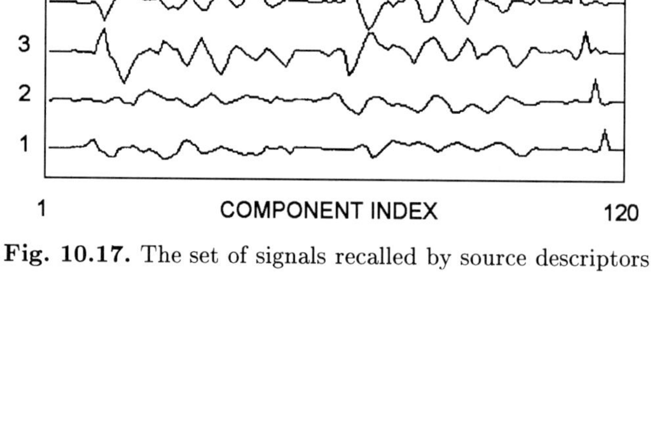

Fig. 10.17. The set of signals recalled by source descriptors

appearing in the auto-correlation of the descriptor portion in the memory matrix. They are formed during learning because of the similarity of various training signals. In addition, the amplitude of the recorded signal vectors is also decreased. We have thus demonstrated that by using the auto-associator we can solve forward as well as inverse problems in elastodynamics, as well as being able to forecast signals.

The examples shown demonstrate that transient ultrasonic phenomena can be modeled by an adaptive system. Its structure is obtained by approximating an exact description of elastodynamic phenomena by an auto-associator. If the elastodynamic description of the excitation of ultrasonic phenomena by forces is strictly taken into account we have either the relation $\boldsymbol{u} = \boldsymbol{G} * \gamma$ or $\gamma = \boldsymbol{G}^{-1} * \boldsymbol{u}$ with $\boldsymbol{G}^{-1}$ indicating the inverse of the Green's tensor. Because the pattern field $\boldsymbol{Z} = (u_1, \ldots, u_k; \gamma_1, \ldots, \gamma_l)$ is comprised of two different components, we conclude that the memory matrix has a structure

$$\boldsymbol{W} = \begin{bmatrix} 0 & \boldsymbol{G} \\ \boldsymbol{G}^{-} & 0 \end{bmatrix} , \tag{10.77}$$

which yields the homogeneous equation

$$(\boldsymbol{u}, \gamma) = (\boldsymbol{G} * \gamma, \boldsymbol{G}^{-} * \boldsymbol{u}) , \tag{10.78}$$

or

$$\boldsymbol{Z} - \boldsymbol{W} * \boldsymbol{Z} = 0 . \tag{10.79}$$

In our approach we obtain non-zero elements along the diagonal in the expression for $\boldsymbol{W}$. One may, therefore, ask what these diagonal terms represent. Since sources of ultrasonic waves are treated only as consequences of forces, we must allow for arbitrary forces and vice versa. Consequently we cannot predict the properties of the force from partial information about it. However, when observing an actual phenomenon there cannot appear arbitrary forces, particularly when they are internally generated in a material, as is the case in acoustic emission. This means that there must exist a certain relationship between the force components in time or space which, in principle, makes it possible to predict the properties of the total force from only partial information about it. The same must hold for the waveform parameter representing the field of the ultrasonic displacements. In our approach we have previously neglected the term corresponding to the solution of the homogeneous equation which does not depend on the forces and makes it possible to predict the ultrasonic wave properties from only partial observation of these waves. This possibility is realized when the diagonal terms in the expression of $\boldsymbol{W}$ are different from zero. The diagonal terms thus represent the solutions of the homogeneous elastodynamic equation for the waveform parameter or the corresponding one for the stresses. The auto-associator is in this aspect more general than the cross-associator.

In applying the auto-associator, two typical problems may arise which are the result of the truncated description of the measured ultrasonic fields. The first arises when the input field is similar for a number of samples comprising the training set. In this case, the estimated field is a mixture of the corresponding prototype vectors and consequently many peaks appear in that portion of the vector corresponding to the characteristics of the source. This case, however, does not correspond to a point-like ultrasonic source. The second problem is related to the time-dependence of the ultrasonic signals. By neglecting the time-dependence of the source forces in our description, only their amplitudes and orientations as well as the positions of the sources can be determined during the recall. Such a description is not time-invariant and can only be successfully applied by a proper triggering of the signal recording instruments when carrying out the measurements. The time-invariant operation of the associator can be obtained by taking into account the shifting of the signals in time and by introducing a multi-dimensional structure of the memory. In this case, the training of the auto-associator forms a Green's matrix that depends only on the difference between times $\Delta t = t - t'$. [14, 32] In a similar way, the invariance with respect to spatial translation can be included into the model.

The structure of the proposed adaptive system is analogous to a simple neural network in which the synaptic joints between neurons represent the elements of the memory matrix. The i-th row of the matrix $\boldsymbol{W}_{ij}$ can be treated as a vector parameter $\boldsymbol{M}_i$ which determines the response $\hat{\boldsymbol{Z}}_i$ of the i-th neuron to the excitation by the input vector $\boldsymbol{Z}$. That is: $\hat{\boldsymbol{Z}}_i = \boldsymbol{M}_i \cdot \boldsymbol{Z}$. The neuron that possesses those parameters most similar to the input vector yields the greatest scalar product and hence the greatest component in the output vector. This interpretation is similar to the corresponding one that was given in relation to the non-parametric regression described in Chap. 9. It is also mentioned there that the completed vector can be reintroduced to the processor as an input vector in order to remove noise which is present in the data. In this case, such a procedure again results in the same completed vector. Since the estimated vector $\hat{\boldsymbol{Z}}$ already lies on the regression line, therefore a repetition of the projection to the same line does not change this vector. A linear auto-associator thus works as a matched filter. [12]

The estimation of oblique forces from the detected ultrasonic signals illustrates a characteristic property of the linear modeler. Although the system was not trained by the corresponding signals it is still capable of finding the correct solution of this inverse problem. This property corresponds to a capability of interpolating or generalizing and can be interpreted as a primitive intelligence. A similar problem arises in relation to the prediction. When a portion of a time-dependent ultrasonic signal is deleted, the auto-associator is capable of supplying the missing part. This corresponds to a linear prediction based on the other portion of signal. Such a capability is one of the most outstanding properties of intelligent behavior of animals and humans and has presumably played a decisive role in their evolution. Therefore it is of importance to examine and develop this property somewhat further.

In our case we have demonstrated only the prediction of the signal in that portion of the signals on which the auto-associator has been trained. For a system of constant structural parameters, the Green's function depends only on the difference of times $\Delta t = t' - t$. When this property is also included in the model, as was done in the first example in this chapter, then the associator is capable of correctly predicting the signal regardless of the time at which it is applied. However, there are many phenomena in Nature that cannot be described by linear models and consequently there arises a question how such phenomena can be predicted. Experience with the linear model suggests that we can use for this purpose a general modeler of natural phenomena and incorporate into it the time invariance. This task will be described in Chap. 11 on the forecasting of chaotic time series.

In addition to an ability to interpolate, the linear modeler exhibits a capability to extrapolate. If the system is trained to reproduce signals of a certain amplitude, it is later capable of repeating the same task on the signals of greater or lower amplitude. This property corresponds to the capability of extrapolation and it is a consequence of the linear structure of the associator. A similar property has not been observed in the general modeler based on the non-parametric regression. These and the previously mentioned properties indicate that it is erroneous to interpret a linear modeler only as a simplification or a first-order approximation of the more general non-linear one. Incorporation of *a priori* information about the properties of the observed phenomena makes the linear modeler more applicable for the modeling of the corresponding class of phenomena but at the same time restricts its generality. A proper selection of the modeler structure thus depends on the properties of the phenomena which are to be modeled. And as has been mentioned previously, it is the task of evolution to find an optimal structure. In this aspect, modeling can generally be represented as a two-stage process. In the first evolutionary stage, a certain class of modeler must be developed and in the second, or training stage, a particular model must be adapted to the phenomenon at hand. The evolutionary stage can proceed by the selection of one particular class of model from the set of available ones by observing the performance of members of the class. The other possibility is to develop by small variations of a class of individuals, a modeler that exhibits the most appropriate performance. The latter possibility resembles the gradient descent method for training and we might expect that it would generally not yield a globally optimal modeler.

11. Modeling and Forecasting of Chaotic Processes

In discussing the modeling and forecasting of chaotic processes we presume that the reader will be familiar with the fundamentals of deterministic chaos. For the interested reader we provide a brief description in Appendix B.

11.1 Modeling of Chaotic Processes

A *process* is generally called a phenomenon that cannot be described by a finite number of data representing the characteristic variable. For example, consider an unending manufacturing of balls characterized by the diameters of the product: $\{x(t);\ t = 1,\ 2,\ \ldots \to \infty\}$, or the drawing of a wire that is characterized by the diameter at a certain position corresponding to the time of production: $\{x(t);\ t \in [0,T]\}$. The first example represents a discrete- and the second one a continuous-parameter process. In this chapter we shall not differentiate between these types of processes and we will generally apply the notation used to describe the continuous case also for the discrete case, provided that a countable set of points is selected in the span $[0,T]$ of the parameter t. We shall restrict our presentation to time-series and consequently we shall henceforth interpret the parameter t as time, although this interpretation is not necessary. At repeated observations of a process we generally obtain various values of x at fixed t and we must treat the process as a stochastic phenomenon. If we want to describe the process empirically we must measure and record the diameter x for various production runs s so that we can obtain the sample functions $x(t,s)$. The ensemble of samples $\{x(t,s);\ t \in [0,T];\ s = 1,\ \ldots,\ N\}$ is then the basis for the empirical characterization of the statistical properties of the phenomenon, as described by the theory of stochastic processes. [20]

However, in performing this task in actual experiments with digital systems, there arise serious questions related to the storing and processing of an infinite number of data contained in the empirical ensembles. This problem is often avoided by directly measuring the statistical parameters of the process, or by representing the continuous process by a finite number of data. In a complementary, analytical approach, the ensemble of samples $\{x(t,s);\ t \in [0,T];\ s = 1,\ \ldots,\ N\}$ is described by a *proper* model. However, the dependence of x on t need not be explicitly described by a function. It is sufficient to specify a generating procedure, such as an iteration, by which the values of x can be generated for an arbitrary time t from some given initial conditions. The expression of the generating procedure is equivalent to a *natural law*. If the parameter t represents time, then the law describes the dynamics of the process.

The specification of such laws is broadly examined in the theory of deterministic chaos. [26, 27, 30] Although very convenient, this method is most frequently applied to cases in which some *a priori* theoretical information about the process dynamics is known. Our intention will be to show how the same method can be utilized for obtaining an empirical description of processes. For this purpose, we must assume that the samples of the observed functions can be generated by some dynamical natural law that is expressible deterministically. The principal task of a general empirical modeling of a process is then to build an information processing system that is capable of extracting information about the model from the samples and to built a model without any *a priori* assumptions about a particular functional form of the corresponding law. As in other examples in this monograph, we use for the empirical modeling the non-parametric regression that was described in Chap. 9.

Among the various problems related to automatic modeling, the prediction of chaotic dynamic phenomena is the most difficult. [5] The principal reason for this is the inherent instability of chaotic phenomena, which results in an irregular variation of the related physical observables. [26, 27, 30] Moreover, chaotic phenomena are often observed in complex systems which cannot be simply modeled by analytical methods. As a typical example from engineering practice, we mention the cutting process that is used in manufacturing. The cutting process involves intricate tool, machine and specimen geometries, the physically complex processes of material deformation and failure as material is cut, tool friction and complicated system dynamics, among others. [6, 7, 8] As there are many important analogous processes of this kind in engineering, technology, economics, biology, and medicine, [18, 22, 25] it is challenging to look for a complement to approaches based on the analytical modeling and prediction. For this purpose, a vast class of linear as well as nonlinear models have been explored. [32, 1]

Modeling by neural networks which has recently been developed [21, 28, 9, 10, 19] appears to be very advantageous, but as in other applications of neural networks, there is a fundamental problem related to the proper selection of the network architecture. The objective of this chapter is to show how such problems can be solved by adapting the non-parametric regression modeler architecture to a description of dynamical phenomena. We shall do this by following an analysis stemming from a general description of chaos.

Although chaotic phenomena exhibit apparently stochastic variability it is a common feature that they usually also exhibit some degree of order, or structure, which makes possible approximate prediction. [26, 27, 30, 5,

21] In this chapter our main task will be to show how this can be done entirely on an empirical basis without applying any prior information about the system model. We obtain guidance for the solution of this problem from the fundamentals of the description of deterministic chaos which are briefly reviewed in Appendix B.

For the sake of simplicity, we consider first a continuous chaotic process that is characterized by the one-dimensional variable $x(t)$. We assume that the signal $x(t)$ is measured in an experiment using an appropriate sensor. This signal represents one of the dynamical variables in a D-dimensional phase space that is needed for the presentation of the complete dynamics of the phenomenon. Without essential loss of generality, we assume that the dynamics develops in a limited region of this space. The corresponding geometrical form which includes the flow of possible trajectories of motion is called a *chaotic attractor*. [26, 27, 30] As is described in Appendix B, the dimensionality of this attractor need not coincide with that of the phase space. The corresponding parameter D_a is most often estimated by calculating the correlation dimension D_2. The value of D_a is usually smaller than D and it may even be fractal. It is generally not known in advance and we must estimate it from the empirical data.

If we cannot measure all D dynamical variables then we must reconstruct the process attractor and extract information about the corresponding dynamics from the measured signal $x(t)$ alone. By shifting the signal $x(t)$ m-times with an arbitrary time shift τ_s we can form a representative vector

$$\boldsymbol{X}(t) = \{x(t),\ x(t-\tau_s),\ \ldots,\ x(t-m\tau_s)\}\ . \tag{11.1}$$

The temporal changes of this vector determine a flow in the corresponding *embedding space*. From this flow the fundamental characteristics of the dynamic phenomenon can be recovered as described in the Appendix B. [26, 27, 30] For this purpose the number $m+1$ components of $\boldsymbol{X}$ must be increased up to some proper value D_e called the *embedding dimension*. [31] From Takens theorem it follows that $D_e = 2D+1$ is generally sufficient for a proper embedding, but in the cases when $D_a < D$ we can expect that smaller values of D_e are also applicable. In addition to the embedding dimension, the proper shifting parameter τ_s must be estimated from the experimental data. We expect intuitively that it should be related to the characteristic period appearing in the time record of the process. [23]

The dynamical law describing the flow in the embedding space is generally described by the differential equation

$$\frac{d\boldsymbol{X}}{dt} = \boldsymbol{\Phi}(\boldsymbol{X})\ . \tag{11.2}$$

in which $\boldsymbol{\Phi}$ represents some nonlinear function called the *chaos generator*. In order to facilitate the modeling of the process by a digital information processing system, we assume that the phenomenon is observed at discrete increments in time, described by a non-dimensional parameter $\{t = 1, 2, \ldots\}$. The dynamical law is then represented by the mapping

$$\boldsymbol{X}(t+1) = \boldsymbol{\Psi}(\boldsymbol{X}(t)) \tag{11.3}$$

in which $\boldsymbol{\Psi}$ is a discrete version of the chaos generator. Expressing this law by the components of the representative state vector, we obtain the expression

$$\{x(t+1),\ x(t),\ \ldots,\ x(t+1-m)\} = \boldsymbol{\Psi}\{x(t),\ \ldots,\ x(t-m)\}\ . \tag{11.4}$$

This equation shows that the complete mapping includes the shifting of m components with respect to the discrete time t and a mapping of the state vector $\boldsymbol{X}(t) = \{x(t),\ \ldots,\ x(t-m)\}$ to the component $x(t+1)$. That is,

$$x(t+1) = \Psi\{x(t),\ \ldots,\ x(t-m)\}. \tag{11.5}$$

The function Ψ describes the dynamics of the representative component of the process. Knowing it we can forecast the forthcoming signal value at time $t+1$ from the past values $\{x(t),\ \ldots,\ x(t-m)\}$. Using the predicted value and the shifted values of x, the vector $\boldsymbol{X}(t+1)$ can be formed and the next value $x(t+2)$ predicted from it, and so on. In such an iterative treatment there is hidden an assumption about the stationarity of the process which is mirrored in the fact that the function Ψ is explicitly time-independent. The system producing the chaotic signal is thus treated as an autonomous system. We shall later describe how non-autonomous systems can be modeled.

The time-shifting of the signal can be carried out using a shift register incorporated in the unit forming the representative vector $\boldsymbol{X}(t)$ while its forecasting must be carried out by a unit capable of modeling the mapping function Ψ. For this purpose, we interpret the joined vectors $(x(t+1),\ \boldsymbol{X}(t))$, observed at past values of time t, as a set of statistical samples. Their joint empirical distribution

$$F = F(x(t+1),\ \boldsymbol{X}(t)) \tag{11.6}$$

in the extended embedding space of dimension $M = m+2$ can be modeled by the self-organization process described previously. With this, the information presented by the empirical samples $\{\boldsymbol{X}'(t); t = 1, 2, \ldots,\}$ is optimally preserved in a finite set of representative samples or prototypes $\{\boldsymbol{Q}_k;\ k = 1,\ \ldots,\ K\}$. With this step, the principal problem of storing an infinite sequence in a system comprised of discrete memory units is solved in an optimal way.

Each prototype represents a typical point in the extended embedding space. It can also be interpreted as a typical mapping of a given vector $\boldsymbol{X}$ to the forthcoming value of x that is in agreement with the previously-observed empirical data. Based on this interpretation, we can then use the prototypes in a non-parametric regression to estimate the forthcoming value $x(t+1)$ from the given vector $\boldsymbol{X}(t)$ that has been constructed from the past and present values of x at an arbitrary moment in time t. In doing this, we implicitly assume that the statistical properties of the described stationary process in the past will continue unchanged into the future.

In order to predict a signal, the modeler must include an element in which the forthcoming signal can be presented in fictitious time. For this purpose we utilize a multi-port, parallel output representing the set $\{x(t+1),\ x(t+2),\ \ldots,\ x(t+\tau)\}$. The index of the output channel τ in this case corresponds to the fictitious time.

In accordance with the above explanation a modeler of chaotic processes must be able to perform the following operations: [9, 10, 11, 12]

1. Sample and time-shift the input signal.
2. From the shifted components form the time-lagged vector $\boldsymbol{X}(t)$.
3. Adapt the prototypes by self-organization to the input vector.
4. Repeat the prediction of the forthcoming signal by conditional average estimation.
5. Present the predicted values using a multi-port parallel output.

The corresponding information processing system consists of two shift registers, a memory, and a conditional average estimator as shown in Fig. 11.1. The signal $x(t)$ detected by a sensor attached to the chaos generator is input to the shift register where a sample of the state vector $\boldsymbol{X}$ is formed by latching successive data values. The sample is then transferred to the self-organizing memory and to the predictor comprised of a shift register and the conditional average estimator. With such a structure the system is similar to a recurrent neural network. [19]

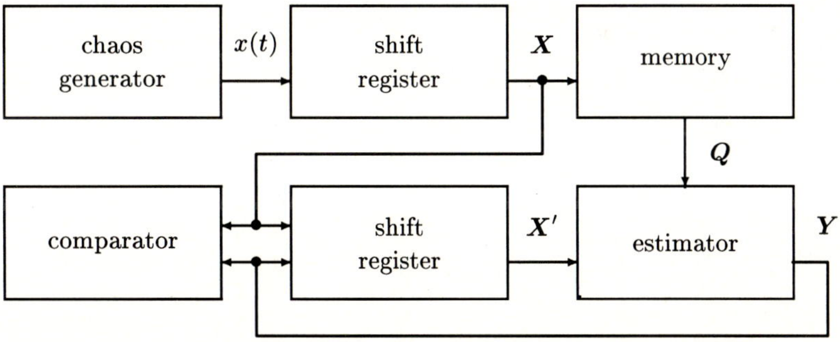

Fig. 11.1. Scheme of the system applicable for the empirical modeling and prediction of chaotic time-series

Numerical experiments were carried out using a predetermined number of prototypes. They were formed by the self-organization in the memory of a computer. The prototypes were transferred to the conditional average estimator. During forecasting, the components of the state vector $\boldsymbol{X}$ are first shifted to obtain a truncated joined vector

$$\boldsymbol{X}'(t) = (\#,\ x(t),\ x(t-1),\ \ldots,\ x(t+1-m)) \ . \tag{11.7}$$

The missing component denoted by # is the forthcoming value $x(t+1)$ of the signal. It is predicted by the conditional average estimator and concatenated to the output vector which will consist of

$$\boldsymbol{Y} = (x(t+1),\ x(t),\ x(t-1),\ \ldots,\ x(t+1-m)) \ . \tag{11.8}$$

This vector is then iteratively fed back to the predictor input in order to forecast the forthcoming term in the time-series. The forecast series is presented as the predictor output.

In order to estimate the performance of such forecasting, a comparator is included into the system. It compares short signal segments of the true and predicted signals. The forecasting performance is described by the relative error defined with respect to the variance of the input signal Var(x) as

$$R_{\mathrm{er}}(\tau) = \frac{1}{\mathrm{Var}(x)\ N} \sum_{t=1}^{N} (y(t+\tau) - x(t+\tau))^2 \ . \tag{11.9}$$

The parameter τ denotes the number of prediction steps. Numerical experiments have shown that for a chaotic signal, the prediction error increases approximately exponentially with the number of steps. The principal source of this instability resides in the chaotic dynamics. Because of this, two adjacent trajectories of a chaotic system originating at non-identical initial points, diverge exponentially in time. In addition, the accumulation of errors caused by the approximate modeling also leads to a divergence of the predicted and the actual, observed trajectories. We conjecture that this effect is less pronounced when the generating function Ψ is well-modeled. One can expect that the characteristic exponent of the prediction error defined by the formula

$$L_{er} = \frac{1}{s} \log \left[\frac{R_{er}(s+1)}{R_{er}(1)} \right] \tag{11.10}$$

should therefore be approximately two times greater than the largest Lyapunov exponent λ of the input chaotic signal. [26, 27, 30, 9, 10, 11, 12] In order to visually estimate the predictor performance, one can compare short time segments and the return map of the predicted and the true chaotic signals. Examples illustrating this are presented in the following section.

11.2 Examples of Chaotic Process Forecasting

In the literature on the prediction of chaotic phenomena, the logistic and Hénon maps as well as the Lorenz system are usually used as prototype examples of the chaos generator. [34] We therefore first demonstrate the results of the numerical prediction of the generated chaotic signals using:

$$x_{n+1} = 3.8\, x_n(1 - x_n) \,, \tag{11.11}$$
Logistic map

$$x_{n+2} = 1 + 0.3\, x_n - 1.4\, x_{n+1}^2 \,, \tag{11.12}$$
Hénon map

$$\begin{aligned} \dot{x} &= 20(y - x) \\ \dot{y} &= (28 - z)x - y \\ \dot{z} &= -8z/3 + x\, y \,. \end{aligned} \tag{11.13}$$
Lorenz system

In the last example the value of $dt = 0.015$ was selected for the discrete numerical treatment while sample values were taken at intervals of $(10\, dt)$.

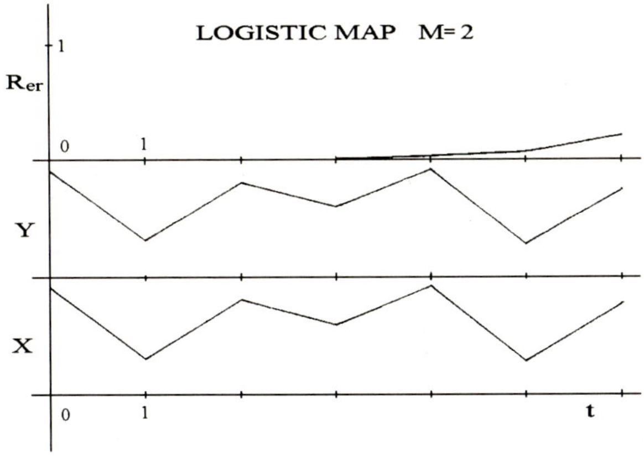

Fig. 11.2. Records of the signal X generated by a logistic map, the predicted signal Y and the relative error R_{er} for $M = 2$

The initial conditions were arbitrarily selected and the initial 500 sample values of the calculated signals, representing the transition from the initial conditions to the chaotic attractor, were discarded in all cases. The self-organization of prototypes was then carried out, using 500 sample values.

The predictor performance is most simply demonstrated for the case of chaos generated by the logistic map for which a two-dimensional return map is sufficient to give a complete graphical presentation. An example of the numerical results obtained with $M = 2$ is shown in Fig. 11.2. The bottom record represents the true signal X obtained from the chaos generator while the middle record shows the signal Y predicted from the first value. With an increasing number of prediction steps, going from left to right, the discrepancy between these two signals becomes evident. The average error, estimated from 500 prediction trials, is shown as a function of the number of prediction steps in the upper record R_{er} of the same picture. The average relative error for the first step of the prediction is $R_{er}(1) = 0.003$, which indicates

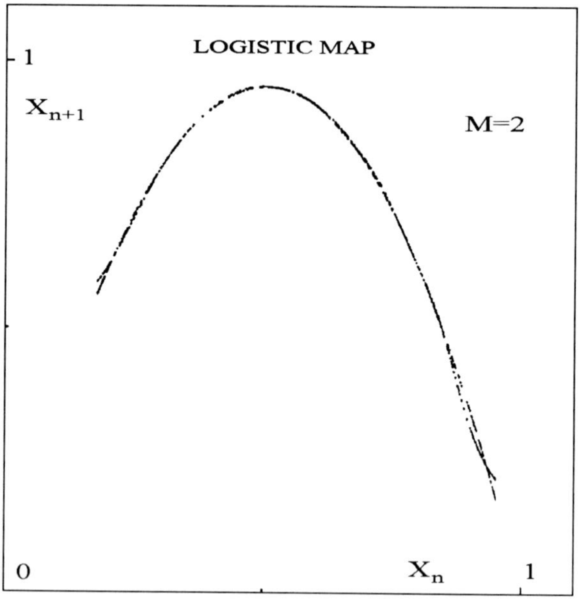

Fig. 11.3. Return maps of the original and the predicted logistic signals for $M = 2$

a good modeling of the chaos. This error then increases with the number of prediction steps in an approximately exponential manner. The exponent estimated from the numerical experiment is approximately $L_{er} \approx 1$. For the same logistic map, the Lyapunov exponent is approximately $\lambda \approx 0.45$ so that the expectation that $L_{er} \approx 2\lambda$ is confirmed. [25] Presumably the higher estimated value is a consequence of the approximate modeling of the logistic map that is obtained with only thirty prototypes. The predictor performance can be visually assessed from the return maps of the predicted and original signals. The return map of the signal that was predicted in five step increments and the return map of the original chaotic signal are shown as two dotted curves in Fig. 11.3. The coincidence of both curves demonstrates the applicability of the system for modeling simple chaos generators.

In addition to the return maps, it is also instructive to graphically present the positions of the prototypes. These are shown in Fig. 11.4. The center of each circle indicates the position of the prototype. The radius describes the range σ of the Gaussian window used in smoothing the delta functions and in estimating the conditional average. As the self-organization is derived from the assumption that an equal probability $1/K$ should correspond to each prototype, the prototypes are more densely distributed in the region of the attractor or on the return map, where the occurrence of the system state is

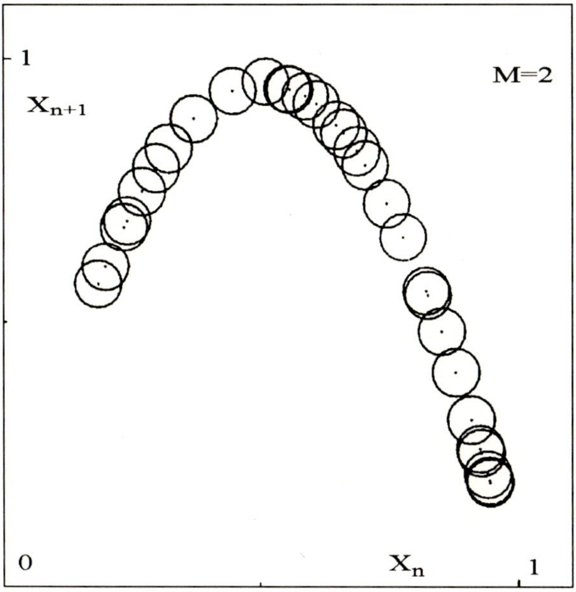

Fig. 11.4. Position of the prototypes in the return map

greater. The distribution of the prototypes thus also indicates the probability distribution.

An individual prototype describes a transition between successive states in the chaotic time-series. A set of prototypes therefore represents various characteristic transitions occurring in the chaotic time-series and thus represents the complete chaos generator. When chaos is generated by the logistic map, a prototype represents just two joined states. Fig. 11.5. shows the transitions $x_n \rightarrow x_{n+1}$ which are obtained by modeling chaos that is generated by the logistic map. A prototype is represented by a line starting at the first value and ending at the second one. By graphically connecting such segments one can construct an approximation to the chaotic signal.

A procedure identical to that described above has been repeated for several prototype dimensions M. The results obtained with $M = 3$ are shown in Figs. 11.6 and 11.7. It is characteristic that the prediction error increases with the dimension. For a simulation of the applied chaotic signal, the two-dimensional state space is therefore most appropriate. A reason for the decrease of the predictor performance which is observed with increasing dimensionality of prototypes is not immediately obvious. A prototype can be interpreted as one of the typical sections or what might be called *strophes* of the chaotic signal. When the predictor forecasts the forthcoming value $x(t+1)$, it

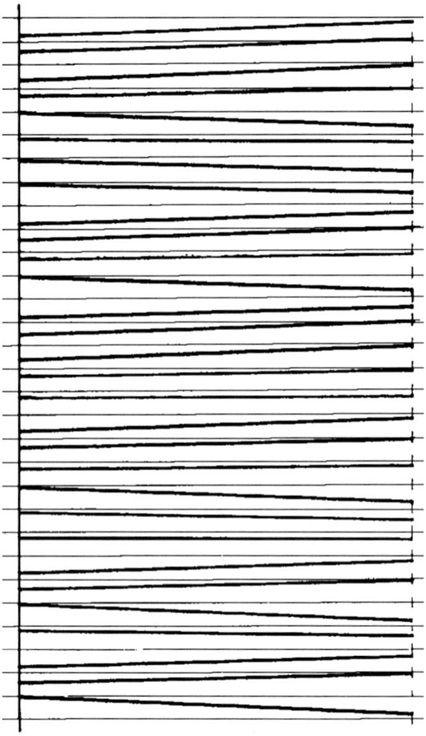

Fig. 11.5. Transitions between successive states represented by prototypes. The graphs of particular prototypes are shifted in the vertical direction in equal increments

compares the given most recent portion of the signal with the corresponding portions of the prototypes. The similarity is most properly estimated when the dimensionality of the prototypes corresponds to the number of components used in the deterministic chaos generator. If one increases the number of components, then the conditional average estimator simultaneously compares more components than is optimally needed. Because of instabilities of the chaotic trajectories, two portions of the chaotic signal that are initially nearly coincident, later diverge and the similarity is therefore generally reduced. If one increases the length of the prototypes and uses them to compare a given segment of the chaotic signal, one can observe several situations. If a short prototype is similar to the given segment, then this similarity becomes less pronounced when the length of the prototype is increasing because of the divergence of trajectories. In contrast, the similarity of the given segment may increase relative to the other prototypes that are in the increased portion

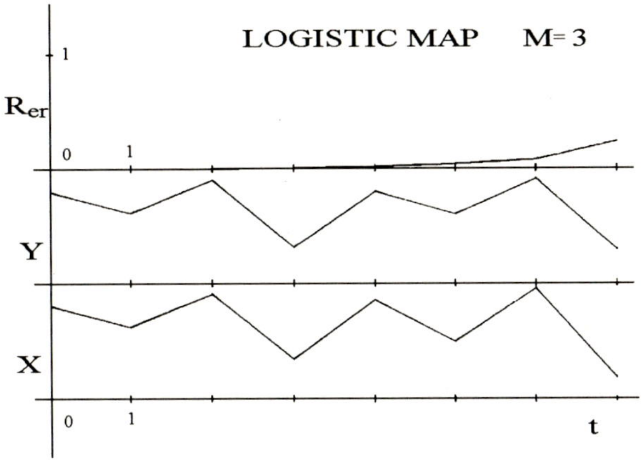

Fig. 11.6. Records of the signal X generated by a logistic map, the predicted signal Y and the relative error R_{er} for $M = 3$

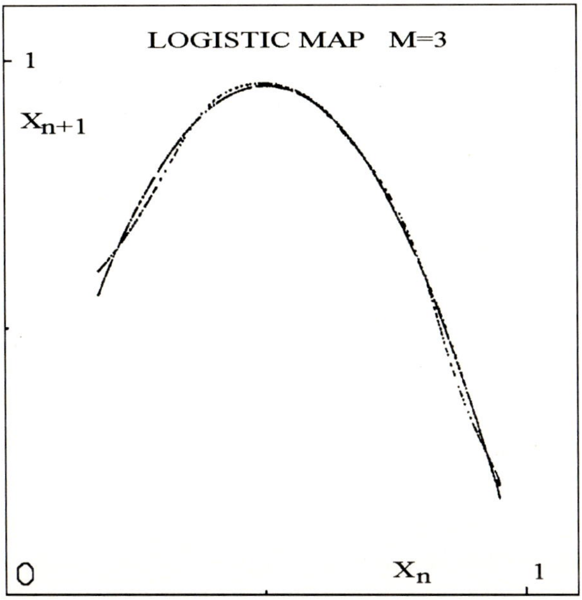

Fig. 11.7. Return maps of the original and the predicted logistic signals for $M = 3$

closer to the given segment. This corresponds to an increasing number of terms that effectively contribute to the conditional average. The resonance-like effect of the similarity is thus smoothed. One obtains a diminished performance of the predictor when one increases the dimensionality of the prototypes above the proper value which is determined by the nature of the chaos generator.

An additional smoothing effect follows from the influence of the dimensionality on the width of prototypes. If one wishes to fill a cube of length L containing the chaotic attractor in an M dimensional state space by K prototypes, then the characteristic width of the prototypes must be scaled as $\sigma \approx L/\ K^{1/M}$. The width σ of the prototypes thus increases with increasing number of vector components M which again corresponds to a smoothing effect. It follows that in order to obtain an optimal performance of the predictor the number of components M should then be as low as permitted by the deterministic chaos generator.

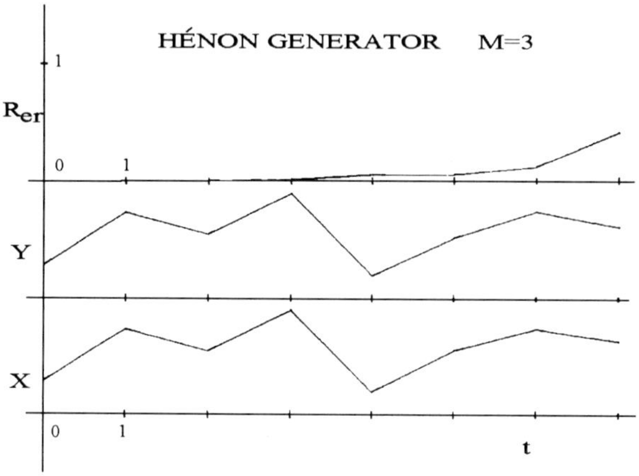

Fig. 11.8. Records of the signal X generated by the Hénon map, the predicted signal Y and the relative error R_{er} for $M = 3$

When using the Hénon chaos generator, a three-dimensional state space is required. By simulation we demonstrate the operation of the modeler with this chaos generator. The representative records and the corresponding return maps are shown in Figs. 11.8 and 11.9. For the applied map, the theory yields $2\lambda \approx 0.84$ while the result of numerical simulation is $L_{er} \approx 0.91$. A good overlapping of the original and predicted return maps is observed. The relative prediction error estimated from 500 steps is $R_{er}(1) \approx 0.01$. This value is higher than that obtained in the simulation with the logistic map. It appears to be the result of the small discrepancy between the return maps in the middle of the graph. This discrepancy can be reduced by increasing the number of prototypes because the system states occupy this portion of the return map with relatively low frequency. The minimal prediction error for the case of the Hénon generator was obtained at $M = 3$, which again corresponds to the correct dimension $D = M - 1 = 2$ of representative vector.

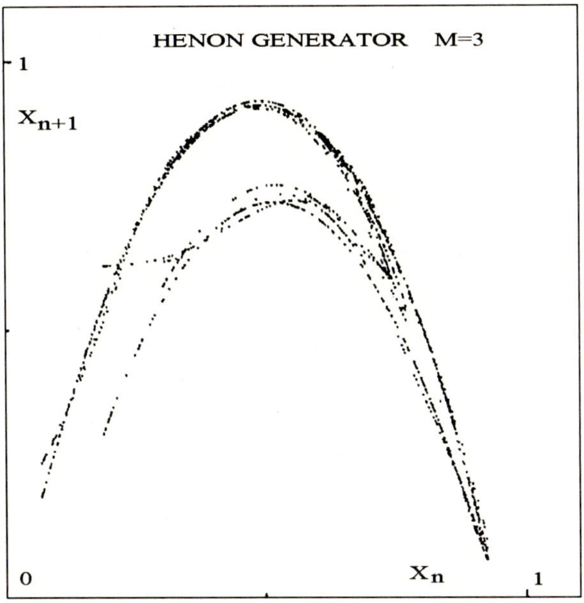

Fig. 11.9. Return maps of the original and the predicted signals of the Hénon map for $M = 3$

The agreement between the dimension of the vector in the chaos generator and the dimension M, for which the minimal prediction error is obtained in discrete cases, suggests that the embedding dimension of the space for presenting the chaotic attractor in the continuous case could be found by searching for the minimum of the prediction error. The estimated value M presumably corresponds to the embedding dimension plus one: $M = D_e + 1$. An argument in favor of this presumption was obtained by predicting the chaotic signals generated by the Lorenz equations, as shown in Figs. 11.10 and 11.11.

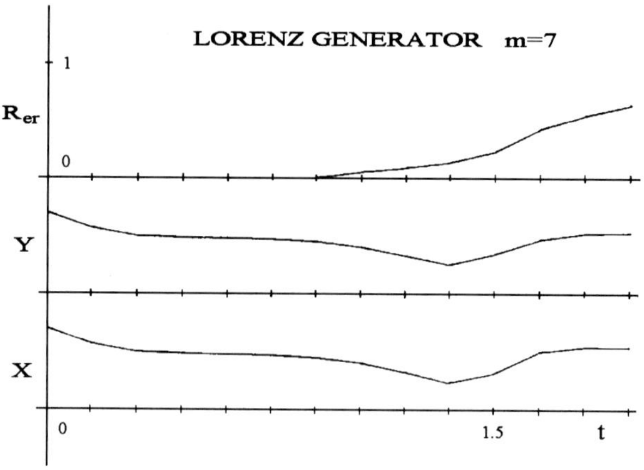

Fig. 11.10. Records of signal X generated by the Lorenz equations, the predicted signal Y and the relative error R_{er} for $M = 7$

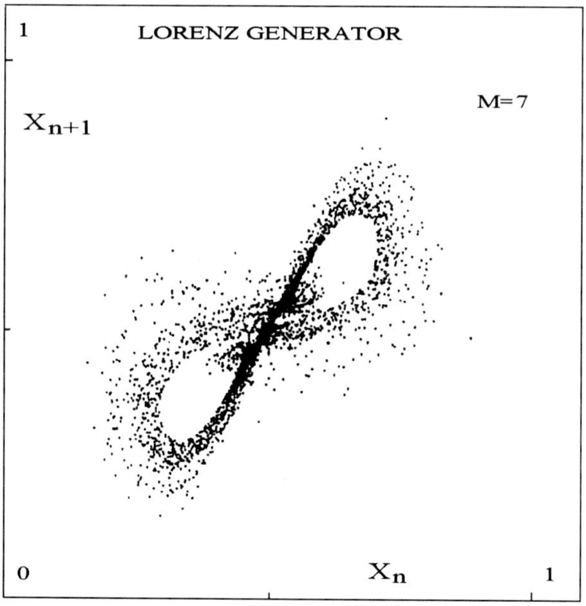

Fig. 11.11. Return maps of the original and the predicted signals of the Lorenz generator for $M = 7$

For the selected parameters in the Lorenz system, the best prediction was obtained with $M = 7$. In this case the phase space is three-dimensional and according to Takens theorem the dimension $D_e = 2D+1 = 7$ assures a proper embedding. This yields $M = D_e + 1 = 8$ which is higher than the observed optimal value $M = 7$. However, the corresponding attractor has a fractal dimension between 2 and 3 which corresponds to the minimal embedding dimension of approximately 6. [17],[27, p.108] We therefore conjecture that the observed optimal value $M = 7$ is related to the minimal embedding dimension.

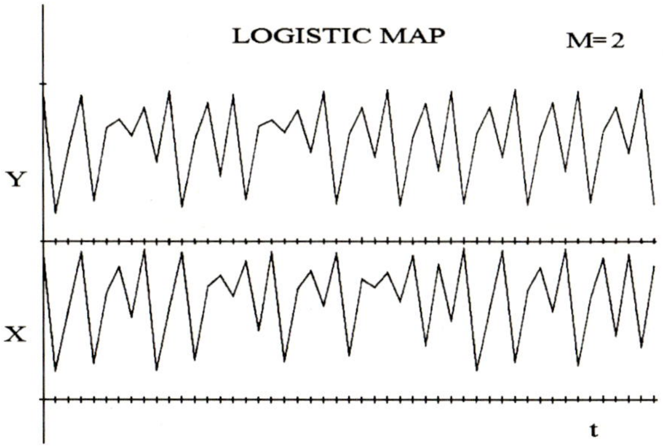

Fig. 11.12. The original signal, X, and the long-term predicted chaotic signal Y from a logistic generator

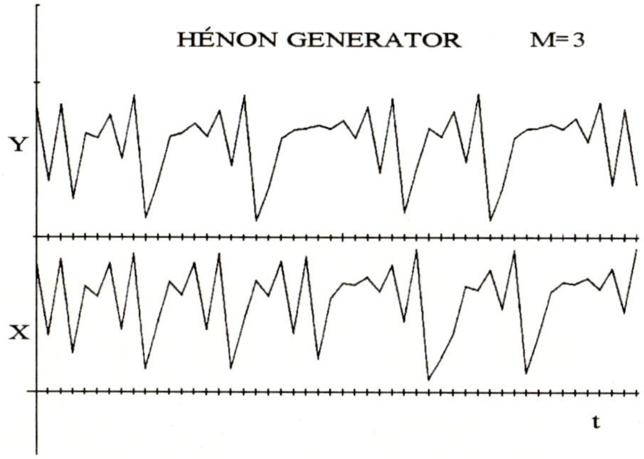

Fig. 11.13. The original signal, X, and the long-term predicted chaotic signal Y from the Hénon generator

The divergence of the trajectories, caused by the instability of the chaotic systems, is the source of the inaccuracy of the long term predictions. Figures from 11.12 to 11.14 show the actual and long-term predicted signals for each of the three chaos generators described above. Although the original and the supplied and predicted signals begin from the same initial values, the difference between them appears obvious after only several steps of prediction. In spite of this divergence, we recognize that both signals exhibit similar properties, which is evidence that the chaos generator is well-modeled.

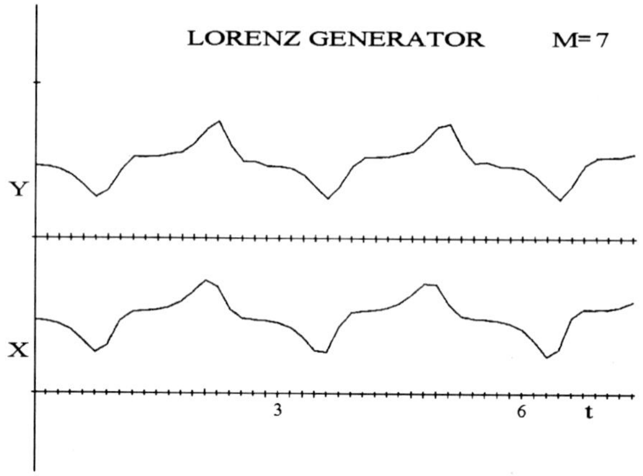

Fig. 11.14. The original signal, X, and the long-term predicted chaotic signal Y from the Lorenz generator

11.3 Forecasting of Chaotic Acoustic Emission Signals

In order to demonstrate the practical applicability of the information processing system described in the previous section, we present here the results of the forecasting of chaotic acoustic emission signals. It was mentioned in Chap. 10 that the physical description of acoustic emission signals generally leads to a modeling of dynamical phenomena. This task is extremely complex when applied to manufacturing processes. A typical example is the cutting process on a lathe or in a drill press. Such processes include: the random properties of the input material, the chaotic dynamics caused by the nonlinear deformation of the cut material, and changing machine parameters because of tool wear or failure. [6, 7, 8] Because of the complexity inherent in manufacturing processes, an exact analytical modeling is often beyond the scope of most theoretical treatments and one must therefore find a proper empirical description in order to enable an optimal control of the process.

As described in the Chap. 10, recent investigations of empirical modeling have shown that it can be applied to transform the properties of AE signals to recover the properties of AE sources. [16] However, for this the modeler must be properly trained using measured AE waveform data which provides empirical information about the observed phenomenon. The same system can also be used to transform one portion of an AE signal into another, which makes the prediction of AE signals possible, based only on previous observations. Such prediction is promising for various applications of acoustic emission measurements in industrial environments, especially in optimal control, and it will therefore be described in more detail. Here we demonstrate the performance of the prediction by forecasting an AE signal generated during the drilling process.

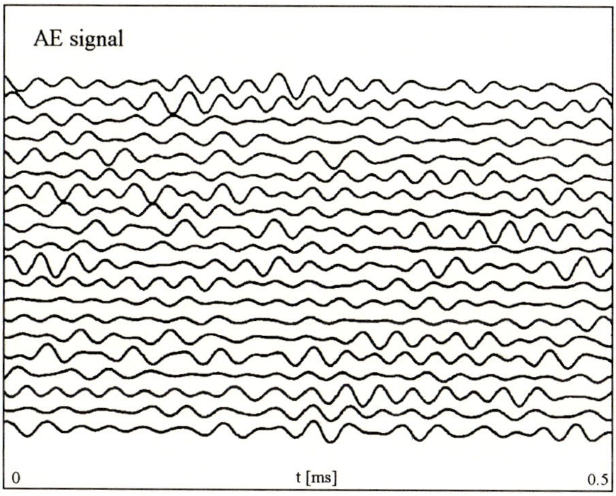

Fig. 11.15. Records of chaotic AE signals generated by a drilling process. Successive records are vertically offset in equal increments. The signal amplitude depends on AE sensor sensitivity and adjustable signal amplifier gain therefore arbitrary units are used on the vertical axis

The experiment was carried out on a steel sample of dimensions $100 \times 100 \times 10\,\mathrm{mm}^3$. A drill of diameter $\phi = 1\,\mathrm{mm}$ was used. A broadband AE sensor was mounted on the edge of the specimen. The AE signal was digitized by a digitizing oscilloscope with a sampling time of $1\,\mu s$ and transferred to a PC in which the information processing was performed by the forecasting system. Fig. 11.15 shows an ensemble of AE waveform records. The corresponding correlation function and the spectrum of one signal are shown in Figs. 11.16 and 11.17 respectively. The correlation function decays with the characteristic correlation time of approximately 3 time periods of the characteristic oscillation. The most expressive frequency of the signal is approximately at 40 kHz and this likely corresponds to a resonant frequency of the sample-tool-machine structure.

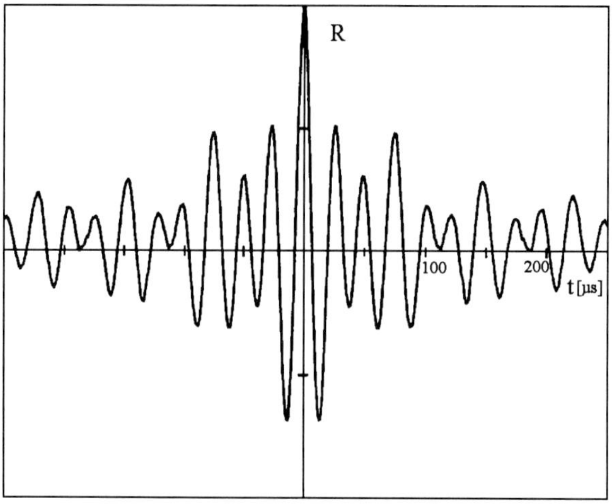

Fig. 11.16. The correlation function estimated from records of the chaotic AE signal. Vertical: arbitrary units

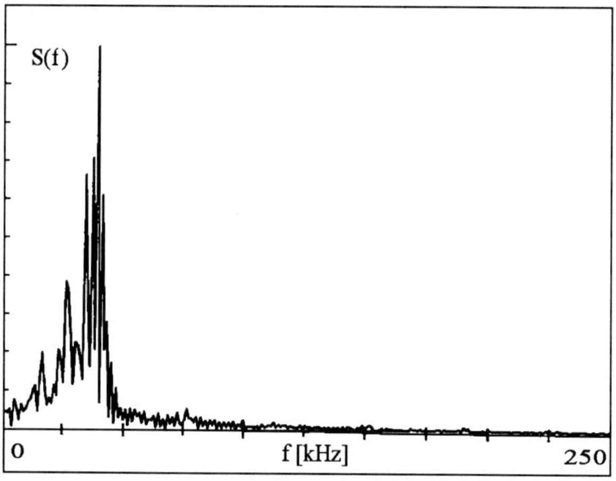

Fig. 11.17. The spectrum of the chaotic AE signal. Vertical: arbitrary units

The sample vectors were formed from the input signals. Seven hundred of these were used in a self-organized adaptation of 100 prototypes which were subsequently used to predict the AE signal. The complete procedure was repeated for the values of the dimension M from 2 to 10. The agreement between the actual and the predicted signals was determined by comparing short signal segments. An example of the input and output signals is shown in Fig. 11.18 by the lower and the middle records, respectively. The average value of the relative prediction error that was estimated from 200 simulations using the dimension $M = 7$ is $R_{er}(1) = 0.035$, which indicates a reasonably good modeling of the chaotic AE phenomenon.

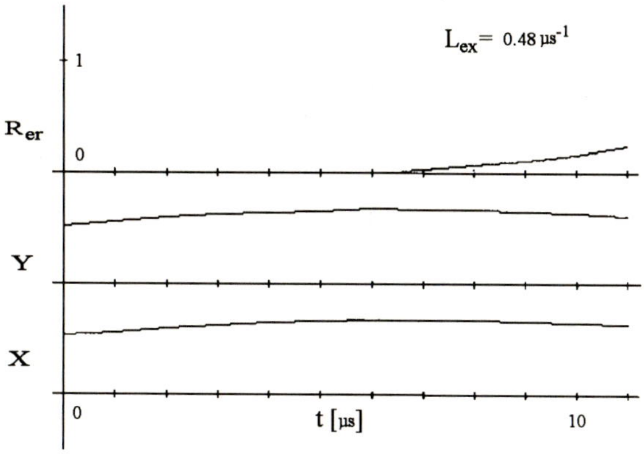

Fig. 11.18. Section of the input signal (*bottom*), the predicted signal (*middle*) and the average relative prediction error (*top*). Vertical: arbitrary units for X and Y

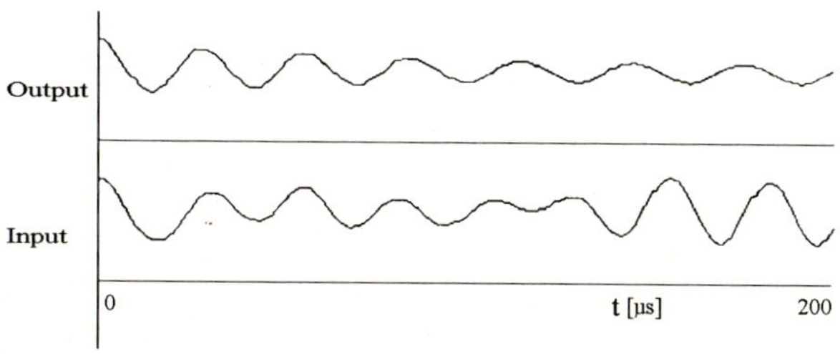

Fig. 11.19. Long-term predicted signal (*top*) and the corresponding true input signal (*bottom*). Vertical: arbitrary units

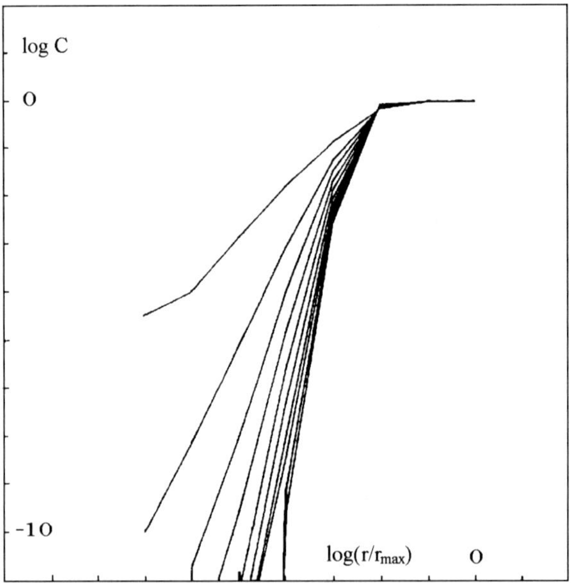

Fig. 11.20. The graph of correlation integral for the embedding dimension increasing from 2 (*left*) to 11 (*right*). With increasing dimension, the value of the slope coefficient stabilizes at approximately 6

The dependence of the prediction error on the number of steps is shown by the upper trace in Fig. 11.18. It grows approximately exponentially with the number of prediction steps. For the AE signal used, the characteristic exponent of relative growth of error is $2\lambda \doteq 0.48\,\mu s^{-1}$, which corresponds to a relatively rapid divergence of chaotic trajectories. Therefore, in this case, one cannot expect good long-term prediction. An example of the input and long-term predicted signals is shown in Fig. 11.19. The agreement between these signals is reasonably close, within the time interval corresponding to the length of the correlation time, which is as expected.

The minimum of the prediction error as a function of embedding dimension can be determined by simulating the adaptation and prediction for several embedding dimensions. It is found that the prediction error has its minimum at the dimension $M = 7$. In studies of chaotic dynamics, the embedding dimension is usually estimated by the correlation exponent. [26, 27, 30] The corresponding graph which depicts the logarithm of the correlation integral versus distance between points on the chaotic attractor is shown in Fig. 11.20. The diagrams from left to right correspond to increasing embedding dimension. The correlation exponent, which is determined from the slope, becomes saturated with increasing dimension of the representative vector at the proper value of approximately $D_e \approx 6$ which corresponds to the minimal embedding dimension that is needed for representing the chaotic attractor. This value is in agreement with the value of the dimension $M = D_e + 1 \approx 7$ of the extended embedding space that is estimated from the best prediction results.

These simulations demonstrate that the modeler is applicable not only for the prediction of synthetic signals which have been generated by deterministic chaos generators but it can also be used to predict chaotic signals which are generated in real systems. This illustrates that very complex phenomena, such as the acoustic emission signals generated during a cutting process, can be modeled, at least approximately, on a completely empirical basis. The same system has also been successfully used to predict the acoustic signals generated by the friction between elastic objects and the prediction of non-chaotic signals representing movements on attractors of finite size. The predicted signal does not deviate appreciably from the actual signal in the succeeding time interval whose duration is approximately equal to the correlation time. The largest Lyapunov exponent can be estimated from the characteristic exponent of the divergence between the input and the predicted signals. These results suggest that the modeler could also be applied to reconstruct other parameters related to the statistics and the dynamics of chaotic attractors.

The accuracy of the modeling depends on the number of applied prototypes. In the approach presented in this chapter, this number is taken as a parameter that is determined by the construction of the system. But there remain many open questions. [2] How can fundamental principles be used to optimize the number of applied prototypes? It is mentioned in Chap. 7 that a properly defined complexity measure could be used for this purpose. Another unanswered question is how one can incorporate *a priori* information or symmetries into a modeler or a predictor. Finally, how can one treat multi-component chaotic phenomena? In the field of elastodynamics, a number of approaches stemming from linear, optimal, multi-component deconvolution have already led to a neural-network-like structure capable of time-invariant operation and prediction [33] but the general method of solving such problems is still not developed.

The properties of the modeler indicate that the development of the described forecasting system can be considered only as a first step in the evolution of intelligent automatic modelers of chaotic phenomena. Prototypes of a modeler correspond to typical transitions occurring in the observed chaotic time-series. The generated model is therefore adapted only to a specific example. An interesting question then arises how the collected knowledge could be generalized and subsequently presented in terms of an abstract natural law. [2] An *ad hoc* replacement of prototypes by various abstract functions only changes a non-parametric treatment into a parametric one but it does not solve the problem because it is not known how one could automatically define an appropriate function that would properly represent the collected knowledge. For this purpose, further levels of modelers, resembling a complex, hierarchically built neural network should be used, but the fundamental physical principles for their development, which could lead to hierarchical scientific reasoning, have not yet been formulated.

11.4 Empirical Modeling of Non-Autonomous Chaotic Systems

In this section we describe the adaptive modeling of a chaotic time-series generated by non-linear dynamical, non-autonomous systems. For illustration, a chaotic economic phenomenon is represented as an example of deterministic chaos generated by a driven dynamic system. We first review the fundamentals of the description of non-autonomous dynamic systems and then describe how empirical modeling similar to that applied previously for the modeling of autonomous chaotic systems can be applied to model non-autonomous systems.

Let the state of a dynamic non-autonomous system be described by the vector $\boldsymbol{s}(t)$ and let the vector $\boldsymbol{v}(t)$ denote the variable which corresponds to the non-autonomous properties of the system. The vector $\boldsymbol{v}$ is generally interpreted as a non-chaotic driving variable which describes the influence of the surroundings on the system. As in the beginning of this chapter, we assume that the system dynamics can be described by a differential equation of the form:

$$\frac{ds}{dt} = \boldsymbol{\Phi}(\boldsymbol{s}(t),\ \boldsymbol{v}(t)) \ . \tag{11.14}$$

For a chaotic phenomenon, the function $\boldsymbol{\Phi}$ is non-linear. We assume that the signal $\boldsymbol{s}(t)$ describes in the phase space a dynamical flow with an attractor of limited extensions. We further assume that the Theorem of Takens is applicable also to non-autonomous systems, which means that the principal properties of the attractor can be estimated from just one time-dependent component of the state vector $\boldsymbol{s}(t)$ which we denote by $x(t)$. Consequently, a new state vector $\boldsymbol{x}(t)$ is formed from $m+1$ samples of this variable: $\boldsymbol{x}(t) = (x(t),\ \ldots,\ x(t-m))$, and the dynamical law is transformed into a discrete mapping

$$\boldsymbol{x}(t+1) = \boldsymbol{\Psi}\left(\boldsymbol{x}(t), \boldsymbol{v}(t)\right) \ . \tag{11.15}$$

Using the components of the state vector, this discrete mapping can be expressed in the form:

$$(x(t+1),\ x(t),\ \ldots,\ x(t+1-m)) = \boldsymbol{\Psi}(x(t),\ \ldots,\ x(t-m), \boldsymbol{v}(t)) \ . \tag{11.16}$$

It includes the shifting of m components by one time increment and the mapping of the state vector $\boldsymbol{x} = (x(t),\ \ldots,\ x(t-m))$ to only one component $x(t+1)$:

$$x(t+1) = \Psi(x(t),\ \ldots,\ x(t-m), \boldsymbol{v}(t)) \ . \tag{11.17}$$

Our task is to model the function $\boldsymbol{\Psi}$ and to determine the dimension $m+1$ of the representative vector $\boldsymbol{x}$ based on empirical data about $x(t)$ and $\boldsymbol{v}(t)$. The principal difference with respect to a non-autonomous system is that the function Ψ now depends on the time-dependent driving variable $\boldsymbol{v}(t)$. The mapping is thus empirically represented by the distribution of sample points $\boldsymbol{X}(t) = \{x(t+1),\ x(t),\ \ldots,\ x(t-m); \boldsymbol{v}(t)\}$, $\{t = 1,\ 2,\ \ldots \to \infty\}$ in the space of dimension $M = m + 2 + d_v$. Here d_v denotes the dimension of the subspace of the driving variable $\boldsymbol{v}(t)$. The set of experimental sample points determines an empirical joint-probability distribution:

$$F(x(t+1),\ x(t),\ \ldots,\ x(t-m); \boldsymbol{v}(t)) = F(x(t+1),\ \boldsymbol{x}(t);\ \boldsymbol{v}(t)) \quad (11.18)$$

which can be modeled by using the self-organization of prototypes ($\boldsymbol{Q}_k$, $k = 1,\ 2,\ \ldots,\ K$) in the M-dimensional state space. Using these, the conditional probability density is determined:

$$F(x(t+1)|x(t),\ \ldots,\ x(t-m); \boldsymbol{v}(t)) = F(x(t+1)|\boldsymbol{x}(t)\ ; \boldsymbol{v}(t))\ . \quad (11.19)$$

Here, the concatenated vectors $(\boldsymbol{x}(t)\,; \boldsymbol{v}(t))$ play the role of a joint condition. Based upon the conditional probability distribution, the function Ψ can then be expressed using a non-parametric regression as an optimal statistical estimator, just as in our earlier treatment of autonomous systems. For a description of non-autonomous systems, the generating function Ψ of the dynamical law is influenced by the vector comprised of chaotic time-lagged values of the representative variable $x(t)$ as well as the driving variable $\boldsymbol{v}(t)$ which is treated deterministically. We are mainly interested in the case in which the driving variable is a periodic function of time. In this case, a self-organization procedure can be carried out in a subspace of the sample space of the variable $\boldsymbol{X}$ that pertains to the chaotic variable $x(t)$ only. However, the complementary components of $\boldsymbol{X}$, corresponding to the driving variable $\boldsymbol{v}(t)$, must be kept fixed and treated as a deterministic condition. The adaptation as well as the prototypes thus become dependent on the condition and we term this *conditional self-organization.*

In order to model the dynamical law corresponding to a non-autonomous system, the prototypes must first be adapted to the input variables $x(t)$ and $\boldsymbol{v}(t)$, and then the forthcoming variable $x(t+1)$ can be estimated in turn from the input data vector and the prototypes. The data vector $\boldsymbol{X}$ is formed from successive values of the time-series shifted in time and the driving variable $\boldsymbol{v}(t)$.

Therefore, as shown in Fig. 11.21, the corresponding forecasting system again includes two shift registers, a memory, and a conditional average estimator. The generated time-series and the driving variable are input into the shift register in which a sample of the state vector $\boldsymbol{X}$ is formed out of M successive data values. The sample is then transferred to the memory of the system and to the predictor which consists of a shift register

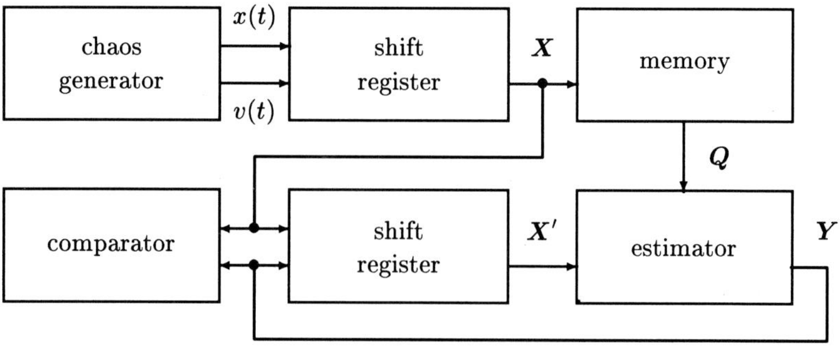

Fig. 11.21. Scheme of the system applicable for the empirical modeling and prediction of non-autonomous chaotic dynamical systems

and the conditional average estimator. The prototypes are formed by self-organization in the computer memory. The process of self-organization proceeds only on the data presenting the chaotic signal $x(t)$ while the driving variable is used as an index which labels the prototypes. This means that the same number of prototypes is assigned to each possible value of the variable $\boldsymbol{v}$. The prototypes are transferred to the conditional average estimator which here plays the role of a predictor. During the prediction a truncated vector $\boldsymbol{X}' = (\#,\ x(t),\ x(t-1),\ \ldots,\ x(t-m)\ ;\boldsymbol{v}(t))$ is led to conditional average estimator where the missing forthcoming value of the signal $x(t+1)$ is estimated. During estimation, the driving variable is again utilized only as an index for labeling the vectors of the subspace pertaining to the lagged components of the signal $x(t)$. The predicted value $\hat{x}(t+1)$ is then concatenated to the output vector

$$\boldsymbol{Y} = (\hat{x}(t+1),\ x(t),\ x(t-1),\ \ldots,\ x(t-m)\ ;\boldsymbol{v}(t))\ . \tag{11.20}$$

This vector is then iteratively fed back to the predictor input in order to obtain a prediction for the next term in the forthcoming time-series. The prediction error is estimated in the comparator which determines the mean square difference between the predicted and the actual input signal. Using this error, the system performance and the optimal dimension M of data vector in the subspace pertaining to $\boldsymbol{x}$ can be described.

11.4.1 Example of Economic Time-Series Forecasting

Here we demonstrate the prediction of a chaotic time-series that is generated by the selling of two kinds of products manufactured in a chemical factory. The first product is fertilizer and the second is a cosmetic agent. Only a one-component driving variable and a one-component state vector are applied. The driving variable denotes the index of the month in a particular year.

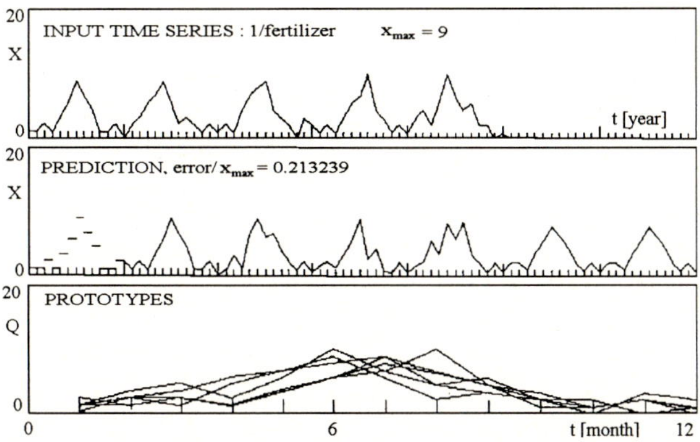

Fig. 11.22. Forecasting of the selling process of a certain chemical product for which a seasonal consumption is characteristic. Top: Input time-series; Middle: Predicted time-series; Bottom: The set of prototypes

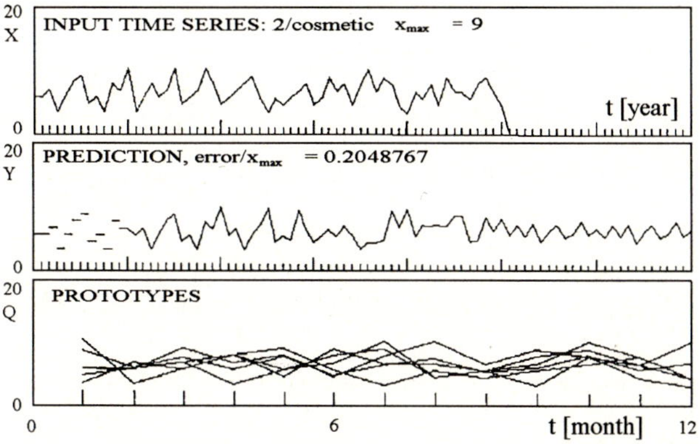

Fig. 11.23. Forecasting of the selling process of a chemical product for which a non-seasonal consumption is characteristic. Top: Input time-series; Middle: Predicted time-series; Bottom: The set of prototypes

Figures 11.22 and 11.23 show two examples of the result of forecasting by three types of records. In each figure the upper record shows the input time-series, the middle one represents the predicted time-series, and the lower record shows the set of prototypes. In the time interval for which the input time-series is specified, the prediction error is estimated from the one-step prediction. Based on this error, the reliability of the forecasting can be estimated in the time interval for which the time-series is not known. The time-series of the first example exhibits strong seasonal variations while the second example shows only chaotic variations around the mean value. These properties are most obvious in the bottom diagram which shows the dependence of the prototypes on the driving variable. Although the characteristics of each example appear different, the performance of the forecasting system does not differ substantially. This indicates that in both cases the generating functions of the corresponding laws of dynamics can be modeled equally well by using only a one-component state and driving variable. However, this is not always the case, especially when the time-series generated by different economical branches are predicted. [13, 14, 15] We have observed in simulations of various examples of synthetic and experimental time-series that the prediction error can be applied to describe the ratio of the random versus the deterministic portions of the input series, which is of importance in economic and technical applications of the forecasting system. As described in the previous section, by minimizing the prediction error, the proper state space dimension D can be estimated.

11.5 Cascade Modeling of Chaos Generators

The principal task related to the prediction of a chaotic time-series is the determination of a proper dimension of the embedding space that is needed for the presentation of the time-series. The formation of prototypes by the self-organization process and a repetition of system performance evaluation by calculating the relative error during forecasting based on the embedding vector of ever greater dimension is time-consuming, although it can, in principle, be carried out as described above. Another problem which is of systematic character, is hidden in the application of the non-parametric regression for forecasting. As mentioned in the previous chapters, the use of a non-parametric regression is an appropriate procedure for the interpolation but not for the extrapolation of various functional dependencies between experimental data. This is however not convenient for the prediction of signals exhibiting an expressive trend. For instance, one cannot successfully predict the signal generated by a linear function $x(t) = t$ outside the region in which it was previously trained. However, this is not a chaotic signal and we could argue that there is a degenerate probability distribution related with it, which is best described by a parametric regression. The question arises how this limitation could be improved by modifying the forecasting system described above that will still permit the use of the convenient non-parametric regression. In this section, we describe the solution of the above-mentioned problems using a proper presentation and an interpretation of the general dynamical law describing a time-series.

Let us again consider a dynamical phenomenon which can be described by the variable $x(t)$ that has been observed at the discrete times $\{t = 1,\ 2,\ \ldots\}$. Related to the fundamental task of forecasting, we now describe the formulation of a method, or a mathematical model, by which forthcoming changes of this variable, denoted by $\Delta x(t) = x(t) - x(t-1)$ can be predicted from a series of past observations $\{x(t),\ x(t-1),\ x(t-2),\ \ldots\}$. For this, we assume that the changes of the variable $x(t)$ can be formally described by the expression:

$$\Delta x(t+1) = x(t+1) - x(t) = G[x(t),\ x(t-1),\ \ldots; \boldsymbol{v}(t)]\ , \tag{11.21}$$

in which the generating function G generally denotes some nonlinear mapping. As will be seen from the examples to be presented, the representation of the dynamical law in terms of changes of the variable x permits a proper accounting of a linear trend in the time-series. In addition to the series of past values of the variable x, we include into the dynamical law, the vector $\boldsymbol{v}(t) = (v_1(t),\ \ldots,\ v_{d_v}(t))$ to provide for the description of non-autonomous phenomena. Our goal is to formulate a procedure by which the generating function G can be modeled from the past series $\{x(t');\ t' \leq t\}$ without any *a priori* information about the embedding dimension of the representative state vector $(x(t),\ x(t-1),\ \ldots)$ that is needed for the description of the dynamical law. This dimension can either be estimated by any of the existing methods of non-linear dynamics and deterministic chaos that has been described earlier or it can be adaptively determined using the approach described in the following paragraphs.

Generally, one-dimensional phenomena are the simplest to model. Such phenomena are determined by a function G that depends only on the first component of the state vector $\boldsymbol{s}(t)$. Numerical simulations show that various higher-order mappings can often be approximated by using only one-dimensional generating functions. [3, 4, 11, 12, 13, 14, 15] A typical example is the modeling of the signal generated by the Hénon map which can be carried out approximately using just two components of the representative vector, even though the original generator includes three components. We therefore conjecture that it might be advantageous to expand the generating function into a hierarchy of terms including an increasing number of components of the state vector $\boldsymbol{s}(t)$. A similar approach has been proposed by Deppisch in his formulation of a neural network. [3, 4] Then the expanded generating function becomes:

$$\begin{aligned} G(x(t),\ x(t-1),\ \ldots; \boldsymbol{v}) &= G_1(x(t); \boldsymbol{v}) \\ &+ G_2(x(t), x(t-1); \boldsymbol{v}) \\ &+ G_3(x(t), x(t-1), x(t-2); \boldsymbol{v}) + \ldots \end{aligned} \tag{11.22}$$

This expansion introduces a cascade model for the chaos generator. After this we also expand the changes of the variable x into a series of terms

$$\Delta x(t+1) = \Delta x_1(t+1) + \Delta x_2(t+1) + \Delta x_3(t+1) + \ldots \tag{11.23}$$

where each term corresponds to an ever more complex dynamical law:

$$\Delta x_1(t+1) = G_1(x(t); \boldsymbol{v})\ ;\ \Delta x_2(t+1) = G_2(x(t),\ x(t-1); \boldsymbol{v})\ ;\ \ldots\ . \tag{11.24}$$

However, such splitting is arbitrary if we do not specify Δx_i and the procedures used for the hierarchical modeling of the generating functions G_i. To avoid this arbitrariness we utilize the previous explanation of modeling a chaotic time-series. We describe the dynamical law

$$\Delta x_i(t+1) = G_i(x(t),\ x(t-1),\ \ldots,\ x(t-i+1); \boldsymbol{v}) \tag{11.25}$$

by a set of points

$$\boldsymbol{x}_i(t) = (\Delta x_i(t+1),\ x(t),\ x(t-1),\ \ldots,\ x(t-i+1); \boldsymbol{v}(t)) \tag{11.26}$$

in the embedding state space of dimension $d_i = i + 1 + d_v$. Using Parzen's window filtering, the corresponding probability density function is again approximately described by:

$$f(\boldsymbol{x}_i) = \frac{1}{K} \sum_{k=1}^{K} w(\boldsymbol{x}_i - \boldsymbol{q}_i(k),\ \sigma)\ . \tag{11.27}$$

This probability density function was previously encountered in relation to self-organization which was discussed in Chap. 8. The prototype vector $\boldsymbol{q}_i(k)$ appearing in this function, describes a multi-dimensional parameter of the k-th formal neuron and w denotes the window or radial basis function. The parameter σ characterizes the width of the receptive field of the neuron. For simplicity we assume that all the neurons possess the same receptive field σ. Information about the phenomenon is hidden in the set of prototype vectors $\{\boldsymbol{q}_i(k),\ k = 1,\ \ldots,\ K\}$ which have been adapted to the past time-series that has been presented to the system, while the model is indirectly specified by the probability density function. The prototypes are formed by the self-organization process in the embedding space. At the start of adaptation the prototypes are equalized with empirical data $\{\boldsymbol{x}_i(t);\ t = 1,\ 2,\ \ldots\}$ and this equalization continues until the number of presented empirical data points, N, exceeds the number of prototypes, K, that has been determined by the formulation of the predictor structure. The corresponding adaptive system, exhibiting the structure of a neural network, at first grows and later adapts by self-organization. This procedure is quite convenient because no training of the system is required, only the inscription of the data into the memory when the number of data points N is less than K. The system is therefore capable of forecasting from the very beginning of its operation. For this purpose, the non-parametric regression is again applied which utilizes the prototypes and partial information about the phenomenon given by the truncated vector

$$\boldsymbol{x}'_i(t) = (\#,\ x(t),\ x(t-1),\ \ldots,\ x(t-i+1)\ ; \boldsymbol{v}(t))\ , \tag{11.28}$$

in which # indicates the missing component $\Delta x_i(t+1)$. Its value is optimally estimated from the vector $\boldsymbol{x}'_i(t)$ by the conditional average estimator of the non-parametric regression as

$$\Delta\hat{x}_i(t+1) = \frac{\sum_{s=1}^{K} \Delta x_i(s+1)\, w(\boldsymbol{x}'_i(t) - \boldsymbol{q}'_i(s),\, \sigma)}{\sum_{r=1}^{K} w(\boldsymbol{x}'_i(t) - \boldsymbol{q}'_i(r),\, \sigma)} \ . \tag{11.29}$$

In accordance with this estimator, we propose the following modeling by a cascade structure which resembles a multi-layer neural network:

1. From a given time-series $\{x(t),\ t = 1,\ \ldots,\ K\}$ we form the prototype vectors $\boldsymbol{q}_1(k)$ representing the parameters of the formal neurons comprising the first layer. Using the conditional average estimator, we predict for each $\boldsymbol{x}'_1(t)$ the corresponding value $\Delta\hat{x}_1(t+1)$ which is treated as the output of the first layer of the predictor. Because of an incomplete modeling of the time-series by the probability distribution function in the embedding space of vectors $\boldsymbol{x}_1$, that correspond to the dynamical law $\Delta x_1(t+1) = G_1(x(t); \boldsymbol{v})$, the estimated, forthcoming value $\Delta\hat{x}_1(t+1)$ does not exactly coincide with the true value $\Delta x(t+1) \equiv \Delta x_1(t+1)$.
2. A crucial step in the development of the cascade predictor is then obtained by assuming that the difference $\Delta x_1(t+1) - \Delta\hat{x}_1(t+1) = \Delta x_2(t+1)$ can be predicted in the next layer of the cascade according to the dynamical law $\Delta x_2(t+1) = G_2[x(t),\ x(t-1); \boldsymbol{v}]$. This law is modeled using the prototype vectors $\boldsymbol{q}_2(t)$ which have been formed from the empirical data $\{\boldsymbol{x}_2(t) = (\Delta x(t+1),\ x(t),\ x(t-1);\ \boldsymbol{v}(t);\ t = 1,\ 2,\ \ldots\}$. The prototype vector $\boldsymbol{q}_2(t)$ represents the parameters of the formal neurons in the second layer. Using these prototypes we can estimate the value $\Delta\hat{x}_2(t+1)$ and define the difference $\Delta x_2(t+1) - \Delta\hat{x}_2(t+1) \equiv \Delta x_3(t+1)$ which is the basis for the formation of the third layer of the network and so on.

By using the outputs from the successive layers, the time-series predictor can be represented as:

$$y(t) = x(t) + \Delta\hat{x}_1(t+1) + \Delta\hat{x}_2(t+1) + \ldots \ . \tag{11.30}$$

The prediction error can be estimated by forming the square of the output of the final layer of the predictor, that is $\Delta\hat{x}_f(t+1)$ and summing over the given time-series. We generally expect that the prediction error will decrease with an increasing number of layers. We will show this result using the numerical examples presented in the following sub-section. Growth of the predictor structure can be stopped when a desired forecasting accuracy is attained. The network thus includes two types of evolutionary adaptations. First, prototypes of an individual layer are formed which correspond to the creation of neurons inside this layer, and secondly, the layers are successively formed until a proper performance of the predictor is established.

11.5.1 Numerical Experiments

The operation of the cascade predictor with the proposed hierarchical structure of a multi-layer neural network was simulated on a personal computer.

The layers of the network were formed from a given time-series using the adaptation described in the previous section. At each level, the prediction performance was estimated and the network growth was halted when the total error of a layer reached a value less than 10^{-3} of the span of x.

Fig. 11.24 shows the results obtained for the case of a regular time-series, of constant trend given by $x(t) = ct$. The upper trace of the figure represents the given time-series, the middle trace shows the time-dependence of the differences for each layer r, and the bottom trace represents the total time-series of which the initial portion is the reproduced input signal and the second portion is that predicted by the network. In this case, the time-series is already well reproduced by only one layer. It is characteristic that the trend is correctly described for $t > K$ which is not the case when using only Parzen's estimator with spherical basis functions and without calculating the differences Δx.

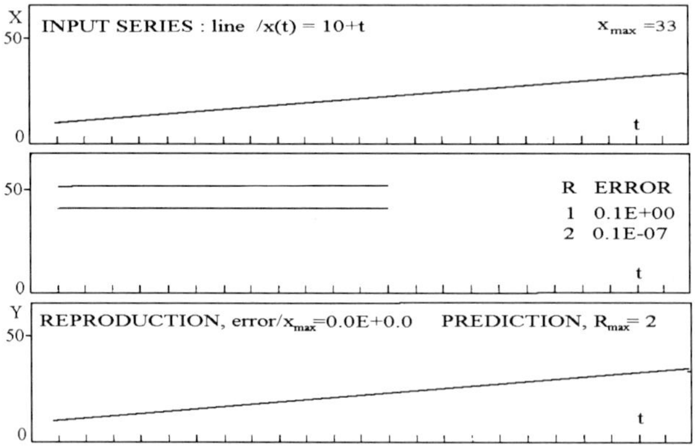

Fig. 11.24. Cascade prediction of a regular time-series generated by $x(t) = 10 + t$. For the modeling, only the initial portion extending over the interval indicated in the second graph was used. Records in the second picture show the difference between the reproduced and the actual time-series for various numbers R of cascade predictor layers. The corresponding average prediction error is written on the right. Successive records are vertically offset in equal decrements

Fig. 11.25 shows the results obtained by a chaotic time-series generated by the logistic map $x(t+1) = 3.9\, x(t)\, [1 - x(t)]$. Also listed is the number of layers in the network that are used to reproduce the given time-series and the corresponding accuracy. The map is one-dimensional and therefore the first layer models it quite accurately. The convergence of the prediction error to zero as well as the good quantitative agreement between the reproduced and input time-series is evident. However, a long-term prediction leads to a

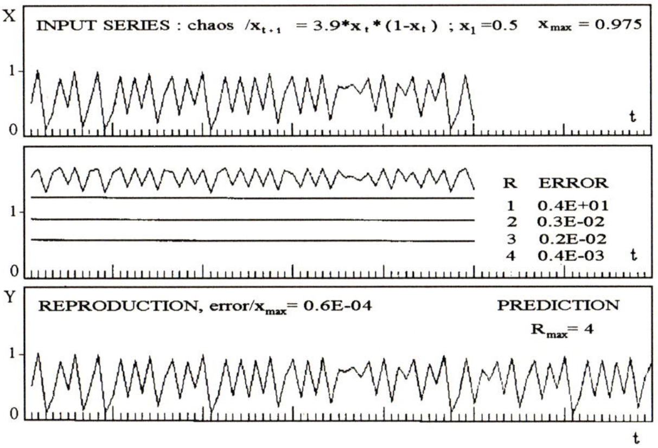

Fig. 11.25. Cascade prediction of a chaotic time-series generated by the logistic map. Records in the middle picture show the difference between the reproduced and the actual time-series for various numbers R of cascade predictor layers. The corresponding average prediction error is written on the right. Successive records in the middle picture are vertically offset in equal decrements

divergence between the given and the predicted time-series at increasing time because of the inherent instability of a chaotic process.

The method proposed in this section is applicable for the prediction of stationary time-series. In this case, one can expect that the model built on the past time-series is not essentially influenced by shifting in time. For the sake of simplicity, when forecasting the signals presented in Figs. 11.24 and 11.25 we have used only the formation of the prototypes without any self-organization. This is possible because of the small number of the presented data points in these examples. However, if an increasing number of sample values is in the presented time-series, a saturation of the computer memory results which must be avoided by limiting the number of prototypes during application of the self-organization procedure. The number of prototypes or formal neurons included in the modeling can be limited by employing the measurement error δx of the variable x. The number of prototypes is then permitted to increase only until their mutual separation in the corresponding phase space is greater than the error δx.

The same multi-layer cascade predictor as used in the above numerical experiments has recently been successfully applied to forecast an economic time-series and energy consumption. In these applications it was shown that the network can be generalized without difficulties to multi-variate time-series. In that case, one needs to convert the input variable x into a corresponding vector. [24] However, in this case, the prototypes are converted into matrices which subsequently require numerical procedures that are computationally more time-consuming.

11.5.2 Concluding Remarks

One may ask why the non-parametric regression is needed at all for the forecasting when there are already so many other available methods. [24, 29, 34] There is a simple answer. Many of the existing methods are based on linear modeling while the non-parametric regression is quite generally applicable to the modeling of non-linear phenomena. A typical class of phenomena for which the non-linearity is essential is represented by examples of deterministic chaos. In these cases, the linear modeling is, in principle, not applicable and one must use a general non-linear modeling. The only comparable approach would be a parametric regression. But in this case, one must provide *a priori* information about the phenomenon modeled in order to select a proper family of nonlinear basis functions which are then subsequently used in the parametric nonlinear regression. But, as mentioned previously, this requirement reduces the generality of the modeler or the predictor.

12. Modeling by Neural Networks

12.1 From Biological to Artificial Neural Networks

Up to now we have considered a number of concepts needed for the development of devices capable of automatic modeling of natural phenomena from the physical description of nature. This approach has already led us to adaptive network systems which consist of a number of relatively simple information processing units. We have called these *formal neurons* because their adaptive properties are similar to those of biological neurons. The aim of this chapter is to explain this similarity in more detail by showing how the adaptation of biological neurons and their networks can be described by dynamical models. Because the corresponding field of research is very broad, we present only those fundamental properties which are of importance for the modeling of natural laws. For the other topics, the reader is advised to consult the literature cited in the bibliography of this chapter.

Some basic functional properties of neurons and neural networks have already been reviewed in Chap. 2 but the most important property, that is related to the adaptive formation of the memory, was left so that it could be better explained here. To begin let us briefly summarize the most essential features related to the characteristics of neurons.

A neuron is an electrically active cell, that receives signals from the other elements of the sensory-neural network over the synaptic joints. The conductivities of the synaptic joints of the i-th neuron are described by the weight vector $\boldsymbol{m}_i = (m_{i1}, m_{i2}, \ldots, m_{ij}, \ldots)$ while the set of synaptic joints in a block of N completely interconnected equal neurons is represented by the weight matrix

$$\boldsymbol{M} = [m_{ij}] . \tag{12.1}$$

The received signals determine the activation state of the i-th neuron which is described by

$$a_i = \sum_{j=1}^{N} m_{ij}\, x_j . \tag{12.2}$$

In the simplest, quasi-static description of a neuron, we express its response to the excitations by some sigmoidal function

$$y_i = \Psi(a_i) = \Psi\left(\sum_{j=0}^{N} m_{ij}\, x_j\right) . \tag{12.3}$$

Here the term with $j = 0$ describes the excitation threshold of a neuron and Ψ denotes its response function that can be properly described using a piecewise linear, logistic or tanh function. Various examples of such a function were given in Chap. 2.

The conductivity of the synapses can be modified by the signals transmitted across them which corresponds to the formation of the memory in the brain. A simple description of the corresponding adaptation can be based on Hebb's hypothesis which maintains *that the efficacy of signal transmission of a neuron is enhanced if the pre- and post-synaptic parts are simultaneously excited.* [1, 13] Let the neuron labeled by index the j transmit the signal x_j to the i-th neuron whose output is denoted by y_i. According to Hebb's hypothesis, the rate of change of the synaptic joint between both neurons caused by the transmission of the signal can be described by the expression

$$\frac{dm_{ij}}{dt} = \frac{1}{\tau}\, y_i\, x_j . \tag{12.4}$$

In this representation of Hebb's learning rule, the coefficient τ denotes the learning or adaptation time.

In the above description the state of a neuron is assumed to be stationary during the adaptation of the synaptic joint. However, a biological neuron is, in fact, not a stationary element, but rather, a living cell possessing an intricate dynamical behavior which is still not completely understood. [26] A quasi-stationary description of a neuron by a smooth nonlinear adaptive response function is, therefore, only a first approximation. [16, 17] In the next approximation, a neuron is treated as a dynamical nonlinear integrating element whose state is governed by the in- and out-flow of chemical substances. A complete dynamical treatment of all flows is difficult to carry out because of its complexity. [11] Therefore, in the modeling, only some of the net effects of the in- and out-flow are usually included. A neuron is then treated as a leaking cell charged by the net in-flow. In the resulting dynamical model, a neuron is represented by an equation describing the rate of change of its activity [16]

$$\frac{dy_i}{dt} = I - L . \tag{12.5}$$

The term I represents the net in-flow which can be approximately described by the flows supplied by the synapses

$$I = \sum_{j=1}^{N} m_{ij}\, x_j . \tag{12.6}$$

Accordingly, the loss term L represents the leaking of the substances through the membrane of the soma and their conversion in the cell. It is assumed to be a nonlinear function of the neural activity y_i

$$L = L(y_i) . \tag{12.7}$$

In a quasi-stationary state the rate of change of neural activity is approximately vanishing and we obtain the nonlinear equation

$$0 \approx I - L(y_i) \tag{12.8}$$

whose solution is

$$y_i = L^{-1}(I) . \tag{12.9}$$

Comparing this equation with Eq. (12.3) which describes the activity of a stationary neuron we conclude that the inverse of the leaking function L^{-1} coincides with the sigmoidal response function. By taking into account the saturating property of the response function we further conclude that the loss term must be an increasing function of the cell activity.

As with the cell activity, the simple Hebb's rule can similarly be viewed as being only a first approximation. The argument for this view is that the Hebb's rule includes only the product of the activities of various cells. Since, by definition, these activities are positive, this means that the synaptic conductances are continually increasing. To obtain a stabilization of the memory, some opposing effect, or *forgetting* must be included into the model. [16, 17] A simple assumption is that the forgetting is proportional to the conductance of the synapse. This can be expressed by the equation

$$\frac{dm_{ij}}{dt} = \frac{1}{\tau} y_i x_j - \frac{1}{\tau_f} m_{ij} . \tag{12.10}$$

Here τ_f describes the forgetting time. It can be expected that the coefficient of forgetting rate $\phi = 1/\tau_f$ is some nonlinear increasing function of the cell activity y. In a quiet state, the activity is zero and in a first approximation, forgetting should also be zero. The forgetting rate coefficient can then be approximated by a linear term

$$\phi = \beta\, y_i , \tag{12.11}$$

with β being a constant. The following adaptation rule

$$\frac{dm_{ij}}{dt} = \frac{1}{\tau} y_i x_j - \beta\, y_i\, m_{ij} \tag{12.12}$$

then corresponds to a generalization of Hebb's hypothesis.

In relation to the last equation, Kohonen formulated the following proposition: [16]

Let $\boldsymbol{x} = (x_1, x_2, \ldots, x_N)^T$ and $\boldsymbol{m}_i = (m_{i1}, m_{i2}, \ldots, m_{iN})^T$ denote the vector of inputs and the synaptic weights of i-th neuron, respectively, where the superscript ^{T} denotes the transpose. If input neural activities x_i are stochastic variables with stationary statistical properties, then the elements m_{ij}, determined by the equation (12.12), converge to asymptotic values that represent the eigenvector of the correlation matrix $\boldsymbol{R}_{xx}$ corresponding to the largest eigenvalue.

According to this proposition, the neuron becomes most sensitive to a particular input vector that represents a certain statistical property or *feature* of all possible input vectors. A further modification of the above learning rule is obtained when making the assumption that learning only occurs in that subset of neurons in the network which are most closely connected to the neuron which is most excited. This leads to the well-known self-organization rule of Kohonen, that is similar to the self-organization algorithm stemming from the absolute maximum entropy principle described in the chapter on self-organization: [16, 17, 8]

$$\frac{d\boldsymbol{m}_i}{dt} = \frac{1}{\tau}(\boldsymbol{x} - \boldsymbol{m}_i)\,\eta_i\ . \tag{12.13}$$

Here the function η_i describes the adaptability of the i-th neuron. Its value is maximal for the neuron that is most sensitive to the input signal and is consequently also most excited by it. Based on biological observations, Kohonen assumed that the adaptation rate decreases with the increasing distance from the most excited neuron, which means that the adaptation essentially occurs only in its vicinity. If the spatial ordering of the neurons in the network resembles the topological ordering of the weight vectors $\boldsymbol{m}_i$ in their sample space then the influence of the excitation on the adaptation of neurons can be described by the function η having the Gaussian form:

$$\eta_i = \exp\left[-\frac{\|\, \boldsymbol{x} - \boldsymbol{m}_i \,\|^2}{2\sigma^2}\right]\ . \tag{12.14}$$

Here the parameter σ describes the typical range of vicinity in which the adaptation occurs. With such an excitation function the adaptation equation (12.13) corresponds to the first approximation of the adaptation rule derived from the absolute maximum entropy principle. However, the sample space of signal vectors generally possesses a dimension that differs from that of the physical space in which the neural network lives. Therefore it is still not completely clear under what conditions Kohonen's approach yields equivalent results as those obtained using the maximum entropy principle.

12.1.1 Basic Blocks of Neural Networks and Their Dynamics

The adaptive and nonlinear character of a neuron leads to difficulties when attempting to understand or to develop an analytical description of the dynamical properties of neural networks. Some properties of the corresponding dynamics can therefore be explained only with the help of numerical experiments which are possible by numerically simulating the operation of a neural network. Nevertheless, the research in this field, which has greatly intensified in the last decade, clears up a number of mysteries and it appears likely that a common physical description of at least some portions of the brain might soon be possible. [16, 17, 10, and references therein] For this effort, the concept of basic network structures will be useful. Using this, a complex neural network with all its peculiarities and features could be represented by a small number of interconnected basic building blocks or operational modules possessing relatively simple structures. A neural building block is then represented by a set of interconnected identical neurons. The block obtains input signals from other blocks and it also supplies signals to other blocks over a set of axons that is usually much less numerous than the number of interconnections inside a particular block. Such a block can be schematically represented by a structure depicted in a simplified drawing that is shown in Fig. 12.1. [16, 17]

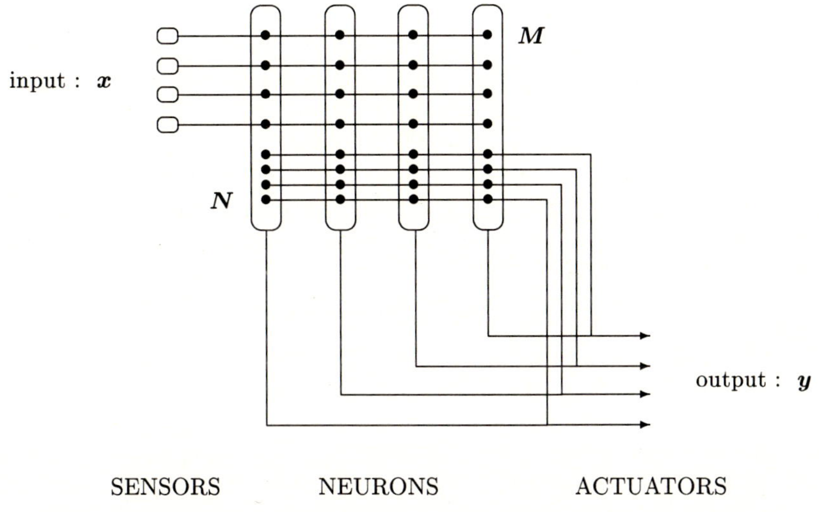

Fig. 12.1. The scheme of a basic building block of neural networks

The block is comprised of K neurons. The input signal to a block is given by the vector $\boldsymbol{x}$ while the output signal is represented by the vector $\boldsymbol{y}$. Both vectors are time-dependent but need not be of the same dimension. The input components are connected to the neurons of the block by a matrix of synaptic joints $\boldsymbol{M} = [m_{ij}]$. Quite often the input components are treated as sensory signals although this certainly is not always the case since various blocks can cooperate as well. The interconnections between the neurons are described by the matrix $\boldsymbol{N} = [n_{ij}]$. In a sense, this matrix represents the feedback in the network. Neuro-anatomical research indicates that the distribution of interconnections generally depends on the distance between neurons and that all neurons are, in fact, not connected to each other. This can be incorporated by setting several elements of the matrix to zero. In this way layered structures, such as a multi-layer perceptron, can be described.

The transformation of the input signal into the output signal can generally be described by some function. It can be treated as a representation of some physical law that relates input and output physical variables. Consequently, a neural network can be further treated as the site of an adaptive modeler of a natural law. It is this property which connects the neural network most directly to our basic task, that is, to the description of systems applicable in autonomous research work.

The fundamental purpose of the block is to transform input signals into output signals and thus to provide for a proper control of the animal organism. For this purpose the matrices $\boldsymbol{M}$ and $\boldsymbol{N}$ must be properly adapted to a specific task. To explain how this can be realized, we must proceed to the theory of adaptation of dynamical systems.

The dynamics of a block can be described by a fundamental system of equations that describe the rate of change of activities of neurons and synaptic weights:

$$\frac{d\boldsymbol{y}}{dt} = \boldsymbol{\Phi}(\boldsymbol{x},\ \boldsymbol{y},\ \boldsymbol{M},\ \boldsymbol{N})\ , \tag{12.15}$$

$$\frac{d\boldsymbol{M}}{dt} = \boldsymbol{\Omega}(\boldsymbol{x},\ \boldsymbol{y},\ \boldsymbol{M})\ , \tag{12.16}$$

$$\frac{d\boldsymbol{N}}{dt} = \boldsymbol{\Theta}(\boldsymbol{y},\ \boldsymbol{N})\ . \tag{12.17}$$

Here $\boldsymbol{\Phi}$, $\boldsymbol{\Omega}$, and $\boldsymbol{\Theta}$ denote vector or matrix functions respectively. The first equation of this system describes the activity of neurons. In a neural network of biological neurons, the activity of neurons can react very quickly, in orders of milliseconds, to an external stimulus. The corresponding relaxation time hidden in the function $\boldsymbol{\Phi}$ must therefore be correspondingly short. In contrast, the learning process generally proceeds much more slowly, and the corresponding relaxation times which are hidden in the functions $\boldsymbol{\Omega}$ and $\boldsymbol{\Theta}$ must be orders of magnitude longer than the relaxation time associated with the neural activity. The first equation can thus be treated as a relaxation equation while the other two equations are adaptation equations. Although the fundamental system of equations appears rather simple, it offers a diversity of solutions because the generating functions $\boldsymbol{\Phi}$, $\boldsymbol{\Omega}$ and $\boldsymbol{\Theta}$ can be rather arbitrarily selected. This variety is still magnified if a multi-layer structure of the network is considered.

According to the characteristic times, we can generally differentiate between two types of problems related to the dynamics of neural networks. In the first type, we consider phenomena on a short time scale over which the matrices $\boldsymbol{M}$ and $\boldsymbol{N}$ can be treated as constants. This generally corresponds to a study of the operational properties of a neural network of fixed parameters and architecture. The operation of the system is described by the first equation of the fundamental system (12.15). The principal difficulty related to its understanding arises from the nonlinear response of neurons. In the second type of problem, we consider phenomena which occur on much longer time scales, that is over which the neural response can be treated as being instantaneous. The focus in this case is on an explanation of the phenomena related to the adaptation of the network parameters or *learning.* The principal difficulties in research on this topic are related to the fact that the site of the memory is a set of adaptable synapses which are distributed in the complete network and can vary in time according to the signals transmitted over them. There is, however, still a third type of problem that is related to the evolution of the network architecture. It should proceed over many generations of similar networks and consequently should have a time scale that is several orders of magnitude longer than those above.

Various specific models of neural networks that appear in the literature of this field can nearly all be derived from those described above by selecting certain combinations of the functions $\boldsymbol{\Phi}$, $\boldsymbol{\Omega}$ and $\boldsymbol{\Theta}$. We shall not proceed further into an analysis of various examples, let us simply mention that there have been published many examples reporting the successful application of neural networks in supervised and unsupervised learning, pattern and speech recognition or synthesis, simulation, prediction, and optimal control of various processes, among others. [16, 10, 24, 23, 12, 14, 5, 21]

When treating an adaptation problem as described above, we usually assume that the characteristic time associated with the evolution of the input signal is much longer than the response time of the neurons. The neurons thus have sufficient time to accommodate their activities to the input and therefore the problem can be treated as being quasi-stationary for which $dy/dt \simeq 0$. In this case, we obtain the implicit nonlinear equation

$$\boldsymbol{\Phi}(\boldsymbol{x}, \boldsymbol{y}, \boldsymbol{M}, \boldsymbol{N}) \simeq 0 \tag{12.18}$$

for the dependence of neural activity on the input stimulus. The corresponding explicit equation can formally be expressed as

$$\boldsymbol{y} = \boldsymbol{y}(\boldsymbol{x}, \boldsymbol{M}, \boldsymbol{N}) \,. \tag{12.19}$$

Relative to the last equation one may ask a general question about the functionality of the neural network in a biological system. If the input components to the network are the signals provided by the biological sensors and the neural activities govern the excitation signals for biological actuators, then we can speculate that nature has developed such systems which are capable of transforming, in some optimal sense, the influences from the surroundings into the reactions of the organism. Unfortunately, because of the complexity of the environments existing in nature, we generally do not know criteria that could be used to define the corresponding functionals by which the optimal behavior could be specified.

The last equation can be interpreted as being a natural law which connects two types of variables: The input $\boldsymbol{x}$ and the output $\boldsymbol{y}$. In view of this interpretation, a neural network can be viewed as an adaptive modeler of empirical natural laws. It was shown in Chap. 2 that a three-layer perceptron is capable of performing a piecewise linear mappings of a one-dimensional input x to the one-dimensional output variable y which corresponds to a given set of sample data. The fundamental theoretical problem is then to examine the generality of such a modeling if the response characteristics of the neurons can be specified by certain kinds of sigmoidal functions and a learning rule. One can proceed in this direction by using Kolmogorov's theorem which states that all continuous functions of n variables have an exact representation in terms of finite superpositions and compositions of a finite number of functions of just one variable. [19, 20] Using this theorem, Sprecher has shown that any continuous function $\boldsymbol{y} = \varphi(\boldsymbol{x})$, that maps from the n-dimensional cube $[0,1]^n$ to the m-dimensional real space $\boldsymbol{R}^m$, can be implemented exactly by using a three-layer neural network structure. [25, 12, 14] However, no description of the neural response function is given in this existence theorem. A step in this direction has been taken by Cybenko who assumed that a single sigmoidal neural response function Ψ is sufficient. [4] He stated that any continuous function $y_i(\boldsymbol{x})$ with support in the unit hypercube $[0,1]^n$ of $\boldsymbol{R}^n$ can be uniformly approximated by linear combinations of the form

$$y_i(\boldsymbol{x}) = \sum_{j=1}^{N} a_{ij}\, \Psi(\boldsymbol{m}_j^T \boldsymbol{x} + \theta_j) \; . \tag{12.20}$$

This transformation of the input vector $\boldsymbol{x}$ into the output component y_i can be implemented by a layer of N neurons all having an identical response function Ψ, synaptic weights $\boldsymbol{m}_j$, thresholds θ_j, and one linear neuron with synaptic weights $\boldsymbol{a}_i = (a_{i1}, a_{i2}, \ldots, a_{iN})$. A transformation of a vector $\boldsymbol{x}$ from a unit cube $[0,1]^n$ to the vector $\boldsymbol{y} = (y_1, \ldots, y_M)$ can thus be performed by a two-layer neural network having in its first layer N neurons with sigmoidal responses and synaptic weights $\boldsymbol{m}_j$, and in the second layer M linear neurons whose synaptic weights are $\mathbf{a}_i$. If the output vector is constrained to be represented in the unit cube $[0,1]^M$ then also the last layer will consist of neurons possessing a sigmoidal response.

Quite often, the distribution of an input vector to various neurons in the first layer is represented by a separate layer of fanout neurons. [12, 14] For the existence of the mapping, the specific properties of the response function Ψ is unimportant, it only needs to satisfy the properties of the sigmoidal functions commonly used in the description of neural activation. That is, it must be continuous, monotonic and limited from above and below. The natural laws that can be expressed by a mapping of one vector variable into another one can therefore be quite generally modeled by a system possessing the structure of a three-layer neural network. The main problem to be answered is then how such a network can be adapted to various specific examples by using empirical data alone and how the number of neurons in the network should be properly selected. The first problem can be solved by the gradient descent method while a general method for the solution of the second problem appears to be still not known. In the next sections of this chapter we review the characteristic features of a linear associator which will be helpful in formulating a back-propagation learning rule that is applicable for training neural networks capable of performing nonlinear mappings and in explaining the performance of a perceptron.

12.2 A Linear Associator

Some of the characteristic properties of intelligent adaptive systems can already be observed on very primitive networks which are described by the simplest generating functions. As a practical example of a primitive network, we present the linear auto-associator whose operation is described by the equations

$$\boldsymbol{y} = \boldsymbol{x} - \boldsymbol{M}\boldsymbol{x} \; , \tag{12.21}$$

$$\frac{d\boldsymbol{M}}{dt} = c\,\boldsymbol{y} \otimes \boldsymbol{x} = c\,\boldsymbol{y}\,\boldsymbol{x}^T \; . \tag{12.22}$$

As before, the input and output vectors are expressed as $\boldsymbol{x}$ and $\boldsymbol{y}$ respectively and $\boldsymbol{M}$ is the memory matrix of the network. The symbol $\otimes$ denotes the outer product operation which transforms two vectors into a matrix, $\boldsymbol{x}^T$ denotes the transposed input vector, and c is a constant controlling the rate of adaptation. A comparison of the last two equations with the equations of the fundamental system of the basic building block shows that the operation of a linear associator is based on the simplest relaxation equation for neural activity. This corresponds to a network consisting of neurons whose response is linear and instantaneous, that is, operating with no time delay. The components of the input vector are simultaneously utilized as threshold parameters. The time-dependence is utilized only in the adaptation equation for developing the memory matrix $\boldsymbol{M}$ that corresponds to Hebb's rule. The structure of the linear auto-associator is shown in Fig. 12.2.

The associator can be interpreted as a network comprised of memory and comparison elements. In an application, the associator must be coupled to an array of sensors that provide input signals x_j. Therefore the matrix $\boldsymbol{M}$ can be interpreted as the set of synaptic joints between the sensory signal channels and the neurons. The input signals are also input to the memory element where they are multiplied by the synaptic weights m_{ij} to yield the output components from the memory $v_i = \sum_j m_{ij}\, x_j$. This is an additional

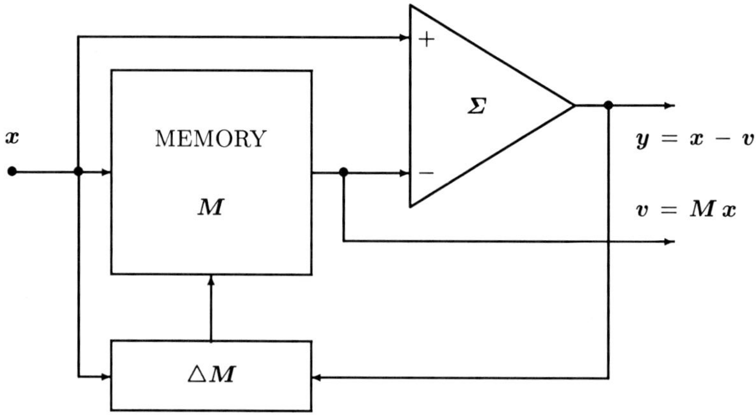

Fig. 12.2. The scheme of a linear auto-associator

output of the network and it is complementary to the output $\boldsymbol{y}$. The adaptive linear element that transforms one input vector into a component of another one in an associator, is usually referred to as an *adaline.* [12, 14] The output vector from the set of adalines $\boldsymbol{v}$, where $\boldsymbol{v} = \boldsymbol{M}\boldsymbol{x}$, is input to the comparison element where the difference $\boldsymbol{y} = \boldsymbol{x} - \boldsymbol{v}$ is determined. This difference is input to the memory where the synaptic joints are adaptively formed according to the adaptation equation given by 12.22 that is usually called the *delta rule.* In a practical simulation on a computer, various samples of input vectors from a certain generator are supplied sequentially to the *associator.* Time is represented by a discrete variable and each presented input vector results in an adaptation of the memory according to

$$\triangle \boldsymbol{M} = \alpha \boldsymbol{y} \otimes \boldsymbol{x} \ , \qquad (12.23)$$

where $\alpha = c\,\Delta t$ is a new learning constant corresponding to the discrete time case. The learning commences with the empty memory, that is, $\boldsymbol{M} = 0$ and it develops sequentially until the vector $\boldsymbol{y}$ vanishes. The initial output from the memory is $\boldsymbol{v} = 0$ therefore $\boldsymbol{y} = \boldsymbol{x}$. If the same input vector is repeatedly fed to the associator, the memory changes until $\boldsymbol{v} = \boldsymbol{x}$. The network thus learns to reproduce the input vector which appears at the output from the memory. If subsequently a new vector is input to the associator, the new output vector $\boldsymbol{y}'$ represents only that part of the input that has not been learned previously and is usually called the *novelty.* [17] In other words, the network is learning to reproduce the input vectors and to diminish the novelties presented to it. When learning is successful, the output vector from the memory $\boldsymbol{v}$ becomes increasingly similar to the input vector and it thus corresponds to an internal representation of the input signals.

It is interesting that the network is capable of simultaneously learning a number of input signals. It is intuitively clear that the number of linearly independent samples is equal to the number of lines in the matrix $\boldsymbol{M}$ which corresponds to the number of neurons. A more detailed analysis, following the steps presented in Chap. 10 on invariant modeling, shows that the learning of the linear auto-associator corresponds to the establishment of a linear regression that acts as a projector to the principal axis of the distribution of input samples. The elements of the memory matrix are therefore determined by the correlations between the components comprising the input vectors. Hence, the network stores the properties of statistical relations between the input components. Because of this, the adapted network is capable of operating associatively. This property is exhibited when a trained network obtains an input sample $\boldsymbol{x}$ which possesses a structure similar to one of the memorized samples. In such a case, the output from the memory $\boldsymbol{v}$ is principally determined by the most similar memorized sample. If, for instance, the input vector includes some components which are equal to zero, or if they are masked by noise with respect to some learned samples, then the vector recalled from the internal representation contains components which approximately equal those of the memorized samples. Such a completion is generally called an auto-associative recall. In the applications presented in Chaps. 9 and 10 dealing with non-parametric regression and invariant modeling, this property is of practical importance because it makes it permits solving problems on the basis of known examples with no analytical modeling. An adapted memory matrix thus represents a model of a natural phenomenon that is represented by the set of training samples. In the auto-associator described here, the network learns to approximate the input samples by recalling the corresponding vectors from the internal representation of the model in the memory. Such learning is called *unsupervised.*

It is interesting that a network whose performance is similar to the linear auto-associator can be described by the following equations [16]

$$\boldsymbol{y} = \boldsymbol{x} - \boldsymbol{N}\boldsymbol{y} , \tag{12.24}$$

$$\frac{d\boldsymbol{N}}{dt} = c\boldsymbol{y} \otimes \boldsymbol{y} . \tag{12.25}$$

Here the matrix $\boldsymbol{N}$ can be interpreted as the set of synaptic joints that represent the feedback in the neural network. In this case, the problem of an unbounded increase of memory weights may arise. But this can be solved by including a term corresponding to a "forgetting" of information into the learning equation. [16]

The linear auto-associator can easily be modified into a linear cross-associator by utilizing two input vectors $\boldsymbol{x}$ and $\boldsymbol{z}$ as shown in Fig. 12.3. In this case, one input is connected to the memory and the other one to the comparator. The network is trained to generate from the first input $\boldsymbol{x}$ another input $\boldsymbol{z}$ as the output $\boldsymbol{v}$ from the memory. In contrast to the auto-associator, such learning is called *supervised* because there exists a desired or target variable $\boldsymbol{z}$ into which the input variable $\boldsymbol{x}$ should be transformed

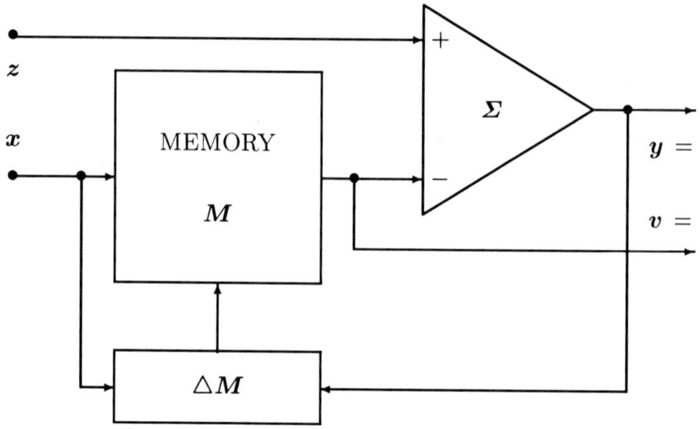

Fig. 12.3. The scheme of a linear cross-associator

or mapped. The cross-associator models a linear relationship between the variables $\boldsymbol{x}$ and $\boldsymbol{z}$.

The linear cross-associator is the simplest example of a mapping neural network which can be used to model natural phenomena and it is instructive to examine its performance in greater detail. To make the cross-associator resemble a neural network, we assume that the output from the memory is determined by the response function of neurons Ψ given by

$$v_i = \Psi\left(\sum_{j=1}^{N} m_{ij}\, x_j\right) . \tag{12.26}$$

If the response function Ψ is described by the unit-step function, the cross-associator corresponds to a binary perceptron. However, for a linear associator, the response function is simply taken to be the identity $\Psi(\xi) \equiv \xi$. Our task is to show how an optimal performance of the cross-associator is achieved under these conditions. For this, we introduce the measure of discrepancy between the desired $\boldsymbol{z}$ and the true output $\boldsymbol{v}$ of the network

$$\mathcal{E} = \frac{1}{2}\,||\boldsymbol{z}-\boldsymbol{v}||^2 = \frac{1}{2}\,||\boldsymbol{y}||^2 \tag{12.27}$$

and we will try to minimize its statistical average by modifying the memory matrix $\boldsymbol{M}$. For this purpose, we apply the gradient descent method. According to Eq. 12.27, the error surface is determined in the space of the weights m_{ij}. When a sample pair $(\boldsymbol{x},\boldsymbol{z})$ is presented to the input, the error will generally not vanish. We try to improve the performance of the associator by changing the memory parameters in such a way that the point representing the memory moves in the direction of the negative gradient on the error surface with respect to the parameters m_{ij}, that is,

$$\Delta m_{ij} = -c \frac{\partial \mathcal{E}}{\partial m_{ij}} . \tag{12.28}$$

The error indirectly depends on the synaptic weights representing the memory. A mediator is the neural response function. We therefore express the gradient by the chain rule

$$\frac{\partial \mathcal{E}}{\partial m_{ij}} = \frac{\partial \mathcal{E}}{\partial v_i} \frac{\partial v_i}{\partial m_{ij}} . \tag{12.29}$$

From the expression for the error function we obtain,

$$\frac{\partial \mathcal{E}}{\partial v_i} = -(\boldsymbol{z} - \boldsymbol{v})_i = -y_i \tag{12.30}$$

which, when substituted into 12.28 yields

$$\Delta m_{ij} = c\, y_i \frac{\partial v_i}{\partial m_{ij}} . \tag{12.31}$$

When the inputs are random variables, we use the mean square error measure $\mathrm{E}[\mathcal{E}]$ for the description of the performance. The last equation is then transformed into

$$\Delta m_{ij} = c\, E \left[y_i \frac{\partial v_i}{\partial m_{ij}} \right] . \tag{12.32}$$

This is the fundamental equation by which the optimal perceptron learning with respect to the mean square error measure is described. It reduces to the simple form for the case of linear associator for which $\Psi(\xi) \equiv \xi$. Further, the expression

$$v_i = \sum_{j=1}^{N} m_{ij}\, x_j \qquad \text{yields} \qquad \frac{\partial v_i}{\partial m_{ij}} = x_j \tag{12.33}$$

and so,

$$\Delta m_{ij} = c\, E[y_i x_j] , \tag{12.34}$$

or in vector notation

$$\Delta \boldsymbol{m} = c\, E[\boldsymbol{y}\boldsymbol{x}^T] = c\{E[\boldsymbol{z}\boldsymbol{x}^T] - \boldsymbol{M}\, E[\boldsymbol{x}\boldsymbol{x}^T]\} . \tag{12.35}$$

This last equation is the well-known Widrow-Hoff delta learning rule for the training of adalines. [29] During adaptation, the adaline sequentially obtains at its input, various joined samples $(\boldsymbol{x}, \boldsymbol{z})_n$ from a certain statistical set. The change of the memory parameters can then be carried out on each presented sample. This corresponds to the so-called momentum learning version. In contrast, in batch learning, the memory is updated only after many presented samples, which are averaged. In this case the resulting correlation matrices $\boldsymbol{R}_{zx} = E[\boldsymbol{z}\boldsymbol{x}^T]$ and $\boldsymbol{R}_{xx} = \mathrm{E}[\boldsymbol{x}\boldsymbol{x}^T]$ represent the variables used in training. The learning process ceases when the memory matrix is equal to

$$M = R_{zx} R_{xx}^{-1} . \tag{12.36}$$

where R_{xx}^{-1} is the inverse auto-correlation matrix. This is also the result to which the sequential learning converges, on average, after a large number of learning samples have been presented to the system. Therefore the linear cross-associator is said to represent a *correlation memory.*

12.3 Multi-layer Perceptrons and Back-Propagation Learning

In the description of a linear associator, there is missing an essential property of neural networks that is related to the nonlinear response characteristic of a neuron. By adding a nonlinear threshold function to the adalines of an associator, as shown in the above derivations of cross-associator, one obtains a network of formal neurons representing a one-layer perceptron. This network has played a crucial role in the research on artificial neural networks. [2, 22, 23] Here we shall not provide a complete description of the properties of the one-layer perceptron because it has been shown that such a network is of limited applicability for modeling natural laws. Rather we turn to the description of its generalized multi-layer form which is sometimes called a *mapping neural network.* [14]

With respect to the work of Cybenko it will suffice to review only a three-layer mapping network. [4] It is described by a function f which maps points from a domain subspace $\boldsymbol{A}$ of n-dimensional Euclidean space $\boldsymbol{R}^n$ to points in a bounded subspace of $\boldsymbol{R}^m$. The exact form of the function f is not known, instead a set of empirical examples of joined input and output data $\{(\boldsymbol{x}_1, \boldsymbol{y}_1), (\boldsymbol{x}_2, \boldsymbol{y}_2), \ldots, (\boldsymbol{x}_N, \boldsymbol{y}_N)\}$ is given, where each sample pair satisfies the mapping relation $\boldsymbol{y}_n = f(\boldsymbol{x}_n)$. Following our treatment of empirical data, we assume that the samples $\boldsymbol{x}_n$ are randomly selected from the domain subspace $\boldsymbol{A}$ in accordance with some fixed probability density $\varphi(\boldsymbol{x})$. The task is finding a learning algorithm for a multi-layer neural network, capable of performing this mapping. The basic structure of a three-layer network is schematically shown in Fig. 12.4. The layers between the first or input layer and the last or output layer are generally called "hidden layers". The first layer is comprised of units that only distribute the input vector components without modification to the nonlinear neurons of the second layer and therefore do not possess the essential non-linear properties of neurons. The mapping is, in fact, performed by the final two layers. They are comprised of neurons that receive the input signals from all the neurons in the previous layer and they generate the output signals to all the neurons of the next layer. Because of the transmission of the signal to layers ahead of it, such a network is referred to as *feedforward.* However, for the purpose of learning, there must be some feedback present in the network to provide information

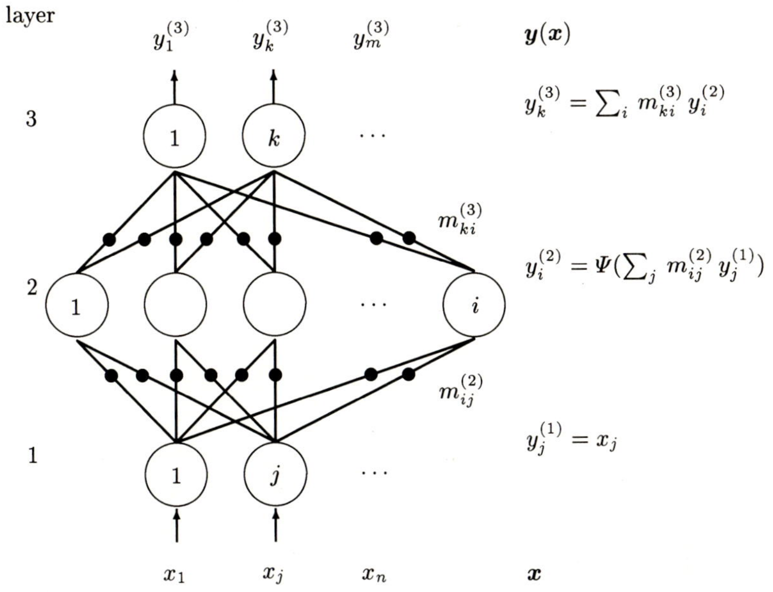

Fig. 12.4. The scheme of a three-layer neural network

needed for correcting the synaptic weights. The properties of the feedback will be discussed in relation to the following derivation of the learning rule.

Let the input to the network be given by the vector $\boldsymbol{x}$ and the output by the vector $\boldsymbol{y}$, while the target output vector is $\boldsymbol{z}$. The output of l-th layer is denoted by the vector $\boldsymbol{y}^l$. One may ask how the synaptic weights of the neurons in the layers should be modified to transform the input into that output which resembles that desired as closely as possible. In transmitting over the l-th layer, the signal is transformed according to the neural response equation:

$$y_i^l = \Psi\left(\sum_{j=0}^{N} m_{ij}^l \, x_j^l\right) . \tag{12.37}$$

Here the neuron response threshold value is represented by the synaptic weight with index 0 and a fixed input $x_0^l = 1$. Note that now $\boldsymbol{y}$ represents the output from the network which previously has been denoted by $\boldsymbol{v}$. Further in this formulation we consider that the input to the l-th layer with index greater than 1 is determined by the output from the previous layer, that is, $\boldsymbol{x}^l = \boldsymbol{y}^{l-1}$. The synaptic weights of the layer have been described by the matrix $\boldsymbol{m}^l$ and the states of neuron activation in the l-th layer are given by the vector $\boldsymbol{a}^l = \boldsymbol{m}^l \boldsymbol{y}^{l-1}$. With these definitions Eq. 12.37 which describes the transformation of a signal through the l-th layer of the network can be rewritten in the form

$$y_i^l = \Psi\left(\sum_{j=0}^{N} m_{ij}^l \, y_j^{l-1}\right) . \tag{12.38}$$

This can be re-written in an abbreviated vector notation as: $\boldsymbol{y}^l = \Psi\left(\boldsymbol{m}^l \, \boldsymbol{y}^{l-1}\right) = \Psi\left(\boldsymbol{a}^l\right)$ which resembles the input-output equation of the linear associator.

We further describe the error of the mapping by the squared norm of the difference between the desired vector and the true output vector from the last layer L:

$$\mathcal{E} = \frac{1}{2}||\boldsymbol{z} - \boldsymbol{y}^L||^2 . \tag{12.39}$$

We will determine the synaptic weights by utilizing the gradient descent method. The adaptive changes in the weights will be made according to the learning rule given by

$$\Delta m_{ij}^l = -c\frac{\partial \mathcal{E}}{\partial m_{ij}^l} . \tag{12.40}$$

To evaluate the right side of the equation, the gradient of the error with respect to a particular weight vector needs to be evaluated. As in the description of the one-layer perceptron, we utilize the chain rule of differentiation with respect to the neuron activation state. We obtain

$$\frac{\partial \mathcal{E}}{\partial m_{ij}^l} = \frac{\partial \mathcal{E}}{\partial a_i^l}\frac{\partial a_i^l}{\partial m_{ij}^l} = \frac{\partial \mathcal{E}}{\partial a_i^l} y_j^{l-1} . \tag{12.41}$$

We further introduce the abbreviated notation $\delta_i^l \equiv -\frac{\partial \mathcal{E}}{\partial a_i^l}$. The learning rule given by Eq. (12.40) can then be expressed as

$$\Delta m_{ij}^l = -c\frac{\partial \mathcal{E}}{\partial m_{ij}^l} = c\,\delta_i^l y_j^{l-1} . \tag{12.42}$$

We assume that the output from each layer $\boldsymbol{y}^l$ is known so that the gradient of the error surface, as described by the Eq.(12.41), can be determined if the values of δ_i^l are first determined. From the definition of δ_i^l we obtain

$$\delta_i^l = -\frac{\partial \mathcal{E}}{\partial a_i^l} = -\frac{\partial \mathcal{E}}{\partial y_i^l}\frac{\partial y_i^l}{\partial a_i^l} = -\frac{\partial \mathcal{E}}{\partial y_i^l}\Psi'\left(a_i^l\right) , \tag{12.43}$$

where Ψ' denotes the derivative of the function Ψ with respect to its argument. As the complete activation state of the network is also known, the remaining task is the evaluation of the derivative $\partial\mathcal{E}/\partial y_i^l$. For the final layer which is denoted by the index L it can be immediately expressed as

$$\frac{\partial \mathcal{E}}{\partial y_i^L} = -(z_i - y_i^L) \tag{12.44}$$

and we obtain

$$\delta_i^L = (z_i - y_i^L)\,\Psi'\left(a_i^L\right) . \tag{12.45}$$

For the hidden layers we take into account that y_i^l influences the activity in the layer $l+1$ and again apply the chain rule for the determination of derivative

$$\frac{\partial \mathcal{E}}{\partial y_i^l} = \sum_{k=0}^{N_{l+1}} \frac{\partial \mathcal{E}}{\partial a_k^{l+1}} \frac{\partial a_k^{l+1}}{\partial y_i^l} . \tag{12.46}$$

By using the previously introduced expressions $\delta_k^l = -\partial \mathcal{E}/\partial a_k^l$ and $\boldsymbol{a}^{l+1} = \boldsymbol{m}^{l+1}\boldsymbol{y}^l$ we obtain

$$\frac{\partial \mathcal{E}}{\partial y_i^l} = \sum_{k=0}^{N_{l+1}} \delta_k^{l+1} m_{ki}^{l+1} \tag{12.47}$$

which further yields for the hidden layers

$$\delta_i^l = \Psi'(a_i^l) \sum_{k=0}^{N_{l+1}} \delta_k^{l+1} m_{ki}^{l+1} . \tag{12.48}$$

Here the variables δ_i^l are determined for the last layer by Eq. (12.45) and for the hidden layers by Eq. (12.48). The last equation shows that the variables δ_i^l of the l-th hidden layer can be determined from the corresponding variables in the layers above it. By starting at the last layer we can successively determine the variables δ_i^l of the preceding layers. The calculation of δ_i^l thus proceeds in the direction opposite to the propagation of the signals in the network and consequently the corresponding algorithm is called the *back-propagation* learning rule. This learning procedure can be carried out for each joined sample that is presented to the system.

The back-propagation rule has been derived independently by several authors. [23, 27] It is fundamental for the application of artificial neural networks in various fields in which a non-linear mapping must be determined from presented data samples. Because of the evaluation of the variables δ_i^l, the training of the network using the back-propagation rule is relatively time-consuming. From application of the gradient descent method, one does not obtain any hints as to the selection of the learning constant, nor does one know if a global or only a local minimum of the mapping error has been achieved. Further, one does not know the number of neurons that should be used in a particular layer or the number of layers that will be optimal. It is also not known what kind of feedback biological neurons could use to learn via a back-propagation type algorithm. These problems have triggered extensive research in the past few years that has been widely published. [15, 28]

One problem related to the application of back-propagation learning rule that is seldom mentioned stems from the following consideration. When measuring various variables on a complex phenomenon, it is often not known *a priori* which data should be treated as being the fundamental ones and which are the derived ones. This means that we do not know in advance how to split the joined data into vectors $\boldsymbol{x}$ and $\boldsymbol{y}$. If we arbitrarily divide the data into the vectors $\boldsymbol{x}$ and $\boldsymbol{y}$ at the start of modeling, then subsequently we cannot utilize the trained network for the description of some other relationship between differently split data into the pair $\boldsymbol{x}'$ and $\boldsymbol{y}'$ having different numbers of components than $\boldsymbol{x}$ and $\boldsymbol{y}$ have. The question thus arises how one can adapt the network in such a way that the selection into given and estimated data can be arbitrarily selected after training. To solve this problem, the modeling should be based on the absolute maximum entropy principle while the estimation can be done by the non-parametric regression that has been described in Chap. 8. The back-propagation connectionist approach to modeling of relations between various variables stems from the analogy with neural networks, while the self-organized and non-parametric regression approach originates from a statistically motivated modeling of natural phenomena and in this aspect appears more fundamental and generally applicable. An additional reason for applying the latter approach stems from the simplicity of model formation when there are a small number of samples. In the self-organized approach, these samples are directly used as prototypes and the learning process is complete. In this case, the time needed for the formation of the model can be orders of magnitude shorter than that needed for the back propagation algorithm. In addition, the self-organized approach yields a more general model.

12.4 Radial Basis Function Neural Networks

To circumvent the above-mentioned problems related to application of multilayer perceptrons comprised of neurons with sigmoidal response functions and trained by a back-propagation algorithm, a neural network has been proposed whose mapping properties closely resemble the conditional average estimator with self-organized prototypes. [3, 7, 18] It is called a *radial basis function network* (RBFN) and consists of neurons with a localized receptive field which is described by a radially symmetric basis function $g(\boldsymbol{x})$ of the input vector $\boldsymbol{x}$. Most generally, the Gaussian function

$$g_k(\boldsymbol{x}) = \exp\left[-\frac{\| \boldsymbol{x} - \boldsymbol{q}_k \|^2}{2\sigma_k^2}\right] \tag{12.49}$$

is used in which the parameters $\boldsymbol{q}_k$ and σ_k, respectively, denote the center and the width of the receptive field of the k-th neuron. The network is generally represented by a layer of K neurons possessing different receptive fields . All the neurons are excited by the same input $\boldsymbol{x}$. The output from the network is denoted by the vector $\boldsymbol{y}$ whose components are a linear superposition of the neural responses determined by the radial basis functions:

$$y_i(\boldsymbol{x}) = \sum_{k=0}^{K} m_{ik}\, g_k(\boldsymbol{x}) . \tag{12.50}$$

This superposition is expressed in terms of the parameters m_{ik} which represent the synaptic weights of the neurons for $k \neq 0$ while m_{i0} represents a bias term with associated $g_0 = 1$. We can recognize in the last equation a generalized description of the conditional average estimator output in which the parameters still need to be determined. For this purpose it is assumed that there is a given training set of $N >> K$ joined samples $\{\boldsymbol{x}_n, \boldsymbol{y}_n; n = 1, \ldots, N\}$. The error of the mapping is represented by

$$\mathcal{E} = \sum_{n=1}^{N} ||\boldsymbol{y}_n - \boldsymbol{y}(\boldsymbol{x}_n)||^2 . \tag{12.51}$$

This error depends on the receptive fields of the neurons and the synaptic weights. The optimal performance of the network is obtained when the following conditions are satisfied:

$$\frac{\partial \mathcal{E}}{\partial q_{kj}} = 0 \quad ; \quad k = 1, \ldots, K; j = 1, \ldots, M_x , \tag{12.52}$$

$$\frac{\partial \mathcal{E}}{\partial \sigma_k} = 0 \quad ; \quad k = 1, \ldots, K , \tag{12.53}$$

$$\frac{\partial \mathcal{E}}{\partial m_{ik}} = 0 \quad ; \quad k = 1, \ldots, K; i = 1, \ldots, M_y . \tag{12.54}$$

Here M_x and M_y denote the dimensions of the input and output vectors, respectively. Note that in this approach, the widths of the receptive fields are generally not assumed to be fixed but are adaptable just like the prototype vectors which determine the centers of the receptive fields. The last three expressions represent a non-linear system of equations for the parameters of the receptive fields of the neurons and their synaptic weights. Its solution can be obtained by using one of the standard procedures such as the backpropagation algorithm.

The approach described above exhibits no advantage of the radial basis function over the multi-layer perceptron architecture. This only becomes evident if we consider also the localized response of the neurons. That is, for each input vector, essentially only one neuron is excited. Therefore, the training can be performed in two stages in which the neural receptive fields are first properly adjusted and after that, the synaptic weights m_{ik} are adapted. [3, 18] The first step resembles the self-organization procedure that is used for the adaptation of the prototype vectors, while the second step corresponds to a generalized formation of the conditional average estimator. As both steps have been described in detail in previous chapters, let us just mention that the structure of the conditional average estimator can be very simply interpreted and that it is very convenient for numerical treatment in many cases of practical interest.

12.5 Equivalence of a Radial Basis Function NN and Perceptrons

As has previously been mentioned in this chapter, the multi-layer perceptrons have played a central role in the research of neural networks. [12, 14] However, as mentioned at the end of Section 12.3, their nonlinear and adaptive characteristics lead to difficulties in the understanding and interpretation of learning and mapping. The most serious difficulty is that training by back-propagation of errors is time-consuming and does not provide for the simple inclusion of *a priori* information. Many of these problems do not appear in simulations of neural networks based on a self-organized representation of empirical probability distributions and a mapping by conditional average in which both can be simply interpreted statistically. *A priori* information can also be simply included by a proper initialization of prototypes. The separation of input signals into independent and dependent variables need not be done prior to training, as it must be for the multi-layer perceptron. Rather, it can be performed when applying a trained network. Because of these advantageous properties, a neural network based on the self-organization and conditional average appears to be more generally applicable than a multi-layer perceptron. However, for historical reasons and the corresponding availability of software, the multi-layer perceptron is currently the most frequently applied paradigm for the simulation of neural networks. It will therefore be advantageous to explore a transition between both paradigms and thus provide a common interpretation for both.

We described in Chap. 2 a procedure for piecewise linear modeling of a regular function $y(x)$ for the case when the sample points x_i are evenly spaced. It was shown that the corresponding mapping can be equivalently represented by sigmoidal or triangular basis functions which when implemented, leads to a perceptron or radial basis function neural network, respectively. We next generalize the analysis to the multi-variate and stochastic cases.

We assume that an observation of a phenomenon can be described by a vector composed of sensory signals $\boldsymbol{x} = (s_1, s_2, .., s_D)$. Its components are of random character and therefore the relations between them are described by the joint probability density function $f_r(x)$. We represent this function by a finite number K of prototype vectors $\{\boldsymbol{q}_1, \boldsymbol{q}_2, \ldots, \boldsymbol{q}_K\}$ as

$$f_r(\boldsymbol{x}) = \frac{1}{K} \sum_{k=1}^{K} w(\boldsymbol{x} - \boldsymbol{q}_k, \sigma) , \tag{12.55}$$

with $w(\boldsymbol{x})$ denoting some approximation of a delta function. In modeling of probability density function the prototypes are first initialized by K samples: $\{\boldsymbol{q}_k = \boldsymbol{x}_k \text{ ,for } k = 1, \ldots, K\}$, which represent *a priori* given information. These prototypes can be later modified by additional samples $\boldsymbol{x}_N$ according to the self-organized adaptation which is described in Chap. 8. We further assume that there is given some partial information, for instance the first

i components of the vector: $\boldsymbol{g} = (s_1, s_2, \ldots, s_i, \emptyset)$. The hidden data, which are to be estimated, are then represented by the complementary truncated vector $\boldsymbol{h} = (\emptyset, s_{i+1}, \ldots, s_D)$.[4] According to the derivations presented in Chap. 9 we describe the relation between both the truncated vectors by the conditional average which can be expressed in terms of the prototype vectors as:

$$\widehat{\boldsymbol{h}} = \sum_{k=1}^{K} C_k(\boldsymbol{g})\, \boldsymbol{h}_k\ , \quad \text{where} \quad C_k(\boldsymbol{g}) = \frac{w(\boldsymbol{g} - \boldsymbol{q}_k, \sigma)}{\sum_{j=1}^{K} w(\boldsymbol{g} - \boldsymbol{q}_j, \sigma)}\ . \tag{12.56}$$

The given vector $\boldsymbol{g}$ represents the condition while the basis functions $C_k(\boldsymbol{g})$ describe the similarity between the condition $\boldsymbol{g}$ and the prototypes. It is important that the splitting of the data into *given* and *hidden* components can be made after training of the network. The model, represented by the prototypes is thus far more fundamental than only a particular version of the conditional average which can be adapted to a specific application.

The conditional average corresponds to a mapping relation $\boldsymbol{g} \to \boldsymbol{h}$ which can be realized by a two-layer radial basis neural network. [3] The first layer consists of K neurons. The k-th neuron obtains the input signal $\boldsymbol{g}$ over synapses which are described by $\boldsymbol{g}_k$. The excitation of this neuron is described by the basis function $C_k(\boldsymbol{g})$. The excitation signal is then transferred to the neurons of the second layer. The i-th neuron of this layer has synaptic weights $h_{k,i}$ and operates as a summation cell generating the output signal $\widehat{h}_i(\boldsymbol{g})$.

To demonstrate how one can proceed to determine the relationship between the conditional average estimator and the multi-layer perceptron, we follow the corresponding presentation of Chap. 2, but in reverse order. We consider the simple, two-dimensional case of a function $y(x)$ which is empirically described by a set of prototypes $\{\boldsymbol{q}_k = (x_k, y_k)\ ;\ k = 1, \ldots, K\}$ with constant spacing $\Delta x = x_{j+1} - x_j$ for $j = 1, \ldots, K-1$. We further introduce a triangular and a piecewise linear sigmoidal basis function:

$$w_i(x) = \left\{1 - \frac{|x - x_i|}{\Delta x} \ldots \text{for } |x - x_i| < \Delta x\ ;\ 0 \ldots \text{elsewhere}\right\}\ , \tag{12.57}$$

$$\Psi_i(x) \equiv \Psi(c_i x - \theta_i) \equiv \begin{cases} 0 & \ldots\ x < x_i \\ (x - x_i)/\Delta x & \ldots\ x_i \le x \le x_{i+1} \\ 1 & \ldots\ x > x_{i+1} \end{cases}\ . \tag{12.58}$$

With these, we represent the function $y(x)$ by straight line segments connecting the sample points. The conditional average is transformed into the expression for a multi-layer perceptron by utilizing the relations:

$$w_{i+1}(x) = \Psi_i(x) - \Psi_{i+1}(x) \qquad \text{and} \qquad \Psi_i(x) = \frac{w_{i+1}(x)}{w_i(x) + w_{i+1}(x)}\ . \tag{12.59}$$

The result is:

$$\widehat{y}(x) = \frac{y_1 w_1(x) + \ldots + y_K w_K(x)}{w_1(x) + \ldots + w_K(x)} \tag{12.60}$$

$$= \frac{y_1 w_1(x)}{w_1(x) + w_2(x)} + y_2 w_2(x) + \ldots + y_{K-1} w_{K-1}(x) + \frac{y_K w_K(x)}{w_{K-1}(x) + w_K(x)} .$$

We retain in the denominator of the first and the last terms of this expression only those basis functions which differ from zero in the region in which the basis function in the numerator also differs from zero. The denominator in the terms of index 2 to $K-1$ equals 1 because of the overlapping of neighboring basis functions. Inserting the relations expressed by Eq. (12.59) into Eq. (12.60) we obtain

$$\widehat{y}(x) = y_1 + \sum_{i=1}^{K-1} (y_{i+1} - y_i)\, \Psi_i(x) . \tag{12.61}$$

By introducing the parameters $\Delta y_i = y_{i+1} - y_i, c_i = 1/(x_{i+1} - x_i), \theta_i = x_i/(x_{i+1} - x_i)$ and a unique, normalized sigmoidal basis function

$$\Psi(x) = \{0 \ldots \text{ for } x < 0 \; ; \; x \ldots \text{ for } 0 \le x \le 1 \; ; \; 1 \ldots \text{ for } x > 1\} \tag{12.62}$$

we can rewrite Eq.(12.61) in the form of a two-layer perceptron mapping relation

$$\widehat{y}(x) = y_1 + \sum_{i=1}^{K-1} \Delta y_i\, \Psi(c_i x - \theta_i) . \tag{12.63}$$

The first layer corresponds to neurons whose synaptic weights are c_i and whose threshold values are θ_i while the second layer contains a linear neuron with synaptic weights Δy_i and a threshold value y_1.

The above derivation demonstrates that for the selected two-dimensional distribution and piecewise linear basis functions, the mapping $x \to y$ that is determined by the conditional average is identical to the mapping relation of a multi-layer perceptron. However, a difference appears when the operations needed for the mapping are executed. The operators involved in both cases are described by different basis functions which correspond to different neurons in the implementation. If the prototypes are not evenly spaced, then the last equation is still applicable, although the transition regions will be of different spans. However, in this case, the basis functions $w_i(x)$ are no longer symmetric which leads to difficulties in the interpretation. The same problem arises when the data are stochastic. For this reason, we shall not continue with the analysis using a one-dimensional function but rather turn to a multivariate case.

We begin our analysis of the correspondence between the multi-variate radial basis function neural network and the multi-layer perceptron by considering the situation with just two prototypes $\boldsymbol{q}_i$ and $\boldsymbol{q}_j$. The conditional average is then described by the function

$$\widehat{\boldsymbol{h}}(\boldsymbol{g}) = \frac{\boldsymbol{h}_i \exp(- \parallel \boldsymbol{g} - \boldsymbol{g}_i \parallel^2 / 2\sigma^2) + \boldsymbol{h}_j \exp(- \parallel \boldsymbol{g} - \boldsymbol{g}_j \parallel^2 / 2\sigma^2)}{\exp(- \parallel \boldsymbol{g} - \boldsymbol{g}_i \parallel^2 / 2\sigma^2) + \exp(- \parallel \boldsymbol{g} - \boldsymbol{g}_j \parallel^2 / 2\sigma^2)} \; . \quad (12.64)$$

We introduce the notation: $\boldsymbol{g}_i \equiv \overline{\boldsymbol{g}} + \Delta\boldsymbol{g}$, $\boldsymbol{g}_j \equiv \overline{\boldsymbol{g}} - \Delta\boldsymbol{g}$, $\boldsymbol{h}_i \equiv \overline{\boldsymbol{h}} + \Delta\boldsymbol{h}$, $\boldsymbol{h}_j \equiv \overline{\boldsymbol{h}} - \Delta\boldsymbol{h}$ in which the bar denotes the average value and $2\Delta\boldsymbol{g}$ is the spacing of the prototypes. If we express the norm by a scalar product and cancel in the numerator and denominator of Eq. (12.64) the term, $\exp[-(\parallel \boldsymbol{g} - \overline{\boldsymbol{g}} \parallel^2 + \parallel \Delta\boldsymbol{g} \parallel^2)/2\sigma^2)$, we obtain the expression:

$$\widehat{\boldsymbol{h}}(\boldsymbol{g}) = \overline{\boldsymbol{h}} + \Delta\boldsymbol{h} \tanh\left[\Delta\boldsymbol{g} \cdot (\boldsymbol{g} - \overline{\boldsymbol{g}})/\sigma^2\right] \; , \quad (12.65)$$

in which "·" denotes the dot product. To obtain the relation between the radial basis function neural network and the multi-layer perceptron, we introduce the weight vector $\boldsymbol{c} \equiv \Delta\boldsymbol{g}/\sigma^2$ and the threshold value $\theta \equiv \overline{\boldsymbol{g}} \cdot \Delta\boldsymbol{g}/\sigma^2$ into Eq. (12.65). The result is:

$$\widehat{\boldsymbol{h}}(\boldsymbol{g}) = \overline{\boldsymbol{h}} + \Delta\boldsymbol{h} \tanh\left[\boldsymbol{c} \cdot \boldsymbol{g} - \theta\right] \; . \quad (12.66)$$

This expression again describes a two-layer perceptron. The first layer is comprised of only one neuron whose synaptic weight is represented by the vector $\boldsymbol{c}$ and the threshold value θ. The second layer is composed of linear neurons whose synaptic weights are Δh_i and threshold values are $\overline{h_i}$. A first-order approximation of the mapping expression is

$$\widehat{\boldsymbol{h}}(\boldsymbol{g}) = \overline{\boldsymbol{h}} + \Delta\boldsymbol{h}\Delta\boldsymbol{g} \cdot (\boldsymbol{g} - \overline{\boldsymbol{g}})/\sigma^2 \; . \quad (12.67)$$

This equation represents a linear regression of $\boldsymbol{h}$ on $\boldsymbol{g}$ which runs through both prototype points provided we set $\sigma^2 = \parallel \Delta\boldsymbol{g} \parallel^2$. Its slope is determined by the covariance matrix $\boldsymbol{\Sigma} = \Delta\boldsymbol{h}\Delta\boldsymbol{g}^{\mathrm{T}}$. However, the nonlinear regression specified in Eq.(12.65) follows a linear regression only in the vicinity of the point determined by $\overline{\boldsymbol{g}}$ and $\overline{\boldsymbol{h}}$ while it becomes saturated when $\boldsymbol{g}$ runs from $\overline{\boldsymbol{g}}$ over the given prototypes to infinity. The saturation is a consequence of function hyperbolic tangent tanh(.) which is basic in the modeling of the multi-layer perceptron.

The analysis presented above for a multi-variate case requires additional consideration when extended to the situation consisting of many prototypes. Let us assume that K prototypes with the indices $1, \ldots, K$ can be found in the hypersphere of radius approximately σ around the given datum $\boldsymbol{g}$ and let these prototypes be approximately evenly spaced. The conditional average can now be expressed in terms of the leading terms and the remainders as

$$\widehat{\boldsymbol{h}}(\boldsymbol{g}) = \frac{\sum_{i=1}^{K} \boldsymbol{h}_i \exp(- \parallel \boldsymbol{g} - \boldsymbol{g}_i \parallel^2 / 2\sigma^2)}{\sum_{i=1}^{K} \exp(- \parallel \boldsymbol{g} - \boldsymbol{g}_i \parallel^2 / 2\sigma^2) + O_w} + O_h \; . \quad (12.68)$$

Here O_h and O_w represent the two remainders which are small in comparison with the two leading terms. We again introduce the average value, but now with respect to K prototypes: $\boldsymbol{g}_i \equiv \overline{\boldsymbol{g}} + \Delta\boldsymbol{g}_i$ and $\boldsymbol{h}_i \equiv \overline{\boldsymbol{h}} + \Delta\boldsymbol{h}_i$ for $i = 1, \ldots, K$. With these, we obtain the approximate expression:

$$\widehat{\boldsymbol{h}}(\boldsymbol{g}) \cong \overline{\boldsymbol{h}} + \frac{\sum_{i=1}^{K} \Delta\boldsymbol{h}_i \exp[\Delta\boldsymbol{g}_i \cdot (\boldsymbol{g} - \overline{\boldsymbol{g}})/\sigma^2]}{\sum_{i=1}^{K} \exp[\Delta\boldsymbol{g}_i \cdot (\boldsymbol{g} - \overline{\boldsymbol{g}})/\sigma^2]} \,. \tag{12.69}$$

For $\boldsymbol{g}$ in the vicinity of the average value, a linear approximation of the exponential function is applicable which yields

$$\widehat{\boldsymbol{h}}(\boldsymbol{g}) \cong \overline{\boldsymbol{h}} + \frac{1}{K} \sum_{i=1}^{K} \Delta\boldsymbol{h}_i \Delta\boldsymbol{g}_i \cdot (\boldsymbol{g} - \overline{\boldsymbol{g}})/\sigma^2 \,. \tag{12.70}$$

This expression represents a linear regression of $\boldsymbol{h}$ on $\boldsymbol{g}$ which is specified by K points. If we express the covariance matrix in terms of two principal vectors $\Delta\boldsymbol{h}_p$ and $\Delta\boldsymbol{g}_p$ as:

$$\boldsymbol{\Sigma} = \frac{1}{K} \sum_{i=1}^{K} \Delta\boldsymbol{h}_i \Delta\boldsymbol{g}_i^{\mathrm{T}} = \Delta\boldsymbol{h}_p \Delta\boldsymbol{g}_p^{\mathrm{T}} \,, \tag{12.71}$$

we obtain a simplified expression of the linear regression

$$\widehat{\boldsymbol{h}}(\boldsymbol{g}) \cong \overline{\boldsymbol{h}} + \Delta\boldsymbol{h}_p \Delta\boldsymbol{g}_p \cdot (\boldsymbol{g} - \overline{\boldsymbol{g}})/\sigma^2 \,, \tag{12.72}$$

which is an approximation of a multi-layer perceptron mapping relation

$$\widehat{\boldsymbol{h}}(\boldsymbol{g}) \cong \overline{\boldsymbol{h}} + \Delta\boldsymbol{h}_p \tanh[\Delta\boldsymbol{g}_p \cdot (\boldsymbol{g} - \overline{\boldsymbol{g}})/\sigma^2] \,. \tag{12.73}$$

The parameters of a single neuron in the perceptron expression thus correspond to the principal vectors of the covariance matrix determining a local regression around the center of several neighboring prototypes. The above expression shows that the transition from a radial basis function neural network to the multi-layer perceptron can be quite generally performed. However, in the multi-variate case, the decomposition of the conditional average into a perceptron mapping is not as simple as in the one-dimensional case because the interpretation of the perceptron parameters is based on a local regression which is determined by the prototypes lying in the neighborhood of the given datum $\boldsymbol{g}$.

13. Fundamentals of Intelligent Control

13.1 Introduction

We describe in this chapter how an empirical modeling of natural phenomena based on learning from examples can be utilized as the basis of an empirical approach for solving control problems. In fact, the importance and advantage of empirical modeling of natural phenomena can best be demonstrated by showing how the difficult problems taking place in the general approach to adaptive and non-linear control can be treated. Our aim is not to become immersed in a detailed description of various specific problems and devices from the broad field of control but rather to show how progress in this field can be facilitated by application of empirical modeling.

Before proceeding with a description of control problems, let us justify the application of the phrase "intelligent control". The word "intelligence" is commonly related to a generalized, flexible, and adaptive capability that we associate with the human brain. We usually assume that it is a consequence of the evolution of animals. On the other hand, the word "control" is in technical language commonly related to influences exerted on various dynamical systems in order to obtain some desired response from a system. "Intelligent control" is thus to be understood as a use of general purpose control systems, that learn over time how to achieve prescribed goals or to optimize their operation in a complex and most often, non-linear environment which can also be very noisy or chaotic. Additionally, we expect that the dynamics must ultimately be learned from examples in real time. An important attribute of an intelligent control system is its ability to use past experiences with its environment to improve its performance as a function of time. In these aspects, an intelligent controller resembles a biological intelligent system capable of self-control in a dynamically changing environment. A number of good review articles on this topic are contained in the books by White and Sofge [23] and Miller, *et al.*. [15]

If we exclude from consideration the evolution of biological control systems, which are the most highly developed and sophisticated on earth, then we can follow the technical development of control procedures beginning in pre-history, when humans began using various processes such as cooking. Stacking wood to build a fire in order to prepare food is not a difficult task, even for children, but if we want to make a device capable of performing this task automatically, we quickly run into difficulties. First, we must develop a manipulator capable of grasping pieces of wood and properly placing them into a fire. If the pieces are of irregular shape, this can lead to difficult problems. Industry has invented a number of devices for a controllable loading of fuel into ovens. And people with their well-developed, complex, sensory-neural network, can usually do this quite readily because of their extraordinary ability to control the movement of their hands. However, in order to grasp a piece of wood, one must first recognize it in its surroundings on the basis of seeing and touching. The corresponding abilities are still not completely achievable by the most sophisticated of today's robots, in spite of all the "intelligent machines" that have already been developed for the analysis of pictures and touch signals.

The next problem in cooking is more chemical in character. It is related to the estimation of the temperature of the pan from the fuel placed in the fire and its effect on the cooking and the final characteristics of the cooked food. People learn from the properties of previous examples, based on the observation of the process and tasting, to predict the result of their manipulation. In the modern food production industry, much effort has been done to develop machines that could prepare food automatically. Yet the best cookies are still those prepared with love by grandmothers for the birthdays of their grandchildren. What is the source of this superiority of cooking skills that some human beings possess? It is, without a doubt, related to the ability to recognize properties of objects or processes, to manipulate them and to continuously learn from them. All of these properties are a consequence of a synergetic operation of a person's sensors, neurons, and actuators. Thus, if we wish to develop intelligent machines that can substitute for human beings in various tasks, we must first develop intelligent controllers which are capable of autonomous sensing or measurement, recognition, learning, and manipulating in a dynamically changing environment. Some of these tasks can be carried out by conventional methods which form the basis of classical control algorithms. The problems of recognition and learning are still not completely solved. Nor can they be fully implemented into control algorithms. Today, recognition and learning are topics under intensive investigation. The research is proceeding first towards the development of a general, intelligent controller that is based on sensing and information processing. It can be expected that this research will be followed with the development of specific devices which will incorporate appropriate electrical, mechanical and other components and tools into a system.

The largest potential users of intelligent controllers are in industry, transportation and medicine. The first group mainly includes manufacturing and chemical process control applications, while the second group includes such tasks as the control of aircraft and space vehicles, the transportation of goods and the generation of energy. In the field of medicine, an intelligent supervision of a person's health and the prediction of diseases, the cloning of surgical procedures, the administration of drugs and the control of various biological processes or bodily functions could be developed based on an intelligent controller. However, in all of the aforementioned fields, there are many cases in which existing regulators and controllers already perform with sufficient accuracy. These are cases in which the system being controlled is well understood.

A typical example is the heating controller for a home in which an intelligent performance in the above-mentioned sense seems a bit superfluous. But in the future, an intelligent heating controller may be used to advantage in optimizing the use of energy in a home. There are also many examples where the processes to be controlled are related to phenomena that cannot be simply modeled and consequently also not easily controlled. Examples include: An accurate machining of a product of irregular shape or made of a material which is anisotropic and heterogeneous, controlling the motions of a boat on a rough sea, or controlling the flight of an airplane under changeable conditions of turbulent air flow. Characteristic in each of these examples is that the underlying physical phenomena, which strongly influence the behavior of the system, can be highly non-linear, unstable and consequently chaotic. The system properties are therefore not static and the desired control actions in different situations may be significantly different and not simply programmable in advance. Because of strongly fluctuating influences acting on the system, which are caused by the underlying dynamics that are essentially non-linear, the controllers used in such applications must also be non-linear and adaptable to changing operating conditions. There are also great difficulties related to the analytical modeling of non-linear phenomena, it is also advantageous if the controller can learn from past experiences or if some *a priori* information on performing in different situations can be incorporated into its structure. For this purpose, the controller must utilize a memory. One of the fundamental problems of an intelligent control system is to develop efficient methods for finding, storing, and recalling the information about the appropriate operation of the controller. We shall show that, for this purpose, the methods of empirical modeling of natural phenomena are applicable. Especially promising are the neural network approaches because they operate in parallel, sometimes with analog signals and are relatively inexpensive to implement electronically. [23] The main task of this chapter is to show how the non-parametric self-organized modeler can be applied as a basis of an intelligent controller. Before doing this, we briefly summarize the basic problems related to the intelligent control and its implementation.

The basic problem encountered in all control applications is the proper dynamical modeling of the system, sometimes called the *plant*, which is to be controlled. In conventional control techniques, an analytical, physical description of system dynamics is commonly used. In more advanced, adaptive control approaches, the specific parameters of a selected dynamical model are left unspecified during the modeling but are later adapted to the controlled plant by some identification technique that is based on sensing of control actions and plant response. Most frequently, linear models are applied, although non-linear modeling has also been developed. [20] An intelligent control is, in fact, just an extension of adaptive, non-linear control, with the principal difference being that a model is not selected by a designer of the control but rather built by an appropriate method in the controller itself. For this purpose, a general approach to modeling of natural phenomena based on learning from examples, which can also be called a *generalized identification*, is applicable. The problem of modeling dynamical phenomena by means of differential equations and corresponding identification of system parameters is thus converted into the design of an information processing system that is capable of mapping the excitation or driving dynamical variable into a plant response based on learning from given examples. The problem of modeling for the purpose of intelligent control is thus similar to the development of a system capable of forecasting that was described in Chap. 11. In the further study of the approaches to the design of intelligent controllers we present here the basic problems related to: (1) Tracking, (2) Cloning, and (3) Empirical Approach to Optimal Control.

In the first class of problems, the operation of a plant is to be controlled and the task is to design a controller capable of generating controlling signals that would result in some prescribed behavior of the plant. The prescription is generally such that results in a particular plant output as function of time. Such a description corresponds to specification of a trajectory in phase space which the plant as a dynamical system should follow. Consequently, this task is called *tracking*. A practically similar problem is encountered when one wishes to keep the system operating in a certain prescribed state. This is known as *stabilization*. The task of generating a corresponding controlling variable from a prescribed system response corresponds to inverse system modeling.

The second class of problems pertains to a plant that is somehow controlled, possibly even manually by a skilled operator or expert, and the task is to develop a controller capable of performing similarly. In this case, the controller must first learn from joined samples of expert actions and system responses to reproduce the manipulating. This problem is similar to a general identification of an operator. By an "intelligent" controller we mean that the controller is capable of reproducing some actions that were previously designed based on the intelligence of the expert. For this purpose, the controller must possess an ability to learn from the expert and to adapt to the characteristics of the plant. In cases where cloning is adequate, one does not expect that there is needed in a strict sense a capability for intelligent, high level reasoning and the creation of strategies based on given facts and properties of the system being controlled. In that case, supervised learning should be sufficient.

The third class of problems pertains to a plant which can be driven by some influences. But in this case, instead of a specified trajectory, an object function related to the state of the plant is entered into the controller. Commonly used may be the cost of the processing or effectiveness of the operation of the plant. The task of the controller is to influence, based on past experiences, the plant such that the extremum of the object function is met. This problem is consequently called a *dynamical optimization.* However, this example can be generalized to those cases in which even the object function is not known but must first be found by an intelligent controller itself and afterwards utilized in the control. By a proper inclusion of the environmental variables into the utility function the problems related to a self-control of a closed system in an open, changing environment can be tackled. The most promising research in this field is focused on the development of controllers possessing brain-like capabilities. [22, 24] It is interesting that tracking and cloning can also be represented as examples of optimal control, and therefore, in this chapter we present the mathematical description of optimal control and show how this description can be changed to include the empirical approach.

In all cases, we assume that the physical phenomena occurring in the plant being controlled may be highly non-linear and to certain extent, intrinsically unstable. This can lead to very complex dynamical behavior which is usually difficult to model analytically. A basis for a successful and fast controller is then an adequate empirical model which stems from given examples that permit the application of a mapping between input and output signals. With this approach, the differential equations modeling the process need not be solved. This can dramatically increase the speed of operation of the controller. Such kind of control has previously been observed in biological control systems and it has also been successfully implemented for controlling the motions of a manipulator. [1] A designer need not have an analytical description on hand to understand the dynamics of a process in order to realize a high-performance modeler. But in that case, the modeler must be capable of modeling the process with sufficient accuracy and solving the corresponding control problem that may be expressed as an inverse problem. However, the standard problems of observability, controllability and the stability of a compound controller and plant system which are well-known in conventional control procedures of linear and nonlinear systems, are also present in an intelligent controller. [19, 20, 24] We shall, however, not consider these problems here. Instead we describe how the empirical modeling can be utilized. In addition to the difficulties related to the strict analytical modeling of complex non-linear dynamical phenomena and the forecasting of their properties, there is also an additional argument in favor of the empirical approach. In order to apply an analytical description such as that given in terms of differential equations, we must also specify the initial and boundary conditions of a process as well as the physical constraints of the system and controller. In real situations, it is often impossible to specify these conditions because of a lack of sufficient information. Most often there are available empirical data about the behavior of the system from which we hope to extract sufficient information for controlling the system. In these cases, the characteristics of the problem at hand dictate the application of the empirical approach. Still another intuitively motivated argument is the following: Nature has developed very efficient biological controllers based on learning rather than on analytical modeling. When chasing a mouse, a cat does not solve the differential equations of mechanics, but rather it acts instinctively based on learning from previously observed hunts. The analytical approach is thus first of all helpful when analyzing a specific problem and making decisions about a proper controller of a given plant, while for the control of complex, highly non-linear and chaotic systems, the use of an empirical approach will likely be advantageous.

The basic design principles of an intelligent control system involves the following components: [22]

1. Construction of learning modelers.
2. Fabricating specific, complex controlling systems comprised of sensors, signal conditioners, learning modelers and actuators.
3. The inclusion of complex controlling systems into other equipment or structures.

The expected benefits from such a systematic approach are as follows:

1. To make feasible the application of high throughput electronic integrated circuits for information processing.
2. To facilitate the development and utilization of software that is based on universal theorems.
3. To realize the ease of use, teaching and the inclusion of additional controllers into applications.
4. To maintain a link to the fields of brain and intelligence exploration that can facilitate further progress of the entire field.

The ultimate goal of the research in this field is, however, to develop the mathematical tools by which the structure and operation of intelligent systems could be described at more sophisticated levels which would in the limit provide a thorough understanding and a functional description of the operation of the brain. At present, methods by which one can explain some of the most fundamental functional properties of the brain and the corresponding intelligent behavior are still missing. [22] It is likely that the development of the brain has been coupled with the ability to control biological organisms, and that, in principle, one cannot separate the functions of control and intelligence in biological systems. We therefore expect that the development of intelligent control will also influence the development of new methods and paradigms for the modeling of natural phenomena.

13.2 Basic Tasks of Intelligent Control

In this section we first describe a transition from a theoretical to empirical description of problems related to control. Following this, we demonstrate an empirical approach to identification, tracking, and cloning.

13.2.1 Empirical Description of a Controlled System

In order to point out the distinction between conventional and intelligent approaches to control and to prepare the foundation for the further work we present a brief review of the basic tasks of control and related mathematical problems. Most of the tasks encountered in the realm of classical control can be described by specific examples of the optimal control theory therefore we present the description of the fundamentals of the latter approach. [8]

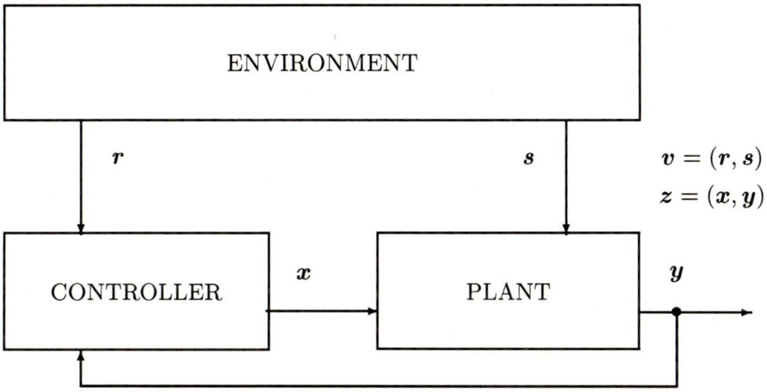

Fig. 13.1. A scheme of a controlled plant

A fundamental object of consideration in the control theory is a discrete system that can most generally be represented by the scheme in Fig. 13.1. Its fundamental units are a system to be controlled, called also a plant, and a controller. Each unit interacts with the other one by some process that is described in terms of physical variables. The interactions are schematically represented by connections between units while the variables representing interactions are indicated as transmitted signals. The basic element of description is the plant state which is generally represented by a time dependent vector $\boldsymbol{y}(t)$. The plant state is influenced by the controller which can obtain also the feedback information about the plant state $\boldsymbol{y}(t)$. The influence is generally described by a time dependent vector $\boldsymbol{x}(t)$ which is commonly called the manipulating variable. The manipulating variable and the plant state are also called input and output respectively. The concatenated vector $\boldsymbol{z}(t) = [\boldsymbol{x}(t), \boldsymbol{y}(t)]$ describes the state of the joint controller-plant system. The controller as well as the plant can both be influenced by the environment of the complete system. The corresponding physical variables are represented by two vectors $\boldsymbol{r}(t)$ and $\boldsymbol{s}(t)$. The first vector $\boldsymbol{r}(t)$ is most commonly called the reference signal, while the second vector $\boldsymbol{s}(t)$ is called the plant disturbance. Both variables can be described in common by a concatenated vector $\boldsymbol{v}(t) = [\boldsymbol{r}(t), \boldsymbol{s}(t)]$. For the sake of simplicity, we first assume that the plant disturbance is negligible, but later we include it again in the consideration. The dynamical properties of the plant are usually described by some dynamical law

$$\dot{\boldsymbol{y}} = \boldsymbol{\Phi}(\boldsymbol{y}(t), \boldsymbol{x}(t), t) \tag{13.1}$$

to which the initial condition $\boldsymbol{y}(t_0)$ and the physical constraints on the plant state and control variable are added. In each controlled system the control's purpose can be clearly described. With this aim we assume that the signals about the state of the controlled plant generally determine the instantaneous utility of the system which can be represented by the variable $u(t) = u(\boldsymbol{x}(t), \boldsymbol{y}(t), t)$. Here the influence of the environment is indicated by the explicit dependence of utility function on time. This variable is applicable for the description of plant performance. Most frequently an integral performance measure is used which is expressed as

$$J = \int_{t_o}^{t_f} u\,(\boldsymbol{y}(t), \boldsymbol{x}(t), t)\, dt\ . \tag{13.2}$$

Here (t_o, t_f) denotes the time interval of the performance description. By a proper specification of the utility function the effects of initial and final controlled plant states can be introduced into the description, therefore they are here not represented by additional terms as is usually done in various approaches to the optimal control. [13] By using different utility functions various goals of control can be achieved. Some characteristic examples will be presented later.

At the design of a controlled system we usually want to provide for a proper description of the plant dynamics first. This task is generally called modeling. It is accomplished by the specification of the dynamical law, or more in particular of the plant function $\boldsymbol{\Phi}(.)$, that can generally be nonlinear. The following approaches are most often used:

1. Analytical modeling - Based on the analysis of the system structure and systematic application of general physical laws to description of the properties of its components.
2. Parametric modeling - Based on synthesis of specific empirical relationships which are expected to correspond to a given structure. In the mathematical expression of the empirical relationships some parameters are left to be adapted using experimental data.

3. Non-parametric modeling - Utilizes an empirical data base comprised of measured input-output data.
4. Physical modeling - Corresponds to building an another system or physical model which has analog properties as the plant.

Cases 1 to 3 are commonly called mathematical modeling. Case 1 is often referred to as theoretical modeling, and cases 3 and 4 as experimental or empirical modeling, while case 2 can be treated as a mixture of both.

Several comments are in order here. In an analytical approach we generally describe the properties of the plant by a dynamical law that is obtained by an analysis of the plant structure. If this structure is not known in advance and we try to avoid its analysis by using empirical approach, then we need some time to observe the plant behavior and a method by which the plant function is determined. In this case the question arises as how to collect experimental data that sufficiently describe the joint properties of the plant input and output. This corresponds to the problem of state space exploration. The problem stems from our unfamiliarity with the properties of the controller prior to its design; consequently, we do not know the properties of the manipulating variable, which will appear later during the optimal control, and how to generate a corresponding signal.

A conventional theoretical approach to the optimal control is usually based on analytical modeling by either linear or non-linear functions. The parametric modeling is usually referred to as a basis of the adaptive optimal control, and often utilizes linear relationships. Analytical as well as parametric modeling can be equivalently utilized in the theoretical methods developed for the solution of the optimal control problems because both yield algebraically expressed plant function. A distinction is only in the specification of the parameters taking place in the expression of this function. Contrary to this the non-parametric modeling is based upon distribution of empirical data points in the state space which cannot be directly utilized in the analytical solution of constrained optimization problem. The intelligent control is based upon learning from examples, therefore we have to develop a proper method by which the empirical probability distribution can be included into the design of an optimal controller. One reason for this is also the observation that with increasing system complexity the analytical and parametric modeling becomes ever more cumbersome so that the non-parametric empirical modeling appears to be the only remaining option of the mathematical modeling that is based on information processing.

In the case of physical modeling the properties of the plant function are transferred to the properties of a real model. For this purpose we can generally apply a system comprised of a network of sensors and actuators, an electronic data acquisition system and a computer with incorporated program for non-parametric modeling of natural laws that corresponds to a simulated neural network. By using such a system one can perform control experiments and eventually proceed towards the development of an optimal controller completely on an empirical basis, eventually also by trial and error. This is the second reason for the study of the empirical approach to the solution of the optimal control problems. An additional reason is the observation that the biological intelligent systems are capable of optimal control based on learning from examples.

13.2.2 General Identification by Non–Parametric Modeling

Let us consider a controlled plant that is schematically shown in Fig. 13.2. Let us further suppose that the system model is not known. Therefore we seek to apply the empirical modeling for this purpose. With this in mind, we make simultaneous recordings of both the input and output variables and thus obtain an ensemble of empirical records. For the sake of simplicity we here assume that the input and output are scalar variables. The resulting joint record, denoted by $\boldsymbol{s}(t) = [x(t), y(t)]$ can be treated as an empirical sample of a two-dimensional vector process which can, in principle, also be of random character.

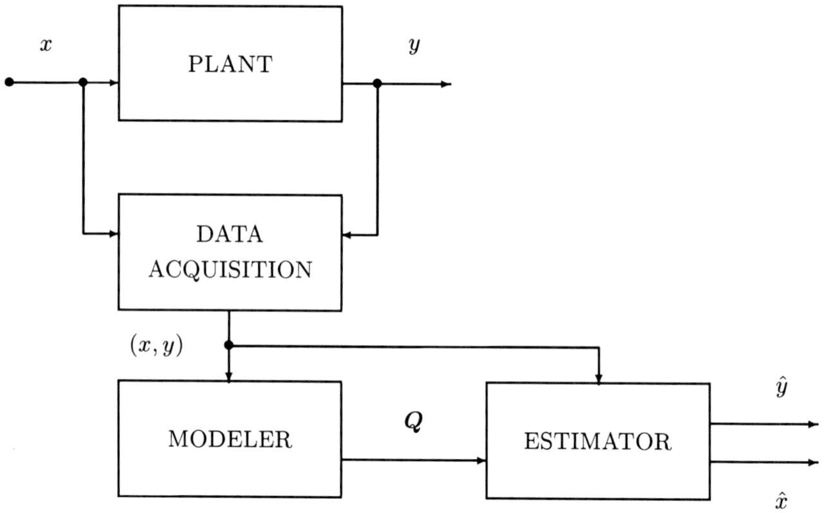

Fig. 13.2. A scheme of a plant and a system for non-parametric identification and prediction

To obtain a description of the process, we will use methods from the theory of random processes. A random process is most generally described if all the joint probability densities of process $\boldsymbol{s}(t)$ for all possible combinations of arbitrary many sampling time values $\{t_1, t_2, t_3, \ldots\}$ are specified. This, however, is possible only in exceptional cases and consequently we are forced to apply a semi-deterministic approach. That is, we assume that a value $s(t)$ can be related to some set of values at previous moments by a functional relation, which we try to estimate on a statistical basis. To proceed along this way, we need to avoid problems related with the presentation of joint probability functions of continuous parameter process. We therefore assume that a discretized time scale $\{0, t_1 = \Delta t, t_2 = 2 \Delta t, \ldots\}$ having a step size Δt is utilized for the presentation of the phenomenon under consideration. For the sake of brevity, we select a time unit to equal the step size, or $\Delta t \to 1$. Such a time scale can be described by the integer indices $\{0, 1, 2, \ldots\}$. In experimental data, the corresponding time step Δt should be approximately an order of magnitude smaller than a characteristic time period of the recorded variables. A continuous time process is then properly represented by a discrete time record which is described in its discrete representation by the sequence

$$\{x(t_1), y(t_1);\ x(t_2), y(t_2); \ldots\} . \tag{13.3}$$

The properties of the phenomenon are described by specifying the joint probability distributions of type

$$\begin{aligned} &F\left(x_1, y_1, x_2, y_2, \ldots; t_1, t_2, \ldots\right) = \\ &P\left\{x(t_1) \leq x_1, y(t_1) \leq y_1, x(t_2) \leq x_2, y(t_2) \leq y_2, \ldots\right\} . \end{aligned} \tag{13.4}$$

If the phenomenon under observation is stationary, then all the joint probability distributions must be invariant with respect to a simultaneous shifting of all times t_n for an arbitrary period τ. If we select $\tau = -t_1$, then the joint probability distribution of a stationary process can be expressed as a function of time differences only

$$\begin{aligned} F(x_1, y_1, x_2, y_2, \ldots; t_1, t_2, \ldots) &= F(x_1, y_1, x_2, y_2, \ldots; t_1 - \tau, t_2 - \tau, \ldots) \\ &= F(x_1, y_1, x_2, y_2, \ldots; 0, t_2 - t_1, \ldots) . \end{aligned} \tag{13.5}$$

This greatly simplifies the modeling of the process because in many cases, we apply the joint probability density function which relates variables at only two instants in time, provided that the process is stationary. The corresponding joint probability density depends only on the time difference which represents just one variable. The result is a tremendous saving in memory.

In a deterministic approach we describe the dynamical law of the plant by a differential equation

$$\dot{y} = \Phi(y, x) \tag{13.6}$$

where $\dot{y}$ denotes dy/dt. After integrating such an equation, one generally obtains a solution by which the response y at time t is related to values of the driving variable as well as the response $y(t)$ in the past time. However, in real dynamical systems, the influence of the excitation usually decreases with time so that for a broad class of dynamical phenomena, a finite effective time of influence can be assumed. If this observation is interpreted in a statistical sense, it means that the joint probability density of the two values $x(t_1)$ and $y(t_2)$ also decreases with increasing time difference (t_2-t_1). We can therefore expect that the number of terms in Eq. (13.3) need not be increased without limit in order to have an appropriate description of the relationship between the input and output. Consequently, we assume that there exists a proper dimension D describing the number of terms that must be taken into account if we want to describe the relation between x and y with sufficient accuracy. If not otherwise specified, the accuracy can be determined by the accuracy of the measurement procedures utilized in the determination of the variables x and y. We also assume that the proper dimension D can be obtained from experimental observations of the variables x and y. A deterministic description of a dynamical phenomenon is then replaced by a specification of the joint probability distribution comprised of a fixed number of terms.

For the sake of simplicity we shall further constrain our treatment to stationary processes, although the approach we are describing can also be generalized to non-stationary phenomena. The principal difference is in the source of information from which the joint probability distribution is estimated. For stationary processes, the invariance with respect to the time shift, makes it possible to determine the probability distribution based on only one record, while for non-stationary processes, an ensemble of records needs to be utilized, which makes the procedures experimentally and computationally much more complicated.

We turn now to the empirical description of joint probability distribution. For this purpose we assume that the dimensionality D is given and that a joint time record of input and output data was acquired. From the recorded data we form a vector sample comprised of only D samples of the vector $\boldsymbol{s}(t) = [x(t), y(t)]$. That is,

$$\boldsymbol{z}(t_1) \;=\; [\,\boldsymbol{s}(t_1), \boldsymbol{s}(t_2), \,\dots\, , \boldsymbol{s}(t_D)\,]\,. \tag{13.7}$$

By lagging this vector in time for values of $t_n = \Delta t, 2\,\Delta\, t, \dots, (N-1)\,\Delta\, t$, we obtain a set of samples

$$\{\boldsymbol{z}(t_1{+}t_n) = [\boldsymbol{s}(t_1{+}t_n), \boldsymbol{s}(t_2{+}t_n), \dots, \boldsymbol{s}(t_D{+}t_n)]\,;\, n = 1, \dots, (N{-}1)\} \tag{13.8}$$

which forms the basis for an empirical statistical description of the process. We have thus reduced the description of the complete phenomenon to a determination of probability distribution from the empirical set of data $\{\boldsymbol{z}_1 = \boldsymbol{z}(t_1), \dots, \boldsymbol{z}_N = \boldsymbol{z}(t_N)\}$ with N being some fixed number of samples of the vector $\boldsymbol{z}$. Using the non-parametric approach, the corresponding representative probability density function is expressed as

$$f(\boldsymbol{z}) \;=\; \frac{1}{N} \sum_{n=1}^{N} w\,(\boldsymbol{z} - \boldsymbol{z}_n)\;, \tag{13.9}$$

where w denotes an appropriate window function. If the number N of samples $\boldsymbol{z}_n$ is too large, a set of smaller number of K prototypes

$\{\boldsymbol{q}_k; k = 1, \ldots, K < N\}$ can be introduced using the self-organization process described in Chap. 8.

Eq. (13.9) represents the basis for an empirical description of the observed dynamical phenomenon. Its specification in terms of empirical records thus corresponds to a generalized identification. The next problem is how this description can be applied in a control application. Let us first describe the so called forward problem. In this case, the input variable $x(t)$ as well as the system state is known for at least $(D-1)$ time steps prior to the moment of interest. In the next time increment, the value $y(t+1)$ can be estimated by using the conditional average estimator and the past values of input and system state as a given joint condition

$$\boldsymbol{g}_y(t) = \{x(t-D), y(t-D); \ldots; x(t), y(t); x(t+1), \emptyset\} \qquad (13.10)$$

in which the symbol $\emptyset$ denotes the value of the system state to be estimated. The corresponding estimator for the system output at time $(t+1)$ is symbolically represented by the expression

$$\begin{aligned} \hat{y}(t+1) &= \mathrm{E}\, y(t+1) \,|\, \boldsymbol{g}_y(t)] \\ &= \mathrm{E}\, [y(t+1) \,|\, x(t-D), y(t-D); \ldots; x(t), y(t); x(t+1)] \; . \end{aligned} \qquad (13.11)$$

This equation corresponds, in fact, to the forecasting the system response to the driving variable $x(t+1)$ and internal dynamics of the system.

An important characteristic of the proposed generalized identification is that for a real system, the response follows the excitation in time. This is the *principle of causality*. In other words, the joint probability distribution will have approximately zero value when the output preceding the input is considered. Thus, it can generally be expected that an output can be estimated on-line from the input but not vice versa. This raises the question of how to determine the input when the desired output is specified. The problem is treated in the subsection on tracking.

The approach of general identification presented here can be generalized to cases where input and output variables are represented by vectors. No essential changes appear in the formalism except that additional components are involved at each time step and thus an expanded memory is needed in order to store all the samples or prototype vectors.

In Chap. 11 on forecasting of chaotic phenomena, the emphasis was on the description of those properties of a system which are primarily a consequence of the internal dynamics of the system and which depend less on the details of the excitation. Therefore, we assumed that the system state could not be represented by just a single time-dependent variable but rather by a vector whose dimensionality was not known *a priori* but that must be estimated from the empirical data. In principle, the same is done here, except that the influence of the excitation variable is more generally described.

Just as the output $y(t)$ can be predicted from the past time series of the input and output variables, so too can the input $x(t)$ be forecast using non-parametric modeling. In this case, the joint condition, expressed by $\boldsymbol{g}_y(t)$ becomes

$$\boldsymbol{g}_x(t) = \{(x(t-D), y(t-D); \ldots; x(t), y(t); \emptyset, y(t+1)\} \tag{13.12}$$

and similar to Eq. (13.12), that was used to estimate the system output, we can estimate the system input, $\hat{x}(t+1)$:

$$\begin{aligned}\hat{x}(t+1) &= \mathrm{E}\left[x(t+1) \,|\, \boldsymbol{g}_x(t)\right] \\ &= \mathrm{E}\left[x(t+1) \,|\, x(t-D), y(t-D); \ldots; x(t), y(t); y(t+1)\right] .\end{aligned} \tag{13.13}$$

However, the information about the forthcoming input value must be introduced into the estimator from the past behavior of this signal, as the output follows the input because of causality. Related to general identification, there should be mentioned two typical control problems called *tracking* and *cloning.*

13.3 The Tracking Problem

A generalized identification that is based on the modeling of the joint probability density is of fundamental importance for control theory. To show this, let us consider the tracking problem which, as has been pointed out earlier, is one of the basic problems in control theory. Let us consider again a system with input $x(t)$ and output $y(t)$ and let us assume that an empirical model has already been obtained from the generalized identification procedure described above. The challenge is to design a controller which will generate such a driving variable $x(t)$ that the system output will follow a certain desired trajectory $r(t)$ that can be treated as a specified reference. The controller should be a mechanistic system and we assume that the corresponding information about the desired trajectory of the process or plant to be controlled must somehow be presented to it. The simplest case is that when the trajectory is presented to the input of the controller as a time-dependent reference signal $r(t)$. The other possibility, which, in fact, turns out to be more interesting, is providing the controller with instructions that can be stored in it as a record $r(t')$ in which the virtual time parameter t' denotes the location or address in the memory where the record is stored. The corresponding reference signal must be transformed in the controller in such a way that it will generate the driving variable $x(t)$ which will lead the system output along the trajectory $y(t) \approx r(t)$.

The corresponding scheme is shown in Fig. 13.3. In terms of optimal control theory, a tracking problem can be specified by describing the utility function as a negative square difference of the plant output and the reference input which yields the performance J:

$$J = -\int_{t_0}^{t_f} (y(t) - r(t))^2 \, dt \; . \tag{13.14}$$

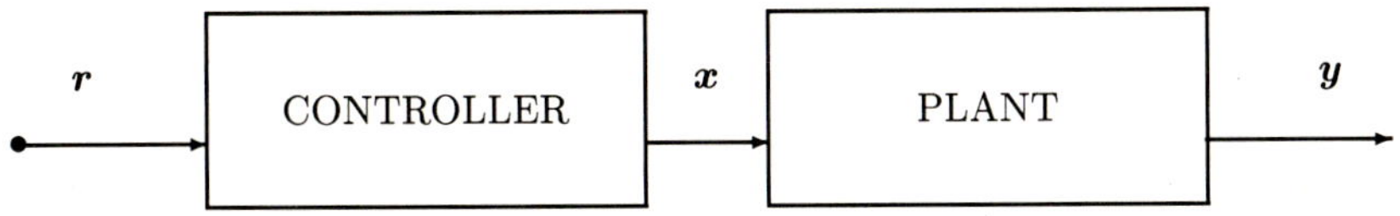

Fig. 13.3. A scheme of tracking controller. The reference signal $r(t)$ describes the desired plant trajectory

It has a maximum if the output equals the reference signal. The problem then, is how to design a controller that will generate such a control signal $x(t)$, that will drive the output $y(t)$ along the trajectory $r(t)$ in accordance to the dynamical law

$$\dot{y}(t) = \Phi(y(t), x(t), t) . \tag{13.15}$$

If the function Φ is given analytically, then we can replace the output variable in the above equation with the reference and solve for the input signal. This, however, is not possible when there are only experimental records available.

Based on linear systems theory, one might think that if the output is determined by some transformation of the input, say $y = T(x)$, then the input can be determined from a given output by the inverse transformation symbolically represented as $x = T^{-1}(y)$.[1] From this viewpoint, a tracking resembles an inversion operation that needs to be performed by a controller.

Let us now ask how the empirical model obtained by the generalized identification can be utilized as a controller. It is represented by the joint probability distribution given by Eq. (13.2). That is,

$$\begin{aligned} F(x_1, y_1, x_2, y_2, \ldots; t_1, t_2, \ldots) &= \\ P\{x(t_1) \le x_1, y(t_1) \le y_1, x(t_2) \le x_2, y(t_2) \le y_2, \ldots\} . \end{aligned} \tag{13.16}$$

By utilizing the conditional average estimator, we can determine from this joint probability density a set of values $\{x(t), x(t+1), \ldots, x(t+D-1)\}$ when the remaining set $\{y(t), y(t+1), \ldots, y(t+D-1)\}$ is given as a condition. At first, this seems to be impossible, because the actual response of the plant is unknown. But we do know the desired response and we can, in principle, apply it as a condition. We could thus design a controller as a system in which the empirical model of the plant is stored and which then utilizes it to estimate the plant input from the given information about the desired trajectory. The corresponding operation needed for a one-step-ahead estimation can be represented by the expression

$$\begin{aligned} \hat{x}(t+1) &= \mathrm{E}[x(t+1) \,|\, \boldsymbol{g}(t)] \qquad (13.17) \\ &= \mathrm{E}[x(t+1) \,|\, x(t-D), r(t-D); \ldots; x(t), r(t); r(t+1)] . \end{aligned}$$

[1] Note that in writing these transformation equations, we have omitted the time-dependence of the variables x and y. The transformation specified by T can represent a general transformation of the process which may be non-local, that is a value $y(t)$ is determined by a set of x values at different times.

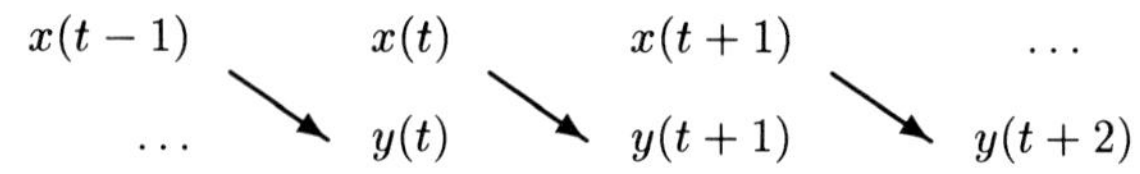

Fig. 13.4. Flow of influence

Here we assume that the past values with respect to time t are all given or could at least have been previously measured and memorized. By performing this estimation sequentially as time progresses, it is possible to predict the input $x(t)$ that will drive the plant along the desired trajectory. When the system obtains the reference signal on-line and in time, then this is the most direct approach to tracking the process. However, this inverse procedure leads to a realizable control only if the relation between input and output is quasi-static and represented by a one-to-one mapping. In a dynamical situation, however, we must assume that the variables are time-dependent and that the response is time-delayed. This, however, represents a serious obstacle for this method because of the causality of real systems.

To explain this in more detail, let us consider the scheme depicted in Fig. 13.4 which shows schematically how an influence in a real system propagates in time from the input to the output. Consider a simple example in which a plant whose input x at time t is only shifted to the output at time $(t+1)$ according to

$$y(t+1) \;=\; x(t)\;. \tag{13.18}$$

This means that we can determine $x(t)$ if we know the output at the time $(t+1)$. But the controller only has the reference signal on-line, that is, it has information about the desired output of the process up to time t. If we want to achieve a tracking, we need to assume that

$$y(t+1) \;=\; r(t+1)\;. \tag{13.19}$$

Using the previous equation, we can then write

$$y(t+1) \;=\; x(t) \;=\; r(t+1)\;. \tag{13.20}$$

This equation shows that at time t, a controller needs the advanced value of the reference signal r at time $(t+1)$ in order to estimate the desired plant input x at t. If a controller obtains information about the desired trajectory on-line and in time, then it cannot be used to control a process since in real dynamical systems, the input signals are delayed because of causality. Only if a time-advanced reference value is used, can the controller properly compensate for the delay in the plant and equalize the plant output with the reference signal. A question thus arises if it is at all possible to control the mentioned plant in such a way that its output $y(t)$ would follow the desired trajectory $r(t)$. The answer is that it is possible, provided that the controller is capable of properly forecasting the reference signal $r(t)$ one step ahead and if the predicted signal is then utilized as a condition from which the driving variable of the plant is estimated.

The forecasting of $r(t)$ can be done based on the past time series of the reference signal only. To realize this, an additional unit called a *forecaster* must be added to the controller. This unit is used to forecast the future reference signal from the supplied reference signal and placed in the memory. For this purpose, the concept of virtual time is needed. It describes the number of prediction time steps $\Delta t'$ ahead from the present time and corresponds to the index of the memory cell in which the forecast signal is stored. It is interesting that the forecasting procedure can be avoided if the desired trajectory is not given on-line but *a priori*. In this case, it can immediately be stored in the memory as mentioned previously. However, in this mode, saturation problems may arise during the memorization of long signals. A proper compromise is obtained by specifying only a certain portion of the desired trajectory in advance. The corresponding time can be experimentally estimated from the typical delay of a propagating influence in the plant. An *a priori* prescription of the trajectory is convenient because in this way the forecasting procedure can be omitted.

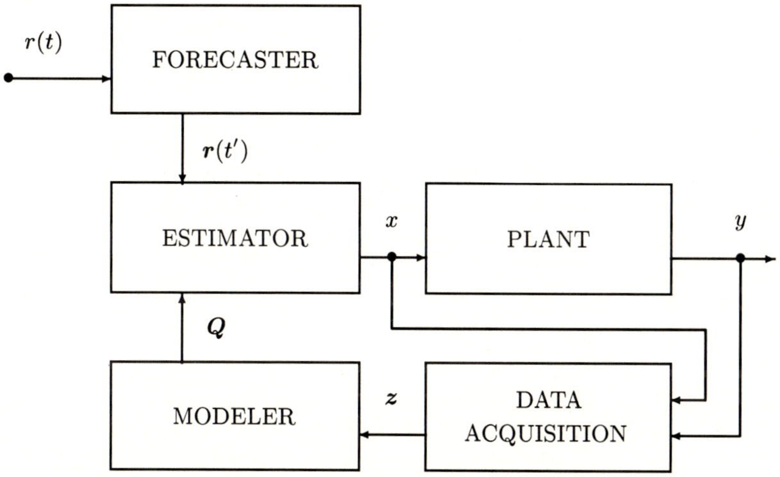

Fig. 13.5. A scheme of intelligent tracking controller

An intelligent tracking controller is thus represented schematically as an information processing system that is depicted in Fig. 13.5. It is comprised of a data acquisition system, a modeler, a forecaster, and an estimator of the driving variable. In order to make such a system feasible, the signal of the driving variable must be properly adapted to the characteristics of the plant. For this purpose, a proper booster should be incorporated in the plant. This is omitted from the figure because we are primarily interested in information processing. The data acquisition system consists of sensors of the input and output plant variables. It also includes the corresponding signal conditioners, and a sampling unit by which the continuous variables x and y are transformed into corresponding discretized sequences of data that are represented by the vector $\boldsymbol{z}$. The modeler consists of a memory and a processor by which a self-organized formation of the prototypes $\boldsymbol{Q} = \{\boldsymbol{q}_k;\ k = 1, 2, \ldots K\}$ is performed. The prototypes are also supplied to the estimator. Here the current value of desired driving variable $x(t)$ is estimated based on the prototypes $\boldsymbol{Q}$ and the reference vector $\{\boldsymbol{r}(t');\ t' = 1, 2, \ldots\}$. Here the index t' denotes the number of forecasting time steps. The reference vector is formed by the forecasting unit using the reference signal $r(t)$. Forecasting is performed in a manner similar to that described for solving the estimation problems for autonomous dynamical systems. It is convenient if the forecaster has direct access to the memory in order to be able to store a desired trajectory $r(t)$ when it is specified in advance.

Let us now point out why such a controller can be called an intelligent controller. First of all, it includes the operation of forecasting based on previous examples. In addition, from the learned examples, it autonomously models the behavior of a plant. It uses this model further, based on association in order to achieve a predetermined goal. All the mentioned operations correspond to basic characteristics of an intelligent being and confirm the application of the name *intelligent tracking controller*. It is characteristic that its design stems from a general empirical description of natural processes. A non-parametric approach makes feasible the inclusion of a self-organization of the input data as well as the utilization of *a priori* information which can be inscribed into the model by initial prototypical data. The intelligent tracking controller requires an initial training procedure. For this purpose, it must be connected to the operating plant for a certain adaptation time. Should this not be possible, then *a priori* information about the system properties must be specified in terms of signal prototypes.

One advantageous feature of the proposed intelligent tracking controller is that it can continuously update the plant model by utilizing current data about plant input and output. This, in fact, represents a feedback in the complex system consisting of plant and tracking system which is so characteristic for a conventional and an adaptive control of a process. The current system state together with its past history can also be utilized for the estimation of the desired plant input variable so that the condition can be written as

$$\boldsymbol{g}(t) = \{x(t-D), y(t-D); \ldots; x(t), y(t); r(t+1)] \,. \tag{13.21}$$

This corresponds to feedback in the system which has short-term influences on the adaptability of the tracking system, while an updating of the prototypes residing in the memory proceeds more slowly and is of long-term character because of the averaging that is inherent in the self-organization process. To differentiate between these two properties, it could be convenient to call the long-term behavior *adaptation* and the short-term behavior *learning.* [5] Without much effort, the proposed scheme of a tracking controller could be generalized to the case where the input and output variables are all represented by vectors.

One problem that should be mentioned in relation to the application of the intelligent tracking controller is the accuracy of the forecasting. It has been already mentioned in Chap. 11 that it is impossible to accurately forecast the long-term behavior of a chaotic system because of the inherent instability of the corresponding trajectories. The same is still true in applications of the forecaster used as a tracking controller. If the reference signal cannot be forecast for a time corresponding to the typical delay of influences in the plant response, then one cannot expect good tracking results. The same is true if the intrinsic dynamics of the plant is such that the influences of the driving variables have little effect on the plant response. In this case, the correlation between input and output variables is low and one cannot strongly influence the behavior of the system by the input variable. An example of such a system is a turbulent jet from a tube where the shape of the tube exhaust can affect the movement of the jet only to a limited extent. The corresponding problems of stability and controllability related to tracking remain under intensive research. [17, 16, 22]

The approach to the tracking problem presented here is based entirely on a statistical approach to the modeling of natural phenomena by means of a self-organized non-parametric regression. The advantage of this approach is that it permits a general treatment of phenomena that results in a feasible physically-based explanation of the operation of a controller. There are, however, another approaches possible. The most common is the connectionist one in which the tracking problem is represented in terms of general mappings which are tried to be realized by designing a neural network which generally includes also time delays and which is trained by some of the existing methods of supervised learning such as back-propagation. [23, 15] Although such an approach can be very promising from the aspect of implementability in terms of commercially available elements, it is unfortunate that the physical background of the system operation often becomes obscured by problems related to the network design and the development of algorithms. A problem that often arises in the application of connectionist approach is how to utilize *a priori* information. When using a statistical approach based on the self-organized non-parametric regression this problem corresponds to specification of initial prototypes which have very clear physical meaning and can be directly related to *a priori* information. It can also be included into the description of joint probability density which has a simple physical interpretation as well. As our task is to give only foundations of the intelligent control based on empirical modeling of natural phenomena we shall not proceed here with the description of specific tracking problems that can be found in the literature [23] but rather show some basic features of the connectionist approach.

Many of the connectionist approaches to intelligent control stem from the assumption that the system input and response can generally be described by some non-linear dynamical model represented in discrete time by the equation

$$y(t) = \Psi\left[y(t-1), \ldots, y(t-n); x(t-1), \ldots, x(t-n)\right] . \qquad (13.22)$$

In the connectionist approach, the function Ψ is modeled by a neural

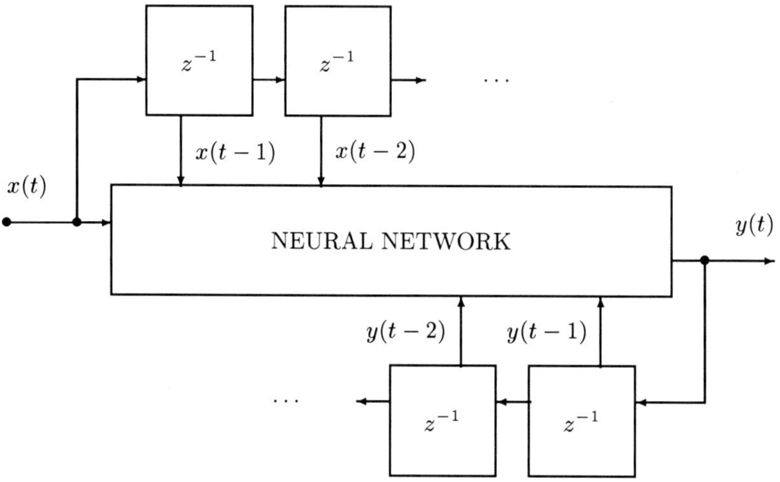

Fig. 13.6. A scheme of a neural network applicable to model a driven nonlinear dynamical system. z^{-1} denotes a unit step time delay of a signal

network represented in the scheme of Fig. 13.6. An identification of the system corresponds to the training of the network by some set of joined samples of $x(t)$ and $y(t)$. For this purpose, a supervised learning of neural networks with sigmoidal neurons by back-propagation algorithm is most frequently used. [16] In a similar way, a simple prediction control problem could be formulated as finding the function Ψ_c such that the control law

$$x(t) = \Psi_c\left[y(t), \ldots, y(t-n); x(t-1), \ldots, x(t-n)\right] \qquad (13.23)$$

when applied to the control described by a previous equation brings the state of the closed loop system to the desired value $y(t)$. [2, 3] With this approach, one hopes that the neural network will automatically learn to forecast the control signals from the past experience but it is seen from this example, that when one relies on the connectionist treatment of the problem using sigmoidal neurons and back-propagation learning, the advantage gained in interpreting the mapping properties in terms of the joint probability density and corresponding empirical samples is obscured. For this purpose self-organized modeling by a set of prototypes and estimation by a conditional average appear more easily interpretable.

13.4 Cloning

Let us consider a plant that is operating under control of an expert as shown in Fig. 13.7. The task of cloning is to design a controller that can be substituted for the expert. To realize the automatic control of the plant, we assume that all the physical variables used in the description can be monitored by an appropriate array of sensors. The expert is assumed to obtain from a director, instructions about the plant operation or signals that correspond to a reference physical variable expressed as $r(t)$. The latter need not directly represent the plant output trajectory. Most often, the orders are quasi-static, but in a flexible production they may be dynamic as well. We assume that under normal operation, the expert obtains also some information from the system by using sensors at the output of the plant. The variable $y(t)$ thus represents feedback to the expert. We avoid the question whether the variable y provides sufficient information to the expert. If the plant operates well and all the channels by which the expert obtains information are known and presented by variable $y(t)$ then we can simply assume that the corresponding information is sufficient. Based on the reference signal $r(t)$ and the current and past state of the plant $y(t)$, the expert manipulates the operation of the plant by setting the control variable $x(t)$.

To permit the replacement of the expert, we install in parallel a controller. This controller obtains the same input signals as the expert, but, in addition, it also obtains the manipulating variable $x(t)$. The controller then works in

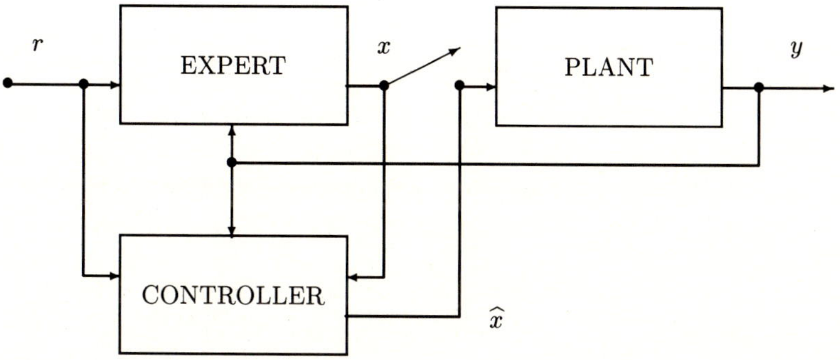

Fig. 13.7. A scheme of cloning

two different modes of operation corresponding to *learning* and *controlling*. During learning, the controller obtains only the input signals. The main task of learning is to construct an internal model of the dynamical phenomenon that is described by the triplet of dynamical variables $\{r(t),\ x(t),\ y(t)\}$. This task corresponds to a modeling of a dynamical system that is here an expert. Because of the delays which occur in the controller and the plant and because of possible nonlinear and chaotic dynamical phenomena of the plant, one needs to pay close attention to the design of the corresponding modeler. We, therefore, assume that during the learning phase all possible instructions described by the reference $r(t)$ appear and that the process is running sufficiently long as to obtain adequate information about all possible combinations of reference, manipulating variable and plant states. The main purpose of the learning is, in fact, not only to form the internal model of the process, but also to train the controller to reproduce by its output $\hat{x}(t)$ the manipulating variable $x(t)$ based on the given signals and the past observations of the expert behavior. In this aspect, cloning corresponds to a general identification of the expert.

In the second, controlling phase of operation, the expert is omitted and only the controller obtains the instructions or reference $r(t)$ and feedback $y(t)$, while its output $\hat{x}(t)$ is used to manipulate the plant. If modeling during learning was accurate, then we can expect that the trained controller will reproduce the "intelligent" control actions of the expert. Even more, we could expect to avoid faults caused by a tired or inattentive expert. Let us mention here that the "expert" need not be a person but instead, it may be a complex controlling device which we want to replace with a learning controller for various reasons such as cost minimization, speed of control, etc.

If the plant operates quasi-statically, no time delay needs to be considered and it is sufficient to compose from the recorded variables a sequence of vector samples representing reference, control and response at particular instants of time

$$\{\boldsymbol{z}_n = \boldsymbol{z}(t_n) = \big(r(t_n),\, x(t_n),\, y(t_n)\big);\ n = 1, 2, \ldots, N\}\,, \tag{13.24}$$

and apply them to obtain an empirical estimation of the joint probability density

$$f(\boldsymbol{z}) \;=\; \frac{1}{N}\sum_{n=1}^{N} w(\boldsymbol{z} - \boldsymbol{z}_n)\,, \tag{13.25}$$

in which w represents a judiciously selected window function. As mentioned previously, when there are too many samples with respect to the available memory space of the controller then we utilize a self-organization procedure by which a set of prototypes

$$\boldsymbol{Q} \;=\; \{\boldsymbol{q}_k = (q_{r,k},\, q_{x,k},\, q_{y,k});\ k = 1, 2, \ldots, K\} \tag{13.26}$$

is formed to represent the sample space and the corresponding joint probability density is

$$f(\boldsymbol{z}) = \frac{1}{K} \sum_{k=1}^{K} w(\boldsymbol{z} - \boldsymbol{q}_k) \,. \tag{13.27}$$

The prototypes must be stored in the controller memory to permit an estimation of the value of the manipulating variable by the conditional average

$$\hat{x} = E\left[x \,|\, r(t),\, y(t)\right] . \tag{13.28}$$

By denoting the condition represented by the sensory data as the vector

$$\boldsymbol{s}(t) = (\, r(t),\, y(t)\,) \tag{13.29}$$

we can express the estimator of the manipulating variable as

$$\hat{x} = \sum_{k=1}^{K} x_k \, C_k(\boldsymbol{s}(t)) \,, \tag{13.30}$$

where the similarity coefficients are expressed as

$$C_k(\boldsymbol{s}) = \frac{w\,(\boldsymbol{s} - \boldsymbol{q}_{s,k})}{\sum_{i=1}^{K} w\,(\boldsymbol{s} - \boldsymbol{q}_{s,i})} \qquad \text{with} \qquad \boldsymbol{q}_{s,k} \equiv (q_{r,k}, q_{y,k}) \,. \tag{13.31}$$

The expression for the conditional average estimator shows that it copies the expert actions described by prototypical values x_k. The coefficients $C_k(\mathbf{s})$ which are based on the present sensory data $\boldsymbol{s}(t)$ determine how much a particular prototypical action x_k influences the resultant one. With little effort, this procedure can again be generalized to the case where vector variables are used to represent the reference and the plant input and output signals.

By assuming a quasi-static operation of the expert and plant in the foregoing, the dynamics of the plant could safely be completely ignored. However, this assumption is not valid when dynamic phenomena with expressive inertial and non-linear effects are on-going. It is known that in such cases an expert incorporates the dynamical behavior of the plant to estimate how it could be manipulated by the input variable based on the past behavior. The assumption that the triplet $\{\, r(t), x(t), y(t) \,\}$ associated with the on-going processes can be adequately described by the joint probability distribution relating just the variables only one at a time time is likely an over-simplification. For a more complete description of the relationship between processes a cascade of probability density functions relating these variables at various time moments should be used instead. However, a general description cannot be obtained and it becomes then a question of how to find a properly truncated description of the dynamical process. We consider this in the next paragraph. For this, we will make some assumptions which arise from the forecasting of the operation of a dynamical process in the future.

First of all, a forecasting of a dynamical process could be done only based on the knowledge of the past long time behavior and the short time development of the phenomenon just before the forecasting. Knowledge of history can be used at the modeling of the phenomenon while the knowledge of the short-time behavior just prior to the forecasting can be used to specify the initial conditions. We, therefore, assume that it could be of advantage to utilize in a controller an approach that is similar to that which can be used to forecast a chaotic time series. We treat the expert and the plant as a complex dynamical phenomenon, driven by the instructions expressed by $r(t)$. The task of the cloning is then to find via a statistical approach, a generator that is capable of describing the dynamical evolution of such a complex system. For this purpose, we assume that the trajectory of the complete state vector $\boldsymbol{z}(t)$ forms in the corresponding phase space an attractor that can be sufficiently well-described by a joint probability distribution relating D successive points. From an extensive empirical record given by

$$\{\boldsymbol{z}(t)\,;\, t = \pm 1, \pm 2, \dots\} \tag{13.32}$$

we form the set of N D–point samples

$$\{\, \boldsymbol{Z}_t \;=\; [\,\boldsymbol{z}(t-D+1), \dots,\, \boldsymbol{z}(t-1),\, \boldsymbol{z}(t)\,];\; t = 1, 2, \dots, N\,\}\,, \tag{13.33}$$

by which we form the joint probability density function

$$f(\boldsymbol{Z}) \;=\; \frac{1}{N} \sum_{n=1}^{N} w(\boldsymbol{Z} - \boldsymbol{Z}_n)\,. \tag{13.34}$$

As in the previous case, but now with many more components involved, we can introduce prototypes representing the process by a corresponding self-organization process. The empirical probability density then provides a basis for estimating the driving variable based on a joint condition represented by the past values of variables. It can be expressed at a given time t by the truncated vector

$$\boldsymbol{g}(t) \;=\; \begin{pmatrix} r(t-D+1), & \dots, & r(t-1), & r(t), \\ x(t-D+1), & \dots, & x(t-1), & \emptyset, \\ y(t-D+1), & \dots, & y(t-1), & y(t) \end{pmatrix} \tag{13.35}$$

in which $\emptyset$ denotes the missing value to be estimated by the conditional average. The result

$$\hat{x} \;=\; E\,[x \,|\, \boldsymbol{g}(t)\,] \tag{13.36}$$

now accounts also for the dynamical behavior of the system and it also resembles the operation of an expert. An additional reason for such an expectation follows from the following consideration. An expert brings intelligent decisions about the manipulation of a plant that are based on past experiences and an ability to forecast the future behavior of the system. The past experience is hidden in the prototypes of the model, while the condition $g(t)$ represents the basis for the forecasting. In fact, the forecasting need not be carried out, because we presume that a mapping of the condition to future plant behavior and utilization of the corresponding forecast signal back to the manipulating variable are equivalent to an estimation of plant input based on previous decisions of the expert. In applying this procedure, one must properly select the dimension D. As in the case of the chaotic time series prediction this can be obtained by increasing the dimension until a proper performance is achieved.

In the foregoing, we have implicitly assumed that the instructions can be treated as being quasi-stationary. Were this not the case, then the problem would resemble, but be generally more difficult than, a tracking problem because the instructions can be treated as encoded information about the desired system trajectory. The task of the expert is to first decode the instructions or orders and then to adapt the control actions to realize the desired trajectory. The problem is thus essentially reduced to tracking. For this purpose, an expert will usually try to forecast instructions which are only based on the previous performance. In order to make feasible cloning of expert operation in such a case it could be of advantage to include a forecaster of orders into the cloning system as well. In this case, cloning is equivalent to *neuro-identification* because we are attempting to model the dynamical operation of an expert's brain. Robust methods being developed for this purpose are still under intensive research. [23]

However, a comment about the control performed by expert operators has to be given here. When bringing decisions an expert often utilizes all the previously learned knowledge about the controlled system as well as tremendous information provided by the human sensors and brain. Therefore, we could hardly expect that an exact cloning would be possible based on a much simpler system comprised of less extensive sensory network and non-parametric modeler.

Cloning of expert operation is similar to the formation of expert systems which are based on methods of artificial intelligence (AI). The principal difference between the method proposed here and the formation of expert systems is that in our case the complete information about the expert operation is provided by sensory signals which can also change dynamically in time. In contrast, expert systems are based predominantly on the transformation of expert reasoning expressed in words or rules into a set of logical functions of AI that are subsequently transformed into control actions. [22]

13.5 An Empirical Approach to Optimal Control

13.5.1 The Theoretical Problem of Optimal Control

In order to proceed with the empirical description of optimal control, let us first give a theoretical description of its fundamental problem. Consider again a controlled plant that can be described by the dynamical law

$$\dot{\boldsymbol{y}} = \boldsymbol{\Phi}(\boldsymbol{y}(t), \boldsymbol{x}(t), t) \tag{13.37}$$

and the initial condition $\boldsymbol{y}(t_0)$. In addition, various physical constraints often exist on the allowable values of plant state and control variable. Let there be also specified the instantaneous utility function $u(t) = u(\boldsymbol{x}(t), \boldsymbol{y}(t), t)$ and a corresponding integral performance measure:

$$J = \int_{t_i}^{t_f} u\,(\boldsymbol{y}(t), \boldsymbol{x}(t), t)\, dt\ . \tag{13.38}$$

The goal of any optimal control theory is generally described by the solution to the following problem:

> **P1**. *Given the plant described by the dynamical law Eq. (13.37) find an admissible control $\boldsymbol{x}(t)$ which causes the plant to follow an admissible trajectory $\boldsymbol{y}(t)$ that optimizes the performance measure Eq. (13.38).*

This is generally a constrained optimization problem. The corresponding functions $\boldsymbol{x}(t)$ and $\boldsymbol{y}(t)$ represent the optimal control signal and the optimal plant trajectory, respectively. By solving the above-stated problem, we seek to obtain a functional relationship

$$\boldsymbol{x}(t) = \boldsymbol{\phi}[\boldsymbol{y}(t), t] \tag{13.39}$$

that yields the optimal control variable for all admissible plant states. It is called the *optimal control law.* A strict mathematical treatment of the optimal control problem is based on the calculus of variation which is often cumbersome. Therefore many methods have been developed in order to find approximate solutions. Among these the dynamical programming is probably the best-known. [13, 8, 23] To execute the optimal control, a controller requires the information about the plant state. The corresponding signal is provided by the feedback from the plant output to the controller.

For a given plant which is to be controlled we are generally faced with the problem of designing a proper controller. A theoretical basis for the design can be extracted from the expression of the optimal control law which reflects the properties of the plant as well as the performance measure. [13] However this is not possible when utilizing an empirical description that is based on measured data of the plant input and output and the corresponding utility. Therefore, in the following paragraph we shall slightly modify the optimal control problem described in **P1**. Accordingly, we shall first consider the empirical description of the utility function and the plant performance which will subsequently lead us to the description of an information processing system which is capable of intelligent control based on an empirical approach.

13.5.2 Experimental Description of Plant Performance and Optimal Control

An experimental characterization of the behavior of a plant is usually obtained using a digital data acquisition system which takes measurements of the plant input and output at evenly spaced sampling times denoted as $t = 1, 2, \ldots, N$. The dynamical law is then represented by a mapping

$$\boldsymbol{y}(t+1) = \boldsymbol{\Psi}(\boldsymbol{y}(t), \boldsymbol{x}(t), t) \qquad (13.40)$$

in which the system function $\boldsymbol{\Psi}(.)$ must be described in terms of the recorded data.

In a theoretical description of the optimal control, the instantaneous utility of the system is expressed by some function of the system state and time, such as $u(t) = u(\boldsymbol{x}(t), \boldsymbol{y}(t), t)$. This function is task-dependent and consequently it cannot generally be specified. In practice, it is usually composed of elementary functions. Suggestion for the composition of the utility function can be obtained from an analysis of the purpose of the control, which is determined by the user of the system to be controlled. Quite different is the situation in an approach based on measured data. In this case, the utility must be experimentally estimated from observations of the system behavior. For this purpose, there must exist an experimental system called *the utility estimator* which transforms the given data of plant input and output into the value of utility variable. This value generally depends on the properties of the environment in which the controlled plant operates. For now we shall not consider the description of the environment, but later we shall consider its influence as well. We can design the utility estimator as an information processing system that obtains signals from the system input and output and transforms them into a signal representing the utility. It can be joined with the data acquisition system.

A real system generally exhibits stochastic properties. Therefore in the experimental approach, we utilize a probabilistic description of the plant input, output and utility. Moreover, if we are observing only one system, then we must assume stationarity of several statistical characteristics in order to prepare an empirical basis for the decisions needed to effect an optimal control. The plant input and output as well as the corresponding utility are then treated as joined stochastic signals. The functional dependence of the plant utility on the control and plant states is empirically described by a measured set of joined data: $\{\boldsymbol{x}(t), \boldsymbol{y}(t), u(t) \; ; \; t = 1, 2, \ldots, N\}$. To provide for the simultaneous description of the dynamical law, we add the data about the plant output at the next time step to the triplet of joined measured data. The data set

$$\mathcal{D} = \{\boldsymbol{X}(t) = (\boldsymbol{x}(t), \boldsymbol{y}(t), \boldsymbol{y}(t+1), u(t)) \ ; \ t = 1, 2, \ldots, N\} \tag{13.41}$$

thus represents the complete empirical information about the system behavior. Our task is to extract the optimal control law from it. In accordance with the statistical treatment of the plant and its utility, in this approach to optimal control we introduce the performance measure that is expressed by the statistically expected value

$$\overline{J(t_\mathrm{i}, t_\mathrm{f})} = E\left[\sum_{t_\mathrm{i}}^{t_\mathrm{f}} u\,(t)\right] . \tag{13.42}$$

The difference between specification of the utility function in the theoretical and experimental approaches leads to a significant distinction between the design of optimal controllers using these approaches. In the theoretical approach, the properties of the utility function, that is specified *a priori* for the complete time interval for which the system is to be controlled, is incorporated into the solution of the general problem represented by the optimal control law. Because the design of the optimal controller is obtained from the optimal control law, the properties of the utility function are consequently also transmitted to the properties of the controller. [13] The structure of the controller is determined in advance. If one possesses information about the utility function then the structure of the controller can be so designed that in the pre-determined time interval it will operate optimally according to the assumed utility. This is, however, not possible in the experimental approach because in this the utility function is not available before the start of experimental work. Rather it needs to be determined experimentally during the operation of the system. We are thus faced with an essentially different problem for optimally controlling a system than that described in **P1**. First of all, we must provide empirical information about the plant properties during some introductory period during which we cannot expect that the system could operate optimally. During this period the phase space, in which the dynamics of the system develops, must be explored and the corresponding empirical data stored. This period corresponds to the specification of the dynamical law and the utility function in the theoretical approach. Only after such an exploration period can we expect that the empirical control could be optimal, provided that the properties of the system do not change in the future, meaning that they are stationary, at least in the statistical sense. In order to delineate between the theoretical and empirical approaches to optimal control, we describe the fundamental problem of empirical optimal control as follows:

P2. *Given a plant whose operation is characterized by a set of joined experimental data* $\mathcal{D} = \{\boldsymbol{X}(t) = (\boldsymbol{x}(t), \boldsymbol{y}(t), \boldsymbol{y}(t+1), u(t)) \ ; \ t =$

$1, 2, \ldots, N\}$, *and a system whose properties are stationary in a statistical sense, one is to design a controller that will drive the system over a trajectory corresponding to the optimal, expected plant performance Eq. (13.42).*

There are several significant differences between the problems stated as **P1** and **P2**. The first is in the description of the system dynamics and utility which, in the latter case is completely based on experimental data. There is no *a priori* information introduced in the experimentally based statement. The second difference is in the specification of the plant performance. The theoretically posed integral performance measure of the first case is replaced in the second case by the statistical average measure that can be experimentally estimated or predicted. For this purpose, reliable experimental data must be provided, which means that statement **P2** implicitly assumes existence of a utility estimator. Because of the stationarity there is a less strict desire about the optimum of the performance in the second case that is a consequence of the probabilistic description of expected performance in the future. By a less stringent desire for optimum a chance is provided for the designer of the controller to include an adaptation procedure by which the system learns how to perform optimally from past examples. Using such a description we proceed towards an empirical treatment of control systems resembling biological organisms living in noisy environments. Such systems exhibit a softer behavior than the rigid mechanistic examples treated by the strict theoretical approach. It is also characteristic that in the above statement nothing is said about the process of data acquisition or the past behavior of the system. This means that the system can be "born" at an arbitrary moment and that any statistically acceptable method of data acquisition can be utilized. However, this ambiguity may lead to problems related to the statistically optimal acquisition of data. The specification of the basic problem of empirical optimal control thus resembles the formulation in the theory of optimal control of stochastic systems based on application of joint probability functions. [4]

13.5.3 Design of an Intelligent Optimal Controller

Our next goal is to show how to proceed to the solution of the experimentally-posed problem of optimal control by using previously developed methods of non-parametric empirical description of dynamical phenomena. Based on the above discussion we assume that the basic tasks of the empirical optimal control can be performed by separate units that are connected by information transmission channels as shown in Fig. 13.8.

The most important unit needed for an empirical optimal control system is an intelligent modeler that is capable of creating an internal empirical model of the complete system to be controlled. For this purpose, it obtains sensory signals that describe the state of the other units of the system. Later, we shall consider also measurements of influences from the environment, but

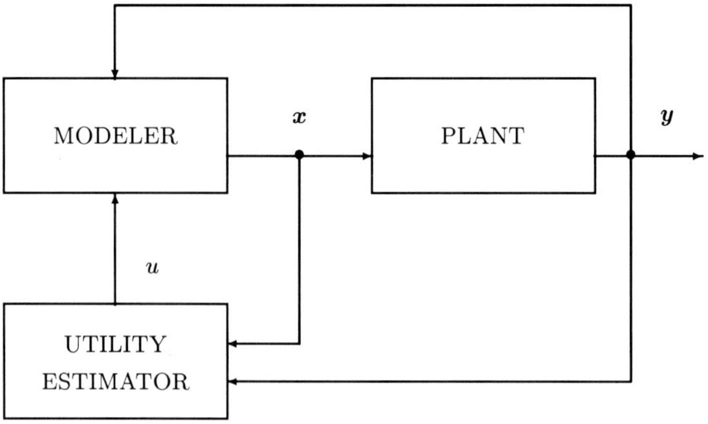

Fig. 13.8. The scheme of an intelligent optimal controller

for the sake of simplicity we first treat the environment as a strictly stationary one. In order to permit the optimization of the control, the utility estimator is included into the system. It obtains signals from the plant input $\boldsymbol{x}(t)$ and output $\boldsymbol{y}(t)$ and generates a signal representing the corresponding instantaneous utility $u(t)$. This signal is transmitted to the modeler which estimates from it, and the system state, an optimal reference control signal $\widehat{\boldsymbol{x}}(t)$. By proper amplification, the reference signal is converted into the manipulating variable $\boldsymbol{x}(t)$ by a booster which can be designed as a classical tracking controller. For sake of simplicity the modeler and booster are represented by the modeler unit in the schematic diagram of the intelligent controller shown in Fig. 13.8.

Let us assume that the modeler of the self-controlled system obtains during its training period complete sensory information in a variety of situations and operating conditions. Using this information, the modeler forms the set of prototypes. We assume that various controls can generally be executed at each possible state which can result in transitions to various successor states with different utilities. This means that the set of prototypes must generally include various prototypes possessing identical initial states and different successor states. In order to emphasize this property of the empirical model, we introduce a two-component labeling of the prototypes using the index $k = (\iota, \kappa)$ in which the maximal value of ι or κ is determined by the number of I possible starting states. The first component denotes the initial and the second the successor state. The corresponding controls and utilities are $x_{\iota\kappa}, u_{\iota\kappa}$. With this notation the prototypes can be represented in the form:

$$\boldsymbol{q}_{\iota\kappa} = (\boldsymbol{y}_\iota, \boldsymbol{y}_\kappa^+, \boldsymbol{x}_{\iota\kappa}, u_{\iota\kappa}) \tag{13.43}$$

and any sum with respect to index k must be considered as a double sum over the indices $\iota\,\kappa$. The superscript $^+$ denotes the successor state.

The prototypes can be used to forecast the system trajectory and the corresponding performance provided that control signal and the initial system state are specified. For this purpose, we determine from each given pair of control and system state variables the utility and the forthcoming state by the conditional average

$$[\widehat{\boldsymbol{y}}(t+1), \widehat{u}(t)] = E[\boldsymbol{y}(t+1), u(t) \mid \boldsymbol{y}(t), \boldsymbol{x}(t)] \\ = \sum_k B_k(\boldsymbol{y}(t), \boldsymbol{x}(t))[\boldsymbol{y}_k^+, u_k]\,, \tag{13.44}$$

$$\text{where} \qquad B_k(\boldsymbol{y}(t), \boldsymbol{x}(t)) = \frac{w(\boldsymbol{x}(t) - \boldsymbol{x}_k, \boldsymbol{y}(t) - \boldsymbol{y}_k)}{\sum_i w(\boldsymbol{x}(t) - \boldsymbol{x}_i, \boldsymbol{y}(t) - \boldsymbol{y}_i)}\,. \tag{13.45}$$

From the state and control variable at time $t+1$ the next successor state and utility can then be predicted and so forth. By summing the predicted values of the utility $\{\widehat{u}(t)\;;\; t = t_{\mathrm{i}}, \ldots, t_{\mathrm{f}}\}$, the predicted performance measure

$$\widehat{J}(t_{\mathrm{i}}, t_{\mathrm{f}}) = E\left[\sum_{t=t_{\mathrm{i}}}^{t_{\mathrm{f}}} \widehat{u}\left(\boldsymbol{x}(t), \widehat{\boldsymbol{y}}(t)\right)\right] \tag{13.46}$$

can be calculated for the trajectory which is determined by the initial condition $\boldsymbol{y}(t_{\mathrm{i}})$ and the control signal $\{\boldsymbol{x}(t)\;;\; t = t_{\mathrm{i}}, \ldots, t_{\mathrm{f}}\}$. Because one uses the utility determined by the conditional average, this performance represents an empirically-estimated statistically expected value.

In order to proceed to the optimal control of the system, the modeler could, in principle, randomly generate various trial control signals, calculate for each one the corresponding performance and memorize the best one for the later application in control of the system. However, such a random search procedure is computationally prohibitive because of the explosive growth of the number of possibilities. Therefore, the question arises as how the empirical model hidden in the set of prototypes could be more efficiently utilized in the search for an optimal control variable. With this goal in mind, let us first explain how such a system can achieve approximately optimal utility in a single step transition from a given state to a possible successor state without extensive search of all possibilities. We then generalize the proposed method to describe the optimization of system performance over many steps.

One-Step Empirical Optimal Control. For the sake of simplicity, we assume that the utility is limited and that its absolute maximum is described by u_o. For this purpose, the optimal utility value from the set of all prototypes can be utilized. Instead of randomly generating various control values and selecting the optimal one among them, the modeler can directly estimate the optimal control variable by the conditional average in which the condition is comprised of the state variable $\boldsymbol{y}(t)$ and the optimal utility u_o:

$$\widehat{\boldsymbol{x}}(\boldsymbol{y}(t), u_o) = E[\boldsymbol{x}|\boldsymbol{y}(t), u_o] = \sum_k B_k(\boldsymbol{y}(t), u_o)\boldsymbol{x}_k \ , \tag{13.47}$$

$$\text{where} \qquad B_k(\boldsymbol{y}(t), u_o) = \frac{w(\boldsymbol{y}(t) - \boldsymbol{y}_k, u_o - u_k)}{\sum_i w(\boldsymbol{y}(t) - \boldsymbol{y}_i, u_o - u_k)} \ . \tag{13.48}$$

To carry out this estimation, no additional optimization procedure is required. Even the exact optimal value of the utility u_o is not needed, only a value that is greater than it. The estimated optimal control variable $\widehat{\boldsymbol{x}}$ need not yield at a given state of the system a utility value equal to u_o but we can expect that when the utility function has a single maximum, the achieved value will be closest to u_o, that is the best possible at this instant. However, this is an expected result only if the system obtains sufficient information about the possible actions during formation of the empirical model in the training period.

Based on the requirement that the determined utility must be similar to the optimal one, we can create from the prototypes $\boldsymbol{q}_{\iota\kappa}$, which include the same starting vector $\boldsymbol{y}_\iota$ and different controls $\boldsymbol{x}_{\iota\kappa}$, just one new prototype $\boldsymbol{q}^*_{\iota,\mu}$. In a simplified treatment this prototype can be represented by that member of the set $\{\boldsymbol{q}_{\iota\kappa}\}$ which represents the transition with the maximal possible utility $u^*_{\iota\kappa}$. A set of such prototypes with different initial state vectors then represents the control law. The number of prototypes that represent the empirical model is $K = I^2$ while the number of prototypes that represent the control law is reduced to I. By application of the control law, the system can be driven in the state space along the trajectory for which it is characteristic that at each transition the utility is the highest momentarily possible. This trajectory is called *the trajectory of optimal utility.* Each point on this trajectory corresponds to a locally optimal behavior of the system. The corresponding states can be estimated by using optimal controls or more directly by successive application of the conditional average

$$\widehat{\boldsymbol{y}}(t+1) = E[\boldsymbol{y}^+|\boldsymbol{y}(t), u_o] = \sum_k B_k(\boldsymbol{y}(t), u_o)\boldsymbol{y}_k^+ \ . \tag{13.49}$$

Multi-Step Empirical Optimal Control. The trajectory of optimal utility may not correspond to a globally optimal system behavior because a locally optimal control can lead the system into a region from which it can later exit only with very low utility. To avoid such a possibility a sequence of dynamic transitions must be considered and the corresponding global performance must be optimized over the sequence. Our task is therefore to describe how the empirical model which is represented by the set of prototypes $\{\boldsymbol{q}_{\iota\kappa}\}$ can be utilized in the statistical estimation of a multi-step optimal trajectory. Let us for this purpose first consider a two-step transition from $\boldsymbol{y}_\iota$ to $\boldsymbol{y}_\lambda$. It can go over various intermediate states $\boldsymbol{y}_\kappa$ with the resulting two-step performance $J_{\iota\kappa\lambda} = u_{\iota\kappa} + u_{\kappa\lambda}$. Information about the corresponding dynamics is represented by the set of double prototypes

$$\mathcal{Q} = \{\mathbf{Q}_{\iota\kappa\lambda} = (\boldsymbol{q}_{\iota\kappa}, \boldsymbol{q}_{\kappa\lambda}) = (\boldsymbol{y}_{\iota}, \boldsymbol{y}_{\kappa}, \boldsymbol{y}_{\lambda}, \boldsymbol{x}_{\iota\kappa}, \boldsymbol{x}_{\kappa\lambda}, u_{\iota\kappa}, u_{\kappa\lambda}); \kappa = 1, \ldots, I\} . \tag{13.50}$$

For fixed initial and final states the performance depends only on the intermediate state. We select among all of them that one which yields the highest performance:

$$J^*_{\iota\lambda} = \max_{\kappa} J_{\iota\kappa\lambda} = \max_{\kappa} (u_{\iota\kappa} + u_{\kappa\lambda}) . \tag{13.51}$$

This condition leads us to the optimal intermediate state with index κ_o and the corresponding controls $\boldsymbol{x}_{\iota\kappa_o}$ and $\boldsymbol{x}_{\kappa_o\lambda}$. However, the execution of the corresponding mathematical procedure requires the comparison of I values of performance. As in the one-step procedure, we therefore propose a simplified treatment that is based on a conditional average which can be determined without comparisons between states. If the absolute maximum of the utility is u_o, then the performance in two steps cannot surpass the value $J^* = 2u_o$. We can therefore employ the condition $u_1 + u_2 = 2u_o$ together with the initial and final states. The corresponding generalized conditional average which yields both optimal controls can be estimated using the set of prototypes $\{\mathbf{Q}_{\iota\kappa\lambda}\}$ as

$$\begin{aligned}(\widehat{\boldsymbol{x}}_1, \widehat{\boldsymbol{x}}_2)_{\iota\lambda} &= E[\boldsymbol{x}_1, \boldsymbol{x}_2 | \boldsymbol{y}_{\iota}, \boldsymbol{y}_{\lambda}, u_1 + u_2 = 2u_o] \\ &= \sum_{\kappa} B(u_{\iota\kappa} + u_{\kappa\lambda} - 2u_o)(\boldsymbol{x}_{\iota\kappa}, \boldsymbol{x}_{\kappa\lambda}) .\end{aligned} \tag{13.52}$$

Here the subscripts $_1$ or $_2$ denote the starting and the succeeding control or utility while the coefficients in the sum are described by the basis function

$$B(u_{\iota\kappa} + u_{\kappa\lambda} - 2u_o) = \frac{w(u_{\iota\kappa} + u_{\kappa\lambda} - 2u_o)}{\sum_{\nu} w(u_{\iota\nu} + u_{\nu\lambda} - 2u_o)} . \tag{13.53}$$

These coefficients are not directly influenced by the values of initial and final state vectors $\boldsymbol{y}_{\iota}, \boldsymbol{y}_{\lambda}$ but only by their indices $\iota\lambda$ which determine the utilities employed in the last expression. As in the one-dimensional case, the estimated sequence of optimal controls $(\widehat{\boldsymbol{x}}_1, \widehat{\boldsymbol{x}}_2)_{\iota\lambda}$ can be described approximately by one prototype $\boldsymbol{Q}_{\iota\kappa\lambda}$ with the most similar control components.

This procedure can be extended to a sequence of m steps. In this case, the performance depends on $s = m - 1$ intermediate states and it can be expressed as

$$J_{\iota\kappa_1\ldots\kappa_s\lambda} = u_{\iota\kappa_1} + u_{\kappa_1\kappa_2} + \ldots + u_{\kappa_s\lambda} . \tag{13.54}$$

The intermediate states must be specified when searching for the optimal performance

$$J^*_{\iota\lambda} = \max_{\kappa_1\ldots\kappa_s} J_{\iota\kappa_1\ldots\kappa_s\lambda} . \tag{13.55}$$

For this purpose Eq. (13.52) can be generalized by using the condition

$$u_{\iota\kappa_1} + u_{\kappa_1\kappa_2} + \ldots + u_{\kappa_s\lambda} = su_o , \tag{13.56}$$

which yields the following empirical statistical estimator of multi-step optimal control

$$\begin{aligned}(\widehat{\boldsymbol{x}}_1, \ldots, \widehat{\boldsymbol{x}}_m)_{\iota\lambda} &= E[\boldsymbol{x}_1, \ldots, \boldsymbol{x}_m | \boldsymbol{y}_\iota, \boldsymbol{y}_\lambda, u_1 + \ldots + u_m = s u_o] \qquad (13.57)\\ &= \sum_{\kappa_1 \ldots \kappa_s} B(u_{\iota\kappa_1} + \ldots + u_{\kappa_s\lambda} - s u_o)(\boldsymbol{x}_{\iota\kappa_1}, \ldots, \boldsymbol{x}_{\kappa_s\lambda}) \; .\end{aligned}$$

The coefficients in the sum are described by the basis function

$$B(u_{\iota\kappa_1} + \ldots + u_{\kappa_s\lambda} - s u_o) = \frac{w(u_{\iota\kappa_1} + \ldots + u_{\kappa_s\lambda} - s u_o)}{\sum_{\nu_1 \ldots \nu_s} w(u_{\iota\nu_1} + \ldots + u_{\nu_s\lambda} - s u_o)} \; . \qquad (13.58)$$

The multiple sum in the last two equations runs over s indices which altogether describe I^s terms. Consequently, an explosive expansion of computation is met with an increasing number of intermediate steps which requires a proper numerical treatment. For this purpose, the method of dynamic programming which was developed by Bellman in the early 1950's can be successfully applied.[6]

Dynamic Programming Based on Prototypes. The multiple sum in Eq. 13.58 indicates that in order to find the optimal trajectory we must search over the complete set of admissible trajectories from the initial to the final state. This significantly complicates the way to the optimal control of a dynamical system. The problem is not a consequence of the empirical description of the dynamics of the system but stems from the character of the optimal control which is based upon the integral performance measure. It is in fact the central problem of optimal control theory and the vast literature of this field is devoted to its study. [7, 4, 8, 13, 23] We shall not proceed with the description of various methods developed for its solution but rather only briefly explain *Bellman's Principle of Optimality* and the fundamentals of the approach to dynamic programming which can be easily related to the non-parametric empirical description of natural phenomena. [6]

The search of optimal controls and corresponding indices in the multi-step case can be arbitrarily split into two steps as indicated by the expression [13, p.54]

$$J^*_{\iota\lambda} = \max_{\kappa_n} \left(J_{\iota\kappa_n} + J_{\kappa_n\lambda} \right) \; . \qquad (13.59)$$

This equation can be given more appropriate form by invoking *Bellman's Principle of Optimality* [6] which can be formulated as follows: [6, 13]

> **Principle of Optimality :** *An optimal policy has the property that whatever the initial state and initial decision are, the remaining decisions must constitute an optimal policy with regard to the site resulting from the first decision.*

In our case the decision is expressed by the generated control variable and the policy is expressed by the control law. Eq. 13.59 can be then expressed as

$$J^*_{\iota\lambda} = J^*_{\iota\kappa_n} + J^*_{\kappa_n\lambda} \, . \tag{13.60}$$

If $n = s/2$ then the total number of terms that must be processed to calculate $J_{\iota\kappa_n}$ and $J_{\kappa_n\lambda}$ is $2I^{s/2}$. This number is smaller by a factor $2I^{-s/2}$ than the number of terms processed in the direct calculation of $J^*_{\iota\lambda}$ which, for large s, represents a significant reduction of calculation. It is, therefore, reasonable to proceed with the indicated splitting still further and to perform the search for the optimal trajectory from an initial to a final state sequentially by starting with a single transition at one terminal and by including ever more transitions. The search can proceed either from the initial state towards the final one or vice versa. The second possibility is most frequently utilized when solving optimal control problems by dynamic programming. By this procedure the explosive growth of the number of trajectories that must be considered in the direct search of the optimal trajectory with an increasing number of steps is suppressed to a linear growth.

In order to proceed to the method of dynamic programming we express the principle of optimality in terms of single transitions as follows:

Let the sequence $\{\boldsymbol{x}^*_{\iota\kappa_1}, \ldots, \boldsymbol{x}^*_{\kappa_n\kappa_{n+1}}, \ldots, \boldsymbol{x}^*_{\kappa_s\lambda}\}$ denote the optimal control for the trajectory starting and finishing in the states $\boldsymbol{y}_\iota$, $\boldsymbol{y}_\lambda$, then the sequence $\{\boldsymbol{x}^*_{\kappa_n\kappa_{n+1}}, \ldots, \boldsymbol{x}^*_{\kappa_s\lambda}\}$ denotes the optimal control for the trajectory starting in an arbitrary intermediate state of optimal trajectory $\boldsymbol{y}_{\kappa_n}$ and finishing at the same final state $\boldsymbol{y}_\lambda$.

Using this principle we can express the optimal performance measure corresponding to the intermediate sequence by the utility in the first succeeding step and the remaining optimal performance by the following *Bellman functional equation*: [4]

$$\overline{J^*_{\kappa_n\lambda}} = \underset{\boldsymbol{x}_{\kappa_n\kappa_{n+1}} \in \mathcal{X}_{ad}}{\text{opt}} \left\{ E[u(\boldsymbol{x}_{\kappa_n\kappa_{n+1}}, \boldsymbol{y}_{\kappa_n})] + \overline{J^*_{\kappa_{n+1}\lambda}} \right\} \, . \tag{13.61}$$

Here $\mathcal{X}_{ad}$ denotes the set of admissible controls and * denotes the optimal value. This equation shows that we can determine at an arbitrary state $\boldsymbol{y}_{\kappa_n}$ the corresponding optimal control $\boldsymbol{x}^*_{\kappa_n\kappa_{n+1}}$ provided we know the optimal performances for all the states that can be achieved from this state by various controls. We therefore do not need to search over all the possible trajectories stemming from the given state to find the optimal performance and control. This generally greatly diminishes the number of computational steps needed in the search of optimal trajectory. For instance, we can begin the computation at the admissible final states and determine the corresponding performance there. Using these performances, we can then find the optimal controls and corresponding optimal performances in the states from which the final states can be achieved. Thus we proceed with calculations backward from the possible final states. This process is identical to Bellman's *the dynamic programming.* The iterative process of backward calculations ends at the initial state. However, the complete procedure is invariant with respect to exchange of labeling of initial and final states and therefore the dynamic programming can also be executed in the forward direction. Because we retain at each step of the search process only the optimal trajectories stemming from the states considered, the number of trajectories that are considered as candidates of becoming the globally optimal one is essentially lower than in the direct search over the set of all possible trajectories and this number increases only linearly with the number of steps s: $N_{\mathrm{tr}} \propto s$. [13] This advantage has led to numerous applications of dynamic programming in optimal control problems. However, because various candidate trajectories must be considered in evaluations it is generally still computationally demanding. [13, 4, 23]

The Bellman's equation provides the basis for the solution of optimal control problems. However, in many cases, an analytical solution of the problem cannot be found. For a numerical treatment the time, the state and the control variables must be quantized into a finite number of levels.,[13] This step is automatically performed in approach described here by utilizing the set of prototypes $\mathcal{Q} = \{\boldsymbol{q}_{\kappa_n \kappa_{n+1}} = (\boldsymbol{y}_{\kappa_n}, \boldsymbol{y}^+_{\kappa_{n+1}}, \boldsymbol{x}_{\kappa_n \kappa_{n+1}}, u_{\kappa_n \kappa_{n+1}});\ \kappa = 1, \ldots, I\}$ to describe the complete dynamical phenomenon. We can carry out the dynamic programming using these prototypes by the following procedure:

1. At the final time t we select as possible admissible system states $\boldsymbol{y}^+_{\kappa_t}$ with κ_t from 1 to I. If a special utility $S(\boldsymbol{y}_{\kappa_t})$ is determined in the specification of the control problem then it is assigned to these states and utilized as the performance of the states at the final time $J^*(\kappa_t, t) = S(\boldsymbol{y}_{\kappa_t})$.
2. To each of the final states there is associated the previous state $\boldsymbol{y}_{\kappa_{t-1}}$ which represents the state at the time $t-1$. The corresponding utility $u_{\kappa_{t-1}\kappa_t}$ is added to the $S(\boldsymbol{y}_{\kappa_t})$ in order to obtain the performances at $t-1 : J^*(\kappa_{t-1}, \kappa_t) = u_{\kappa_{t-1}\kappa_t} + J^*(\kappa_t, t)$. These performances must be memorized for all κ_{t-1} and $\kappa_t \in (1, I)$.
3. The application of Bellman's equation begins at time $t-2$ for which we select an arbitrary state $\boldsymbol{y}_{\kappa_{t-2}}$. We assume that the system transits from this state to a state $\boldsymbol{y}^+_{\kappa_{t-1}}$ at the time $t-1$ by an appropriate control $\boldsymbol{x}_{\kappa_{t-2}\kappa_{t-1}}$ and with the corresponding utility $u_{\kappa_{t-2}\kappa_{t-1}}$. There are generally various possible paths for going from the state $\boldsymbol{y}_{\kappa_{t-2}}$ to the state at the final time, each corresponding to different κ_{t-1}. With each of them is associated the performance measure

$$J(\kappa_{t-2}, \kappa_{t-1}, \kappa_t) = u_{\kappa_{t-2}, \kappa_{t-1}} + J^*(\kappa_{t-1}, \kappa_t) \,. \tag{13.62}$$

According to Bellman's equation, we select among all of them that one for which $J(\kappa_{t-2}, \kappa_{t-1}, \kappa_t)$ is optimal. This condition determines the value of optimal κ_{t-1} and the corresponding optimal path to the final state. With it the optimal control $\boldsymbol{x}^*_{\kappa_{t-2}, \kappa_{t-1}}$

and the corresponding performance measure $J^*(\kappa_{t-2}, \kappa_t)$ is determined at the time $t-2$. This procedure must be repeated for all values of κ_{t-2} and the corresponding performances and the optimal controls memorized.

4. Using the memorized optimal performances at the time $t-2$, the next step of iteration starts at time $t-3$, etc. The iteration ends at the initial time $t_i = 1$.

In using Bellman's equation in the above procedure we have omitted the statistical average because we have assumed that the prototypes have been obtained by the self-organization process which is related to empirical averaging. During execution of the dynamic programming procedure we determine at the time t' in each particular state $\boldsymbol{y}_{\kappa_{t'}}$ the optimal performance measure $J^*(\kappa_{t'}, \kappa_t)$ and the optimal control $\boldsymbol{x}^*_{\kappa_{t'},\kappa_{t'+1}}$. We can join with this triplet, the optimal successor state of $\boldsymbol{y}_{\kappa_t}$ which is denoted as $\boldsymbol{y}^{*+}_{\kappa_{t'}}$. If we divide the optimal performance by the remaining number of steps to reach the final time $\Delta t = t - t'$, we can define the average utility for the determined optimal path as

$$u^*_{\kappa_{t'},\kappa_{t'+1}} = \frac{J^*(\kappa_{t'}, \kappa_t)}{\Delta t} . \tag{13.63}$$

Using these data we can describe the optimal control trajectory by the set

$$\mathcal{Q}^*(t') = \{\boldsymbol{y}_{\kappa_{t'}}, \boldsymbol{y}^{*+}_{\kappa_{t'}}, \boldsymbol{x}^*_{\kappa_{t'}\kappa_{t'+1}}, u^*_{\kappa_{t'}\kappa_{t'+1}} \; ; \; \kappa_{t'} = 1, \ldots, I\} . \tag{13.64}$$

It describes in an empirical way the optimal system dynamics and the corresponding optimal control law that is extracted from the fundamental set

$$\mathcal{Q} = \{\boldsymbol{y}_\iota, \boldsymbol{y}^+_\kappa, \boldsymbol{x}_{\iota\kappa}, u_{\iota\kappa} \; ; \; \iota, \kappa = 1, \ldots, I\} \tag{13.65}$$

by which the set of admissible trajectories of the complete system was initially characterized. The set of prototypes given by Eq. 13.64 represents an empirically estimated optimal control law that leads the system from the initial to the final state. It is still not clear how this procedure can be extended to the case of an unlimited number of steps and an unspecified final state. As has previously been mentioned, for many phenomena of practical interest, the optimal control law converges with the number of steps to a steady law which is independent of t' and can be represented by the set of prototypes

$$\mathcal{Q}^* = \{\boldsymbol{y}_\kappa, \boldsymbol{y}^{*+}_\kappa, \boldsymbol{x}^*_{\kappa\kappa+}, u^*_{\kappa\kappa+} \; ; \; \kappa = 1, \ldots, I\} . \tag{13.66}$$

In some cases the optimal trajectory corresponds, or is at least close to the trajectory of optimal utility that has been described by the conditional average $\widehat{\boldsymbol{x}} = E[\boldsymbol{x} \,|\, \boldsymbol{y}, u_0]$ in Eq. (13.47). Such an estimator is also applicable in cases when only an approximately optimal trajectory must be estimated.

In the computational procedure for the empirical estimation of optimal control that is described above it was assumed that the trajectories run only over states represented by prototypes. This restriction can be relaxed by allowing for arbitrary states and using the conditional average to describe the transitions between them.

Related to the search of optimal trajectories there arises a question how one can modify the self-organization process such that the resulting prototypes would directly yield the result presented in Eq. (13.64) rather than the fundamental set described by Eq. (13.65). In principle this could be achieved during the exploration of the state space during the development of the model. This option is still an object of intense research in the field of neural networks and we shall turn to it when discussing the methods of state space exploration. [23] Before doing this however, we shall first generalize our description by considering the effects of the environment.

13.5.4 The Influence of the Environment on Optimal Control

In our treatment of optimal control in the previous section we have assumed for the sake of simplicity that the environment is unchanging. However, dynamical systems are a part of nature and generally cannot be isolated from their surroundings. Moreover, the influences from the environment often critically affect the behavior of the system. We must therefore consider the environment of a system as an essential object of the description in the domain of optimal control. This is especially true when systems such as mobile robots or living beings are considered. We expect that, by including the environment into the modeling of a system's dynamics and its performance, we can proceed towards a description of the fundamentals of natural intelligence. With this goal, let us consider a closed, controllable, dynamical system in an open environment.

We again represent a system by a controller and plant as shown schematically in Fig. 13.9. We further describe the influence of the environment on the controller and plant by the time-dependent vector $\boldsymbol{v}(t)$. We assume that the vector $\boldsymbol{v}(t)$ describes the signals detected by a set of sensors, such as eyes, ears, etc. This vector therefore represents information about the properties of the environment that are related to mechanical forces or constraints, thermal flow and other influences that determine the gross interaction between the system and the environment. We also suppose that the modeler in the controller has a set of internal sensors to obtain the feedback signals describing the state $\boldsymbol{y}$ of the plant. In a deterministic treatment, the system dynamics is described by the difference equation for the system state:

$$\boldsymbol{y}(t+1) = \boldsymbol{\Psi}_y[\boldsymbol{x}(t), \boldsymbol{y}(t), \boldsymbol{v}(t)] \tag{13.67}$$

in which the dynamics generating, vector function $\boldsymbol{\Psi}_y$ is generally nonlinear. In the case when the vectors $\boldsymbol{x}$, $\boldsymbol{y}$, $\boldsymbol{v}$ adequately represent all the degrees of freedom needed to describe the dynamics of a system the function $\boldsymbol{\Psi}_y$ depends only on the local value of these vectors at time t, otherwise it must be

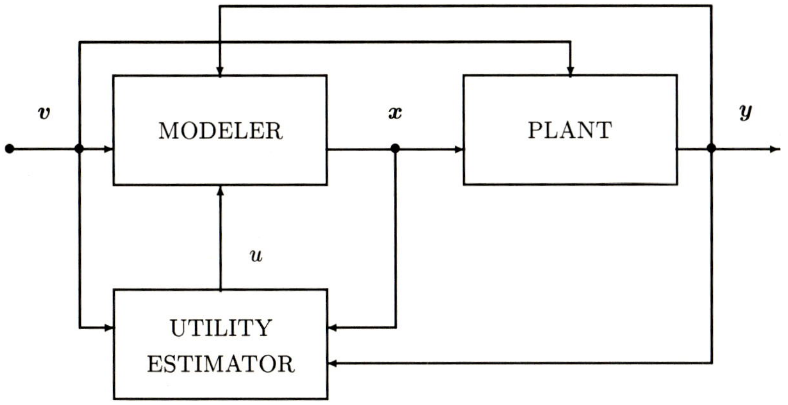

Fig. 13.9. The scheme of an intelligent optimal controller coupled to a dynamic environment

non-local, which means that it connects arguments at various past times. The treatment is thus apparently reduced to a study of driven, non-linear, multi-component systems. However, this is not exactly the case, because changes of the driving variable $\boldsymbol{v}(t)$ do not only depend on the dynamics of the environment but may generally also be influenced by the actions of the system itself. The corresponding dynamics is formally represented by the equation

$$\boldsymbol{v}(t+1) = \boldsymbol{\Psi}_v[\boldsymbol{v}(t), \boldsymbol{y}(t)] \, , \tag{13.68}$$

where again the vector function $\boldsymbol{\Psi}_v$ may generally be nonlinear and non-local. The above system of equations does not describe the dynamics completely because the equation by which the control variable is determined is missing. For this purpose, we look for the optimal control law that is based on a specification of the system's performance.

Our ultimate goal is to describe how the empirical modeling of natural phenomena can be utilized to improve the behavior of a system in a given environment. For this purpose, we shall assume that the performance of the system in a given state of the environment can be quantified based on sensing. The system is supposed to have for this purpose a set of specifically adapted sensors. Such sensors in biological systems provide signals of pain, temperature, odor, taste, etc. We include these signals in the set of components of vectors $\boldsymbol{v}$ and $\boldsymbol{x}$, $\boldsymbol{y}$ which together describe the compound system–environment state $\boldsymbol{c} = (\boldsymbol{x}, \boldsymbol{y}, \boldsymbol{v})$. In a deterministic approach, the quantification is described by some utility function $u(t) = u(\boldsymbol{x}(t), \boldsymbol{y}(t), \boldsymbol{v}(t)) = u(\boldsymbol{c}(t))$. It is determined empirically by the utility estimator that provides the corresponding signal which can subsequently be utilized in the estimation of the system performance

$$\overline{J(t_\mathrm{i}, t_\mathrm{f})} = \int_{t_\mathrm{i}}^{t_\mathrm{f}} \overline{u\left(\boldsymbol{c}(t')\right)}\, dt' \,. \tag{13.69}$$

However, this functional is not known in advance and certainly not to an animal in nature. We therefore treat it merely as an aid for the description of the optimal behavior. In other words, we suppose its existence and assume that it can be empirically estimated, based on a measurement of the utility. For this purpose, the intelligent controller includes in addition to the modeler a utility estimator. The modeler obtains complete information about the environment, system, and utility and must provide an estimation of the optimal control variable $\widehat{\boldsymbol{x}}(t)$ which is amplified in the booster of the controller.

Related to the scheme we have described, one might ask how the utility estimator should be constructed. There is no unique answer to this question, at least natural evolution has not found it. This conclusion is based on the simultaneous existence of various animals in the same environment. The brains of members of various species differ essentially as do the multitudes of behavior of various animals in the same environment. In the sense of control theory this means that the utility functions essentially depend on the construction of the system. Depending on the structure of the system, the action that is good for one system need not be optimal or even good for another in the same circumstances. Therefore, the properties of the utility estimator must be system-dependent. To facilitate the subsequent discussion, we avoid specifying the utility estimator in detail, but rather simply assume that for a given system it exists and provides for each compound state $\boldsymbol{c}(t)$ the corresponding signal $u(t)$. This signal can be associated in the modeler with the state vector and utilized in the formation of the memory. As in the previous subsection, the basis for the modeling is then a more complex vector which is now denoted by

$$\boldsymbol{X}(t) = [\boldsymbol{x}(t), \boldsymbol{y}(t), \boldsymbol{y}(t+1), \boldsymbol{v}(t), \boldsymbol{v}(t+1), u(t)] \,. \tag{13.70}$$

By utilizing empirical samples of such vectors, the modeler can form the set of prototypes that describe the coupled dynamics of the controlled system and the environment. The optimal control law can be extracted from such a model by the method of dynamic programming in a manner similar to that described in the previous subsection. The principal difference here is a greater complexity because of the inclusion of the environmental variables. Because the complete system is considered to be closed and situated in an open environment, the corresponding control can be also called *self-control.*

In the previous sections of this chapter the general identification, tracking, and cloning of dynamical systems were discussed. These can all be formulated as specific problems of the empirical optimal control. Identification can be introduced by assuming that there exists in the environment a system with input $\boldsymbol{v}_\mathrm{i}(t)$ and output $\boldsymbol{v}_\mathrm{o}(t)$ for which we want to design a modeler that is capable of generating the signal $\boldsymbol{x}(t)$ which is similar to the signal $\boldsymbol{v}_\mathrm{o}(t)$ when the signal $\boldsymbol{v}_\mathrm{i}(t)$ is supplied to the modeler. With this goal in mind, we define the utility as the unit that determines the negative square discrepancy between $\boldsymbol{x}(t)$ and $\boldsymbol{v}_o(t)$ and require that the performance $\overline{J(t_i, t_f)} = -\int_{t_i}^{t_f} \overline{(\boldsymbol{x}(t) - \boldsymbol{v}_o(t))^2}\, dt$ should be maximal. This is, in fact, a degenerate optimal control problem because the plant which should be controlled need not be specified.

Tracking can be described by interpreting the reference signal $\boldsymbol{v}(t)$ as the influence of the environment and defining the utility by $u = -(\boldsymbol{y}(t) - \boldsymbol{v}(t))^2$. Specification of the empirical optimal control problem is then completed by the joint measurement of the system input and output. This is equivalent to a specification of the system's dynamical law $\boldsymbol{y}(t+1) = \boldsymbol{\Psi}(\boldsymbol{y}(t), \boldsymbol{x}(t))$ which permits treatment of the system using the methods described previously in this section.

Just as general identification and tracking, cloning can also be represented as an optimal control problem. For this purpose, the general scheme of the intelligent controller operating under the influence of an environment is applicable. The expert obtains the reference $\boldsymbol{v}$ from the environment and converts it into the control variable $\boldsymbol{x}$ that is further transformed by the plant into the response $\boldsymbol{y}$. The expert also provides information about the utility corresponding to the control and plant state $u(t) = u(\boldsymbol{x}(t), \boldsymbol{y}(t))$. All these variables are then available to the modeler which can utilize them in the formation of the model during cloning of the expert. In this case it must also learn to estimate the utility from the given reference, control and plant output signals so that it can later simultaneously play the role of the utility estimator which is the characteristic of an expert. A trained modeler can later find optimal control solutions based on the methods presented in this section.

13.5.5 The Problem of Phase Space Exploration

In the description of the empirical approach to optimal control presented in this section we have assumed that the modeler of the intelligent controller obtains information about the system behavior in an open environment during some exploration phase. It was also assumed that during this phase, which can be also called *learning*, the modeler forms in its memory the prototype vectors which represent typical dynamical transitions of the system state. However, in order to obtain information about various possible actions, the intelligent controller must create during the exploration phase control signals and there arises a question how this can be properly accomplished. The problem of *exploration learning* is specific for the empirical approach to solving the optimal control task and it is more complex than the learning that was mentioned when we described the empirical approach to modeling of natural phenomena. There, the modeler only accepts signals from the phenomenon observed, but it does not generate them. A suggestion for the approach to exploration learning cannot be extracted from the theoretical methods used in the analytical treatment of optimal control, because information about the dynamics is given *a priori* in the analytical formulation of the problem. However, the problem of modeling of system dynamics is also related to a proper experimental and analytical exploration of the system under consideration, but it is not included in the treatment of optimal control because it is solved separately, in advance, by an intelligent explorer. In the empirical approach to optimal control, the modeler of the controller plays the role of an intelligent explorer that must solve this problem by a proper strategy of state space exploration.

To obtain some insight towards a heuristic approach for achieving a solution of the control problem using exploration learning, let us imagine how an intelligent operator learns from experience to control an initially unknown system in an unknown, possibly even noisy environment. For example, let us consider a person attempting to paddle a kayak in a stream. Using a paddle, a kayaker begins by exploring the properties of the system and the response of the kayak to the particular environment by a small random generation of control actions. Based on these actions, the kayaker collects information about the performance of the system being controlled. Among various actions, those which correspond to increased performance are then selected to *reinforce* the control until an optimal operation in a particular situation is obtained. While exploring the characteristics of the system, an intelligent operator usually also remembers which actions are most favorable in a particular situation and later uses this information to predict the proper actions which are associatively based on the perception of the state of the system and the environment. Good operators also predict the consequences of their actions and try to choose from all the possible ones those, that will likely lead to the best performance. In the field of dynamic programming, the approach to optimal control in noisy and random environments has been studied in relation to development of various iterative and stochastic methods. [7, 12]

A strategy similar to that described above is applicable to the design of intelligent controllers capable of exploration learning. This can be described as follows: To obtain proper information about the system dynamics the modeler initially generates random actions and remembers them together with the corresponding system response and the utility. It then gradually tries to optimize, at least on average, the instantaneous utility function at each particular moment by utilizing previous information. The optimization being carried out uses the utility of past action to find a better one. For this purpose, the previously memorized actions of the controller must be slightly varied and those actions that increase the performance of the system must be properly reinforced. In order to obtain a reinforcement of the proper actions of the controller, the utility is repeatedly measured and the corresponding information is subsequently further utilized to adjust the actions of the controller. Such a treatment corresponds to a local minimization principle and it need not lead to a global extremum of performance measure. The likelihood of obtaining a global extremum can be enhanced by gradually including ever more steps into the estimation of the performance that corresponds to a step-wise development of the dynamic programming that is based on locally, approximately optimal, prototype transitions.

The strategy described above can be achieved by the intelligent control system represented by the block diagram shown in Fig 13.9. as follows. Using its model, the controller estimates at a particular instant the best control signal represented by the vector $\widehat{\boldsymbol{x}}$ and determines the corresponding system state and utility. This vector is then disturbed by a small random trial variation $\delta\boldsymbol{x}$ and the resulting signal is again utilized for the manipulation that drives the system into a new state $\boldsymbol{y}+\delta\boldsymbol{y}$ corresponding to the utility $u+\delta u$. The variation of the utility is then used for the final correction of the control variable by enforcing favorable variations. If the test variation $\delta\boldsymbol{x}$ results in an increased utility i. e. $\delta u > 0$, then this disturbance is further amplified by some properly estimated reinforcement factor $\alpha(\delta u)$. When there is a decreased utility $\delta u < 0$, the sign of the disturbance is reversed and the correction is done in the opposite direction, using again a proper reinforcement factor. The corresponding correction term is then expressed as

$$\Delta\boldsymbol{x} \;=\; \alpha(\delta u)\,\delta\boldsymbol{x}\;. \tag{13.71}$$

The reinforcement rate can be estimated provided that the gradient of the utility as a function of control $\boldsymbol{x}$ and the position of the utility maximum can somehow be estimated. Quite commonly, when approaching the optimal utility u_0 ever smaller variations of the utility are observed and therefore, corresponding smaller changes of the control variable are also needed. With this aim we include into the reinforcement factor the ratio of the difference we wish to achieve $\Delta u = u_0 - u$ and the variation of the utility δu. The corresponding expression of the reinforcement rate which also accounts for the proper sign of reinforcement is then given by

$$\alpha(\delta u) \;=\; c\,\frac{\Delta u}{\delta u} \tag{13.72}$$

with c being a constant. For a one-dimensional case and a quadratic dependence of the utility function on the control variable $\boldsymbol{x}$ close to its maximum the proper value is $c = 2$. The control variable changed for $\Delta\boldsymbol{x}$ is expected to drive the system with approximately optimal utility.

In using the reinforcement learning procedure, the question arises how to select the span of the variation $\delta\boldsymbol{x}$. A natural measure for this purpose is the width of the window function utilized in the estimation of the conditional average. It corresponds to the mean distance between prototypes in the state space of the control variable.

The corrected control variable $\boldsymbol{x}_c = \widehat{\boldsymbol{x}} + \Delta\boldsymbol{x}$ results in a new state $\boldsymbol{y}_c$ and the corresponding utility u_c which is expected to be better than that in the initially estimated state $\widehat{\boldsymbol{x}}$. It is, therefore, reasonable to use the corrected values and their successors in the self-organization process by which the set of prototypes

$$\mathcal{Q} = \{\boldsymbol{q}_k = (\boldsymbol{x}_k, \boldsymbol{y}_k, \boldsymbol{y}_k^+, \boldsymbol{v}_k, \boldsymbol{v}_k^+, \overline{u_k}) \; ; \quad k = 1, \ldots, K\} \tag{13.73}$$

is adapted to the dynamical phenomenon occurring in the controlled system and the environment. The corresponding changes $\Delta \boldsymbol{q}_k$ are expected to drive the prototypes into the region of the state space with a high average utility $\overline{u_k}$. The memory of the modeler is thus increasingly filled with information from which the optimal behavior of the system can be estimated. The reinforcement procedure thus corresponds to a training process of an intelligent controller. Using a random variation of sample vectors, a refined set of learning vectors is found by which the memory is then formed. It can be expected that with an increasing number of tests the values of the reinforcement factor decrease. When training commences, the memory is thus mainly filled with prototype vectors which are randomly distributed in the state space, but later they become increasingly more concentrated in the region of maximal average utility. At the start of training, the estimation of the optimal control is approximate and inaccurate, but is later refined as the number of tests increases.

The reinforcement procedure described above corresponds to the gradient descent learning described in Chap. 7 relative to the adaptive modeling of natural laws and it will therefore not be discussed in additional detail. However, in its implementation for control application, the corrections are created by the modeler itself. Because the reinforcement is subsequently utilized in the modeler it is also reminiscent of a back-propagation algorithm which is widely applied in the supervised training of neural networks. Moreover, the reinforcement learning resembles the properties of the calculus of variations that is carried out in finding the solution of optimal control problems in an arithmetic way. The complete problem of reinforcement learning stems from the fact that the gradient of the utility in the state space of control cannot be properly estimated from measured data samples.

The goal of reinforcement learning described above is a refinement of optimal utility estimation. This procedure corresponds to a local or one-step performance optimization. One may ask how this procedure could be generalized to a non-local or multi-step performance optimization. With this as our goal, the next level of learning and training can be introduced by which the model, that has been trained to be optimal in one-step performance, is further modified by a multi-step optimization. To do this, we consider that the dynamics of the system trained in one-step procedure is represented by a set of prototypes Eq. (13.73). To proceed towards the multi-step performance optimization, we must first measure the integral $\overline{J}$ and estimate with it the average utility $\overline{u_k}$. The measurement is generally time-consuming because of the averaging procedure. We denote the characteristic time in the interval of performance estimation by the time index τ. The set of prototypes during this time denoted by $\mathcal{Q}(\tau)$ is associated with the estimated multi-step performance $\overline{J(\tau)}$ instead of the momentary utility. We then obtain the new set of prototypes

$$\mathcal{Q}(\tau) = \{\boldsymbol{q}_k(\tau) = (\boldsymbol{x}_k, \boldsymbol{y}_k, \boldsymbol{y}_k^+, \boldsymbol{v}_k, \boldsymbol{v}_k^+, \overline{J(\tau)}) \ ; \ k = 1, \ldots, K\} \tag{13.74}$$

in which the same average performance is assigned to each prototype compound state because during a longer time interval the performance is determined by all prototypes in common.

This set of prototypes characterizes the joint behavior of the controller and the plant in a given environment. It is interesting, that with respect to modeling, this compound environment–system can be treated as a single unit possessing a complex dynamical behavior. However, from the optimal control viewpoint, the variables corresponding to the controller or plant output possess quite different meanings. The model is represented by the set of prototypes. Thus, the complete system dynamics can be changed by a variation of this set. This procedure corresponds to a deterministic correction of the parameters of the controller. We further suppose that the properties of the controller can be changed by influencing those components of the prototype vector of the model that pertain to the controller. For this purpose, a trial variation $\delta\boldsymbol{x}$ is created by the modeler. A variation of the controller parameters further changes the behavior of the plant, which after a certain time of operation, is reflected by a change of the average performance. The procedure of self-organized reinforcement learning can be then continued by using the multi-step performance $\overline{J(\tau)}$ and its variation rather than the utility $\overline{u_k}$. For this purpose, longer time intervals must be used to permit observing the consequences of the varied control variable. Correspondingly, when using the multi-step procedure the self-organization process proceeds more slowly. To correct the operation of the controller, the modeler must be capable of estimating the performance variation $\overline{\delta J(\tau)}$ caused by random changes of the control variable $\delta\boldsymbol{x}$. For this purpose, it must memorize the past performance for the time needed to complete the averaging and then compare it with the new value to determine the reinforcement term

$$\Delta\boldsymbol{x} = \alpha(\overline{\delta J(\tau)})\,\delta\boldsymbol{x} \ . \tag{13.75}$$

The search for optimal performance is thus similar to a random walk in a multi-dimensional state space caused by a random disturbance of the control variable. In this context, the rate corresponds to the mobility of the controller state in the field of the generalized force $\mathcal{F} = \alpha(\overline{\delta J(\tau)})\,\delta\boldsymbol{x}$. [11]

The time index τ in the above procedure describes the present set of prototypes. It was intentionally denoted differently from discrete time to emphasize an important property which can be intuitively understood on the basis of everyday experience with improvement of control of various systems. Consider again an intelligent operator controlling a system in an unknown environment. When performing a variation of the plant manipulation, the operator pauses for a certain time interval so that the system can properly adapt to the modified control and that the performance estimation can be made over a time interval which is normally appreciably longer than the response time of the plant. Numerous examples illustrate the onset of instability and the impaired performance that results when an operator reacts too quickly with respect to the response time of the plant. It is characteristic for novice athletics, as for example when paddling a kayak on rough water, skiing downhill, skating, etc., that their performance in a new environment is impaired if they react too quickly to every disturbance of the environment. One can, therefore, expect that a step in the index τ should generally include many steps in discrete time to realize a stable operation of the optimal controller. In other words, the characteristic time scale of the dynamics of the system should, in general, be shorter than the characteristic time needed for the optimization of the performance. Experimentally, the proper proportion must be found by trial and error.

After an initial exploration, the presence of the random variation of control may seem superficial. But simple reasoning indicates that it is, in fact, essential for a good optimization of strategy. Assume, for instance, that in a particular time interval τ, the trial variation $\delta \boldsymbol{x}$ has left the performance unchanged. In this case, one could erroneously expect that there the extremum of the utility function was achieved. But this is only true under the condition that one proceeds in the direction of $\delta \boldsymbol{x}$, while nothing is said about the effect of steps in the other directions. In the algorithm for adjusting the prototype vectors described above, because of the random character of the trial term $\delta \boldsymbol{x}$ various directions are explored. If, however, a trial in the side step leads to an improved performance it is then further reinforced as described by the Eq. (13.75). The necessity of a permanent random exploration of the state space is also needed when, at a certain time interval, the set of prototypes is indeed the optimal one, because subsequently the properties of the environment may again change. For a corresponding change of the optimal controller, the modeler must re-explore the state space. The random trial variation of the control variable thus resembles the *creation of a new idea* in the brain of an intelligent being. It is characteristic for an intelligent modeler that the random exploration of the state space and the self-organization process can be permanently active during the operation of the system. Such a compound process could be termed *reinforced self-organization.* It is the basis of exploring the state space on which an improvement of an intelligent controller can be obtained. [21]

13.5.6 Numerical Simulations of Optimal Control

In order to demonstrate the applicability of empirical modeling by a set of self-organizing prototypes for the solution of optimal control problems, we present three examples: the estimation of an optimal parameter in a chaotic manufacturing process; the self-stabilization of a randomly influenced system and a vehicle backing operation.

Control of Chaotic Cutting Process. The purpose of this example is to demonstrate how the conditional average can be efficiently applied to

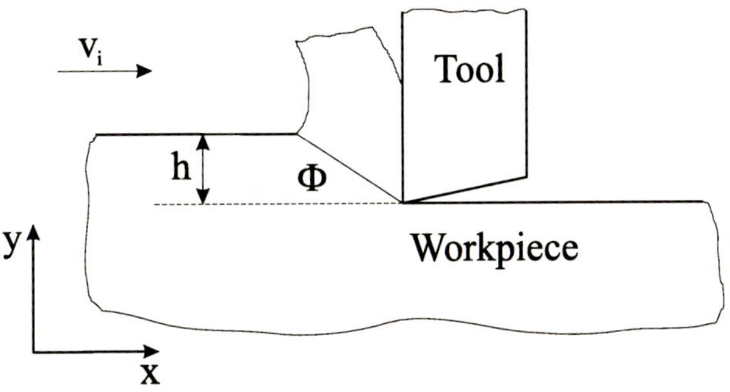

Fig. 13.10. Schematic diagram of an orthogonal cutting system

specify an optimal cutting depth in a manufacturing process. The process is schematically shown in Fig. 13.10. For its description we utilize two dimensions (x, y).[10] A homogeneous workpiece with a straight surface moves in the x direction with a constant velocity v_i towards the cutting tool. The parameter H describes the depth at which the free tool edge is set by the controller of the process. The material flow generates a cutting force component F_x and a friction component F_y of the total cutting force vector $\boldsymbol{F}$. This force is a non-linear function of the material flow velocity and the cutting depth. This force results in an elastic deformation of the tool which influences the actual cutting velocity and the actual cutting depth h. The elastic and inertial properties of the tool and the workpiece can be described by a coupled two-dimensional nonlinear oscillator. The nonlinear dependence of the cutting force on the cutting velocity causes tool oscillations which may be chaotic at particular values of the cutting parameters. [9, 10] The evolution of the amplitude of the tool oscillations with increasing cutting depth H is illustrated by the bifurcation diagrams shown in Figs. 13.11 and 13.12.

At $H \sim 0.23$, the oscillations become irregular, such that many different amplitude values appear at a particular cutting depth. The records of the tool displacement in the characteristic segments of the bifurcation diagrams show a transition from harmonic to quasi-harmonic oscillations - I, chaotic oscillations - II, inharmonic periodic oscillations - III, and quasi-periodic oscillations - IV.

A cut made with an oscillating tool edge results in a rough surface as depicted in Fig. 13.13. The surface roughness generally increases with cutting depth, but the time needed to remove a certain volume of material decreases. From an economic standpoint it is advantageous to select that cutting depth at which the cutting time would be appropriately short while the surface would be sufficiently smooth. For the definition of the optimal depth as function of cutting depth, we introduce a cost function which incorporates these two criteria:

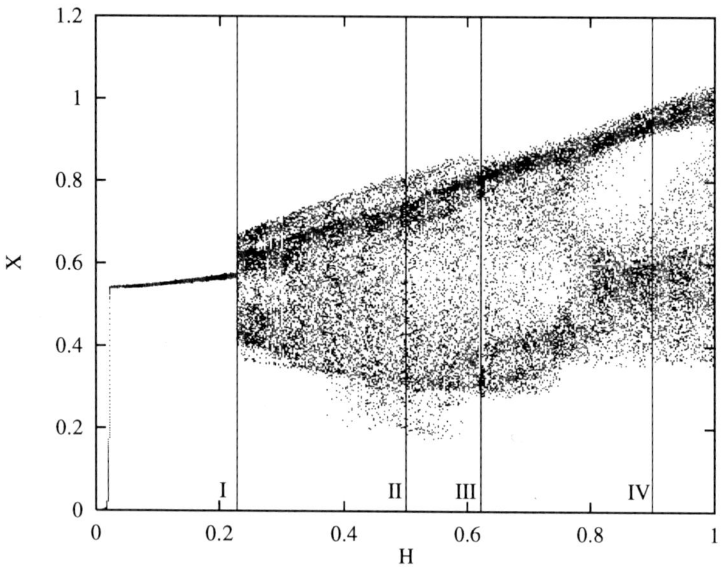

Fig. 13.11. Dependence of relative oscillation amplitude in the x direction on the cutting depth H

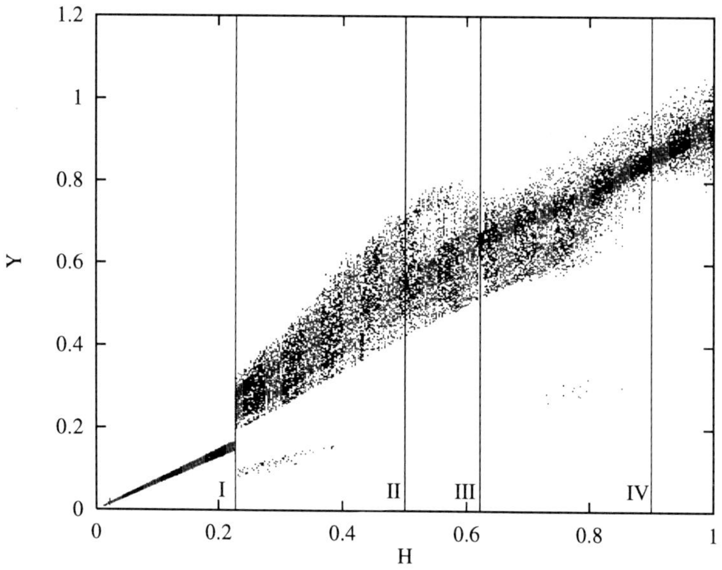

Fig. 13.12. Dependence of relative oscillation amplitude in the y direction on the cutting depth H

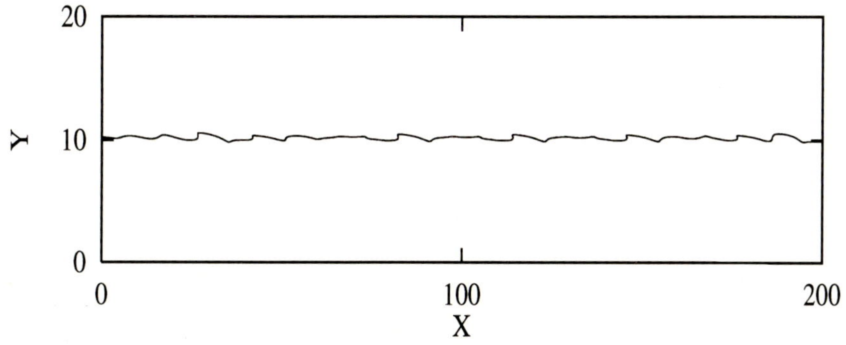

Fig. 13.13. A sample with a rough surface profile

$$C(H) = \left(\frac{\sigma(H)}{\sigma_{max}}\right)^2 + \frac{A}{1 + \left(\frac{H}{H_{max}}\right)^2} \, . \qquad (13.76)$$

The first term in this equation describes the effect of the surface roughness, where $\sigma(H)$ is the standard deviation of the surface profile after machining at the set cutting depth H, and σ_{max} denotes the maximum surface profile deviation after cutting at different cutting depths in the range $[0, 1]$ of H/H_{max}. The second term describes the effectiveness of the machining. The constant A is an arbitrary positive weight which adjusts the importance of the second term in the equation. The cost function can be treated as a negative performance measure. Because an analytical expression for $\sigma(H)$ is not available we cannot perform the optimization analytically. Instead, we must use an empirical method consisting of two steps: Empirical modeling of the cost function and an empirical estimation of its minimum.

Figure 13.14 shows the record of the cost function as a function of the cutting depth and a set of $K = 18$ prototype points which were adapted to the cost function by the self-organized process. Since a prototype vector consists of joint data $\boldsymbol{q}_k = (C_k, H_k)$, the set of prototypes represents an empirical model of the cost function $C(H)$. From this definition it follows that the cost function is positive for all H, and hence among all its values the minimum value lies most closely to 0. The position of the minimum can therefore be determined by the conditional average

$$\widehat{H}(C) = E[H|C] = \sum_k B_k(C) H_k \qquad (13.77)$$

in which $C = 0$ is used as the condition even though this value cannot be reached. The estimated value of the optimal cutting depth is $H = 0.77$. The optimal value determined from the recorded data in Fig. 13.14 was $H^* = 0.76$ which is in good agreement with the estimated value. It is emphasized that

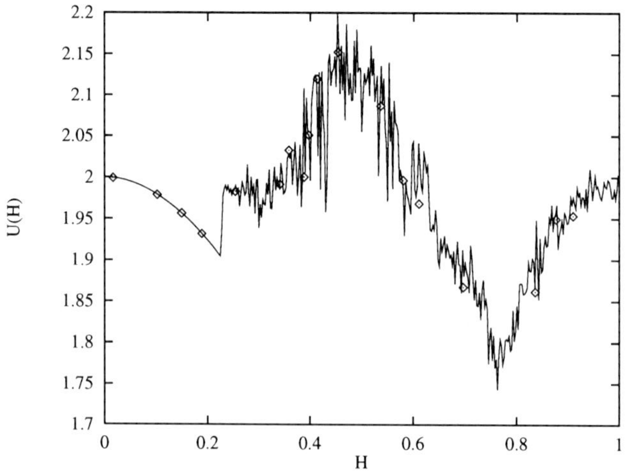

Fig. 13.14. The cost function and prototypes (o)

for the estimation of the optimal value no comparison of various values of the cost function is needed. Comparison is automatically made during calculation of the conditional average because of the similarity measure that is expressed by the basis function B_k.

Self-Stabilization of a Randomly Excited System. The purpose of the following numerical simulation is to demonstrate the applicability of the non-parametric method for the empirical modeling and control of a system in a random environment and to show the effect of reinforcement learning on the performance. An example of a stabilizer has been simulated because it is simple to model and the corresponding trivial solution can be analytically expressed. The problem is to find the controlling variable x that compensates the random influence of the environment v such that the output y of the system remains at some fixed point, which, in this example, we have arbitrarily chosen to be the origin, that is, $y = 0$. The output of the plant is therefore described by $y = v - x$. In this simplified case, the signals representing the plant and environment state, utility and control variable are thus represented by scalar functions $y(t), v(t), u(t)$ and $x(t)$. In this example, the utility is defined as the negative compensation error that is specified by the function $u = -y^2$. To simulate the presence of the utility estimator, the utility was calculated separate from the control variable. In this demonstration, the influence of the environment was randomly varied.

In the first example only randomly generated samples of the control variable were utilized without reinforcement learning in the formation of the model. Figure 13.15 shows the distribution of 200 random points representing values of the environmental and control variables that were generated

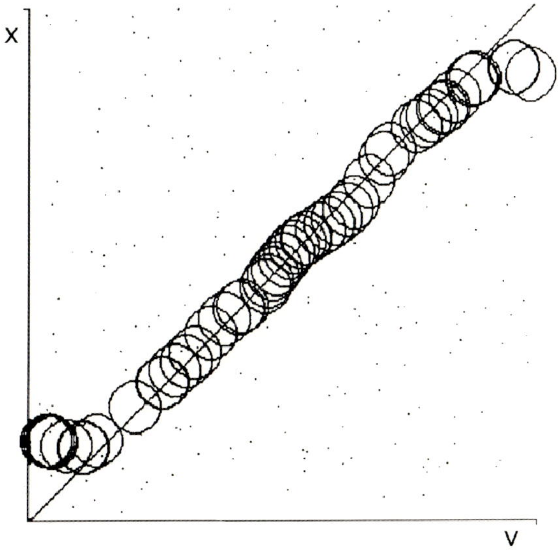

Fig. 13.15. Distributions of random learning samples obtained by random influence from the environment and independent random control variable (shown as *points*) and the generated optimal control of a trained system (determined by centers of *circles*). The diagonal line corresponds to the optimal control

mutually independent during the exploration phase. The points representing the influence of the randomly changing environment and the control variable that has been generated in 50 test steps subsequent to the exploration phase by the one-step optimal control are shown in Fig. 13.15 by the centers of the circles. The radius of each circle corresponds to the width σ of the window function utilized in the estimation of the conditional average which also describes the statistical measure of the smoothing effect in the determination of the utility maximum. The centers of the circles lie close to the analytical solution described by the line $x = v$ which corresponds to the case in which the output y is zero. It is statistically more acceptable to say that the circles with radius σ practically cover the theoretically given optimal solution $x = v$. In other words, the plant has achieved nearly maximal utility, or $u_0 \approx 0$. The average utility estimated from 50 test points after learning was of the order -10^{-3} which is less than, but close to, the optimal utility $u_0 = 0$. The exact value depends slightly on the selection of the seed of the random number generator and the corresponding train of samples utilized in the calculation. The small systematic error that appears in this demonstration at the edges of the distribution of the environmental variable is a result of our choice of representing the probability density by spherical Gaussian window functions. This error can be avoided by utilizing more complex, elliptical multivari-

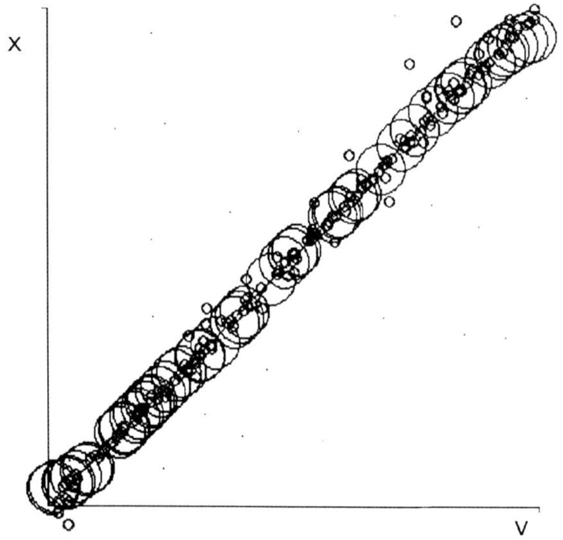

Fig. 13.16. Distributions of random learning samples (shown as *points*), samples obtained by reinforcement learning (shown by *small circles*) and data obtained by random influence from the environment and corresponding generated optimal control of a trained system (determined by centers of *large circles*). The line corresponds to the optimal control of the system

ate approximations of Gaussian functions instead of constant prototypes in forming the conditional average.

In the second numerical experiment, the same model was applied but with only 20 random samples acquired initially which were subsequently utilized in the reinforcement learning in the following 180 steps. The complete number of presented samples in both examples was thus 200 to permit comparison of the performance. In the reinforcement learning the trial variation of the control variable δx was generated by the random number generator with the uniform probability distribution in the interval $\pm\sigma_x/2$ around the mean value 0, where the parameter σ_x denotes the mean distance between the prototypes in the space of the control variable. The results are shown in Fig. 13.16. In this figure the small points denote the initially randomly generated samples, the small circles designate the positions of the samples generated by reinforcement learning while the great circles denote the positions of the estimated points that correspond to optimal control after the training was completed. This numerical experiment confirms our expectation that the samples obtained by the reinforcement learning should lie predominantly in the neighborhood of states with optimal utility. In this case the average utility of the order -10^{-5} was achieved by the trained system. This indicates a significant refinement of the training process with respect to the previous case resulting in an improvement of the utility by about two orders of magnitude.

In both examples, for the sake of simplicity only the initialization of the prototypes was utilized. Because of non-correlated influences of the environment and the momentary response of the system to the control variable, the optimal utility trajectory represents the optimal control trajectory in both cases. Consequently a multi-step optimization cannot improve the performance of the system and it has therefore not been carried out in this example.

The samples of the control variable $\boldsymbol{x}_c = \widehat{\boldsymbol{x}} + \Delta\boldsymbol{x}$ are obtained in the reinforcement learning in two steps: First, an approximately optimal value $\widehat{\boldsymbol{x}}$ is determined by the conditional average and then this value is disturbed by the reinforced trial variation. The numerical experiments have revealed that utilization of the random trial variation introduced in the reinforcement learning in order to obtain favorable samples with respect to the utility is essential for the improvement of the learning process. If, for instance, the new samples are generated only by application of the conditional average $\widehat{\boldsymbol{x}}$ that is based on previous samples, then the new samples always fall into that region of the state space that is covered by the window functions centered on previous samples. New portions of the state space which are separated from the previous prototypes are in this case not explored and the learning process does not lead to an improvement in the estimation of the optimal control variable. This is especially critical near the margins of the sample point distribution in the state space as is evident from the comparison of the figures in both numerical examples. Creation of prototypes by the self-organization process must be interpreted as a memorization process that optimally preserves the information provided by the samples, while the reinforcement process represents the search for appropriate samples. The reinforcement thus leads to *creativity* which is a characteristic of intelligence of the proposed empirical modeler.

Optimal Vehicle Backing. In the previous two examples the applicability of empirical modeling and the direct estimation of optimal values by a conditional average was demonstrated using a quasi-static description of system behavior. However, the characteristic optimal control problems can be demonstrated only by the inclusion of system dynamics. With this aim, we present the solution of the well-known problem of the backing of a vehicle using the non-parametric modeling of system dynamics and reinforcement learning. The problem is to design an intelligent system capable of learning to steer a vehicle such as a truck while backing up to a loading dock from any initial position [18, 24]. Fig. 13.17 shows the geometry of the truck and the loading dock. The coordinates ξ, η determine the position of the truck, while ζ represents the angle between the truck and the normal to the dock. The state is thus represented by the vector $\boldsymbol{y} = (\xi, \eta, \zeta)$. The motions of the vehicle can be deterministically described by the equation

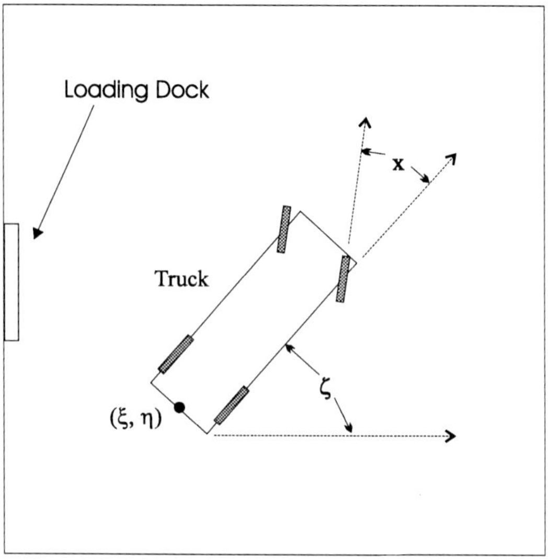

Fig. 13.17. Scheme of a truck in the vicinity of a dock

$$\boldsymbol{y}(t+1) = \boldsymbol{\Psi}(\boldsymbol{y}(t), x(t)) \tag{13.78}$$

in which the function $\boldsymbol{\Psi}$ is determined by the geometrical properties of the vehicle. The truck is placed at some random initial position and it is backed up with constant speed while being steered via its front wheels by the controller. The control variable x represents the angle of the wheels with respect to the axis of the truck. The motion of the truck stops when the truck reaches the line parallel to the dock. The goal is to reach the state $\boldsymbol{y} = (0, 0, 0)$ as close as possible. This corresponds to the maximization of the performance described by the objective function

$$J = -\left(c_1 \xi_f^2 + c_2 \eta_f^2 + c_3 \zeta_f^2\right) \tag{13.79}$$

in which the constants c_1, c_2, c_3 determine the importance of the deviation represented by each state component.

The function $\boldsymbol{\Psi}$ is determined by a set of empirical data about the control variable, the system state and its successor. It could be represented using an extensive set of joint prototypes. However, our goal is to select among them a reduced set by which the optimal control variable could be estimated at each system state $\boldsymbol{y}(t)$ using the conditional average estimator. We denote this optimal set by

$$\{\boldsymbol{q}_k^* = (x_k^*, \boldsymbol{y}_k) \ ; \ k = 1, \ldots, K\} \ . \tag{13.80}$$

This set empirically describes the optimal control law

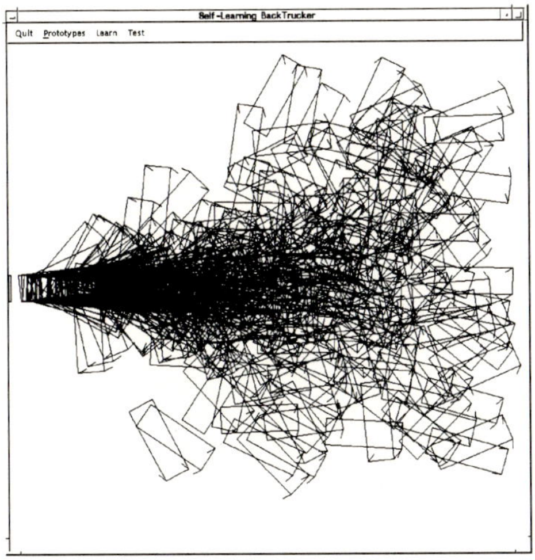

Fig. 13.18. The set of learning samples for the truck backing application

$$x^*(t) = F(\boldsymbol{y}(t)) \ . \tag{13.81}$$

The estimated control variable $x^*(t)$ determines the steering-angle which optimally leads the system from the given state $\boldsymbol{y}(t)$ to the successor state $\boldsymbol{y}(t+1)$. Our goal in this numerical experiment is to create the optimal set based on reinforcement learning without extensive search of all possible trajectories.

During creation of the empirical model the samples were generated randomly in the state space. The positions of the samples are shown in Fig. 13.18. To permit a more precise steering in close vicinity of the dock, the density of samples was higher there. In the learning process, the prototypes with the positions approximately one step away from the dock were formed. From these positions single steps were evaluated according to the dynamic law with randomly chosen control. The objective function was determined in the final state. The initial random control was then improved by the reinforcement procedure until the best control at each position was obtained. The data obtained in this way were stored as prototypes. In the next step of learning the samples displaced from the previously formed prototypes by approximately one step were again varied by the reinforcement procedure. In this case, only the control of the first step was modified while the succeeding step was estimated from the previously formed optimal prototypes. The value of the objective function achieved in the final state was used as the criterion for improving the control during the first step. After formation of the prototypes with the vehicle position two steps away from the dock, the procedure was iteratively continued by including ever more distant positions of the vehicle. At each distance the set of previously formed prototypes was utilized to ex-

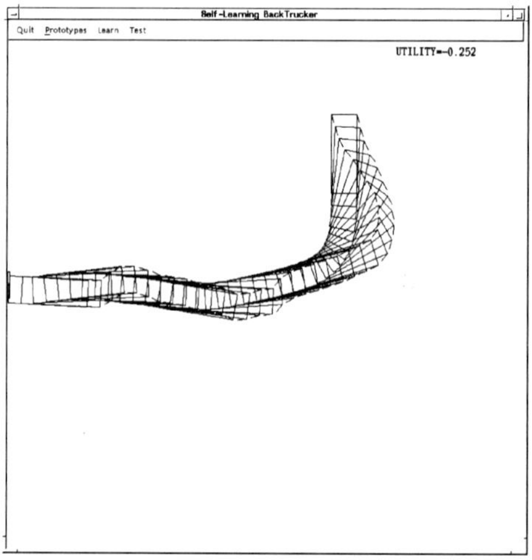

Fig. 13.19. Truck backing test case: Initial position inside the zone of learning

ecute the steering of the truck after the first step, while the optimal control of the first step was found by reinforcement of the most favorable variations. This procedure, in fact, corresponds to dynamic programming for the case in which the number of steps is increased.

After the formation of the prototypes the empirical model of the optimal control law was tested for various random initial states which were placed either in the region of the state space for which the controller was previously trained or outside it. It was found that the controller is able to successfully steer the truck back to the loading dock in either case. Three typical test examples are presented in Figs. 13.19 – 13.21. They indicate that a controller with a conditional average estimator is even capable of extrapolating to a certain extent. If the initial state is outside of the domain of learning, the controller is able to steer the truck first from this position to one of the nearby states for which control of the vehicle was previously improved by reinforcement and then from there along the optimal route to the loading dock.

13.5.7 Summary and Conclusions

The examples presented in this section demonstrate that the empirical modeling of natural phenomena by a non-parametric approach and the optimal statistical estimation based on a conditional average is applicable for the treatment of optimal control problems. An advantage of this approach is that the stored contents of memory cells can be simply interpreted as prototype data. When their number is not too large, the prototypes can be determined directly by measured samples. As the number of samples increases, a self-organized adaptation process must be utilized. The selection of the number of prototypes was left here to the user, but it can also be optimized based on the prescribed accuracy of the modeling that is sought.

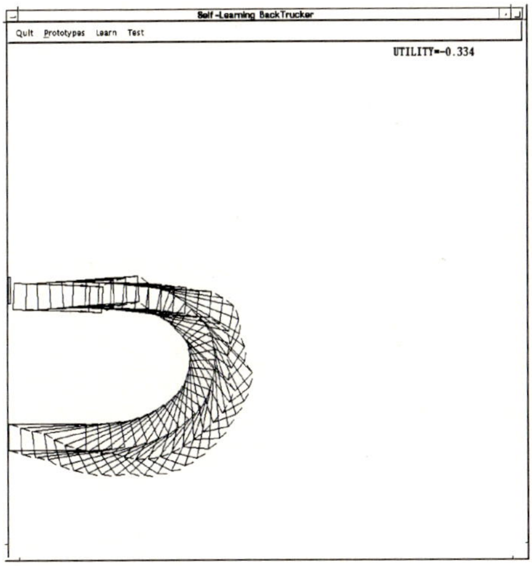

Fig. 13.20. Truck backing test case: Initial position outside the zone of learning

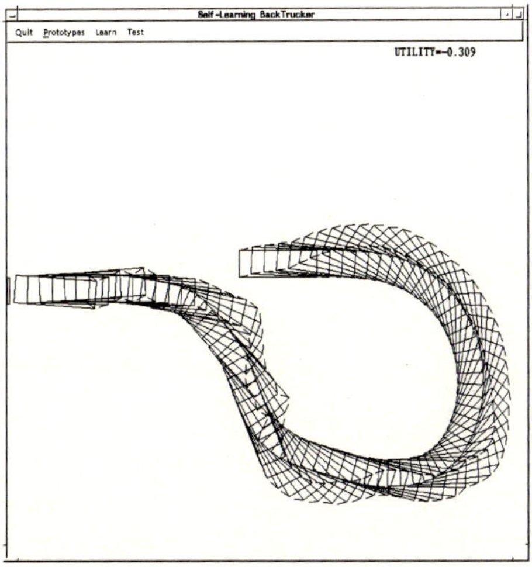

Fig. 13.21. Truck backing test case: Initial position outside the zone of learning

For applications of the modeler the estimation of missing components from given partial information about the system state by a conditional average is convenient. With no specific programming it can also be used to estimate the optimal control variable. In the examples presented in this chapter, spherical basis functions were used to obtain the estimation. The performance of the estimation can be increased by replacing the spherical functions by multivariate basis functions that yield in the conditional average estimator the linear functions of the condition rather than the fixed prototype values.[14] Optimal control laws can be modeled using the same approach. For this purpose the optimization problem can be solved by a generalized conditional average. This can be efficient, provided that there is available exhaustive information about all possible states and controls. Extraction of an optimal control law then represents the reduction of the number of prototypes. This generally appears to be an overwhelming task and it is probably advantageous to include in the learning phase a search for favorable control actions which are based on reinforcement learning. Examples of intelligent control that are found in biology indicate the existence of both paradigms, although the latter appears to be more direct.

14. Self-Control and Biological Evolution

14.1 Modeling of Natural Phenomena by Biological Systems

Fundamentals of the intelligent control offer an interesting interpretation of the link between natural intelligence and the capability of modeling natural phenomena. Let us consider the organism of an animal to be a closed, self-controlled dynamical system that lives in an open environment. If its nervous system is a center in which the dynamical model of the organism and the environment is created and stored then this model can be utilized to obtain an optimized behavior of the animal. For this purpose, the controlling signals in the center should be generated by reinforced variations that help find an improved system performance and create a proper model of the environment and the dynamical system. It is expected that the result would be an optimized movement of the animal in its environment which corresponds to a knowledge-based or intelligent behavior. According to this view, thinking can be interpreted as an internal representation of various situations and events. An imagination of possible future events and their consequences thus corresponds to the forecasting of system trajectories in the environment based on knowledge stored in the model. Making decisions, then is equivalent to taking such actions as are expected to lead to some optimal trajectory or favorable effect. We know from experience that the treatment of various tasks and problems in everyday life is, in fact, based on our expectations or predictions of benefits. A similar conclusion can also be drawn for the intelligent behavior of animals although we do not know operationally how they make decisions.

The basis of intelligent behavior is the empirical information which has been obtained by experience. This information is stored in a memory which can therefore be generally interpreted as the site of the empirical model. A model in itself has no value unless it has been obtained in relation to an optimization of the performance of the system. For instance, some disabled persons with injured brains may recall certain situations or events but they cannot make decisions based on this memorized information. Hence, this information is of little use to them. According to this view, the value of the stored information or its meaning corresponds to an increase of a system's performance which is determined by the behavior of an animal in a particular environment. We can conjecture that the ability of an individual to optimize his behavior has played a decisive role in the selection process during biological evolution which led to the development of biological, intelligent information processing systems capable of abstract reasoning, an ability that is needed to forecast events and plan behavior. In accordance with this conjecture, we conclude that the evolution of natural intelligence has been closely connected with the development of self-control and consequently both properties should be studied and described together. However, we do not yet know how to analytically describe in terms of physical variables and laws the complex environmental properties and the corresponding influences on a system which have played the dominant role in the evolution of the species. Nevertheless, based on the above reasoning, we can at least use our intuition to elucidate the fundamental structural properties of a closed system which is capable of intelligent self-control in an open environment. For this purpose, we could apply the scheme of the intelligent control system previously presented in Chap. 13. We presume that the properties of the environment, and especially its dynamics, have a critical influence on the behavior of the system and therefore they must also be included in any description of the system. In addition, we presume that the performance of the system is not specified *a priori* by some deterministic utility function which depends only on the system dynamics but rather that it emerges as an empirical statistical characteristic which describes the interaction between the system and the environment. According to this view, the survival of a particular animal depends essentially on the occurrence of various situations and events and not primarily on the dynamical properties of natural laws. We expect that a joint statistical treatment of a closed controllable system and its environment could finally lead to an operational description of some basic properties of biological intelligence.

From a philosophical point of view, this approach, at a first glance, represents a serious deviation from the commonly accepted physical description of natural phenomena which is based on the specification of natural laws. These laws, expressed in terms of relations between physical variables, represent the *necessity* or general properties of the Nature. It is characteristic that the initial and boundary conditions of the natural phenomena imposed on the physical variables are most often excluded from the description of natural laws and thus they do not represent an important ingredient in the physical description of Nature. In accordance with this view, the complex properties that characterize an environment can be interpreted only as one particular realization from the menu of various possibilities. Contrary to this view, the approach proposed above treats an environment as part of a unique Nature. Its structure and the corresponding information that physically corresponds to the specification of a certain combination of the boundary and initial conditions, is in fact the most important ingredient of its description because it essentially influences the evolution of Nature. By using the information and combinations we introduce the necessity and also the concept of *chance* that appears as the basis for the description of evolution. In such an approach, necessity is treated as an emergent statistical property of Nature, one which represents some common properties of chance. According to this view, in a description of the evolution of living beings, information about the boundary and initial conditions, or more generally about the existing structure of Nature is as important as the natural laws themselves. For example, it is equally important for an animal to know where food is, as it is to know how to reach for it from a certain position. The ability to process information about the state of environment and the system itself is thus considered the most important property of living beings and it should be of primary importance for the description of evolution. It is, therefore, not surprising that the empirical modeling of natural phenomena on the basis of sensory information, appears to be more fundamental for an understanding of the evolution of natural intelligence than the physical description of natural phenomena based on abstract analytical laws.

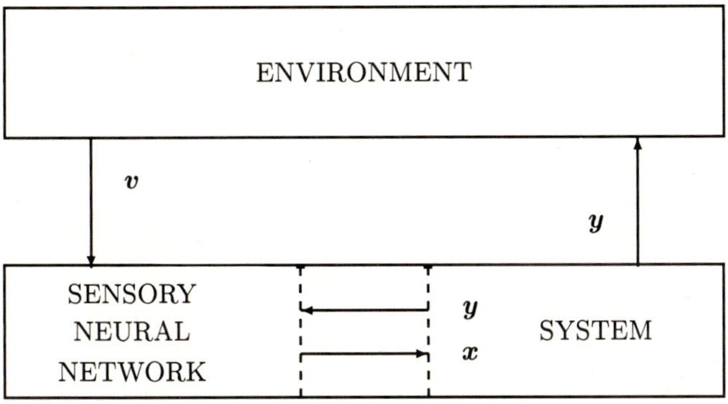

$$\boldsymbol{z} = (\boldsymbol{x}, \boldsymbol{y})$$

Fig. 14.1. The scheme of a controllable organism in an open environment

14.2 Joint Modeling of Organism and Environment

Let us consider a closed, controllable organism in an open environment. We formally represent it by two coupled blocks called a sensory-neural network (SNN) and a body as shown by the scheme in Fig. 14.1. These blocks correspond to the controller and the plant, respectively, of the dynamical system considered in the Chap. 13. [3] The SNN provides signals that control the behavior of the body. As an example, we mention an autonomous robot in which the SNN is an information processing system and the body is a mechanical system capable of exerting various operations and actions. Let the controlling signals be described by the time-dependent vector $\boldsymbol{x}(t)$, and the body state by the vector $\boldsymbol{y}(t)$. Additionally, we describe the influence of the environment on the organism by the time-dependent vector $\boldsymbol{v}(t)$. We assume that the vector $\boldsymbol{v}$ describes the signals provided by a set of sensors of the external influences and it therefore represents the interaction between the system and the environment. We also suppose that, from a set of internal sensors, the SNN obtains the feedback signals which describe the current state $\boldsymbol{y}$ of the body. According to this scheme, the complete state of the environment is characterized by the vector $\boldsymbol{v}$, while the state of the closed organism is described by the concatenated vector $\boldsymbol{z} = (\boldsymbol{x}, \boldsymbol{y})$.

In a deterministic treatment, the dynamics of a body can be represented by the differential equation describing the rate of change of its state :

$$\dot{\boldsymbol{y}}(t) = \boldsymbol{F}_y\left[\boldsymbol{y}(t), \boldsymbol{x}(t), \boldsymbol{v}(t)\right] . \tag{14.1}$$

The vector function $\boldsymbol{F}_y$ is generally non-linear. When the SNN performs an optimal control of the organism, the corresponding optimal control law can be expressed as:

$$\boldsymbol{x}(t) = \boldsymbol{\phi}\left[\boldsymbol{y}(t), \boldsymbol{v}(t)\right] \tag{14.2}$$

which, at least in principle, can be extracted from some performance functional by a proper optimization technique. However, the sensory neural network is itself a dynamical system and we can therefore expect that the optimal control law that governs its response to the body and the state of environment can also be expressed in terms of a differential equation that is analogous to Eq. (14.1)

$$\dot{\boldsymbol{x}}(t) = \boldsymbol{F}_x\left[\boldsymbol{x}(t), \boldsymbol{y}(t), \boldsymbol{v}(t)\right] . \tag{14.3}$$

The vector function $\boldsymbol{F}_x$ is generally non-linear and also non-local. By using this equation, the dynamics of the organism can be formally represented by a single differential equation for the rate of change of the complete state vector, $\boldsymbol{z}(t)$. This is

$$\dot{\boldsymbol{z}}(t) = \boldsymbol{F}_z\left[\boldsymbol{z}(t), \boldsymbol{v}(t)\right] . \tag{14.4}$$

The complete description of the organism is thus apparently reduced to a specification of the dynamics of a non-linear, multi-component system which is driven by the influences from the environment. However, this is not exactly the case, because changes of the driving variable $\boldsymbol{v}(t)$ may also depend on the actions of the organism. Therefore, the corresponding dynamics is formally represented by the equation

$$\dot{\boldsymbol{v}}(t) = \boldsymbol{F}_v\left[\boldsymbol{z}(t), \boldsymbol{v}(t)\right] \tag{14.5}$$

where, as before, the vector function $\boldsymbol{F}_v$ can generally be non-linear. It might be expected that it is also non-local because the sensory signals of the external influences cannot completely represent the state of an open environment. By introducing the joint vector which represents the state of the organism-environment,

$$\boldsymbol{S}(t) = [\boldsymbol{z}(t), \boldsymbol{v}(t)] \tag{14.6}$$

all dynamical equations can be combined into a single, more general equation that can be written as

$$\dot{\boldsymbol{S}}(t) = \boldsymbol{F}[\boldsymbol{S}(t)] \,. \tag{14.7}$$

This equation represents a joint dynamical description of the coupled organism and its environment.

While we generally do not know the vector function $\boldsymbol{F}$ that appears in the above dynamical equation, we assume that the corresponding dynamical law can be modeled at least empirically in the sensory neural network of the organism. The theory of non-parametric modeling provides an explanation how this might occur, while research into artificial neural networks has already revealed how the related tasks can be implemented. At present, we still do not know exactly how the empirical modeling in living organisms proceeds, but the approach to the empirical optimal control that is presented in Chap. 13 shows that the performance of an organism in a dynamically changing environment can be improved by self-control that is based on empirical modeling and reinforced learning. However, for this purpose the sensory neural network must somehow estimate the utility or performance of the organism in the environment from the sensory signals. The modeler and utility estimator appearing in the scheme of an intelligent controller that is presented in Fig. 13-9 could thus be interpreted as the sensory neural network of an organism. From a theoretical viewpoint the possibility of performance improvement appears self-evident but it leads to a basic question of natural sciences: *"How do the organisms capable of self-control evolve?"* The answer to this question is expected to be obtained from research into the field of self-organizing dynamical systems which should explain what kinds of functions $\boldsymbol{F}$ can emerge from the fundamental physical properties of Nature. [4]

14.3 An Operational Description of Consciousness

Let us consider the SNN of the self-controlled organism, as depicted in Fig. 14.1 and assume that it is capable of non-parametric modeling. With this aim it obtains the sensory information about the state of the environment as well as about the state of the body and utilizes this in the formation of its memory. After the exploration period, it possesses empirical information about the expected dynamics of the environment and its own admissible behavior. In addition, it also possesses empirical information about the utility that can be further applied to achieve an optimization of the organism's future behavior. All of the mentioned properties are usually jointly represented by the word *consciousness* which is one of the most difficult terms to define although it is intuitively well-understood. [12] Therefore we conjecture that specification of consciousness, at least in its most primitive form, could be supported by the description of the fundamental operational properties of an intelligent, self-controlled, closed organism. It appears that the most fundamental characteristic of consciousness stems from the feedback of sensory information from the body to the modeler present in SNN where an internal joint representation of the organism and the environment is formed. According to our scheme of non-parametric modeling there is, in fact, no picture but rather a set of memorized prototypes that represent the model of Nature. The second characteristic stems from the available empirical information that can be applied for the control of the organism as explained in the Chap. 13. The difficult question as to how the modeler knows where a particular item that is needed for forecasting of the organism's optimal trajectories is stored in the memory need not be addressed because the items are associatively recalled based on a conditional average and current sensory signals. Using the empirical information stored in prototypes, that control variable is formed which corresponds to the optimally estimated intention of the organism. The state that is achieved following this control from the SNN may, in fact, be different from that which is sought, but it is expected that, on average, the states obtained by such a control will correspond to an optimal performance. The last property is related to the ability of the modeler of the SNN to forecast the dynamics of the environment and the behavior of the complete organism. For this purpose, there must exist a sub-network in the SNN where the optimal virtual states can be estimated and temporarily stored. The corresponding process in the SNN might be thought to correspond to a *self-imagination* of the organism in the predicted states of the environment. From the analogy between a self-controlled system and a conscious organism, consciousness can be thought of as the activity of the sensory neural network in a self-controlled organism by which the information about the real and virtual states of the organism and its environment is processed. It can be thus represented by the set of operations needed to

1. Accept sensory signals
2. Create and adapt the prototypes in the memory and eventually, pay attention to a particular item among them
3. Utilize the stored empirical information for the associative recall based on the present sensory information which is needed for the estimation of the complete organism–environment state
4. Utilize the stored empirical information for forecasting of the future sequence of organism–environment states
5. Find optimal virtual trajectories of the organism in the environment
6. Generate an admissible control of the organism

All the above stated items can be nearly completely simulated by utilizing the methods based on empirical non-parametric modeling of natural phenomena.

For its implementation, an information processing system consisting of a sensory network and a computer is needed.

The most characteristic property of a conscious organism stems from the internal representation of the complete organism–environment state which can be described by the activity of the neurons storing the prototypes. We usually say that such an organism is aware of itself.

According to our presentation, we conjecture that consciousness in the biological world has developed to optimize the performance of organisms in changeable environments and it represents a basic dynamical property of self-controlling organism. When the organism is destroyed, or when a biological subject dies, this property ceases.

14.4 The Fundamental Problem of Evolution

The self-controlled organism presented by the scheme depicted in Fig. 14.1 includes only those elements that are inevitably required to achieve its self-control. It is characteristic that an organism is comprised of elements that perform specific tasks. However, an intelligent behavior of the organism stems principally from the SNN that is capable of modeling and forecasting optimal future actions. In Chap. 13 we formulated the optimal utility trajectory by the optimization of the momentary utility that is determined only by the current state of the organism. This appears to be sufficient for the planning of very primitive behavior in simple environments. However, we know from experience that Nature is very complex, and that this complexity increases with the growth of scales on which the environment is observed. For an optimal self-control of an organism moving in a complex environment we should formulate a modeler that is capable of optimizing its behavior by utilizing information that is presented to it by a long sequence of states. This means that the modeler should eventually include an additional unit in which long-term forecasting and performance optimization can be performed.

According to the classical concept of *determinism* the dynamical behavior of closed systems is completely determined by the extremum of the action integral

$$J(t) = \int_0^t L(t')\,dt' \tag{14.8}$$

where the Lagrangian function is determined by the difference of the kinetic and potential energies of the observed system $L = T - U$. Such a description permits no additional optimization because a single trajectory of a dynamical system is completely determined by the action integral alone. The question thus arises how to avoid this rigidity of mechanical determinism. Such a determinism is difficult to adapt so that it can provide a description of living organisms which optimize their behavior in open environments according to apparently freely selected internal measures of performance. Such an optimization can only be performed when in a certain state there exist various possible next states that can be attained by a self-control of the organism. The various possibilities are usually related to the inherent instability of dynamical phenomena which can be explained by the instabilities of dynamical trajectories and the corresponding, sensitive dependence of the dynamics of the organism on its initial conditions. It is known from the theory of deterministic chaos that infinitesimally small disturbances of an inherently unstable system can result in an observable effect on its trajectory in a finite time. [2, 13] This further corresponds to an advantage with respect to self-controlling organisms, because small influences can lead to appreciable effects. By a proper generation of small disturbances, a self-controlling organism can therefore achieve states which are favorable with respect to its utility. Consider the following remarkable example. Just a few hundred photons, detected by the eye of an elephant may excite him to reach speeds of several m/s. This represents a gain in the system energy in excess of 10^{22}. We may then ask, how can biological systems such as elephants which possess such a high system gain, cope with it, without being chaotic? In fact, can we conclude from examples as this, that life exists on the edge of chaos while, at the same time, it is well-organized? [9] The inherent instability of natural dynamical phenomena thus brakes the rigidity of the mechanistic determinism and enables the planning of behavior and the execution of intentions by self-control that is one of the most characteristic properties of life. [11] The planning of behavior corresponds to a "free will" and we can conclude that it is possible only in those cases in Nature in which there exists an inherent instability and a corresponding, sensitive dependence on initial conditions. Related to the inherent instability of dynamical organisms is usually a development of structure which represents a kind of order in Nature. [6, 7, 11] It is interesting to note that if Nature were completely random and did not possess its present structure, it would not permit self-control because, just as in the deterministic case, there would be no planning possible. In this case, any state of the environment could be followed by any other state without any preference and there would be no benefit obtained by planning and self-control. According to this observation, it is necessity as well as chance that is related to the instability of Nature which makes self-control possible and hence the existence of life itself. [6] Thus the theory of deterministic chaos appears to be of fundamental importance for the explanation of the properties of life and the development of natural intelligence. [4]

However, even more fundamental than forecasting itself there is the planning of future behavior that is based on an optimization of the utility function. Human beings and probably also animals, in general, use abstract reasoning based on ideas or notions to plan their future behavior. Abstract notions and ideas could be treated as encoded data about particular aspects of prototype states and their sequences. But exactly how they might be created in a self-organized organism is still not completely clear. Based on research on problems of feature extraction [10] an additional optimal processing of information is needed for this that eventually also needs specific sub-networks in the complete SNN where abstract notions and ideas are processed in an organism. A similar requirement for additional sub-networks appears if the organism must perform various specific tasks such as moving, feeding, communicating, etc. The problems related to the construction and interconnection of such sub-networks must be solved during the evolution of the organism, which, it would appear, could not be described for the general case, but rather only with respect to a particular organism and its environment. In order to understand the properties of the hierarchical structure of the brains of animals, we should first understand the fundamental laws of biological evolution and life itself, which have not, as yet, been successfully described in terms of physical laws. Nevertheless, it appears that the hierarchical structure of information processing organisms is favorable for self-controlling larger organisms. This does, however, raise the question of whether there exist some simple general laws by which the corresponding properties of biological organisms could be described without a detailed understanding of natural evolution. [10]

If we wish to increase the applicability of a self-controlled organism that has adapted to perform particular tasks in a specified environment, then we must provide for the modeling of more complex environments and the planning and execution of more complicated tasks. We can expect that, for this purpose, the properties of an organism should in general adapt in both their physical and their information processing aspects. Nature does not appreciably change the structure of organisms that presently exist in it. Organisms grow older, adapt, and then die. Evolutionary changes in their structure appear only in their descendants. The performance of the parents results in the number of descendants and their fit to the environment. Similar processes can be observed in the industrial evolution of various machines and electronic information processing systems. The fundamental problem of such an evolutionary strategy is how to preserve the favorable properties of the parent organism and how to include those novelties of the organism which eventually increase its performance. In the biological world, mutations and the exchange of genetic information are the primary mechanisms for this purpose. Genes of parent cells can be treated as extremely concentrated pools of information corresponding to the properties of the parents. A descendant obtains from both parents an amount of information that is sufficient for the growth of the complete structure of the organism, provided that it evolves in a proper environment. The growth of a biological organism thus resembles a complex manufacturing process which is based on a specific technology. The exchange of genetic information is in this aspect similar to the mixing of design schemes, ideas and the technologies of two workshops. Only the mutations represent the creation of essentially new properties of products. The performance of an individual product is evaluated during its use in its lifetime and that which requires less energy to operate, that which is more versatile and possesses greater endurance with less service, etc. will more likely be manufactured. One product thus competes during its lifetime with all other similar products in the marketplace. It appears that there might exist quite general principles or laws by which the evolution of the biological world as well as the development of the industrial or economic spheres can be understood. The description of evolution is thus related to two basic processes: the exchange of the information at birth of an offspring, and the selection of organisms that is based on the performance of a particular subject during the life in a certain environment. With these two processes, we do not describe *why*, but *how* evolution proceeds. The complete description can also be mathematically formulated by a so-called genetic algorithm. [8, 5, 1]. At present, this algorithm appears as the only other option by which one can quantitatively describe the optimization that occurs in evolutionary processes. We therefore conjecture that this algorithm represents a quite general basis for the solution of previously-mentioned problems related to the development of proper architectures of various artificial neural network-like information processing systems. Work in this field of research is currently very active and has already led to various applications of practical importance. [4, 5, 8]

A. Fundamentals of Probability and Statistics

The purpose of this appendix is to briefly present the basic concepts of probability theory and statistics which are relevant for a quantitative empirical modeling of natural laws.

A.1 Sample Points, Sample Space, Events and Relations

In formulating a quantitative description of a natural phenomenon we assume that it can be characterized experimentally. Related to every experiment is the specification of its plan and its execution. In planning an experiment, a portion of nature or an environment is selected according to some experimental set-up that can be intuitively or operationally described. During execution of the experiment, the examined natural phenomenon causes changes in the selected environment that can be described as a transition from the initially prepared state into the final one. We generally assume that an experiment can, at least in principle, be repeated arbitrarily many times under identical initial conditions of the experiment. Consequently, the initial state is assumed always to be the same and it is therefore usually omitted from further description. The result of an experiment is thus described only by specifying the outcome. A particular transition from the initial to the final state is generally called an *elementary experimental event* and this is described symbolically by an abstract element s. It is treated as a single sample point in an abstract *sample space* S which denotes the set of all possible realizations of experimental outcomes. The main property of a sample space is described by the number of sample points; if it is countable, the sample space is discrete, otherwise it is continuous.

An observed phenomenon is called a deterministic one if a repetition of experiments under given, fixed set of initial conditions always yields the same result. If this is not the case, then the phenomenon is called a stochastic or a random one. It is worth noting that this classification strongly depends on the selected experimental procedures and therefore the same phenomenon can often be identified as being either deterministic or stochastic.

For the description of natural phenomena it is convenient to introduce the concept of a *compound event* consisting of set of elementary events, that is, $A = \{s_1, s_2, \ldots\}$. Such a compound event is realized whenever the outcome of the experiment is represented by an element of this set. In agreement with this definition, we conclude that the complete sample space S corresponds to an event that is realized at each execution of an experiment because it includes all possible elementary outcomes; it therefore represents a certain event. On the other hand, an empty set $\emptyset = \{\}$ represents a non-realizable event.

Introduction of compound events establishes the link between an experimental description of natural phenomena and set theory. The operations defined in this theory are applicable in defining relations between events. The following relations between events are of importance to us:

1. *Equality*: $A = B$
   Two events are equal if and only if both sets A and B have the same elements.
2. *Implication or inclusion*: $A \subset B$
   The event A implies (is included within) the event B if for every $s \in A$ it follows that $s \in B$.
3. *Union*: $C = A \cup B$
   An event C is a union of events A and B if it has the elements that are either in A or in B or in both simultaneously.
4. *Intersection*: $C = A \cap B$
   An event C is intersection of events A and B if it has elements that are simultaneously in A and B.
5. *Complement*: A^c
   An event A^c is a complement of A if it is comprised of all the elements from S that are not in A.

Two events are mutually *exclusive* or *disjoint* if they have no common elements or $A \cap B = \emptyset$. Other properties of compound events are described in detail in the standard literature of elementary courses on probability theory and will not be further considered here. [1, 2] Let us only mention that in the case when a continuous sample space is used in the description of a natural phenomenon, then it is most often unavoidable that one uses compound events representing intervals rather than elementary sample points in giving a description of its stochastic properties. For a class σ of such events it is characteristic that:

I. To each event $A \in \sigma$ there corresponds a complement $A^c \in \sigma$.
II. To any pair of events $A \in \sigma$ and $B \in \sigma$ there corresponds a union $C = A \cup B \in \sigma$.

Such a collection of events σ is often called a *sigma algebra* and it is closed with respect to operations 1-5 defined for events.

A.2 Probability

As mentioned previously, a repetition of experiments on a stochastic phenomenon does not always yield the same result. Let us assume that we apply for the description of the experimental outcomes, a finite set of mutually disjoint events $A_1, A_2, \ldots, A_K$ such that $A_1 \cup A_2 \ldots \cup A_K = S$ and $A_i \cap A_j = \emptyset$ for each pair $i \neq j$. When we perform an experiment, then one and only one event from this set is realized. After N repetitions of an experiment there are $N_1, N_2, \ldots, N_K$ realizations of corresponding events which satisfy the relation $N_1 + N_2 + \ldots + N_K = N$. The *relative frequency* of occurrence of i-th event is defined by

$$\Phi_i = N_i/N . \tag{A.1}$$

The frequency satisfies the conditions

$$0 \leq \Phi_i \leq 1 , \tag{A.2}$$

$$\sum_{i=1}^{N} \Phi_i = 1 . \tag{A.3}$$

The relative frequency is the basic variable applicable for a quantitative empirical description of random phenomena. One of the most important properties of this description is its statistical regularity. The relative frequency generally fluctuates with repetition of experiments, but it has been empirically observed by numerous investigations in the history of science that the fluctuations generally decrease with increasing N. Therefore one can expect with practical certainty that Φ_i asymptotically converges to a fixed value p_i which can be treated as a characteristic quantity characterizing the random properties of the phenomenon under investigation. This quantity is called the empirical probability of the event, while the set $\{p_i,\ i = 1, 2, \ldots, N\}$ describes the empirical probability distribution.

Experience with the relative frequency leads us to the conclusion that it is reasonable to expect that associated with each random phenomenon there is a specific probability distribution which characterizes its properties in a manner similar as is done by deterministic physical laws. It is therefore important to specify the fundamental properties of probability in a mathematically consistent manner by a set of axioms.

Axiom A.1. With an experiment there is related a sample space S representing all possible elementary experimental outcomes and a collection σ of subsets A of S called events.

Axiom A.2. To each event A from the collection σ, there can be assigned a non-negative real number $P[A] \geq 0$ called the probability of this event.

Axiom A.3. Any assignment of probability P must satisfy the condition $P[S] = 1$.

Axiom A.4. For any pair of mutually exclusive events, A and B, satisfying the condition $A \cap B = \emptyset$, the probability assignment must satisfy the relation $P[A \cup B] = P[A] + P[B]$.

These axioms suffice for describing the probability properties in cases with a sample space comprised out of finite number of sample points. If this number is not finite then Axiom A.4 must be extended. For this we have the following additional axiom:

Axiom A.5. For any set of mutually exclusive events $A_1, A_2, A_3, \ldots$ satisfying the conditions $A_i \cap A_j = \emptyset$ for $i \neq j$, the assignment of probability must satisfy the relation $P[\bigcup_{i=1}^{\infty} A_i] = P[A_1] + P[A_2] + P[A_3] + \ldots = \sum_{i=1}^{\infty} P[A_i]$.

A rule which assigns the probability to the events from sigma algebra represents a function defined on a set and it is therefore called a *set function.* The set function that satisfies Axiom A.4 is finitely additive while that which satisfies Axiom A.5 is countably additive. The triplet $\{S, \sigma, P\}$ is usually called a *probability space.*

When the sample space S consists of a finite number of sample points, then one can assign a probability to each of the elementary events represented by sample points. In contrast to this, in cases possessing a continuous sample space, the probability cannot always be properly assigned when the collection σ is comprised of elementary events, but only if larger events represented by intervals are applied. This property is a consequence of the discreteness of the repetition of experiments by which the relative frequency is introduced. Simply stated, a continuum is "too large" for properly assigning to each sample point in it a relative frequency that is determined by a countable repetition of experiments.

The quoted set of axioms is sufficient to reproduce the general characteristic properties of relative frequency by using only standard arithmetic techniques. Specifically it follows: $P[A^c] = 1 - P[A]$; $P[A] \leq 1$; $P[\emptyset] = 0$.

Let A and B be two arbitrary events from the collection σ. A new event C can be defined by their intersection, $C = A \cap B$. The probability corresponding to this event is called a *joint probability.* It is often convenient to express the joint probability by the probability of one basic event either A or B. For this purpose the *conditional probability* is defined by the relation:

$$P[A \cap B] \;=\; P[B|A] \cdot P[A] \,. \tag{A.4}$$

For $P[A] \neq 0$ the conditional probability of event B at given condition represented by event A can be explicitly expressed as

$$P[B|A] \;=\; P[A \cap B]/P[A] \,. \tag{A.5}$$

If this conditional probability does not depend on the condition A then $P[B|A] = P[B]$ and it follows from the preceding expression

$$P[A \cap B] = P[B] \cdot P[A] , \tag{A.6}$$

which further implies that $P[A|B] = P[A]$. In this specific case, the conditional probability of A is independent of condition B, and vice versa, and consequently both events are called *statistically independent.*

A.3 Random Variables and Probability Distributions

Our ultimate goal is to apply probability to a quantitative description of natural phenomena. For this purpose the results of experiments must be quantitatively described by measurements. This means that a rule is specified by which each sample point $s \in S$ is associated to one or more numbers denoted by a scalar or vector $X(s)$ respectively. Such a rule is mathematically represented by a function $s \to X(s)$ and it defines a *random variable.* A particular *realization* of this variable is denoted by $x = X(s)$. In this text large letters are used to denote various random variables while small leters denote their particular realized values. A totality of all possible realizations

$$S_x = \{x : x = X(s) \quad \text{for } s \in S\} \tag{A.7}$$

represents a new sample space which is an *image* of S and it determines the *range* of the variable X. By the function X, an event $A \subset S$ is mapped into a new event $A_x \subset S_x$ specified in the range of the random variable: $A \overset{X}{\to} A_x$. In the following, we always assume that the probability assignment defined on S leads to a new probability assignment in the range of random variable:

$$\forall A \overset{X}{\to} A_x \;\Rightarrow\; P[A \subset S] = P[A_x \subset S_x] . \tag{A.8}$$

Let us now consider a scalar, random variable X and let us compose a particular event A_x in the collection σ_x in the interval from $(-\infty, x]$ or $A_x = \{X(s) \le x\}$. The *cumulative probability distribution function* of the random variable X is then defined as

$$F_X(x) \equiv P[A_x] \equiv P[X(s) \le x] . \tag{A.9}$$

This function has the following characteristic properties:

Limits:	$F_X(+\infty) = 1 ;\; F_X(-\infty) = 0$
Monotonicity:	$b \ge a \;\Rightarrow\; F_X(b) \ge F_X(a)$
Continuity on the right:	$F_X(x + 0) = F_X(x)$
Interval probability:	$P[X(a, b]] = F_X(b) - F_X(a)$
Decomposition:	Every cumulative probability function can be expressed as a sum of a continuous and discontinuous function

This function written in terms of its continuous and a discontinuous parts is:

$$F_X(x) = C(x) + D(x) . \tag{A.10}$$

The discontinuous part may be expressed as

$$D(x) = \sum_i p_i U(x - x_i) \tag{A.11}$$

where U indicates the unit step function, that is,

$$U(x) = \{1 \ldots \text{ for } x \geq 0 ; \; 0 \ldots \text{for } x < 0\} \tag{A.12}$$

and $p_i = P[X = x_i]$. For $C = 0$ or $D = 0$, the random variable is called discrete or continuous respectively.

The cumulative distribution function can be expressed by a Lebesgue-Stieltjes integral as:

$$F_X(x) = \int_{-\infty}^{x} dP[x] = \int_{-\infty}^{x} f(x)\, dx . \tag{A.13}$$

If the cumulative distribution function is differentiable then its inversion leads to the definition of the probability density function pdf:

$$f_X(x) = \frac{dF_X(x)}{dx} . \tag{A.14}$$

This definition can be generalized to the discontinuous distribution functions if the unit *step function* is described by the Dirac *delta function*

$$U(x) = \int_{-\infty}^{x} \delta(x)\, dx \tag{A.15}$$

so that we can formally write

$$\delta(x) = \frac{dU(x)}{dx} . \tag{A.16}$$

The density function of an arbitrary probability distribution can be then formally expressed as

$$f_X(x) = \frac{dC(x)}{dx} + \sum_i p_i\, \delta(x - x_i) . \tag{A.17}$$

For the probability density function the following properties are characteristic:

$$\textit{Non-negativeness:} \quad f_X(x) \geq 0 . \tag{A.18}$$

$$\textit{Normalization to unit area:} \quad \int_{-\infty}^{+\infty} f_X(x)\, dx = 1 . \tag{A.19}$$

The probability of an event $A \subseteq S_x$ can be expressed by the density function as

$$P[X \in A] = \int_A f(x)\,dx \,. \tag{A.20}$$

The most common *normal density* is defined by the Gaussian function as

$$f_X(x) = \frac{1}{\sqrt{2\pi}\sigma} \exp\left[-\frac{1}{2}\left(\frac{x-m}{\sigma}\right)^2\right] \tag{A.21}$$

where the parameters m and σ denote the *mean value* and the *standard deviation* respectively.

Let us further consider the multivariate case when the result of an experiment is described by a n-dimensional vector:

$$\boldsymbol{X}(s) = [X_1(s), X_2(s), \ldots, X_n(s)] = (x_1, x_2, \ldots, x_n) \,. \tag{A.22}$$

The properties of this random variable are described by the *joint-probability distribution function*

$$F_{\boldsymbol{X}}(\boldsymbol{x}) = P[X_1(s) \le x_1,\ X_2(s) \le x_2, \ldots,\ X_n(s) \le x_n] \,. \tag{A.23}$$

As in the previous case, this distribution function is bounded:

$$F_{\boldsymbol{X}}(-\infty, -\infty, \ldots, -\infty) = 0, \tag{A.24}$$

and

$$F_{\boldsymbol{X}}(+\infty, +\infty, \ldots, +\infty) = 1 \,. \tag{A.25}$$

The *joint probability density function* is defined in the multivariate case by the partial derivative

$$f_{\boldsymbol{X}}(\boldsymbol{x}) = \frac{\partial^n F_{\boldsymbol{X}}(\boldsymbol{x})}{\partial x_1 \ldots \partial x_n} \,. \tag{A.26}$$

This probability density function is also non-negative and normalized to 1. In terms of this, the cumulative distribution is expressed as

$$F_{\boldsymbol{X}}(\boldsymbol{x}) = \int_{-\infty}^{x_1} \cdots \int_{-\infty}^{x_n} f_{\boldsymbol{X}}(\boldsymbol{x})\, dx_1 \ldots dx_n \,. \tag{A.27}$$

Finally, the i-th *marginal probability distribution* is defined as

$$\begin{aligned} F_{X_i}(x_i) &= P[X_1(s) \le \infty, \ldots, X_i(s) \le x_i, \ldots, X_n(s) \le \infty] \\ &= F_{\boldsymbol{X}}(\infty, \ldots, x_i, \ldots, \infty) \,, \end{aligned} \tag{A.28}$$

which is:

$$F_{X_i}(x_i) = \int_{-\infty}^{\infty} \cdots \int_{-\infty}^{x_i} \cdots \int_{-\infty}^{\infty} f_{\boldsymbol{X}}(\boldsymbol{x})\, dx_1 \ldots dx_n \,. \tag{A.29}$$

And the corresponding density function is given by

$$f_{X_i}(x_i) = \frac{dF_{X_i}(x_i)}{dx_i} \; . \tag{A.30}$$

Two components of a random vector are *statistically independent* if their joint probability density can be expressed as

$$f_{X_1 X_2}(x_1, x_2) = f_{X_1}(x_1) \cdot f_{X_2}(x_2) \; . \tag{A.31}$$

The probability of an event $A \subseteq S_x$ can be expressed by the multivariate density function as

$$P[\boldsymbol{X} \in A] = \int \overset{\cdots}{_{\boldsymbol{x} \in A}} \int f_{\boldsymbol{X}}(\boldsymbol{x}) \, dx_1 \ldots dx_n \; . \tag{A.32}$$

Using the above definitions, we can further define the conditional probability distribution and its density. Let us select $(n-1)$–dimensional subspace $S_{\boldsymbol{X}''}$ described by components $\{x_2, x_3, \ldots, x_n\}$ and define a condition by $C = [X_2 < x_2, \ldots, X_n < x_n]$. The distribution function of the conditional probability of the first component x_1 at the specified condition C is then

$$\begin{aligned} F(x_1|C) &= F_{\boldsymbol{X}}(x_1, x_2, \ldots, x_n)/P[C] \\ &= \frac{\int_{-\infty}^{x_1} \int \overset{\cdots}{_{\boldsymbol{x} \in C}} \int f_{\boldsymbol{X}}(\boldsymbol{x}) \, dx_1 \, dx_2 \ldots dx_n}{\int \overset{\cdots}{_{\boldsymbol{x} \in C}} \int f_{\boldsymbol{X}}(\boldsymbol{x}) \, dx_2 \ldots dx_n} \; . \end{aligned} \tag{A.33}$$

If the condition is specified by an infinitesimally small region arround the point $\boldsymbol{X}'' = (x_2, x_3, \ldots, x_n)$ in the subspace $S_{\boldsymbol{X}''}$ then this equation reduces to

$$F(x_1|x_2, x_3, \ldots, x_n) = \frac{\int_{-\infty}^{x_1} f_{\boldsymbol{X}}(\boldsymbol{x}) \, dx_1}{f_{\boldsymbol{X}''}(x_2, x_3, \ldots, x_n)} \; . \tag{A.34}$$

The corresponding density of conditional probability of x_1 relative to the remaining coordinates is then

$$f(x_1|x_2, x_3, \ldots, x_n) = \frac{f_{\boldsymbol{X}}(\boldsymbol{x})}{f_{\boldsymbol{X}''}(x_2, x_3, \ldots, x_n)} \; . \tag{A.35}$$

In a similar way the following density of conditional probability is derived when the condition is specified by the truncated vector $\boldsymbol{X}'' = (X_{r+1}, \ldots, X_n)$:

$$f(x_1, \ldots, x_r|x_{r+1}, \ldots, x_n) = \frac{f_{\boldsymbol{X}}(\boldsymbol{x})}{f_{\boldsymbol{X}''}(x_{r+1}, \ldots, x_n)} \; . \tag{A.36}$$

This density is of fundamental importance for the non-parametric estimation of a truncated vector $\boldsymbol{X}' = (x_1, x_2, \ldots, x_r)$ from a given vector $\boldsymbol{X}'' = (x_{r+1}, \ldots, x_n)$. This is the basic task for developing empirical models of natural phenomena.

A.4 Averages and Moments

Let X be a scalar random variable and y its measurable function. The *average* or *expected value* of this function is defined by the Lebesgue-Stieltjes integral

$$\mathrm{E}\,[y(X)] = \int_{-\infty}^{+\infty} y(x)\,dP[x] = \int_{-\infty}^{+\infty} y(x) f_X(x)\,dx\,. \tag{A.37}$$

For a discrete, random variable this integral can be written as a sum

$$\mathrm{E}\,[y(X)] = \sum_i y(x_i)\,p_i\,. \tag{A.38}$$

Average values of various functions can be used to represent the properties of a random variable. Often, the expected value of the k-th power of X is used, corresponding to the *k-th moment*:

$$\mathrm{E}\,[X^k] = \int_{-\infty}^{+\infty} x^k f_X(x)\,dx \equiv m_k\,. \tag{A.39}$$

For $k=1$ we have the *mean value*

$$\mathrm{E}\,[X] = \int_{-\infty}^{+\infty} x\, f_X(x)\,dx \equiv m\,. \tag{A.40}$$

For the k-th power of the deviation from the mean value, the *central moments* are defined as

$$\mathrm{E}\,[(X-m)^k] = \int_{-\infty}^{+\infty} (x-m)^k f_X(x)\,dx \equiv \mu_k\,. \tag{A.41}$$

From $k=2$ we obtain the *variance* of X

$$\mathrm{var}(X) = \mathrm{E}\,[(X-m)^2] = \int_{-\infty}^{+\infty} (x-m)^2 f_X(x)\,dx \equiv \mu_2\,. \tag{A.42}$$

Using this moment, we define the *standard deviation* as

$$\sigma \equiv \sqrt{\mathrm{var}(X)}\,. \tag{A.43}$$

In a manner similar to that for a scalar random variable, the expected value of a measurable function y of a random vector $\boldsymbol{X} = (X_1, X_2, \ldots, X_n)$ is defined:

$$\begin{aligned}\mathrm{E}\,[y(\boldsymbol{X})] &= \int_{-\infty}^{+\infty} \cdots \int_{-\infty}^{+\infty} y(\boldsymbol{x})\,dP[\boldsymbol{x}] \\ &= \int_{-\infty}^{+\infty} \cdots \int_{-\infty}^{+\infty} y(\boldsymbol{x})\, f_{\boldsymbol{X}}(\boldsymbol{x})\,dx_1 \ldots dx_n\,. \end{aligned} \tag{A.44}$$

For vector random variables, *joint moments* are defined by the average of the product of various powers of the vector components $\mathrm{E}[X_1^k X_2^l \dots X_n^m]$. Among these, the most frequently used are the correlations of two components

$$R_{ij} = \mathrm{E}[X_i \cdot X_j] \,. \tag{A.45}$$

The corresponding central moments are the covariances

$$\sigma_{ij} = \mathrm{E}[(X_i - m_i) \cdot (X_j - m_j)] \tag{A.46}$$

which, together with the variances $\sigma_i^2 \equiv \sigma_{ii}$, comprise the covariance matrix $\boldsymbol{\Sigma} = [\sigma_{ij}]$. Using this matrix, the probability density of a normal distribution of a multivariate random variable is written as

$$\begin{aligned} f(\boldsymbol{x}) &= \frac{\exp[-\frac{1}{2}(\boldsymbol{x}-\boldsymbol{m})^T \boldsymbol{\Sigma}^{-1}(\boldsymbol{x}-\boldsymbol{m})]}{(2\pi)^{n/2}|\boldsymbol{\Sigma}|^{1/2}} \\ &= (2\pi)^{-n/2}|\boldsymbol{B}|^{1/2} \exp\left[-\frac{1}{2}(\boldsymbol{x}-\boldsymbol{m})^T \boldsymbol{B}\,(\boldsymbol{x}-\boldsymbol{m})\right] . \end{aligned} \tag{A.47}$$

Here $|\boldsymbol{\Sigma}|$ is the determinant of the matrix $[\boldsymbol{\Sigma}]$, and ^{T}, $\boldsymbol{\Sigma}^{-1} = \boldsymbol{B}$ denote the transposed vector and inverse covariance matrix respectively.

In any practical application involving the empirical modeling of natural phenomena, the conditional average is of central importance. Suppose that the vector of the random variable $\boldsymbol{X}$ is split into two truncated vectors written as $\boldsymbol{X}'$ and $\boldsymbol{X}''$ which are specified by $\boldsymbol{X}' = (X_1, \dots, X_r)$ and $\boldsymbol{X}'' = (X_{r+1}, \dots, X_n)$. Then, the conditional average of $\boldsymbol{X}'$ at given condition $\boldsymbol{X}'' = \boldsymbol{x}'' = (x_{r+1}, \dots, x_n)$ is defined by

$$\begin{aligned} \mathrm{E}[\boldsymbol{X}'|\boldsymbol{x}''] &= \int_{-\infty}^{+\infty} \dots \int_{-\infty}^{+\infty} \boldsymbol{x}' f(x_1, \dots, x_r | x_{r+1}, \dots, x_n)\, dx_1 \dots dx_r \\ &= \int_{-\infty}^{+\infty} \dots \int_{-\infty}^{+\infty} \boldsymbol{x}' f(\boldsymbol{x}'|\boldsymbol{x}'')\, d^r \boldsymbol{x}' \,. \end{aligned} \tag{A.48}$$

A.5 Random Processes

In practice, we often encounter phenomena that must be described by an infinite sequence of data. A typical example is the description of the weight of bricks produced by a sequential manufacturing process. For this purpose the concept of a random vector can be extended to infinitely many components.

$$\boldsymbol{X} = (X_1, X_2, \dots, X_n, \dots) = \{X(n)\,;\ n = 1, 2, \dots \to \infty\} \,. \tag{A.49}$$

In this case, the phenomenon is usually called a *discrete-parameter, random process*. A sample of the corresponding random variable is determined by an infinite sequence of data $(x_1, x_2, \dots, x_n, \dots)$. It is generally denoted by

$$\{X(n,s)\,;\quad n=1,2,\ldots\} \tag{A.50}$$

with the index n describing the position of the component in the sequence and the second index denoting the experiment which yielded the data or the abstract sample point. The random properties of such a variable are sufficiently described by specifying a rule which determines the joint-probability distribution function for an arbitrarily selected finite subset of components

$$\boldsymbol{X}'(s) = (X(i,s), X(j,s), \ldots, X(k,s))\ , \tag{A.51}$$

$$F_{\boldsymbol{X}'}(\boldsymbol{x}') = P[X(i,s) \le x_i, X(j,s) \le x_j, \ldots, X(k,s) \le x_k] \tag{A.52}$$

with $i, j, \ldots, k$ an arbitrary combination of indices. However, this treatment can be extended also to the description of continuous random processes such as a turbulent jet of water from a pipe. In this case, the diameter of the jet, specified by X, depends continuously on the time t in the selected time interval of observation $(0, T)$ and on the sample s obtained by the repetition of the experiment. The corresponding random variable is then denoted by

$$\{X(t,s)\,;\quad t \in (0,T)\} \tag{A.53}$$

and is said to represent a *continuous-parameter random process.*

In order to characterize the random properties of a contiuous, random proces by probability distributions, we first convert a continuous process into a finite-dimensional random vector. For this, we arbitrarily select a finite number of time values $\{t_1, t_2, \ldots, t_N\}$ and specify the random variable $X(t,s)$ at the corresponding moments. By this step, we form an auxiliary random vector

$$\boldsymbol{X}' \;=\; (X(t_1,s), X(t_2,s), \ldots, X(t_N,s))\ . \tag{A.54}$$

The continuous-parameter random process is then said to be sufficiently described from the probabilistic point of view if a rule is specified by which all the joint probability distributions of the arbitrarily formed auxiliary random vector

$$F_{\boldsymbol{X}'}(\boldsymbol{x}') \;=\; P[X(t_1,s) \le x_1, X(t_2,s) \le x_2, \ldots, X(t_N,s) \le x_N] \tag{A.55}$$

can be determined. Such a probability distribution is generally a function of the selected times $\{t_1, t_2, \ldots, t_N\}$ as well as of the values $\{x_1, x_2, \ldots, x_N\}$.

$$F_{\boldsymbol{X}'}(\boldsymbol{x}') \;=\; F(x_1, x_2, \ldots, x_N; t_1, t_2, \ldots, t_N)\ . \tag{A.56}$$

A continuous-parameter random process is said to be strictly stationary when all the probability distributions formed in such a way are invariant with respect to an arbitrary translation τ of the time parameter t, that is,

$$\begin{aligned} F_{\boldsymbol{X}'}(\boldsymbol{x}') &= F(x_1, x_2, \ldots, x_N; t_1, t_2, \ldots, t_N) \\ &= F(x_1, x_2, \ldots, x_N; t_1+\tau, t_2+\tau, \ldots, t_N+\tau)\ . \end{aligned} \tag{A.57}$$

If this is not the case, then the process is said to be nonstationary in the strict sense.

Using the probability distributions defined above, we can define various moments that are of practical interest for the characterization of random processes. Most frequently the first and the second moments are applied. These are defined by the expected values

$$m_X(t_1) = \mathrm{E}\left[X(t_1)\right] = \int x_1 \, dF(x_1, t_1) \; , \tag{A.58}$$

$$R_{XX}(t_1, t_2) = \mathrm{E}\left[X(t_1)X(t_2)\right] = \int x_1 x_2 \, dF(x_1, x_2; t_1, t_2) \; . \tag{A.59}$$

These are called the *mean value* and the *auto-correlation function*, respectively. For a stationary random process, the mean value is a constant and the correlation function depends only on the difference $\Delta t = t_2 - t_1$. If a random process has a constant mean value and the correlation function depends only on the time difference, the process is said to be stationary in the wide sense. Every random process that is strictly stationary is included in this class as well.

In describing random processes we often encounter the situation in which only one variable is insufficient for describing the phenomenon. For example, let us consider a water jet which is to be characterized by its diameter and flow velocity where both of these quantities can vary randomly. In this case, we generalize the description presented above by assuming that the variable $X(t,s)$ can be a vector of k components $\{X_1(t,s), \ldots, X_k(t,s)\}$. In order to specify the random properties of such a process from the probabilistic aspect we must specify all possible joint probability densities of the type

$$\begin{aligned} F = P[&X_i(t_{i,1}, s) \le x_{i,1}, X_i(t_{i,2}, s) \le x_{i,2}, \ldots \\ &\ldots, X_j(t_{j,1}, s) \le x_{j,1}, X_j(t_{j,2}, s) \le x_{j,2}, \ldots] \; . \end{aligned} \tag{A.60}$$

By using such functions, we can then define moments such as the cross-correlation function

$$R_{X_1X_2}(t_1, t_2) \;=\; \mathrm{E}\left[X_1(t_1) \cdot X_2(t_2)\right] \tag{A.61}$$

which can be used to obtain an approximate description of the relationship between two processes.

In addition to multicomponent random processes we often encounter phenomena in which one parameter is insufficient for describing the random variable. For example, we can imagine the description of the air temperature which depends randomly on the time t and the position specified by vector $\boldsymbol{r}$. In this case we are referring to a random field variable. A generalization of the probabilistic description of random processes to random fields does not represent an insurmountable obstacle. Instead of the single parameter t, we must now use a multicomponent set of parameters denoted as $\Theta(t, \boldsymbol{r})$.

With these, the concept of stationarity can be transferred to the description of random fields. It can be further extended to the concepts of homogeneity and isotropy. A random field is stationary in time or homogeneous in space if all the corresponding probability distributions are invariant to arbitrary translations in the temporal and spatial portions of the parameter $\Theta(t, \boldsymbol{r})$. Similarly, the random field is isotropic if the probability distributions are invariant under arbitrary rotations in the spatial portion of the parameter Θ.

A.6 Sampling, Estimation and Statistics

Whenever we are dealing with a random phenomenon we can hypothesize that some probability distribution function corresponds to it. Generally there are two possibilities for drawing inferences about the corresponding hypothetical distribution function: an analytical approach and an empirical one. In the analytical approach some information about the properties of the phenomenon is utilized in a theoretical determination of the probability distribution. For example, in a coin tossing experiment, the symmetry of the coin is presumed, therefore equal probabilities $p_1 = p_2$ are assigned to both possible outcomes, which together with the probability axioms leads to the assignment that $p_1 = p_2 = 1/2$. In contrast, in the empirical approach, the probability distribution is estimated from the relative frequency determined directly from a series of experiments. However, often only some averages are needed. It is therefore advantageous to define the empirical average directly and to relate it to the hypothetical, expected value.

Let us consider a random phenomenon characterized by the variable X and let us assume that an experiment performed N times yielded the outcomes $x_1, x_2, \ldots, x_N$. Each result represents a sample from the sample space S_x. The complete sequence of experiments is treated as a single, combined experiment yielding the *sample vector* $\boldsymbol{v} = (x_1, x_2, \ldots, x_N)$. The components of this vector are treated as realizations of different random variables X_i that are identically distributed according to some hypothetical probability distribution $F_X(x)$ and are mutually statistically independent. Successive combined experiments generally yield different realizations of the sample vector which is therefore treated as a random vector variable $\boldsymbol{V} = (X_1, X_2, \ldots, X_N)$. The empirical average, or the sample mean, defined by

$$\langle x \rangle = \frac{1}{N} \sum_{i=1}^{N} x_i \tag{A.62}$$

is then interpreted as a realization or a sample of a function of the random vector $\boldsymbol{V}$:

$$\langle X \rangle = \frac{1}{N} \sum_{i=1}^{N} X_i \, . \tag{A.63}$$

Such functions of sample vector are generally called *statistics.*

Using these concepts, we can define and determine the hypothetical expected value of the sample mean. That is,

$$\mathrm{E}\left[\langle X\rangle\right] = \frac{1}{N}\sum_{i=1}^{N}\mathrm{E}\left[X_i\right] = \frac{1}{N}\sum_{i=1}^{N} m_i = m\,. \tag{A.64}$$

Here we have taken into account that the averaging is a linear operation and that all the components X_i possess the same hypothetical mean value m. Because of the final relation, the sample mean is called the *estimator* of the hypothetical mean value. This estimator is called *unbiased* because the expectation value of the sample mean is equal to the mean value of the random variable being estimated.

The above interpretation illustrates how the hypothetical predictions can be related to experimental observations and this consequently represents the basic link between the probability theory and empirical sciences or statistics. The corresponding bridge is the sample mean and it is therefore reasonable to examine its properties in more detail.

The statistical dispersion of the sample mean around the hypothetical mean value m is described by the variance:

$$\mathrm{var}(\langle X\rangle) = \mathrm{E}\left[(\langle X\rangle - m)^2\right] = \frac{1}{N^2}\,\mathrm{E}\left[\left(\sum_{i=1}^{N}(X_i - m)\right)^2\right]\,. \tag{A.65}$$

Taking into account that all the components X_i possess an identical distribution and are mutually statistical independent, we obtain the result

$$\mathrm{var}(\langle X\rangle) = \frac{1}{N}\,\mathrm{var}(X)\,. \tag{A.66}$$

Consequently, we can generally expect that the dispersion of the sample mean X_i around the hypothetical mean value m decreases to zero with an increasing number of samples N

$$\lim_{N\to\infty}\mathrm{E}\left[(\langle X\rangle - m)^2\right] = 0\,. \tag{A.67}$$

The sample mean is therefore said to converge in mean square to the mean value being estimated and it is consequently called a *consistent estimator.* This means that with a repetition of combined experiments and with an increasing number of samples N, we can expect ever better agreement between the sample mean and the estimated value. The probability that the estimation error exceeds an arbitrarily small value ϵ therefore tends to zero with an increasing number of samples N:

$$\lim_{N\to\infty} P\left[|\langle X\rangle - m| \geq \epsilon\right] = 0\,, \qquad \text{for any } \epsilon > 0\,. \tag{A.68}$$

This result indicates that it is possible to infer with unlimited certainty about the hypothetical probabilistic properties of random phenomenon by using the sample mean estimated in repeated combined experiments.

For the modeling of natural phenomena, it is of interest that the relative frequency of an event can also be expressed as a sample mean. This means that all the convenient properties possessed by the sample mean are characteristic for the relative frequency as well. Let us consider an event A and a N-time successive random sampling of variable X. Let us further describe whether the event A occurs at the realization of a random sample X_i or not by the *set-indicator function*

$$I_A(X_i) = \begin{cases} 1 & \text{if} \quad X_i \in A \\ 0 & \text{if} \quad X_i \in A^c \end{cases} . \tag{A.69}$$

The expected value of this function equals the probability of event A

$$\mathrm{E}\left[I_A(X)\right] = \int I_A(x)\, dP = \int_A 1\, dP + \int_{A^c} 0\, dP = P[A] \tag{A.70}$$

while its sample mean corresponds to the relative frequency

$$\langle I_A \rangle = \frac{1}{N} \sum_{i=1}^{N} I_i = \frac{N_A}{N} = \Phi_A . \tag{A.71}$$

We therefore conclude that the relative frequency of an event obtained by independent random samples is an unbiased and consistent estimator of the corresponding probability.

B. Fundamentals of Deterministic Chaos

B.1 Instability of Chaotic Systems

This Appendix provides a review of the theoretical fundamentals of chaotic dynamics which are needed for the development of a forecasting system. With this as our goal, we consider a discrete dynamical system, whose state can be described by a set of physical variables $\boldsymbol{X} = \{x_1, \ldots, x_D\}$ representing a point in a D-dimensional phase space. The dynamics of the system is described by a physical law, in which the velocity of a point in the phase space and its corresponding state vector $\boldsymbol{X}$ are related to the set of system parameters $\boldsymbol{p} = \{p_1, p_2 \ldots\}$ [10, 16, 18, 19]

$$\frac{d\boldsymbol{X}}{dt} = \boldsymbol{F}(\boldsymbol{X};\, \boldsymbol{p}) \,. \tag{B.1}$$

We further assume that the system is autonomous. That is, the dynamics of the system is completely determined by its structure and therefore the function $\boldsymbol{F}(\boldsymbol{X};\, \boldsymbol{p})$ does not depend explicitly on time. With the above dynamical law and an initial state $\boldsymbol{X}(0) = \boldsymbol{X}_0$, a particular trajectory $\boldsymbol{X}(t; \boldsymbol{X}_0)$ is uniquely determined in the phase space. In many cases of practical interest, the motion develops in a limited portion of the phase space, so that the condition $|\boldsymbol{X}| \leq L$ is fulfilled, where L denotes the span of the trajectory .

Let us consider two trajectories $\boldsymbol{X}_1$ and $\boldsymbol{X}_2$ originating from two nearby initial points and let us analyze positions for a sequence of times $(t_1, t_2, \ldots)$ as indicated in Fig. B.1 With respect to the behavior of the distances $d_{12}(t_i) = |\boldsymbol{X}_1(t_i) - \boldsymbol{X}_2(t_i)|$, the dynamical systems can be divided into two groups. The first group includes examples in which the distance on average increases with increasing time exponentially or more quickly, while the second group includes those cases in which the average increases more slowly with time. Typical examples are various self-excitable oscillators in unstable equilibrium, and a damped pendulum respectively. [13] Let us further assume that the initial state is not determined precisely. In this case, we only specify a confining region in which the system state can exist. This region changes with increasing time as determined by the flow of trajectories. For a system of the first group, the width of the confining region spreads exponentially or more quickly with time and reaches the trajectory span in a finite time interval. This means that after this time, the system state can occur anywhere

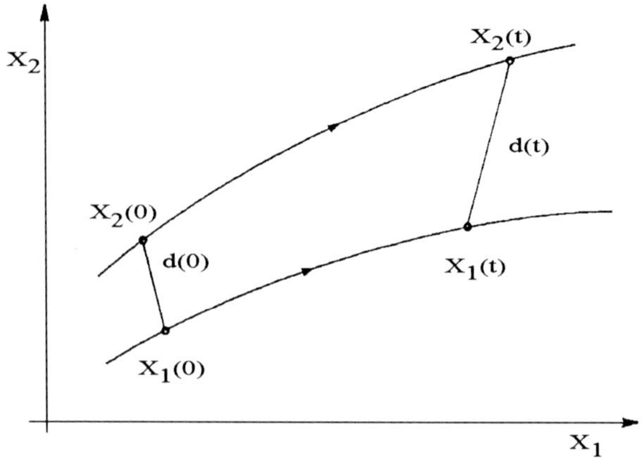

Fig. B.1. Examples of trajectories in the two-dimensional phase space

in the region of all possible states. It is thus practically undetermined, although the corresponding dynamical law is deterministically described. Such a system is called *unstable.* The accuracy of the experimental determination of the system state is generally limited. Because of experimental errors, an initial state cannot be exactly described or reproduced. If experiments on the same unstable system are repeated, then essentially different states can be observed after finite time intervals even though the initial states differ by less than the experimental error. Because of the divergence of the trajectories, the behavior of an unstable system is thus unpredictable for long prediction times. If the trajectories diverge everywhere in the phase space, then experimentally unobservable influences will generally lead to unpredictable or chaotic behavior. [6, 13, 16, 18, 19]

The instability and with it the possibility of the development of chaos can be quantitatively characterized by the rate of trajectory divergence in the phase space. Let us select a point $\boldsymbol{X}_0$ in the phase space and a swarm of possible initial states which are distributed on a hyper-sphere of radius d_0 around it, as shown in Fig. B.2. By following the trajectories from the initial points, a new set of points is obtained which, after a short time, is distributed on an approximately ellipsoidal surface. Using its principal axes d_k, the spectrum of characteristic or Lyapunov exponents λ_k is defined by [15, 16, 18, 19]

$$\lambda_k = \lim_{d_0 \to 0} \lim_{t \to \infty} \frac{1}{t} \ln \left[\frac{d_k(t)}{d_0} \right] . \tag{B.2}$$

It is convenient to represent the spectrum in decreasing order ($\lambda_{\max}, \ldots, \lambda_{\min}$). These exponents characterize the divergence of the trajectories originating from the vicinity of the point $\boldsymbol{X}_0$ in the phase space and generally depend on the position of the selected point. In order to characterize a

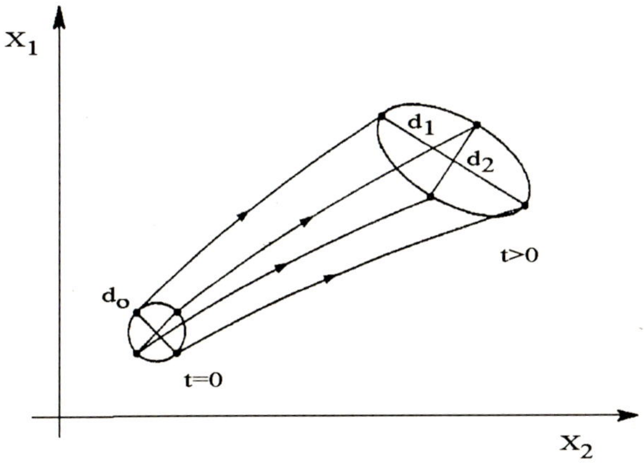

Fig. B.2. The flow in the phase space changes the form of the confining region of a swarm of representative system states

global property, the average values are usually used. A necessary condition for chaotic behavior is that at least one Lyapunov exponent is positive, which describes an unstable dynamical system.

Observation of distances between simultaneously selected points on trajectories offers an additional possibility of quantitatively characterizing dynamical systems by specifying changes in the differentially small volume element of the phase space. Let us for this purpose imagine a swarm of points inside a small D-dimensional cube of side $\triangle x_{io}$ and volume $\triangle V_0 = \prod_{i=1}^{D} \triangle x_{io}$ which includes them. Following the flow of points in the phase space during a time interval t, this volume changes into

$$\triangle V'(t) = \sum_{i=1}^{D} \frac{\partial x_i'}{\partial x_i} \triangle V_0 \ . \tag{B.3}$$

The average rate of change of the volume element is then defined by

$$\Lambda = \lim_{t\to\infty} \frac{1}{t} \ln \left| \frac{\triangle V'(t)}{\triangle V_0} \right| \ . \tag{B.4}$$

The system is conservative if $\Lambda = 0$ and non-conservative or dissipative if $\Lambda < 0$. It has been shown elsewhere [10, p. 383] that the average rate of divergence is equal to the sum of Lyapunov exponents. That is,

$$\Lambda = \sum_{k=1}^{N} \lambda_k \ . \tag{B.5}$$

Many technically important dynamical systems are dissipative, therefore we further limit our treatment to this class. [13, 21, 4] The chaotic behavior of a dissipative system can be observed if at least one of the Lyapunov exponents is positive, but at the same time their sum must be negative because of a dissipative property. This indicates that a one-dimensional system cannot be chaotic. Moreover, it has been found by Poincarè and Bendixon [9] that chaos is not possible in a two-dimensional system so that continuous chaotic flows can only develop in three- or more-dimensional dissipative systems. However, the dynamics described by discrete maps can exhibit chaotic behavior at lower dimensions as well. [16, 18, 19] It is apparently paradoxical that unpredictable irregular behavior can appear in systems where dynamics is described by the deterministic natural law given by Eq. (B.1). In order to emphasize this property, the phenomenon is called deterministic chaos.

When following the flow of trajectories originating from a widespread set of initial points, one can often observe that they asymptotically tend to some specifically formed region called the *attractor*. Well-known examples in three-dimensional space are: A point attractor or a focus corresponding to damped oscillations, a cycle corresponding to undamped oscillations, and a torus for a case of multi-dimensional, cyclical motion. [10, 18] The corresponding spectra of the Lyapunov exponents can be described by the sign–schemes: $(-,\ -,\ -)$, $(0,\ -,\ -)$, $(0,\ 0,\ -)$ in which '$-$' denotes negative value.

For the case of dissipative chaotic systems, the attractor exhibits rather strange properties. Because of the instability the trajectories must diverge in the phase space. But at the same time, differentially small volumes must on average contract. This can occur simultaneously if at least one dimension of the volume element increases, while the others dimensions decrease at such a high rate that the volume of the element also decreases in time. An example of the corresponding spectrum of Lyapunov exponents is represented by the scheme $(+,\ 0,\ -)$. Because of the divergence of trajectories one could expect that the motion cannot be limited. But the divergence is just an average local property. [10] The typical span of a strange attractor can still be limited if its geometrical form is stretched and folded over onto itself. This stretching and folding is reminiscent of a baker kneading dough and leads to an infinitely leaved structure of strange attractors.

A layered, leafed structure, in a sense, preserves some general properties of motion in a phase space and therefore chaotic motion appears qualitatively self-similar and to some extent ordered and is therefore also somewhat predictable. [7] Another argument for predictability is that the relevant physical laws can be deterministically described by the function appearing in Eq. (B.1). We can therefore conclude that *the principal goal of modeling an observed chaotic phenomenon should not be an exact reproduction of the trajectories, which are in every case unstable, but to obtain as far as possible a correct description of the function* $\boldsymbol{F}(\boldsymbol{X},\ \boldsymbol{p})$ *which is called the chaos generator.* Because of the instability of the trajectories we can also expect that the predicted and the observed trajectories should generally diverge, as expected from the maximum Lyapunov exponent. Therefore the correctness of the model can be estimated from the agreement between true and predicted trajectories only in short time intervals.

The simplest example of a dynamical system with diverging trajectories is described by a system of linear differential equations with positive coefficients

$$\frac{dx_k}{dt} = \lambda_k x_k \ . \tag{B.6}$$

If k indicates the principal axis corresponding to the Lyapunov exponent λ_k, then such a description can be used to characterize the local divergence of trajectories around a selected point in a phase space. Trajectories described by the solution

$$x_k(t) = x_k(0) \exp(\lambda_k \, t) \tag{B.7}$$

diverge but they are not confined to a limited domain of the phase space. Confinement can be obtained only by including nonlinear terms in Eq. (B.6). From this simple example we can conclude that a nonlinear model is inevitable when considering chaotic phenomena in dissipative systems. If the model for a system is known, the solution of a nonlinear dynamic problem cannot generally be determined by analytical but by numerical methods. This is also the reason why many properties of deterministic chaos have been discovered only recently by numerical simulation on computers. A similar conclusion also applies for the modeling of chaotic systems that relies on experimental observation, where the irregularity and complexity of experimental data usually calls for a computer-controlled data organization and numerical information processing system.

B.2 Characterization of Strange Attractors

A strange attractor is a geometrical structure determined by the set of all possible asymptotic trajectories occurring in a limited domain of the phase space. Because of the contraction of volume elements that occurs in the dynamics of dissipative systems, the dimension of the corresponding geometrical structure must be smaller than the dimension of the phase space. Determination of a proper parameter, whether from the dynamical law or from the records of trajectories, is thus a fundamental task in characterizing strange attractors. A qualitative estimation of attractor dimension can be obtained by observing the set of points at which the trajectories pierce an arbitrarily selected hyper-surface in the phase space. This set, called the Poincarè section, is a general aid for characterizing dynamical phenomena. [10] As an example imagine a three-dimensional phase space cut by a simple plane, as shown in Fig. B.3. If the flow develops on a limit cycle or torus, then a typical Poincarè section is a point or a closed loop respectively.

A strange attractor in a three-dimensional space cannot lie entirely on a two-dimensional surface because in this case chaos is not possible. Neither

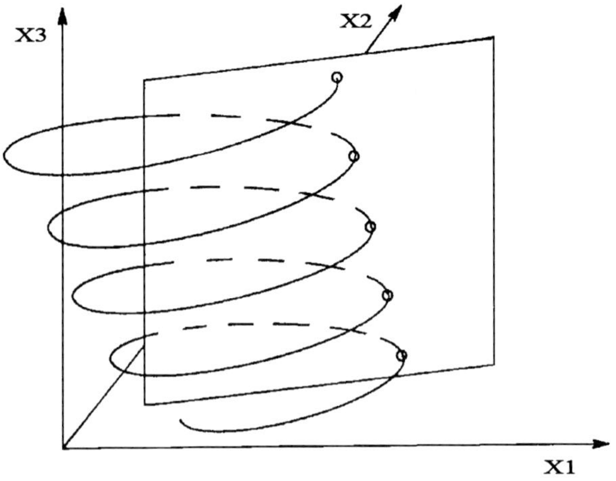

Fig. B.3. An example of the Poincarè section

can it fill the phase space because the system is dissipative. Therefore its structure must be inhomogeneous and the Poincarè section can show an isolated point, lines, and continuously distributed points in two dimensions. This indicates that the dimensionality of a strange attractor can describe only its average property. [16, 18, 19] Therefore, the value of a corresponding parameter need not correspond to an integer number and for a three-dimensional phase space, it should lie between 2 and 3. Very often a proper selection of the cross section reveals a rather simple Poincarè section. [13] By estimating the relation between successive positions of a trajectory piercing, $\boldsymbol{X}_n$, one obtains a transition from a continuous flow description to a discrete map. An application of the corresponding map

$$\boldsymbol{X}_{n+1} = \boldsymbol{f}\,(\boldsymbol{X}_n) \tag{B.8}$$

then reduces the dimensionality of the problem and it greatly simplifies the analysis of the dynamical behavior of chaotic systems. [10] For example, the approximately quadratic dependence between successive values of a properly selected representative variable can be expressed by the logistic map [13, 16, 18, 19]

$$X_{n+1} = a\,X_n(1 - X_n)\ . \tag{B.9}$$

The generic properties of this discrete map have been extensively investigated and can be successfully used to describe a number of chaotic phenomena. [3] The map therefore serves as a fundamental example that demonstrates deterministic chaos. Relations such as Eq. (B.9) are usually graphically represented as a return map, as shown in Fig. B.4.

A quantitative parameter which can be used to obtain a description of attractor dimensionality can be described in the following way. [16, 18,

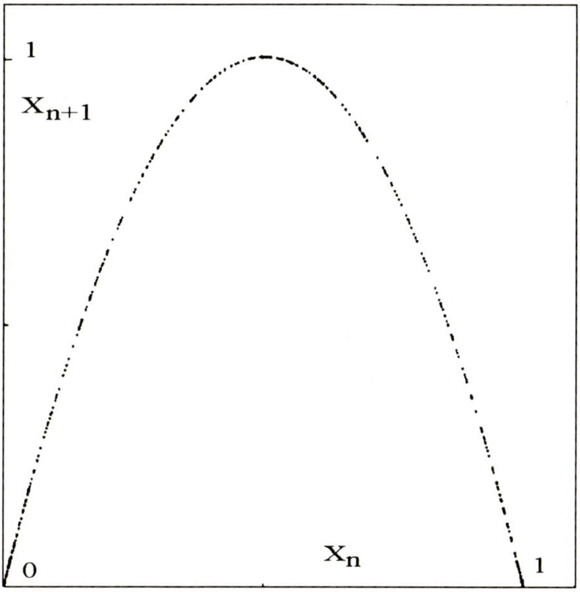

Fig. B.4. The return map of the logistic chaos generator: $X_{n+1} = 4X_n(1 - X_n)$

19] Let us cover the attractor by a net of D-dimensional cube cells of side s and let us represent a trajectory by a sequence of sample points $\{\boldsymbol{X}(t_k)\ ;\ t_k = kT\ ,\ k = 1,\ 2\ ,\ldots,\ D\}$. The distribution of the attractor over the phase space is then described by the probability that a trajectory point lies in a selected cell

$$P_i = \lim_{N\to\infty} \frac{N_i}{N}\ , \quad i = 1,\ 2,\ \ldots,\ M\ . \tag{B.10}$$

Here N_i is the number of sample points in the cell with index i and M is the number of cells used to cover the attractor. In order to describe the inhomogeneous attractor structure quantitatively, we define a set of dimensions by the expression

$$D_n = \lim_{s\to 0} \left(\frac{1}{n-1}\right) \frac{\ln\left(\sum_{i=1}^{M} P_i^n\right)}{\ln s}\ . \tag{B.11}$$

For $n = 0$ we obtain the Hausdorff dimension

$$D_0 = -\lim_{s\to 0} \frac{\ln M}{\ln s}\ . \tag{B.12}$$

The number of cells needed to cover the attractor generally grows with decreasing side length s. The last equation shows that we can expect for small s a scaling relation

$$M = s^{-D_0} \qquad \text{for} \quad s \to 0 \; . \tag{B.13}$$

Simple geometrical structures such as a focus, cycle, or torus are characterized by the integer values 0, 1, 2 of the Hausdorff dimension, respectively. However, there are many geometrical objects with non-integer or *fractal* dimension and it is also characteristic for various strange attractors. [12]

For $n = 1$ the dimension is determined by the L' Hôpital rule as the limit for $n \to 1$

$$D_1 = - \lim_{s \to 0} \frac{S(s)}{\ln s} \; . \tag{B.14}$$

The quantity $S(s)$ is defined by

$$S(s) \equiv - \sum_{i=1}^{M(s)} P_i \ln P_i \tag{B.15}$$

and it denotes the entropy of information. [8] It quantitatively describes the information needed to specify the position of an average point of a trajectory in the network of cells if the distribution of probabilities is given by P_i. Therefore, D_1 is called the *information dimension.* For a homogeneous attractor, $P_i = 1/M(s)$, $S(s) = \ln M(s)$ and $D_1 = D_0$. Both dimensions coincide in this case, but with the transition to an inhomogeneous structure, the entropy of information can only decrease and therefore, more generally, $D_1 \leq D_0$. A similar relation $D_m \leq D_n$ is also valid for higher indices m and n where $m > n$. Equality holds for homogeneous attractors and therefore attractor inhomogenity can be characterized from a comparison of various dimensions.

Generally, the values of the dimensions D_0 and D_1 must be determined numerically. The corresponding limit process $s \to 0$ requires a tedious repetition of computations. Less involved is the estimation of D_2, which is simplified using the following formula [5, 16, 18, 19] in which D_2 is defined as

$$D_2 = \lim_{s \to 0} \frac{\ln \sum_{i=1}^{M} P_i^2}{\ln s} \; . \tag{B.16}$$

Here P_i^2 denotes the probability that two independently selected points on the attractor occur in the i-th cell. The sum $\sum_{i=1}^{M} P_i^2$ then corresponds to the probability that two points anywhere on the attractor are in the same cell. This value is approximately equal to the probability that the distance between two arbitrarily selected points of the attractor is less than s or

$$C(s) = \sum_{i=1}^{M} P_i^2 = \lim_{N \to \infty} [\text{number of pairs } ij \text{ with } |\boldsymbol{X}_i - \boldsymbol{X}_j| < s] \; . \tag{B.17}$$

This quantity is called the *correlation integral* and it corresponds to the number of correlated pairs of points inside a hyper-sphere of radius s in the phase space. If the points are homogeneously distributed on a line or a surface or another object inside a sphere, then D_2 scales as s, s^2, etc, respectively. The correlation exponent obtained from the scaling relation specified by $C(s) = s^{D_2}$ is given by

$$D_2 = \lim_{s \to 0} \frac{\ln C(s)}{\ln s} \tag{B.18}$$

which is simply evaluated. Because of its relation to the correlation integral C , the exponent D_2 is called the *correlation dimension.* It is most widely used for estimating the dimensionality of strange attractors. [2]

The dimensions D_n characterize the static properties of attractors. The partition of the phase space into cells is also useful for describing the dynamic properties of the motion on an attractor. For this purpose we first define the joint probability $P_{i_0,\, i_1,\, \ldots,\, i_n}$ that the points $(\boldsymbol{X}(t_k)\ ;\ k = 1,\ \ldots,\ N)$ are in cells with indices $i_0,\ i_1,\ \ldots,\ i_n$. Similar to Eq. (B.15) the entropy of joint information is defined by

$$K_n = -\sum_{i_0} \sum_{i_1} \cdots \sum_{i_n} P_{i_0,\, i_1,\, \ldots,\, i_n} \ln P_{i_0,\, i_1,\, \ldots,\, i_n} \ . \tag{B.19}$$

The difference $\Delta K_n = K_{n+1} - K_n$ corresponds to information that is required in order to specify the state in the $(n+1)$-st cell, provided that the sequence of states resides in the cells with indices $i_0,\ i_1,\ \ldots,\ i_n$. ΔK_n measures the loss of information or the increase of uncertainty resulting from dynamical changes in the system during the time interval $(nT,\ (n+1)T)$. But more than this loss of information, it is interesting to estimate the average rate of its change, defined by the Kolmogorov entropy

$$K = \lim_{T \to 0} \ \lim_{s \to 0} \ \lim_{N \to 0} \frac{1}{NT} \sum_{n=0}^{N-1} K_n \ . \tag{B.20}$$

This entropy is 0, greater than 0, and ∞ for a deterministic, a chaotic and a system following random motions, respectively. [16, 18, 19]

Chaotic behavior is a consequence of the divergence of trajectories, which can be described by positive Lyapunov exponents. It was shown elsewhere [17, 2] that the sum of positive Lyapunov exponents averaged over the attractor is equal to the Kolmogorov entropy. A proper quantity for an average characterization of chaotic motion on the attractor is thus the Kolmogorov entropy. This is demonstrated by the following example. Let us assume that a strange attractor is limited to a region of the typical span L and let r denote a typical width of the region where an inexactly determined initial state of the system occurs in the phase space. Because of the divergence of trajectories, the system state can move in a characteristic time T_m from the initial region to any point of the complete span. This time can be estimated from the relation

$$L \approx r \ \exp(K\, T_m) \qquad \text{or} \qquad T_m \approx \frac{1}{K} \ln \frac{L}{r} \tag{B.21}$$

in which K denotes an estimated average rate of divergence of trajectories that is described by the sum of positive Lyapunov exponents or the Kolmogorov entropy. Its inverse value thus determines in large measure the typical time interval during which the behavior of the system can be predicted. Eq. (B.21) shows that the uncertainty of the position of the initial state influences this time only logarithmically or negligibly. Because of the complicated numerical procedure, the Kolmogorov entropy is only rarely determined. This is in spite of its importance for characterizing the dynamic properties of chaotic phenomena.

B.3 Experimental Characterization of Chaotic Phenomena

In a system described by a continuous flow, chaos can appear if the phase space is at least three-dimensional. It is generally difficult to follow fluctuations of various variables simultaneously. Therefore most frequently only one variable, say x, is selected for a representative observation or recording. From the form of the corresponding records, which are obtained by an oscilloscope or a similar recording device, one can qualitatively estimate if the behavior of the system is stationary, periodic or chaotic. [13, 2] However, it is difficult to distinguish between quasi-periodic and chaotic motions. A quasi-periodic motion is composed of many cyclical motions of incommensurable frequencies and it is often apparently chaotic. A quantitative method is thus needed to make the distinction. For this purpose, the covariance function

$$\Sigma(t) = \lim_{T\to\infty} \frac{1}{T} \int_0^\infty (x(t') - m)\,(x(t+t') - m)\; dt' \tag{B.22}$$

is used. Here m denotes the average value of the signal

$$m = \lim_{T\to\infty} \frac{1}{T} \int_0^\infty x(t')\, dt' \; . \tag{B.23}$$

Often the origin of x is chosen so that $m = 0$ and in that case, the covariance function is identical to the correlation function. For periodic or quasi-periodic phenomena the covariance function oscillates between certain limits while for chaotic motions it converges to 0 with increasing t. The last property arises because the system trajectories diverge so that the values of variable x at two separate moments become, on average, increasingly less correlated with increasing time lags t. Based on intuition, it is expected that a typical correlation time corresponds approximately to the time interval during which the behavior of the system can be predicted. [1] By the Fourier transform of the covariance function the spectral density is determined. For a quasi-periodic signal the spectral density differs from zero only at discrete values of frequency. Contrary to this, a chaotic phenomenon is characterized by a continuous type of spectral distribution. [2] Both types of motions are therefore usually most simply distinguished by observation of the spectral density which is easily measured. More difficult, however, is the determination of other statical and dynamical characteristics of strange attractors. For this purpose, we can either simultaneously measure various time-dependent variables or we can try to reconstruct the properties of the strange attractor from observations of only one representative variable. The first possibility is experimentally more demanding so we present here the arguments made for the second approach. [2, 20]

From the law of system evolution described by Eq. (B.1) it follows that the system state at some time $t+T$ is uniquely related to the state at the previous time t. If the representative variable $x(t)$ is applied to form the time-shifted components such that

$$x_1(t) = x(t) \; ; \; x_2(t) = x(t-T) \; ; \; \ldots \; ; \; x_n(t) = x(t-(n-1)T) \qquad \text{(B.24)}$$

then it can be expected that the system dynamics is mirrored in the time dependence of the vector, which is comprised of the components

$$\boldsymbol{X} = (x_1, \; \ldots, \; x_n) = (x(t), \; x(t-T), \; \ldots, \; x(t-(n-1)T)) \; . \qquad \text{(B.25)}$$

This conclusion further suggests that chaotic properties could also be estimated from the observation of the attractor of the time-dependent vector $\boldsymbol{X}$. A more rigorous analysis of this approach has shown [20] that under relatively mild restrictions, the vector composed of $n = 2D+1$ shifted variables can indeed be applied to represent a smooth embedding of flow with the same metric properties as the original strange attractor in D-dimensional space. In experimental work the dimensionality of strange attractors is therefore most often estimated from just one representative variable by using the correlation exponent. [5, 2]

However, an additional problem needs to be mentioned. From the observation of just one characteristic variable, it is possible to estimate the correlation dimension of the attractor. However, the dimension of the phase space which is needed for a complete representation of the phenomenon, cannot be easily determined. In the existing literature on chaotic phenomena, it is usually mentioned [2, 14] that this dimension can be estimated by repeating the calculation of the correlation exponent with an ever greater number N_d of shifted components. With increasing N_d the correlation exponent D_2 reaches its saturation at the particular value $N_{d,s}$. A proper dimension N of the phase space is then estimated from the definition of the embedding dimension which is given by the relation $N_{d,s} = 2N+1$. This further indicates the minimum number of variables which are needed to model the observed phenomenon. In Chapt. 11 of this monograph, we proposed a complementary procedure for estimating the proper dimension of the embedding and of the phase space from the minimum of the prediction error. In addition to this problem, there usually also arises the question of how to choose a proper time shift. An acceptable answer to this question can be found in the literature. [11]

References

Chapter 1

1. *The Handbook of Artificial Intelligence*, ed. by A. Barr, P. R. Cohen, and E. A. Feigenbaum (Addison-Wesley Publ. Co., Reading, MA 1989)
2. *Self-Organization and Life: From Simple Rules to Global Complexity*, Proc. Int. Conf. ECAL '93 (Center for Non-Linear Phenomena and Complex Systems, CP 231, Université Libre de Bruxelles, Brussels 1993)
3. *Evolution, Games and Learning, Models for Adaptation in Machines and Nature*, Proc. 5th Annual Int. Conf. (Center for Nonlin. Stud., Los Alamos, NM 1985); ed. by D. Farmer, A. Lapedes, N. Packard, B. Wendroff (North-Holland, Amsterdam), Physica, **22D**, (1986)
4. I. Grabec: "Self-organization of neurons described by the maximum entropy principle", Biol. Cyber., **63**, 403–409 (1990)
5. I. Grabec and W. Sachse: "Automatic Modeling of Physical Phenomena: Application to Ultrasonic Data", J. Appl. Phys., **69**(9), 6233–6244 (1991)
6. H. Haken: *Synergetic Computers and Cognition, A Top-Down Approach to Neural Nets* (Springer, Berlin 1991)
7. S. A. Kauffman: *The Origins of Order, Self-Organization and Selection in Evolution* (Oxford University Press, New York 1993)
8. T. Kohonen: *Self-Organization and Associative Memory* (Springer, Berlin 1989)
9. C. Mead: *A Silicon Model of Early Visual Processing*, Neural Networks, **1**, 91–97 (1988)
10. D. E. Rumelhart, J. L. McClelland, and PDP Research Group: *Parallel Distributed Processing, Explorations in Microstructure of Cognition* (MIT Press, Cambridge, MA 1988)
11. *Handbook of Intelligent Control: Neural, Fuzzy, and Adaptive Approaches*, ed. by A. White, D. A. Sofge (Van Nostrand Reinhold, New York 1992)

Chapter 2

1. J. A. Anderson, E. Rosenfeld Eds: *Neurocomputing, Foundations of Research* (MIT Press, Cambridge, MA 1988)
2. C. M. Bishop: "Neural networks and their applications", Rev. Sci. Instr., **65**, 1830–1832 (1994)
3. *ECAL '93*, Proc. Int. Conf. Self–Organization and Life: From Simple Rules to Global Complexity (Center for Non–Linear Phenomena and Complex Systems, CP 231 Université Libre de Bruxelles, Brussels 1993)

4. *Evolution, Games and Learning, Models for Adaptation in Machines and Nature*, Proc. 5th Annual Int. Conf. (Center for Nonlin. Studies, Los Alamos, NM 1985), ed. by D. Farmer, A. Lapedes, N. Packard, B. Wendroff (North-Holland, Amsterdam), Physica, **22D**, (1986)
5. P. Glansdorf, I. Prigogine: *Thermodynamic Theory of Structure, Stability and Fluctuations* (Wiley–Interscience, London 1971)
6. I. Grabec and W. Sachse: "Automatic Modeling of Physical Phenomena: Application to Ultrasonic Data", J. Appl. Phys., **69**(9), 6233–6244 (1991)
7. I. Grabec: "Self-Organization of Neurons Described by the Maximum-Entropy Principle", Biol. Cyb., **63**, 403–409 (1990)
8. *Handbook of Intelligent Sensors for Industrial Automation*, ed. by N. Zuech (Addison–Wesley, Reading, MA 1992)
9. *Handbook of Measurement Science* (John Wiley and Sons, Chichester 1982)
10. *Pattern Formation by Dynamic Systems and Pattern Recognition', Proc. Int. Symp. Synergetics*, ed. by H. Haken (Springer, Berlin 1979)
11. H. Haken: *Synergetics, An Introduction, Nonequilibrium Phase Transitions and Self-Organization in Physics, Chemistry, and Biology*, Springer Ser. Synergetics, Vol. 1 (Springer, Berlin 1983)
12. H. Haken: *Advanced Synergetics, Instability Hierarchies of Self-Organizing Systems and Devices*, Springer Ser. Synergetics, Vol. 20 (Springer, Berlin 1983)
13. H. Haken: *Synergetic Computers and Cognition, A Top-Down Approach to Neural Nets*, Springer Ser. Synergetics, Vol. 50 (Springer, Berlin 1991)
14. S. Haykin: *Neural Networks* (McMillan, New York 1994)
15. D. O. Hebb: *The Organization of Behavior, A Neurophysiological Theory* (Wiley, New York 1949)
16. R. Hecht-Nielsen: *Neurocomputing* (Addison-Wesley, Reading, MA 1990)
17. S. A. Kauffman: *The Origins of Order; Self-Organization and Selection in Evolution* (Oxford University Press, New York 1993)
18. T. Kohonen: *Self-Organization and Associative Memory* (Springer-Verlag, Berlin 1989)
19. T. Kohonen: "An Introduction to Neural Computing", Neural Networks, **1**, 316 (1988)
20. C. G. Langton: "Studying Artificial Life with Cellular Automata", in *Evolution, Games and Learning, Models for Adaptation in Machines and Nature*, Proc. 5th Annual Int. Conf. (Center for Nonlin. Studies, Los Alamos, NM 1985), ed. by D. Farmer, A. Lapedes, N. Packard, B. Wendroff (North-Holland, Amsterdam), Physica, **22D**, 120–149 (1986)
21. W. S. McCulloch, W. A. Pitts: "A Logical Calculus of the Ideas Immanent in Nervous Activity", *Bulletin of Mathematics and Biophysics*, **5**, 115–133 (1943)
22. J. von Neuman: *The General and Logical Theory of Automata* (Hixon Symposium, Pasadena, CA 1948)
23. G. Nicolis, I. Prigogine: *Self–Organization in Nonequilibrium Systems, From Dissipative Structures to Order Through Fluctuations* (John Wiley & Sons, New York 1977)
24. F. Rosenblatt: "The Perceptron: A Probabilistic Model for Information Storage and Organization in the Brain", Psychoanalytic Review, **65**, 386-408 (1958)
25. B. Russell: *The Principles of Mathematics* (W. W. Norton, Co., New York 1937)
26. P. H. Sydenham: *Introduction to Measurement Science and Engineering* (John Wiley and Sons, Chichester 1989)
27. D.D. Swade: "Redeeming Charles Babbage's Mechanical Computer", Scientific American, **268**, 86–91 (1993)
28. R. F. Thompson: *The Brain* (W. H. Freeman Co., New York 1985)

29. *Theory and Applications of Cellular Automata*, ed. by S. Wolfram (World Scientific, Singapore 1986)

Chapter 3

1. N. Zuech, Ed., *Handbook of Intelligent Sensors for Industrial Automation*, Addison-Wesley Publishing, Reading, MA (1992)
2. S. Deutsch and A. Deutsch, *Understanding the Nervous System*, Inst. Electr. Electron. Eng., New York (1993), Chapt. 2
3. S. Deutsch and E. Micheli-Tzanakou, *Neuroelectric Systems*, New York University Press, New York (1987), Chapt. 2
4. A. Roberts and B. M. H. Bush, *Journal of Experimental Biology*, **54**, 515–524 (1971)
5. S. E. Blackshaw and S. W. Thompson, *Journal of Physiology*, **396**, 121–137 (1988)
6. A. J. Hudspeth, M. M. Poo and A. E. Stuart, *Journal of Physiology*, **272**, 25–43 (1977)
7. R. M. White, "A sensor classification scheme", *IEEE Trans. Ferroelec. Freq. Contr.*, **UFFC-34**(2), 124–126 (1987)
8. E. O. Doebelin, *Measurement Systems, Application and Design*, Third Edition, McGraw-Hill Book Company, New York (1983), Chapt. 2
9. C. Chang and W. Sachse, "Analysis of elastic wave signals from an extended source in a plate", *J. Acoust. Soc. Am.*, **77**(4), 1335–1341 (1985)
10. C. Chang and W. Sachse, "Separation of spatial and temporal effects in an ultrasonic transucer", in *Review of Progress in Quantitative Nondestructive Evaluation*, Vol. 5A, D. O. Thompson and D. E. Chimenti, Eds., Plenum Press, New York (1986), pp. 139–143
11. K. Y. Kim and W. Sachse, "Self-aligning capacitive transducer for the detection of broadband ultrasonic displacement signals", *Rev. Sci. Instrum.*, **57**(2), 264–267 (1986)
12. K. Y. Kim, L. Niu, B. Castagnete and W. Sachse, "Miniaturized capacitive transducer for detection of broadband ultrasonic displacement signals", *Rev. Sci. Instrum.*, **60**(8), 2785–2788 (1989)
13. H. K. P. Neubert, *Instrument Transducers, An Introduction to their Performance and Design*, Second Edition, Clarendon Press, Oxford (1975), pp. 5
14. J. P. Bentley, *Principles of Measurement Systems*, Second Edition, Longman Scientific and Technical, Harlow, Essex, UK (1988), Chapt. 2
15. W. Sachse and N. Hsu, "Ultrasonic Transducers for Materials Testing and their Characterization", Chapt. 4, in *Physical Acoustics*, Vol. XIV, W. P. Mason and R. N. Thurston, eds., Academic Press, New York (1979), pp. 277–406
16. C. B. Scruby and L. E. Drain, *Laser Ultrasonics: Techniques and Applications*, Adam Hilger, Bristol (1990)
17. J. Merhaut, *Theory of Electroacoustics*, McGraw-Hill Book Co., New York (1981), Chapt. 6
18. E. K. Sittig, "Design and Technology of Piezoelectric Transducers for Frequencies above 100 MHz", Chapt. 5, in *Physical Acoustics*, Vol. IX, W. P. Mason and R. N. Thurston, Eds., Academic Press, New York (1972), pp. 221–275
19. E. A. Robinson, *Multichannel Time Series Analysis with Digital Computer Programs*, Holden-Day, Inc., San Francisco (1967)
20. E. A. Robinson and M. T. Silva, *Digital Signal Processing and Time Series Analysis*, Holden-Day, Inc., San Francisco (1978)

21. E. A. Robinson and S. Treitel, *Geophysical Signal Analysis*, Prentice-Hall, Englewood Cliffs, NJ (1980), Chapts. 6, 7 and 8
22. J. E. Michaels, *Fundamentals of Deconvolution with Applications to Ultrasonics and Acoustic Emission*, M. S. Thesis, Cornell University, Ithaca, NY (1982)
23. *The Neural Network Toolbox*, The MathWorks, Inc., Natick, MA, USA (1994)
24. S. M. Zse, Ed., *Semiconductor Sensors*, John Wiley and Sons, New York (1994)
25. T. M. Canh, *Biosensors*, Chapman and Hall, London (1993)
26. N. C. MacDonald, L. Y. Chen, J. J. Yao, Z. L. Zhang, J. A. McMillan, D. C. Thomas and K. R. Haselton, "Selective chemical vapor deposition of tungsten for microelectromechanical structures", *Sensors and Actuators*, **20**, 123–133 (1989)
27. K. A. Shaw, Z. L. Zhang and N. C. MacDonald, "SCREAM I: A single mask, single-crystal silicon, reactive ion etching process for MicroElectroMechanical structures", *Sensors and Actuators A*, **40**, 63–70 (1994)
28. M. T. A. Saif and N. C. MacDonald, "A milli-Newton microloading device", *The 8th International Conference on Solid State Sensors and Actuators (Transducers'95)* and *Eurosensors IX*, Vol. 2, (Stockholm, Sweden, 1995), pp. 60–63
29. S. G. Adams, F. M. Bertsch and N. C. MacDonald, "Independent tuning of the linear and non-linear stiffness coefficients of a micromechanical system", *IEEE Micro Electro Mechanical Systems'96*, (San Diego, 1996). In Press
30. T. Yasuda, I. Shimoyama and H. Miura, "Microrobot actuated by a vibration energy field", *7th International Conference on ...*, pp. 42–45
31. T. Fukuda, A. Kawamoto, F. Arai and H. Matsuura, "Mechanism and swimming experiment of micro mobile robot in water", *Proc. IEEE MEMS'94 Workshop*, IEEE, (New York, 1994), pp. 273–278
32. K. I. Arai, W. Sugawara and T. Honda, "Magnetic small flying machines", *The 8th International Conference on Solid State Sensors and Actuators (Transducers'95)* and *Eurosensors IX*, Vol. 1, (Stockholm, Sweden, 1995), pp. 316–319
33. C. Mead, *Analog VLSI and Neural Systems*, Addison-Wesley Publishing Co., Reading, MA (1989)
34. C. Mead and M. A. Mahowald, "A Silicon Model of Early Visual Processing", *Neural Networks*, bf 1, (1988)
35. M. A. Mahowald and C. Mead, in *Analog VLSI and Neural Systems*, Addison-Wesley Publishing Co., Reading, MA (1989), Chapt. 15
36. J. Tanner and C. Mead, in *Analog VLSI and Neural Systems*, Addison-Wesley Publishing Co., Reading, MA (1989), Chapt. 14
37. R. F. Lyon and C. Mead, *IEEE Trans. Acoust. Speech and Signal Proc.*, **36**(7), (1988)
38. R. F. Lyon and C. Mead, in *Analog VLSI and Neural Systems*, Addison-Wesley Publishing Co., Reading, MA (1989), Chapt. 16

Chapter 4

1. P. Révész: "Density Estimation", in *Handbook of Statistics* ed.by P. R. Krishnaiah, P. K. Sen (North-Holland, Amsterdam 1991), Vol. 4, pp. 531-549
2. E. A. Nadaraya: *Non-parametric Estimation of Probability Densities and Regression Curves* (Kluwer Academic Publ., Dordrecht 1989)
3. M. Rosenblatt: "Remarks on Some Non-parametric Estimators of a Density Function", Ann. Math. Stat., **27**, 832-835 (1956)
4. E. Parzen: "On Estimation of Probability Density Function and Mode", Ann. Math. Stat., **35**, 1065-1076 (1962)

5. R. O. Duda, P. E. Hart: *Pattern Classification and Scene Analysis* (J. Wiley and Sons, New York 1973), Ch. 4
6. K. Fukunaga: *Introduction to Statistical Pattern Recognition*, (Academic Press, New York 1972), Ch. 6
7. K. Fukunaga, L. D. Hosteler: "Optimization of k-Nearest-Neighbor Density Estimates", IEEE Trans. Inf. Theory, **IT-19**, 320-326 (1973)
8. D. O. Loftsgaarden, C. P. Quesenbery: "A Non-parametric estimate of a Multivariate Density Function", Ann. Math. Stat., **36**, 1049-1051 (1965)
9. Y. Mack, M. Rosenblatt: "Multivariate K-nearest Neighbour Density Estimates", J. Multivariate Anal., **9**, 1-5 (1979)
10. I. Grabec, W. Sachse: "Automatic Modeling of Physical Phenomena: Application to Ultrasonic Data", J. Appl. Phys., **69**, 6233-6244 (1991)
11. I. Grabec: "Self-Organization of Neurons Described by the Maximum Entropy Principle", Biol. Cyb., **63**, 403-409 (1990)
12. I. Grabec: "Optimization of Kernel-Type Density Estimator by the Principle of Maximal Self-Consistency", Neural, Parallel and Scientific Computation, **1**, 83-92 (1993)

Chapter 5

1. H. Haken: *Synergetics, An Introduction*, Springer Series in Synergetics, Vol. 1 (Springer, Berlin 1983)
2. H. Haken: *Information and Self-Organization, A Macroscopic Approach to Complex Systems*, Springer Series in Synergetics, Vol. 40 (Springer, Berlin 1988)
3. J. N. Kapur: *Maximum Entropy Models in Science and Engineering* (John Wiley and Sons, New York 1989)
4. S. Kullback: *Information Theory and Statistics* (J. Wiley and Sons, New York 1959)
5. G. Jumarie: *Relative Information, Theories and Applications*, Springer Series in Synergetics, Vol. 47 (Springer, Berlin 1990)
6. C. E. Shannon, W. Weaver: *The Mathematical Theory of Communication* (University of Illinois Press, Urbana, Chicago 1949 and later editions)
7. R.L. Stratonovitch: *Teoriya Informacii* (Sov. Radio, Moskva 1975), in Russian

Chapter 6

1. A. Gersho: "On the Structure of Vector Quantizers", IEEE Trans. Inf. Theory, **IT-28**, 157-166 (1982)
2. J. W. Gibbs: *Elementary Principles in Statistical Mechanics* (Yale University Press, New Haven, Conn. 1902)
3. I. Grabec: "Self-Organization of Neorons Described by the Maximum-Entropy Principle", Biol. Cyb., **63**, 403–409 (1990)
4. H. Haken: *Synergetics, An Introduction*, Springer Series in Synergetics, Vol. 1 (Springer, Berlin 1983)
5. H. Haken: *Information and Self-Organization, A Macroscopic Approach to Complex Systems*, Springer Series in Synergetics, Vol. 40 (Springer, Berlin 1988)
6. E. T. Janes: "Where do we stand on maximum entropy?" in *The Maximum-Entropy Formalism*, ed. by R. D. Levine, M. Tribus (MIT Press, Cambridge, MA 1978)

7. G. Jumarie: *Relative Information: Theories and Applications*, Springer Series in Synergetics, Vol. 47 (Springer, Berlin 1990)
8. J. N. Kapur: *Maximum-entropy Models in Science and Engineering* (John Wiley and Sons, New York 1989)
9. J. N. Kapur, H. K. Kesavan: *Entropy Optimization Principles with Applications* (Academic Press, Boston 1992)
10. T. Kohonen: "Learning Vector Quantization for Pattern Recognition", Report TKK-F-A601, ISBN 915-753-950-9, (Helsinki University of Technology, Dpt. Techn. Physics, SF-02150, Espoo, Finland 1986)
11. G. A. Korn, T. M. Korn: *Mathematical Handbook for Scientists and Engineers, Definitions, Theorems, and Formulas for Reference and Review* (Mc Graw Hill, New York 1968)
12. S. Kullback: *Information Theory and Statistics* (J. Wiley and Sons, New York 1959)
13. Y. Linde, A. Buzo, R. M. Gray: "An Algorithm for Vector Quantizer Design", IEEE Trans. Com., **Com. 28**, 84-95 (1980)
14. J. Makhoul, S. Roucos, H. Gish: "Vector Quantization in Speech Coding), Proc. IEEE, **73**, 1551-1588 (1985)
15. N. M. Nasrabadi, R. A. King: "Image Coding Using Vector Quantization, A Review", IEEE Trans. Com., **36**, 957- 971 (1988)
16. *Maximum-Entropy and Bayesian Methods in Inverse Problems*, ed. by C. Ray Smith and W.T. Grandy, Jr. (D. Reidel Publishing Company, Dordrecht 1985)
17. C. E. Shannon, W. Weaver: *The Mathematical Theory of Communication*, (University of Illinois Press, Urbana, Chicago 1949 and later editions)
18. R.L. Stratonovitch *Teoriya Informacii* (Sov. Radio, Moskva 1975), in Russian

Chapter 7

1. A. E. Albert, L. A. Gardner: *Stochastic Approximation and Nonlinear Regression* (MIT Press, Cambridge, MA 1967)
2. S. T. Alexander: *Adaptive Signal Processing* (Springer, New York 1986)
3. D. J. Bell: *Mathematics of Linear and Nonlinear Systems, An Introduction for Engineers and Applied Scientists* (Clarendon Press, Oxford 1990)
4. A. Dvoretzky: "On Stochastic Approximation", Proc. 3rd Berkeley Symposium on Mathematical Statistics and Probability, Vol. 1, pp. 39–55, 1956
5. R. O. Duda, P. E. Hart: *Pattern Classification and Scene Analysis* (John Wiley and Sons, New York 1973), Ch. 5
6. K. Fukunaga: *Introduction to Statistical Pattern Recognition* (Academic Press, New York 1972)
7. I. Grabec: "Self-Organization Based on the Second Maximum-Entropy Principle", 1st IEE Conference on "Artificial Neural Networks", London 1989, Conf. Publication No.313 , pp 12-16
8. I. Grabec: "Self-Organization of Neurons Described by the Maximum-Entropy Principle", Biol. Cyb., **63**, 403-409 (1990)
9. I. Grabec, W. Sachse: "Automatic Modeling of Physical Phenomena: Application to Ultrasonic Data", J. Appl. Phys., **69 (9)**, 6233-6244 (1991)
10. S. Grossberg: "Nonlinear Neural Networks: Principles, Mechanisms, and Architectures", Neural Networks, **1**, 17-61 (1988)
11. H. Haken: *Information and Self-Organization, A Macroscopic approach to Complex Systems*, Springer Series in Synergetics (Springer, Berlin 1988)
12. S. Haykin: *Adaptive Filter Theory* (Prentice-Hall Int., London 1991)

13. R. Hecht-Nielsen: *Neurocomputing* (Addison-Wesley, Reading, MA 1990)
14. J. Kiefer, J. Wolfowitz: *Stochastic Estimation of the Maximum of a Regression Function*, Annals of the Mathematical Statistics, **23**, 462-466 (1952)
15. D. E. Kirk: *Optimal Control Theory, An Introduction* (Prentice-Hall, Englewood Cliffs, NJ 1970)
16. T. Kohonen: "Self-Organized Formation of Topological Correct Feature Maps", Biological Cybernetics, **43**, 59-69 (1982)
17. T. Kohonen: *An Introduction to Neural Computing*, Neural Networks, **1**, 3-16 (1988)
18. T. Kohonen: *Self-Organization and Associative Memory* (Springer, Berlin 1989)
19. G. A. Korn, T. M. Korn: *Mathematical Handbook for Scientists and Engineers, Definitions Theorems and Formulas for Reference and Review* (McGraw Hill, New York 1961, 1968 and later editions)
20. B. Kosko: *Neural Networks and Fuzzy Systems, A Dynamical systems Approach to Machine Intelligence* (Prentice-Hall, Englewood Cliffs, NJ 1992)
21. S. J. Orfandis: *Optimum Signal Processing: An Introduction* (MacMillan, New York 1988)
22. H. Robbins, S. Monro: *A Stochastic Approximation Method*, Annals of Mathematical Statistics, **22**, 400-407 (1951)
23. D. E. Rumelhart, J. McClelland and The PDP Research Group: *Parallel Distributed Processing* (MIT Press, Cambridge, MA 1986)
24. G. N. Saridis: *Stochastic Approximation Methods for Identification and Control - A Survey*, IEEE Transactions on Automatic Control, **AC 19**, 798-809 (1974)
25. R. J. Schalkoff: *Pattern Recognition: Statistical, Structural and Neural Approaches* (John Wiley and Sons, New York 1992)

Chapter 8

1. *Neurocomputing, Foundations of Research*, ed. by J. A. Anderson, E. Rosenfeld (MIT Press, Cambridge, MA 1988), p. 209 and 509
2. R. O. Duda, P. E. Hart: *Pattern Classification and Scene Analysis* (John Wiley and Sons, New York 1973) Ch. 4
3. J.W. Gibbs: *Elementary Principles in Statistical Mechanics* (Yale University Press, New Haven, Conn. 1902)
4. I. Grabec: "Self-Organization Based on the Second Maximum-Entropy Principle", 1st IEE Conference on "Artificial Neural Networks", London 1989, Conf. Publication No. 313, pp. 12-16
5. I. Grabec, W. Sachse: *Automatic Modeling of Physical Phenomena: Application to Ultrasonic Data*, J. Appl. Phys., **69 (9)**, 6233-6244 (1991)
6. I. Grabec: *Self-Organization of Neurons Described by the Maximum-Entropy Principle*, Biol. Cyb., **63**, 403-409 (1990)
7. I. Grabec: *Modeling of Chaos by a Self-Organizing Neural Network*, in *Artificial Neural Networks*, Proc. ICANN, Espoo, Finland, ed. by T. Kohonen, K. Mäkisara, O. Simula, J. Kangas (Elsevier Science Publishers B. V., North-Holland, Amsterdam 1991), Vol. 1, pp. 151-156
8. S. Grossberg:*Nonlinear Neural Networks: Principles, Mechanisms, and Architectures*, Neural Networks, **1**, 17-61 (1988)
9. H. Haken: *Synergetics, An Introduction*, Springer Series in Synergetics, Vol.1 (Springer, Berlin 1983)
10. H. Haken: *Information and Self-Organization, A Macroscopic Approach to Complex Systems*, Springer Series in Synergetics, Vol. 40 (Springer, Berlin 1988)

11. D. O. Hebb: *The Organization of Behavior, A Neurophysiological Theory* (John Wiley, New York 1948)
12. E. T. Jaynes: *The Maximum-Entropy Formalism*, ed. by R.D. Levine, M.Tribus (MIT Press, Cambridge, MA 1978)
13. J. N. Kapur: *Maximum-Entropy Models in Science and Engineering* (John Wiley and Sons, New York 1989)
14. T. Kohonen: *Self-Organized Formation of Topologically Correct Feature Maps*, Biol. Cybernetics, **43**, 59-69 (1982)
15. T. Kohonen: *An Introduction to Neural Computing*, Neural Networks, **1**, 3-16 (1988)
16. T. Kohonen: *Self-Organization and Associative Memory* (Springer, Berlin 1989)
17. M. Kokol: *Modeling of Natural Phenomena by a Self-Organizing Regression Neural Network*, MSc Disertation (Faculty of Natural Sciences and Technology, University of Ljubljana 1993)
18. M. Kokol, I. Grabec: "Training of Elliptical Basis Function NN", World Congress on Neural Networks, San Diego, CA 1994
19. R. Linsker: *Self-Organization in a Perceptual Network*, Computer, **21**, (3), 105-117 (1898)
20. Chr. von der Malsburg: *Self-Organization of Orientation Sensitive Cells in the Striate Cortex*, Kybernetik, **14**, 85-100 (1973)
21. *Maximum Entropy and Bayesian Methods in Inverse Problems*, ed. by C. R. Smith, W.T Grandy, Jr. (Reidel, Dordrecht 1985)
22. D. E. Rumelhart, J. McClelland, and The PDP Research Group: *Parallel Distributed Processing*, (MIT Press, Cambridge, MA 1986)

Chapter 9

1. W. B. Davenport, JR.: *Probability and Random Processes, An Introduction for Applied Scientists and Engineers* (McGraw-Hill, New York 1970)
2. R. O. Duda, P. E. Hart: *Pattern Classification and Scene Analysis* (John Wiley & Sons, New York 1973)
3. I. Grabec, W. Sachse: "Experimental Characterization of Ultrasonic Phenomena by a Learning System", J. Appl. Phys., **66**, 3993-4000 (1989)
4. I. Grabec, W. Sachse: "Application of an Intelligent Signal Processing System to AE Analysis", J. Acoust. Soc. Am., **85**, 1226-1235 (1989)
5. I. Grabec, W. Sachse: "Automatic Modeling of Physical Phenomena: Application to Ultrasonic Data", J. Appl. Phys., **69**, (9), 6233-6244 (1991)
6. I. Grabec, W. Sachse: "Automatic Modeling of Ultrasonic Phenomena", Conf. Proc. *Ultrasonics International '91*, La Touquet, France, 1991 (Butterworth, London 1991), pp. 633-636
7. I. Grabec: "Automatic Modeling of Acoustic Emission Phenomena by Neural Networks", in *Dynamic, Genetic and Chaotic Programming, The Sixth Generation*, ed. by B. Souček and IRIS Group (John Wiley & Sons, New York 1992), pp. 144-163
8. I. Grabec: "Prediction of Chaotic Dynamical Phenomena by a Neural Network", in *Dynamic, Genetic and Chaotic Programming, The Sixth Generation*, ed. by B. Souček and IRIS Group (John Wiley & Sons, New York 1992), pp. 470-500
9. I. Grabec: "Self-Organization of Neurons Described by the Maximum-Entropy Principle", Biol. Cyb., **63**, 403-409 (1990)

10. I. Grabec: "Modeling of Chaos by a Self-Organizing Neural Network, Artificial Neural Networks", *Proc. ICANN*, Espoo, Finland, ed. by T. Kohonen, K. Mäkisara, O. Simula, J. Kangas (Elsevier Science Publishers B. V., North-Holland 1991), Vol. 1, pp. 151-156
11. I. Grabec, D. Grošelj: "Prognosis of Periodontal Disease Healing Process", *Proc. CADAM-95 Workshop*, Bled, Slovenia, 1995, ed. by N. Lavrač (IJS Sci. Pub., IJS-SP-95-1), pp. 146-154
12. D. Grošelj: *Evaluation of Periodontal Tissue Therapy Using Tooth Mobility Measurements*, PhD Disertation (Medical Faculty, Department of Dentistry, University of Ljubljana 1990)
13. D. Grošelj, I. Grabec: "Modeling of a Periodontal Disease Healing Process", *Abstracts Booklet of* 32nd *Annual Meeting CED/IADR*, Ljubljana, Slovenia, 1995, (Continental European Division / International Association for Dental Research), Rep. No. 082
14. H. Haken: *Synergetics, An Introduction*, Springer Series in Synergetics, Vol. 1 (Springer, Berlin 1983), p. 181
15. H. Haken: *Synergetic Computers and Cognition, A Top-Down Approach to Neural Nets*, Springer Series in Synergetics, Vol. 50 (Springer, Berlin 1991), Ch. 13
16. J. J. Hopfield: "Neural Networks and Physical Systems with Emergent Collective Computational Abilities", Proc. Natl. Acad. Sci. USA, **79**, 2554-2558 (1982)
17. J. J. Hopfield: "Neurons with Graded Response have Collective Computational Properties like those of Two State Neurons", Proc. Natl. Acad. Sci. USA, **81**, 3088-3092 (1984)
18. T. Kohonen: *Self-Organization and Associative Memory* (Springer, Berlin 1989)
19. I. D. Lefas, M. D. Kotsovos, N. N. Ambraseys: "Behavior of Reinforced Concrete Structural Wals: Strength, Deformation Characteristics, and Failure Mechanism", ACI Structural Journal, **87**, 23-31 (1990)
20. *Nondestructive Testing Handbook*, ed. by R. K. Miller, P. McIntire (American Society for Nondestructive Testing, Columbus, OH 1987), Vol. 5
21. G. Nicolis, I. Prigogine: *Self-Organization in Nonequilibrium Systems, From Dissipative Structure to Order through Fluctuations* (John Wiley & Sons, New York 1977)
22. Y. H. Pao, W. Sachse: J. Acoust. Soc. Am., **56**, 1478 (1974)
23. I. Peruš, P. Fajfar, I. Grabec: "Prediction of the Seismic Capacity of RC Structural Walls by a Neural Network", the article in print at ACI Structural Journal
24. W. Sachse, S. Golan: in *Elastic Waves and Non-Destructive Testing of Materials*, ed. by Y. H. Pao (ASME, New York, 1978), AMD Vol. 29, pp. 11-31
25. D. E. Rumelhart, J. McClelland, and The PDP Research Group: *Parallel Distributed Processing* (MIT Press, Cambridge, MA 1986)
26. H. Schiøler, U. Hartmann: "Mapping Neural Network Derived from the Parzen Window Estimator", Neural Networks, **5**, 903-909 (1992)
27. D. F. Specht: "Probabilistic Neural Networks", Neural Networks, **3**, 109-118 (1990)
28. D. F. Specht: "A General Regression Neural Network", IEEE Trans. on Neural Networks, **2**, No. 6, 568, (1991)
29. A. Tarantola: *Inverse Problem Theory, Methods for Data Fitting and Model Parameter Estimation* (Elsevier, Amsterdam 1987)
30. S. L. Wood: "Shear Strength of Low-rise Reinforced Concrete Walls", ACI Structural Journal, **87**, 99-107 (1990)

31. S. L. Wood: "Observed Behavoir of Slender Reinforced Concrete Walls Subjected to Cyclic Loading", in *Earthquake-resistant Concrete Structures, Inelastic Response and Design*, ed. by S. K. Ghosh, ACI SP 127-11, 453-477 (1991)

Chapter 10

1. A. E. Albert, L. A. Gardner: *Stochastic Approximation and Nonlinear Regression* (MIT Press, Cambridge, MA 1967)
2. S. T. Alexander: *Adaptive Signal Processing* (Springer, New York 1986)
3. J. A. Anderson, E. Rosenfeld: *Neurocomputing, Foundations of Research*, (The MIT Press, Cambridge, MA 1988), Vol. 1, 2
4. J. A. Anderson, A. Pellionisz, E. Rosenfeld: *Neurocomputing 2, Directions for Research* (The MIT Press, Cambridge, MA 1990)
5. M. Bertero: "Linear Inverse and Ill-Posed Problems", *Report of Instituto Nazionale Di Fisica Nucleare*, Frascati, Italy, INFN/TC-88/2
6. I. Grabec: "Application of Deconvolution in Description of Operation of Linear Systems", Mechanical Engineering Journal, **27**, 1-7 (1981)
7. I. Grabec: "Description of Operation of Noisy Linear Systems", Mechanical Engineering Journal, **29**, E 1-5 (1983)
8. I. Grabec: "Optimal Filtering of Transient AE Signals", Proc. *Ultrasonics International '85*, London 1985, (Butterworth, Guildford, UK 1985), pp. 219-224
9. I. Grabec: "Chaos Generated by the Cutting Process", Physics Letters, **117**, 384-386 (1986)
10. I. Grabec, E. Elsayed: "Analysis of AE During Martensitic Transformation in the CuZnAl Alloy by the Measurement of Forces", Physics Letters, **113**, 376-378 (1986)
11. I. Grabec, E. Elsayed: "Quantitative Analysis of AE During Martensitic Transformation in the CuZn Alloy", J. Phys., D - Appl. Phys., **19**, 605-614 (1986)
12. I. Grabec, W. Sachse: "Application of an Intelligent Signal Processing System to Acoustic Emission Analysis", J. Acoust. Soc. Am., **85**, 1226-1235 (1989)
13. I. Grabec, W. Sachse: "Experimental Characterization of Ultrasonic Phenomena by a Learning System", J. Appl. Phys., **66**, 3993-4000 (1989)
14. I. Grabec, K. Zgonc, W. Sachse: "Application of a Neural Network to Analysis of Ultrasonic Signals", Proc. *Ultrasonics International '89*, Madrid 1989 (Butterworth, Guildford, UK 1989), pp. 796-802
15. H. Haken: *Synergetic Computers and Cognition, A Top-Down Approach to Neural Nets*, Springer Series in Synergetics, Vol. 50 (Springer, Berlin 1991)
16. S. Haykin: *Adaptive Filter Theory* (Prentice-Hall, London 1991)
17. R. Hecht-Nielsen: *Neurocomputing* (Addison-Wesley, Reading, MA, 1990)
18. D. O. Hebb: *The Organization of Behavior, A Neurophysiological Theory* (Wiley, New York, 1948)
19. T. Kohonen: *Self-Organization and Associative Memory* (Springer, Berlin 1989)
20. Y. W. Lee: *Statistical Theory of Communication* (J. Wiley & Sons, New York 1960)
21. G. I. Marchuk: *Methods of Numerical Mathematics* (Springer, Berlin 1975)
22. J. E. Michaels, T. E. Michaels, W. Sachse: "Application of Deconvolution to Acoustic Emission Signal Analysis", Materials Evaluation, **39**, 1032-1036 (1981)
23. S. J. Orfandis: *Optimum Signal Processing: An Introduction* (Macmillan, New York 1988)
24. Y. H. Pao: *Elastic Waves and Nodestructive Testing of Materials* (American Society of Mechanical Engineers, New York 1987), Vol. 29, pp. 107-128

25. H. Robbins, S. Monro: "A Stochastic Approximation Method", Annals of Mathematical Statistics, **22**, 400-407 (1951)
26. D. E. Rumelhart, J. McClelland, and The PDP Research Group: *Parallel Distributed Processing* (MIT Press, Cambridge, MA 1986)
27. W. Sachse, S. Golan: in *Elastic Waves and Non-Destructive Testing of Materials*, ed. by Y. H. Pao (AMD, ASME, New York 1978), Vol. 29, pp. 11-31
28. W. Sachse, N. N. Hsu: "Ultrasonic Transducers for Materials Testing and Their Characterization", in *Physical Acoustics*, ed. by W. P. Mason and R. N. Thurston (Academic, New York 1979), Vol. 14, pp. 177-405
29. W. Sachse: "Application of Quantitative AE Methods", in *Solid Mechanics for Quantitative NDE*, ed. by J. D. Achenbach, Y. Rajapakse (Martinus Nijhoff, Dordrecht 1987), pp. 41-64
30. A. Tarantola: *Inverse Problem Theory, Methods for Data Fitting and Model Parameter Estimation* (Elsevier, Amsterdam 1987)
31. B. Widrow, M. E. Hoff: *Adaptive Switching Circuits*, 1960 IRE WESCON Convention Record, 96-1054, New York, 1960
32. K. Zgonc: *PhD Dissertation* (Faculty of Mechanical Engineering, University of Ljubljana 1992)
33. K. Zgonc, I. Grabec: "A Multidimensional Optimal Deconvolution Applied to Hetero-associative Recall of Acoustic Emission Signals", Proc. ECPD *Neurocomputing*, Dubrovnik, 1990, Vol. 1, No.1, pp. 206-212
34. K. Zgonc, I. Grabec: "Multidimensional Deconvolution Applied to Acoustic Emission Analysis", 1st Symp. Eval. Adv. Mat. by AE, Tokyo (JSNDI, Tokyo 1990)

Chapter 11

1. H. D. I. Abarbanel, Phys. Rev. A, **41**, 1782-1807 (1990)
2. J. P. Crutchfield, B. S. McNamara: "Equations of Motion from Data Series", Complex Systems, **1**, 417-452 (1987)
3. J. Deppisch: "Vorhersagen chaotischer Zeitreihen mit neuronalen Netzen", (Phys. Inst., Theoret. Phys., Univ. Würzburg 1990), Diplomarbeit
4. J. Deppisch, H. U. Bauer, T. Geisel: "Hierachical Training of Neural Networks and Prediction of Chaotic Time Series", Phisics Letters A, **158**, 57-62 (1991)
5. J. D. Farmer: Report No. LA-UR-87-1502 and No. LA-UR-88-901 (Los Alamos Nat. Lab., Los Alamos, NM)
6. I. Grabec: "Chaos Generated by the Cutting Process", Physics Letters, **117**, 384-386 (1986)
7. I. Grabec: "Chaotic Vibrations and AE Caused by the Cutting Process", in *Progress in AE III*, Proc. 8th Int. AE Symp., Tokyo 1986, (Jap. Soc. NDI, Tokyo), pp. 87-94
8. I. Grabec: "Chaotic Dynamics of the Cutting Process", Int. J. Machine Tools & Manufacturing, **28**, 19-32 (1988)
9. I. Grabec: "Modeling of Natural Phenomena by a Self-Organizing Neural Network", in *Neurocomputing*, Proc. ECPD, Dubrovnik, 1990, Vol.1, pp. 142-150
10. I. Grabec: "Prediction of a Chaotic Time Series by a Self-Organizing Neural Network", *Dynamics Days*, Düsseldorf, 1990, poster
11. I. Grabec: "Prediction of Dynamical Phenomena by a Neural Network", Proc. Conf. "Melecon '91", Ljubljana, Slovenia, ed. by B. Zajc, F. Solina, IEEE Cat. No. 91CH2964-5, 1991, pp. 18-23

12. I. Grabec: "Modeling of Chaos by a Self-Organizing Neural Network, Artificial Neural Networks", Proc. ICANN, Espoo, Finland, ed. by T. Kohonen, K. Mäkisara, O. Simula, J. Kangas (Elsevier Science Publishers B. V., North-Holland 1991), Vol.1, pp. 151-156
13. I. Grabec: "Prediction of Chaotic Dynamical Phenomena by a Neural Network", in *Dynamic, Genetic and Chaotic Programing, The Sixth Generation*, ed. by B. Souček and IRIS Group (John Wiley & Sons, New York 1992), pp. 470-500
14. I. Grabec: "Prediction of a Chaotic Economic Time Series by a Self-Organizing Neural Network", Neural Network World, **2**, 607-614 (1992)
15. I. Grabec: "Prediction of Chaos in Non-Autonomous Systems by a Neural Network", Proc. ICANN, Brighton, UK, 1992, ed. by I. Aleksander, J. Taylor (Elsevier Sci. Pub., Amsterdam 1992), pp. 379-382
16. I. Grabec, W. Sachse: "Automatic Modeling of Physical Phenomena: Application to Ultrasonic Data", J. Appl. Phys., **69 (9)**, 6233-6244 (1991)
17. P.Grassberger, I. Procaccia: "Characterization of Strange Attractors", Phys. Rev. Lett., **50**, 346-349 (1983)
18. H. Haken: *Synergetics, An Introduction*, Springer Series in Synergetics, Vol. 1 (Springer, Berlin 1983)
19. R. Hecht-Nielsen: *Neurocomputing* (Addison-Wesley, Reading, MA 1990)
20. G. A. Korn, T. M. Korn: *Mathematical Handbook for Scientists and Engineers* (McGraw-Hill, New York 1961, 1968 and latter editions)
21. A. Lapedes, R. Farber: Report No. LA-UR-87-2662 (Los Alamos Natl. Lab., Los Alamos, NM 1987)
22. A. J. Lichtenberg, M. A. Lieberman: *Regular and Stochastic Motion* (Springer, New York 1983)
23. W. Liebert, H. G. Schuster: "Proper Choice of the Time Delay for the Analysis of Chaotic Time Series", Phys. Lett., **142**, 107-111 (1989)
24. H. Lütkepohl: *Introduction to Multiple Time Series Analysis* (Springer, Berlin 1991)
25. F. C. Moon: *Chaotic Vibrations, An Introduction for Applied Scientists and Engineers* (J. Wiley & Sons, New York 1987)
26. E. Ott: *Chaos in Dynamical Systems* (Cambridge University Press, Cambridge 1994)
27. H. G. Schuster: *Deterministic Chaos* (Physik-Verlag, Weinheim 1984)
28. K. Stokbro, D. K. Umberger, J. A. Hertz, Complex Systems, **4**, 603-622 (1990)
29. P. Strobach: *Linear Prediction Theory, A Mathematical Basis for Adaptive Systems* (Springer, Berlin 1990)
30. S. Strogatz: *Nonlinear Dynamics and Chaos* (Addison–Wesley, Reading MA 1994)
31. F. Takens: "Detecting Strange Attractors in Fluid Turbulence", Lecture Notes in Mathematics, **898**, 366-381 (1981), ed. by D. A. Rand, L. S. Young (Springer, Berlin 1981)
32. M. West, J. Harrison: *Bayesian Forecasting and Dynamic Models*, (Springer, Berlin 1989)
33. K. Zgonc, I. Grabec: "A Multidimensional Optimal Deconvolution Applied to Hetero-Associative Recall of AE Signals", Proc. ECPD *Neurocomputing*, Dubrovnik, 1990, Vol.1, pp. 206-212
34. *Time Series Prediction: Forecasting the Future and Understanding the Past*, Proc. NATO Advanced Res. Workshop on Comparative Time Series Anal., Santa Fe, NM 1992, ed. by A. S. Weigend, N. A. Gershenfeld (Addison-Wesley, Reading, MA 1994)

Chapter 12

1. *Neurocomputing, Foundations of Research*, ed. by J. A. Anderson, E. Rosenfeld (MIT Press, Cambridge, MA 1988)
2. R. Beale, T. Jackson: *Neural Computing, An Introduction* (Adam Hilger, Bristol 1990)
3. C. M. Bishop: "Neural Networks and their Applications", Rev. Sci. Instr., **56**, 1803-1832 (1994)
4. G. Cybenko: *Approximations by Superpositions of a Sigmoidal Function*, Mathematics of Control, Signals & Systems, **2**, 303-314 (1989)
5. E. Domany, J. L. van Hemmen: *Models of Neural Networks* (Springer, Berlin 1991)
6. R. O. Duda, P. E. Hart: *Pattern Classification and Scene Analysis* (John Wiley, NY 1973)
7. E. Girosi, T. Poggio: "Networks for Learning: A View from the Theory of Approximation of Functions", in *Neural Networks: Concepts, Applications, and Implementations*, ed. by P. Antognetti, V. Milutinović (Prentice Hall, Englewood Cliffs, New Jersey 1991), Vol. 1, pp. 110-154
8. I. Grabec: "Self-Organization Based on the Second Maximum-Entropy Principle", 1st IEE Conference on *Artificial Neural Networks*, London 1989, Conf. Pub. No. 313, pp. 12-16
9. I. Grabec, W. Sachse: "Automatic Modeling of Physical Phenomena: Application to Ultrasonic Data", J. Acoust. Soc. Am., **69**, (9), 6233–6244 (1991)
10. S. Grossberg: "Nonlinear Neural Networks: Principles, Mechanisms, and Architectures", Neural Networks,**1**, 17-61 (1988)
11. J. Guckenheimer, S. Gueron, R.M. Harris-Warrick: "Mapping the dynamics of a bursting neuron", Phil Trans. R. Soc. Lond. B, **341**, 345-359 (1993)
12. S. Haykin: *Neural Networks* (McMillan, New York 1994)
13. D. O. Hebb: *The Organization of Behavior, A Neurophysiological Theory* (Wiley, New York 1949)
14. R. Hecht-Nielsen: *Neurocomputing* (Addison-Wesley, Reading, MA 1990)
15. *Neural Networks.* Vols. 1-6 (1988-1994)
16. T. Kohonen: "An Introduction to Neural Computing", Neural Networks **1**, 3-16 (1988)
17. T. Kohonen: *Self-Organization and Associative Memory* (Springer, Berlin 1989)
18. M. Kokol: "Modeling of Natural Phenomena by a Self-Organizing Regression Neural Network", Msc. Disertation (Faculty of Natural Sciences and Technology, Department of Physics, University of Ljubljana 1993)
19. A. N. Kolmogorov: "On the Representation of Continuous Functions of Many Variables by Superpositions of Continuous Functions of one Variable and Addition", Dokl. Ak. Nauk SSSR, **114**, 953-956 (1957)
20. G. G. Lorentz: "The 13th Problem of Hilbert", in *Mathematical Developments Arising from Hilbert's Problems*, ed. by F. Browder, (Am. Math. Soc., Providence, RI 1976)
21. R. K. Miller, CMfgE, T. C. Walker, A. M. Ryan: *Neural Net Applications and Products*, SEAI Techn. Publ. Madison GA 30650 (Graeme Publ., Amherst NH 1990)
22. F. Rosenblatt: "The Perceptron: A Probabilistic Model for Information Storage and Organization in the Brain", Psychoanalytic Review, **65**, 386-408 (1958)
23. D. E. Rumelhart, J. McClelland, and The PDP Research Group: *Parallel Distributed Processing* (MIT Press, Cambridge, MA 1986)
24. P. K. Simpson: *Artificial Neural Systems, Foundations, Paradigms, Applications, and Implementations* (Pergamon Press, New York 1990)

25. D. A. Sprecher: "On the Structure of the Continouous Functions of several Variables", Trans. Am. Math. Soc., **115**, 340-355 (1965)
26. R. F. Thompson: *The Brain* (W. H. Freeman, New York 1985)
27. P. J. Werbos: "Beyond Regression: New Tools for Prediction and Analysis in the Behavioral Sciences", PhD Disertation (Appl. Math., Harvard University 1974)
28. P. J. Werbos: *The Roots of Back-Propagation* (John Wiley and Sons, New York 1994)
29. B. Widrow, M. E. Hoff: "Adaptive Switching Circuits", 1960 IRE WESCON Convention Record, 96-104, New York 1960

Chapter 13

1. J. S. Albus: "A New Approach to Manipulator Control: The Cerebellar Model Articulation Controller (CMAC)", Trans. ASME, J. of Dyn. Sys., Meas. and Control, **97**, 220–227 (1975)
2. K. J. Åström: *Introduction to Stochastic Control Theory* (Academic Press, New York 1970)
3. K. J. Åström, T. J. McAvoy: "Intelligent Control: An Overview and Evaluation", in *Handbook of Intelligent Control, Neural, Fuzzy and Adaptive Approaches*, ed. by D. A. White, D. A. Sofge (Van Nostrand Reinhold, New York 1992)
4. A. Bagchi: *Optimal Control of Stochastic Systems* (Prentice–Hall, New York 1993)
5. W. L. Baker, J. A. Farrell: "An Introduction to Connectionist Learning Control Systems", in *Handbook of Intelligent Control, Neural, Fuzzy, and Adaptive Approaches*, ed. by D. A. White, D. A. Sofge (Van Nostrand Reinhold, New York 1992), pp. 35–63.
6. R. Bellman, R. E. Kalaba: *Dynamic Programming and Modern Control Theory* (Academic Press, New York 1965)
7. D. P. Bertsekas: *Dynamic Programming: Deterministic and Stochastic Models* (Prentice–Hall, Englewood Cliffs, NJ 1987)
8. W. L. Brogan: *Modern Control Theory* (Prentice–Hall, Englewood Cliffs, NJ 1985)
9. I. Grabec: "Chaos Generated by the Cutting Process", Phys. Lett., **117**, 384-386 (1986)
10. J. Gradišek, E. Govekar, I. Grabec: "Chaotic Cutting Process and Determining Optimal Cutting Parameter Values Using Neural Networks", Int. J. Machine Tools & Manuf. , (1996) in print
11. H. Haken: *Synergetics: An Introduction*, Springer Series in Synergetics, Vol. 1 (Springer, Berlin 1983)
12. R. Howard: *Dynamic Programming and Markov Processes* (MIT Press, Cambridge, MA 1960)
13. D. E. Kirk: *Optimal Control Theory – An Introduction* (Prentice–Hall, Englewood Cliffs, NJ 1970)
14. M. Kokol: "Modeling of Natural Phenomena by a Self-Organizing Regression Neural Network", MSc Disertation (Faculty of Natural Sciences and Technology, University of Ljubljana 1993)
15. *Neural Networks for Control*, ed. by W. T. Miller, III, R. S. Sutton, P. J. Werbos (The MIT Press, Cambridge, MA 1990)

16. K. S. Narendra: "Adaptive control of Dynamical Systems Using Neural Networks", in *Handbook of Intelligent Control, Neural, Fuzzy, and Adaptive Approaches*, ed. by D. A. White, D. A. Sofge (Van Nostrand Reinhold, New York 1992), pp. 141–184
17. K. S. Narendra, A. M. Annaswamy: *Stable Adaptive Syestms* (Prentice–Hall, Englewood Cliffs, NJ 1989)
18. H. Nguyen, B. Widrow: "Neural Networks for Self-Learning Control Systems", IEEE Control Systems Magazine, **4**, 18-23 (1990)
19. J. G. Reid: *Linear System Fundamentals, Continuous and Discrete, Classic and Modern* (McGraw–Hill, New York 1983)
20. Jean-Jacques E. Slotine, Weiping Li: *Applied Nonlinear Control* (Prentice–Hall, Englewood Cliffs, NJ 1991)
21. S. B. Thurn: "The Role of Exploration in Learning Control" in *Handbook of Intelligent Control, Neural, Fuzzy and Adaptive Approaches*, ed. by D. A. White, D. A. Sofge (Van Nostrand Reinhold, New York 1992) pp. 527–559
22. P. J. Werbos: "Neural Control and Elastic Fuzzy Logic: Capabilities, Concepts, and Applications", Industrial Electronics, **40**, 170–180 (1993)
23. *Handbook of Intelligent Control, Neural, Fuzzy, and Adaptive Approaches*, ed. by D. A. White, D. A. Sofge (Van Nostrand Reinhold, New York 1992)
24. B. Widrow, E. Walach: *Adaptive Inverse Control* (Prentice–Hall PTR, Upper Saddle River, NJ 1996)

Chapter 14

1. Proc. 4th Int. Conf. *Genetic Algorithms*, ed. by R. Below, L. Booker, (Morgan Kaufman, San Mateo, CA 1991)
2. G. Chen, X. Dong: "Control of Chaos – A Survey", in Proc. *1993 IEEE Control and Decision Conference*, San Antonio, TX, USA, 1993, pp. 469-474
3. I. Grabec and W. Sachse: "Intelligent Controller", Record of Invention, Cornell Research Foundation, Cornell University, Ithaca, NY, February, 1994
4. Proc. Int. Conf. *ECAL '93 , Self-Organization and Life: from Simple Rules to Global Complexity* (Center for Non-Linear Phenomena and Complex Systems, CP231, Université Libre de Bruxelles, Brussels 1993), Vols. I , II
5. D. E. Goldberg: *Genetic Algorithms in Search, Optimization and Machine Learning* (Addison-Wesley, Reading, MA 1989)
6. H. Haken: *Synergetics*, Springer Series in Synergetics, Vol. 1 (Springer, Berlin 1983)
7. H. Haken: *Advanced Synergetics*, Springer Series in Synergetics, Vol. 20 (Springer, Berlin 1983)
8. J. H. Holland: *Adaptation in Natural and Artificial Systems* (MIT Press, Cambridge, MA 1992)
9. S. A. Kauffman: *The Origins of Order; Self-Organization and Selection in Evolution* (Oxford University Press, New York 1993)
10. T. Kohonen: *Self-Organization and Associative Memory* (Springer, Berlin 1989), Chapter 7.6
11. G.Nicolis, I. Prigogine: *Self-Organization in Nonequilibrium Systems, from dissipative Structures to Order through Fluctuations* (John Wiley & Sons, New York 1977)
12. R. Penrose: *The Emperors New Mind, Concerning Computers, Minds and the Laws of Physics* (Vintage, London 1989), Chapter 10

13. T. Shinbrot, C. Grebogi., E. Ott, J. A. Jorke: "Using Small Perturbations to Control Chaos", Nature, **363**, 411-417 (1993)
14. S. Thurn: "The Role of Exploration in Learning Control", in *Handbook of Intelligent Control, Neural, Fuzzy and Adaptive Approaches*, ed. by D. A. White, D. A. Sofge, (Van Nostrand Reinhold, New York 1992), p. 527

Appendix A

1. W. B. Davenport, JR: *Probability and Random Processes, An Introduction for Applied Scientists and Engineers* (McGraw–Hill, New York 1970 and later editions)
2. G. A. Korn, T. M. Korn: *Mathematical Handbook for Scientists and Engineers; Definition, Theorems and Formulas for Reference and Review* (McGraw–Hill, New York 1968 and later editions)

Appendix B

1. H. Atmanspacher, H. Scheingraber: "A Fundamental Link Between System Theory and Statistical Mechanics", Found. Phys., **17**, 939 (1987)
2. J. P. Eckmann, D. Ruelle: "Ergodic Theory of Chaos and Strange Attractors", Rev. Mod. Phys., **57**, 617 (1985)
3. M. Feigenbaum: *Dynamical Systems and Chaos* (Springer, Berlin 1983)
4. I. Grabec: "Sources of Chaos in Mechanical Machines", in Proc. of *IMACS*, Paris 1989, Numerical and Applied Mathematics, (W.F. Ames Ed., J.C. Baltzer AG), pp. 309-314
5. P. Grassberger, I. Procaccia: "Characterization of Strange Attractors", Phys. Rev. Lett., **50**, 346-349 (1983)
6. J. Guckenheimer, P. Holmes: *Nonlinear Oscillations, Dynamical Systems and Bifurcations of Vector Fields* (Springer, New York 1983)
7. *Evolution of Order and Chaos*, ed. by H. Haken (Springer, Berlin 1982)
8. H. Haken: *Synergetics*, Springer Series in Synergetics, Vol. 1 (Springer, Berlin 1983)
9. M. W. Hirsch, S. Smale: *Differential Equations, Dynamic Systems and Linear Algebra* (Academic Press, New York 1965)
10. A. J. Lichtenberg, M. A. Lieberman: *Regular and Stochastic Motion* (Springer, New York 1983)
11. W. Liebert, H. G. Schuster: "Proper Choice of the Time Delay for the Analysis of Chaotic Time Series", Phys. Lett., **142**, 107-111 (1989)
12. B. B. Mandelbrot: *The Fractal Geometry of Nature* (Freeman, San Francisco 1982)
13. F. C. Moon: *Chaotic Vibrations, An Introduction for Applied Scientists and Engineers* (J. Wiley & Sons, New York 1987)
14. G. Nicolis: "Dissipative Systems", Rep. Prog. Phys, **49**, 873-949 (1986)
15. V. I. Oseledec: Trans. Moscow Math. Soc., **19**, 197 (1968)
16. E. Ott: *Chaos in Dynamical Systems* (Cambridge University Press, Cambridge 1994)
17. Ya. B. Pesin: Usp. Math. Nauk, **32**, 55 (1977)
18. H. G. Schuster: *Deterministic Chaos* (Physik-Verlag, Weinheim 1984)

19. S. Strogatz: *Nonlinear Dynamics and Chaos* (Addison–Wesley, Reading MA 1994)
20. F. Takens: "Datecting Strange Attractors in Fluid Turbulence", Lecture Notes in Mathematics, **898**, 366-381 (1981)
21. J. M. T. Thompson, H. B. Stewart: *Nonlinear Dynamics and Chaos, Geometrical Methods for Engineers and Scientists* (John Wiley & Sons, New York 1986)

Index

Printing: Mercedesdruck, Berlin
Binding: Buchbinderei Lüderitz & Bauer, Berlin